Lecture Notes in Artificial Intelligence 8239

Subseries of Lecture Notes in Computer Science

LNAI Series Editors

Randy Goebel
  University of Alberta, Edmonton, Canada
Yuzuru Tanaka
  Hokkaido University, Sapporo, Japan
Wolfgang Wahlster
  DFKI and Saarland University, Saarbrücken, Germany

LNAI Founding Series Editor

Joerg Siekmann
  DFKI and Saarland University, Saarbrücken, Germany

Guido Herrmann   Martin J. Pearson
Alexander Lenz   Paul Bremner
Adam Spiers   Ute Leonards (Eds.)

Social Robotics

5th International Conference, ICSR 2013
Bristol, UK, October 27-29, 2013
Proceedings

 Springer

Volume Editors

Guido Herrmann
University of Bristol, Faculty of Engineering
Bristol BS8 1TR, UK
E-mail: g.herrmann@bristol.ac.uk

Martin J. Pearson
Alexander Lenz
Paul Bremner
Adam Spiers
University of the West of England
Bristol Robotics Laboratory
Bristol BS16 1QY, UK
E-mail:{martin.pearson, alex.lenz, paul.bremner, adam.spiers}@brl.ac.uk

Ute Leonards
University of Bristol, School of Experimental Psychology
Bristol BS8 1TU, UK
E-mail: ute.leonards@bristol.ac.uk

ISSN 0302-9743   e-ISSN 1611-3349
ISBN 978-3-319-02674-9   e-ISBN 978-3-319-02675-6
DOI 10.1007/978-3-319-02675-6
Springer Cham Heidelberg New York Dordrecht London

Library of Congress Control Number: 2013949860

CR Subject Classification (1998): I.2, K.4, H.5, J.4, I.5

LNCS Sublibrary: SL 7 – Artificial Intelligence

Typesetting: Camera-ready by author, data conversion by Scientific Publishing Services, Chennai, India

Printed on acid-free paper

Springer is part of Springer Science+Business Media (www.springer.com)

Preface

These proceedings contain the papers presented at 5th International Conference on Social Robotics (ICSR 2013), which was held in the city of Bristol, in the United Kingdom, during October 27th–29th, 2013. Three workshops as part of the conference took place on Sunday, 27 October 2013, while the conference presentations were held on Monday, 28 October 2013, and Tuesday, 29 October 2013.

The ICSR first assembled in 2009 at Incheon in Korea as part of the FIRA RoboWorld Congress. Since then, it has grown in size while maintaining its founding principle of bringing researchers together to report and discuss the latest progress in the field of social robotics in a collegial, supportive, and constructive atmosphere. The theme of the conference this year was "Companionship," which was to both inspire interest in the emerging roles that robotic systems may play in our society in the near future, and to encourage researchers from a diverse range of backgrounds to engage with the conference. Toward this end, the conference attracted submissions from researchers with a broad range of interests, for example, human–robot interaction, child development, and care for the elderly. There were also a large number of submissions describing progress in the more technical issues that underlie social robotics, such as visual attention and processing, motor control and learning, emphasizing the fascinating cross-disciplinary nature of this field of research. We received 108 papers and extended abstracts this year from research institutes in Europe, Asia, North/South America, and Australasia, reflecting the truly international community that has developed and grown through social robotics research. Of these submissions, 55 were accepted as full papers with a further 13 accepted as extended abstracts. The Program Committee consisted of an experienced team of active researchers in the field of social robotics, many of whom have supported the ICSR series since its inception. We gratefully acknowledge the contribution made by the reviewers this year in selecting what we see as the definitive "snap-shot" of the social robotics research landscape in 2013.

There were two invited speakers this year delivering keynote speeches on each of the working days of the conference; Professor Rolf Pfeifer, director of the Artificial Intelligence Laboratory at the University of Zurich; and Professor Qiang Huang from the Intelligent Robotics Institute at the Beijing Institute of Technology. A highly successful dinner speech enlightened the conference participants during the conference banquet. Moreover, two competitions were held at the conference which inspired future researchers in this highly multi-disciplinary field of social robotics: The Robot design competition was sponsored by the Technical and Professional Network of the Institution of Engineering and Technology (IET) in Robotics and Mechatronics and co-organised together with the Pervasive Media Studio from Bristol. The NAO Programming Competition was organized together with Aldebaran Robotics, and their UK distributor Active Robots who also provided the prizes for this highly exciting competition. We acknowledge the effort of all who helped in the organization of the conference, from the founding members of the standing committee for their guidance and momentum, through to the student helpers and caterers on the day.

We hope that the readers will find within this volume papers that stimulate their interest in social robotics, further their knowledge base in this field, and inspire fruitful work and collaboration in the future.

October 2013

Guido Herrmann
Martin J. Pearson
Alexander Lenz

Organization

ICSR 2013 was organized by the University of Bristol and the University of the West of England, Bristol, working together through the joint venture of the Bristol Robotics Laboratory. The Pervasive Media Studio from Bristol co-organized the Robot Design Competition of the conference.

Standing Committee

Shuzhi Sam Ge
    The National University of Singapore and University of Electronic Science and Technology of China

Oussama Khatib   Stanford University, USA

Maja Mataric   University of South California, USA

Haizhou Li   A*Star, Singapore

Jong Hwan Kim   KAIST, Korea

Paolo Dario   Scuola Superiore Sant'Anna, Italy

Ronald Arkin   Georgia Tech., USA

Organizing Committee

General Chairs

Guido Herrmann   University of Bristol, UK

Alois Knoll   Technische Universität München, Germany

Program Chairs

Martin Pearson   Bristol Robotics Laboratory, UK

Peter Ford Dominey   Inserm-Stem Cell and Brain Research Institute, France

Organizing Chairs

Alexander Lenz   Bristol Robotics Laboratory, UK

Hongsheng He   University of Tennessee, USA

Honorary Chair

Chris Melhuish   Bristol Robotics Laboratory, UK

Publicity Chairs

Praminda Caleb-Solly   Bristol Robotics Laboratory, UK

Martin Saerbeck   Institute of HPC, Singapore

Awards Chairs

Kerstin Dautenhahn   University of Hertfordshire, UK
Reid Simmons   Carnegie Mellon University, USA

Design Competition Chairs

Adam Spiers   Bristol Robotics Laboratory, UK
Verity McIntosh   Pervasive Media Studio, Bristol, UK
Ho Seok Ahn   University of Auckland, New Zealand

Special Sessions Chairs

Paul Bremner   Bristol Robotics Laboratory, UK
Giorgio Metta   Italian Institute of Technology

Panel Discussions Chairs

Tony Pipe   Bristol Robotics Laboratory, UK
Verena Hafner   Humboldt-Universität zu Berlin, Germany

Workshop Chairs

Ute Leonards   University of Bristol, UK
Mary-Anne Williams   Univeristy of Technology, Sydney, Australia

International Program Committee

A. Agah	G. Dissanayake	K. Kühnlenz
R.Akmeliawati	S. Dogramadzi	S. Lauria
F. Amirabdollahian	Z. Doulgeri	H. Lehmann
M. Ang	F. Dylla	A. Lenz
K. Arras	K. Eder	N. Lepora
T. Assaf	F. Eyssel	Z. Li
I. Baroni	M. Falahi	B. Liu
P. Baxter	I. Georgilas	K. Lohan
N. Bellotto	J. Ham	M. Lohse
T. Belpaeme	M. Hanhede	J. Ma
L. Brayda	M. Harbers	B. MacDonald
P. Bremner	W. He	A. Meghdari
E. Broadbent	M. Heerink	T. Meng
F. Broz	G. Herrmann	B. Mitchinson
E. Cervera	B. Johnston	J. Na
R. Chellali	A. Jordi	M. Paleari
X. Chen	H. Kamide	A. Pandey
R. Chongtay	T. Kanda	M. Pearson
S. Costa	M. Kim	T. Pipe
R. Cuijpers	K. Koay	K. Pitsch

S. Poeschl
S. Sabanovic
M. Salem
M. Salichs
K. Severinson-Eklundh
R. Silva
W. So

K. Suzuki
J. Tan
Y. Tan
K. Tee
C. Wang
F. Wallhoff
M. Walters

A. Weiss
M. Williams
G. Wolbring
A. Wykowska
H. Zhang

Sponsoring Organizations

The Technical and Professional Network of the Institution of Engineering and
    Technology (IET) in Robotics and Mechatronics
Active Robots
Aldebaran Robotics
Bristol Robotics Laboratory
The University of Bristol
The University of the West of England

Table of Contents

Extended Abstracts

Building Companionship through Human-Robot Collaboration

Yanan Li[1], Keng Peng Tee[2], Shuzhi Sam Ge[1], and Haizhou Li[2]

[1] Social Robotics Laboratory (SRL), Interactive Digital Media Institute,
and Department of Electrical and Computer Engineering,
National University of Singapore, Singapore, 117576
[2] Institute for Infocomm Research (I²R), Agency for Science,
Technology and Research, Singapore, 138632

Abstract. While human-robot collaboration has been studied intensively in the literature, little attention has been given to understanding the role of collaborative endeavours on enhancing the companionship between humans and robots. In this position paper, we explore the possibilities of building the human-robot companionship through collaborative activities. The design guideline of a companion robot Nancy developed at SRL is introduced, and preliminary studies on human-robot collaboration conducted at SRL and I²R are elaborated. Critical issues and technical challenges in human-robot collaboration systems are discussed. Through these discussions, we aim to draw the attention of the social robotics community to the importance of human-robot collaboration in companionship building, and stimulate more research effort in this emerging area.

1 Introduction

Population ageing has become an extremely serious issue affecting almost every country across the globe. The increasing shortfall of caregivers, together with the increasing occurrences of elderly disability, loneliness and depression, can put a strain on a country's healthcare and economic system, as well as create deep social problems. It is thus envisaged to be highly beneficial to create intelligent robots that provide companionship to ease loneliness and prevent depression.

Most works on robot companionship take an approach focusing on communications and interactions, such as providing cognitive support, personalized chats, or pet-like affective therapy. To expand the application scope and fulfil more general needs, research in robot companions should also look into goal-oriented collaborative activities between human and robot. This is motivated by the observation that people who work well together enjoy good bonding experience. Likewise, participating in a joint activity with a robot to achieve a practical goal (e.g. house cleaning) can increase feelings of companionship in the human towards the robot. At the same time, the robot fulfils its role to assist with activities of daily living, and due to its collaborative nature, imparts a sense of empowerment and personal autonomy to the human since he/she still needs to

G. Herrmann et al. (Eds.): ICSR 2013, LNAI 8239, pp. 1–7, 2013.

be actively involved in the joint task instead of just being a passive beneficiary of robot assistance.

In fact, robots still need human help to operate in unstructured environments like homes, since human beings and robots have complementary advantages. On the one hand, robots equipped with the state-of-art sensors, computing power, and powerful actuators are capable of carrying out repetitive, high-load tasks tirelessly and with high precision. On the other hand, human beings possess excellent cognitive skills and are generally superior in situational awareness and decision making. An effective human-robot collaboration system will free human beings from heavy loads and achieve more robust operability in unstructured environments by maximizing complementary capabilities of human beings and robots. In the context of elderly companionship, an added benefit for the elderly person comes from the cognitive stimulation associated with learning how to interact and work with a robot with an intelligence of its own.

While human-robot collaboration has been studied intensively in applications such as search and rescue, healthcare, and construction (see the surveys [1,2]), little attention has been given to understanding the role of collaborative endeavours on enhancing feelings of acceptance and companionship towards robots. Our position is that there is a strong fundamental link between collaboration and companionship, which is worth investigating to unlock new but pragmatic ways of alleviating loneliness and accomplishing tasks at hand. But first, technical challenges must be solved to enable human and robots to work together organically. Designing such a system requires interdisciplinary studies of robot design, sensor fusion, control design, and system integration, among others. Much effort has been made in the literature. However, in many situations robots are still not "intelligent" enough to collaborate with human beings in a natural and effective way. What are the reasons that have caused such a gap, and what can we do to fill it? In this position paper, we will attempt to shed light on key aspects of this question by discussing critical technical challenges in human-robot collaboration and results of preliminary studies conducted at SRL and I^2R. We aim to draw the attention of the social robotics community to the importance and yet-to-be-realized value of human-robot collaboration in companionship building, and stimulate more research effort from the community in this emerging area.

2   Robot Companion Nancy for Human-Robot Collaboration

A social robot named Nancy has been developed (Fig. 1) as a platform for investigations and experiments in human-robot collaboration and its role in building companionship. Details of the robot design can be found in [3]. Here, we outline how Nancy was designed to be suitable for human-robot collaboration.

2.1   Design

The overarching design guideline for Nancy has been to make it able to move in a human-like manner with high performance, flexibility, and safety. As shown

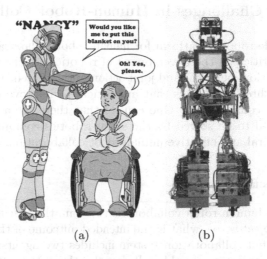

Fig. 1. (a) An envisaged scenario for Nancy the companion robot. (b) An actual first prototype of Nancy.

in Fig. 1, Nancy is a humanoid 160.7cm in height and has a total of 32 degrees-of-freedom (DOFs) throughout the whole body. Similar to a human, the upper body of Nancy includes the modules of hand, wrist, forearm, elbow, upper arm, shoulder, scapula, neck, head and torso. In particular, each arm has 8 DOFs to achieve human-like reaching and manipulating motions, and it has a 2-DOF neck and 3-DOF eyes with built-in cameras. A cable-pulley driving mechanism has been developed to ensure a light-weight upper body and to provide more natural physical human-robot interaction. Nancy's lower body has been designed to facilitate smooth and multi-direction movements with two legs and a wheel base.

2.2 Sensing and Perception: Visual, Vocal and Haptic

Nancy has been equipped with an intelligent scene-understanding engine to sense and perceive the environments through computer vision [4]. It is able to track and recognize objects, perceive the emotion, feeling and attitude of the human beings, and determine its attention autonomously in a scene. Another essential modality for human-robot collaboration is speech. Perception modules for recognition and understanding of spontaneous speech have been embedded in Nancy [3]. Besides vision and speech, haptic sensing and perception have also been built into the robot to enable studies in physical human-robot collaboration. In particular, force feedback is measured by a 6-axis force/torque sensor mounted on Nancy's wrist, enabling human motion intention to be estimated for the purpose of human-robot co-manipulation [5]. Artificial skin has also been integrated to provide distributed tactile sensing to the robot as well as to enhance tactile sensation for the user [6].

3   Technical Challenges in Human-Robot Collaboration

Building a suitable robotic platform for human-robot collaboration is only half the story. More critical is the development of robotic intelligence that controls the robot's perception, behavior and joint action in the presence of the user. This section analyzes three key issues that are critical to improve the performance of a human-robot collaboration. One can see that there are a number of open problems that need to be solved by the social robotics community in order to achieve more natural and effective human-robot collaboration.

3.1   Task Objective

When designing a human-robot collaboration system, the first thing to come up with is the *task objective*, or, what is the intended outcome of the collaboration? Since a human-robot collaboration system includes two agents: a human being and a robot, this objective should well describe the coherence and conflict of individual objectives of two agents. This can be a specific application of multi-agent systems [7], where the weight of the individual objective of the human being is larger than that of the robot in most cases. Our previous work [5] considered human-robot co-manipulation where the objective is to physically free the human from the heavy load. To prescribe an objective for a human-robot collaboration system is vital, but it can be challenging since the collaboration is dynamic. The human may change his/her strategy for the task, so the robot needs to be able to detect the excursion from the original plan and infer the new objective, or there should be an easy and intuitive way for the human to specify the task objective on-the-fly. It may also be essential to have prior knowledge of the physical limits of both human and robot, as well as models of the operation process and environment.

3.2   Task Allocation and Adaptation

After the objective is determined, the task needs to be allocated to the human and robot. *Task allocation* is based on the task objective and determined by analyzing the complementary advantages of human and robot. As discussed in Introduction, humans are generally better at understanding the circumstances and making decisions, and robots at carrying out regular tasks with the guaranteed performance. Therefore, it is natural to develop a leader-follower model, where the human being plays the role of a leader and executes a higher-level task. The leader-follower model has been employed in a vast number of research studies. In our previous work [5], we have shown that the robot becomes a load to the human being when the motion intention of the human being changes, and it can be estimated to allocate a larger share of the task to the robot. However, even if the motion intention of the human being can be predicted by the robot, the ideal case is that the robot follows the human "perfectly". We assume implicitly that the motion intention of the human being is stationary with respect to the actual robot trajectory. In other words, the adaptation of the robot trajectory is assumed to have no effect on the motion intention of the human being. However, human motion is also an output of the neuromuscular control system, so the dynamic interaction with the robot could well result in *concurrent adaptation* in the motion intention of the human being. This concurrent adaptation should be carefully investigated to ensure that the human-robot collaboration is stable.

Through learning and communication, the robot is able to increase its knowledge about the environment and task. Based on the acquired knowledge, it is possible for the robot to be allocated a higher level task. In robot programming by demonstration (see e.g. [8]), the human programs a robot to perform a certain complex task by physically holding a robot arm. After several iterations of operations, the robot is able to complete the task with less interference from the human being. In this simple example, the task allocation to the human being becomes less when the robot acquires more knowledge. It can be expected to develop a mechanism of *adaptive task allocation*, such that the proportion of the task allocation of the human being is increased only when necessary.

3.3   Communication and Interface

The *communication* between the human being and robot is important during the task execution. It can be used to resolve the conflict, disambiguate uncertainties, refine the task allocation and even the task objective. It relies on the sensing and perception capabilities of both human and robot. In most cases, communication is limited on the robot side due to limited sensing and perception capabilities, and this is the original motivation of human-robot collaboration. Technologies such as intention estimation [5], speech recognition [9], human detection and tracking [10,11], and gesture recognition [12,13] have been developed to enhance the perception capability of the robot. Augmented reality has been studied to make the human being and robot share the same reference frame [1]. Besides these, many sensors have been designed and developed for human-robot collaboration systems. Human beings understand the circumstances by listening, looking, smelling, and touching. It is natural to expect that robots have equivalent capabilities so a natural and efficient collaboration can be achieved. Nevertheless, most of state-of-art human-robot collaboration systems are embedded with only one or two of the above modules/technologies. A natural and efficient collaboration can be anticipated only when these modules/technologies are integrated to form a *multi-modality interface*. Such an interface goes beyond straightforward system integration, as sensor fusion needs to be carefully investigated and hierarchical architecture designed to handle differing levels of complexity in perception.

Many state-of-art human-robot collaboration systems are designed to make the robot understand the human being, while very few works in the literature address the issue of *mutual understanding*. On the one hand, there is still a long way to go for the robot to "understand" the human being, due to the limitation of sensing and perception capabilities, and learning skills. On the other hand, it is important to consider the acceptance and understanding of the robot by the human, especially when the human is a non-expert in robotics. For example, transparency and intuitive interface are essential in tele-operation. We have developed human-aided grasp assistance for easy and fast communication of grasp intent to robot [14] as well as robot-assisted grasping based on local proximity sensors to simplify the tele-operation task for the user [15]. A robot will be better understood and accepted if its behavior emulates that of human beings [1], and this can be taken into account when developing an *interface* for a human-robot collaboration system.

4   Conclusion

In this position paper, we have postulated and explored the intriguing possibility of building human-robot companionship through collaborative activities. Preliminary studies on human-robot interaction and collaboration conducted at SRL and I^2R show promising results in terms of individual components such as motion intention estimation, human detection and tracking, speech and gesture recognition. However, much more work needs to be done, especially in integrating the various components into a stable, robust human-robot collaboration system, as well as designing and carrying out experimental trials to investigate the effect on human-robot companionship. Key issues in human-robot collaboration, including i) task objective, ii) task allocation and adaptation, and iii) communication and interface, have been discussed, and technical challenging problems proposed: adaptive task allocation, concurrent adaptation, mutual understanding, and multi-modality interface. More effective and natural human-robot collaboration leading to stronger companionship can be expected by resolving these problems. We see this topic as providing abundant scope for multi-disciplinary, collaborative research in social robotics.

References

1. Green, S.A., Billinghurst, M., Chen, X., Chase, J.G.: Human-robot collaboration: A literature review and augmented reality approach in design. International Journal of Advanced Robotic Systems 5(1), 1–18 (2008)
2. Bauer, A., Wollherr, D., Buss, M.: Human-robot collaboration: A survey. International Journal of Humanoid Robotics 5(4), 47–66 (2008)
3. Ge, S.S., Cabibihan, J.J., Zhang, Z., Li, Y., Meng, C., He, H., Safizadeh, M.R., Li, Y.B., Yang, J.: Design and development of nancy, a social robot. In: Proceedings of the 8th International Conference on Ubiquitous Robots and Ambient Intelligence, Incheon, Korea, November 23-26, pp. 568–573 (2011)
4. He, H., Ge, S.S., Zhang, Z.: Visual attention prediction using saliency determination of scene understanding for social robots. International Journal of Social Robotics 3, 457–468 (2011)
5. Li, Y., Ge, S.S.: Human-robot collaboration based on motion intention estimation. IEEE/ASME Transactions on Mechatronics (2013), doi:10.1109/TMECH.2013.2264533

6. Cabibihan, J.J., Pradipta, R., Ge, S.S.: Prosthetic finger phalanges with lifelike skin compliance for low-force social touching interactions. Journal of NeuroEngineering and Rehabilitation 8(1), 16 (2011)
7. Ge, S.S., Fua, C.H., Lim, K.W.: Agent formations in 3D spaces with communication limitations using an adaptive q-structure. Robotics and Autonomous Systems 58(4), 333–348 (2010)
8. Billard, A., Calinon, S., Dillmann, R., Schaal, S.: Robot programming by demonstration. In: Siciliano, B., Khatib, O. (eds.) Springer Handbook of Robotics. Springer (2008)
9. Li, H., Lee, K.A., Ma, B.: Spoken Language Recognition: From Fundamentals to Practice. Proceedings of the IEEE 101(5), 1136–1159 (2013)
10. Li, L., Yu, X., Li, J., Wang, G., Shi, J.Y., Tan, Y.K., Li, H.: Vision-based attention estimation and selection for social robot to perform natural interaction in the open world. In: Proceedings of the ACM/IEEE International Conference on Human-Robot Interaction (HRI), pp. 183–184 (2012)
11. Li, L., Yan, S., Yu, X., Tan, Y.K., Li, H.: Robust multiperson detection and tracking for mobile service and social robots. IEEE Transactions on Systems, Man and Cybernetics, Part B 42(5), 1398–1412 (2012)
12. Ge, S.S., Yang, Y., Lee, T.H.: Hand gesture recognition and tracking based on distributed locally linear embedding. Image and Vision Computing 26(12), 1607–1620 (2008)
13. Yan, R., Tee, K.P., Chua, Y., Li, H., Tang, H.: Gesture recognition based on localist attractor networks with application to robot control. IEEE Computational Intelligence Magazine 7(1), 64–74 (2012)
14. Chen, N., Chew, C.M., Tee, K.P., Han, B.S.: Human-aided robotic grasping. In: Proceedings of the IEEE International Symposium on Robot and Human Interactive Communication (RO-MAN), pp. 75–80 (2012)
15. Chen, N., Tee, K.P., Chew, C.M.: Assistive grasping in teleoperation using infrared proximity sensors. In: Proceedings of the IEEE International Symposium on Robot and Human Interactive Communication, RO-MAN (2013)

Older People's Involvement in the Development
of a Social Assistive Robot

Susanne Frennert, Håkan Eftring, and Britt Östlund

Department of Design Sciences, Lund University, Sweden
Susanne.Frennert@certec.lth.se

Abstract. The introduction of social assistive robots is a promising approach to enable a growing number of elderly people to continue to live in their own homes as long as possible. Older people are often an excluded group in product development; however this age group is the fastest growing segment in most developed societies. We present a participatory design approach as a methodology to create a dialogue with older people in order to understand the values embodied in robots. We present the results of designing and deploying three participatory workshops and implementing a subsequent robot mock-up study. The results indicate that robot mock-ups can be used as a tool to broaden the knowledge-base of the users' personal goals and device needs in a variety of ways, including supporting age-related changes, supporting social interaction and regarding robot aesthetic. Concerns that robots may foster inactivity and laziness as well as loss of human contact were repeatedly raised and must be addressed in the development of assistive domestic robots.

Keywords: Participatory Design, Older People, Social Assistive Robots.

1 Introduction

Older people may become a major user group of social assistive robots. These robots have the potential to transform what is currently understood as normal routines in older people's everyday lives. The development and design of assistive robots is situated between what exists and what could exist. From a broader perspective, every assistive robot design process can be seen as a unique, ethical, aesthetic, political and ideological process that will affect older people's practices [13]. *Unique* since the designer, situation and resources will be different in the specific design situation at hand. *Ethical* since the design decisions will affect older people's practices and possible choices of actions. *Aesthetic* since every design we decide to adopt signals "who we are" and "who we want to be" and has to fit into its context. *Political and ideological* since social assistive robots will carry a set of assumptions about what it means to be old and to live in society as an elderly person. According to Bijker, no technological system or artefact has just one potential use; instead, the users mediate its use or non-use [2]. Schön argues that there are no right or wrong answers, only actions and consequences when designing artefacts [15]. Schön introduced the concepts *reflection-in-action* and *reflection-on-action* [15]. *Reflection-in-action* helps

G. Herrmann et al. (Eds.): ICSR 2013, LNAI 8239, pp. 8–18, 2013.

us to reshape what we are doing while we are doing it. *Reflection-on-action* helps us to evaluate what we have done and understand our tacit knowledge [15]. As a consequence, a participatory design process can be seen as an on-going conversation between the designers, the users and the situation [13].

In this paper we describe the implementation of three participatory workshops that were followed by a robot mock-up study. In the workshops, sketches and mock-ups are used to explore ideas and discuss them with older people. In order for assistive robots to be adopted by older people and since design is about the *not-yet-existing*, we need to understand the underlying problems of ageing as well as the meaning and possibilities assistive robots have for older people. Older people are experts on their lives and we, as researchers, need to gain an understanding of what might be meaningful for older people in regard to robotic assistance. We do not consider only functionality; we also need to understand their attitudes and thoughts. As Krippendorff eloquently put it, "No artefact can survive within a culture without being meaningful to those that can move it through its defining process" [11: 413]. The meaning of an artefact is not inherent or inscribed in the artefact; it is mediated by its users [11]. The approach of this study draws on theories from anthropology and design science. It aims to draw out the subjective and personal feelings an older person has about the experience of working with sketches of robots (*reflection-in-action*) and having a robot mock-up at home (*reflection-on-action*) [13].

The robot mock-up study is inspired by Design Noir – the Placebo Project [5]. Dunne and Raby conducted the placebo project as an experiment to test their ideas on electronic technology and its aesthetic meaning. The design project has its origin in critical theory and their design aims to challenge people's preconceptions and provoke new ways of thinking about an electronic artefact [10]. To our knowledge no study has reported the use of full-scale robot mock-ups in the domestic environment of elderly people.

Evidence indicates that involving the user in the development of technological systems has positive effects, such as increased sales, increased user productivity, decreased training costs and decreased user support [12]. This paper describes the methods used to involve older people in the development of a social assistive robot, and presents a summary of the findings from the first part of our participatory research project: HOBBIT. The HOBBIT Project is funded by the EU's Seventh Framework Programme. It aims to develop, in close contact with older people, the kind of help they would like from a robot. The purpose of the robot is to enable older people to continue living and functioning in their homes. The target users are older people, aged 70+, who perceive themselves to have moderate vision, hearing and mobility impairments. The reason for targeting this user group is that by the age of 70+, age-related impairments starts to occur [9]. We also believe that the HOBBIT needs to be introduced to the user in their home before the decision is taken that they can no longer live independently in their home. This research project was reviewed and approved by the Swedish Regional Review Board.

Concept Exploration by the Prospective Older Users: Workshops

As a preliminary to the concept development, and to identify main design features, a workshop, questionnaires (n=36) and interviews (n=14) were undertaken with older people (aged 70+). The findings indicated that older people wanted a robot that was easy to clean, not too small or too big, round in shape with no sharp edges, soft to touch and, preferably, neutral in colour to blend in with the home furniture [6].

The first phase in the conceptual development consisted of deploying an industrial designer, Johan Rosberg, who made sketches of the robot appearances based on the user and technological requirements. Concept generation activities were organised around the appearance of a prospective robot, the challenges of growing older, as well as, what kind of help was desired and needed. These ideation activities were facilitated by experience prototyping [3] and participatory workshops. Sketches of different looks, colours and forms of robots were shown to prospective users and the participants selected the ideas they thought offered the most interesting prospect for a future robot.

Workshop 1

Participants: Ten people (one man and nine women), between 70 and 84 years of age (mean age=77), participated in the first workshop (Fig. 1). They perceived themselves to have moderate vision (8), hearing (5) and mobility impairments (2). All the participants had signed a consent form.

Fig. 1. Participants in workshops 1 and 2

Procedure: The workshop was conducted and recorded in a usability lab at the Department of Design Sciences at Lund University, Sweden. The participants were divided into two groups and sat at two tables with one researcher at each. The participants were shown different sketches of robots and asked to discuss them. The researchers fed the discussion by asking what they liked and did not like about the different sketches. The participants were also asked to fill in a form with "three wishes" regarding the three most important things they would like to see in an assistive robot. They were also given a sheet of paper showing all the different shapes of the robot and were asked to choose the one they liked the most (Fig. 2). The researchers took field notes during the whole workshop.

Result: The most common wishes from the participants in workshop 1 were that the robot should look happy, be easy to clean, not too big, easy to use, and stable so it could be used for mobility support These results are in line with the ones we found during our previous workshop [6]. The favourite choice of robot shape was C, which we think looks like a butler (Fig. 2). The second favourite was F, which we call the "Scandinavian model" because of its minimalistic shape.

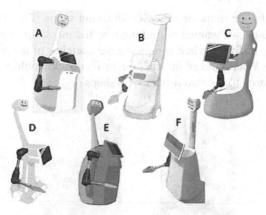

Fig. 2. Robot sketches

Workshop 2
To validate and confirm the findings from the first workshop a second one was arranged (Fig. 1)

Participants: Eight people (four men and four women), between 69 and 81 years of age (mean age=75), participated in the workshop. They perceived themselves to have moderate vision (6), hearing (4) and mobility impairments (3). The second workshop was also conducted and recorded in the usability lab at the Department of Design Sciences at Lund University. None of the participants received any monetary reward but they were given free coffee and cakes.

Procedure: As in workshop 1, the participants were invited to a workshop regarding robot appearance and they had all signed a consent form. This time, the participants first saw the sketches of the different robot appearances in black and white, and after they had discussed the different shapes (while the researchers took notes) they were shown the same sketches but in different colours. The reason for showing the sketches first in black and white was that we wanted the participants to focus on the shape of the robot without being influenced by its colour. We wanted to validate the findings from the first workshop. Was the same shape popular even if it was in black and white (fig 2)? And which colour would they like their favourite shape to be in? The participants were also asked to describe what they considered to be the three most important things about a robot and its appearance.

Result: The favourite choices in the second workshop were the same as in the first. For the three wishes, most of the participants wanted, just as in workshop 1, a robot that was easy to use, neutral in colour, happy in appearance and reliable. They also wanted it to have a big screen (at least 17 inches) that was adjustable.

Feedback from the Prospective Older Users: Mock-Up Workshop
The results from the two workshops fed into the next phase, the mock-up phase. Two full-scale mock-ups were produced in polystine style foam. The shapes of the two

mock-ups were the butler shape and the Scandinavian shape (Fig. 3). We organised a third workshop because we wanted to find out how the mock-ups were perceived. Did the subjects feel any different when comparing the sketches of the robot shapes to the real-size mock-ups? What kind of information or feedback would we get by showing the mock-ups compared to the two former workshops?

Fig. 3. The butler and the Scandinavian mock-ups

Workshop 3
Participants: Nine people (two men and seven women), between 72 and 81 years of age (mean=76), participated in the third workshop. They perceived themselves to have moderate vision (7), hearing (4) and mobility impairments (3). The workshop was conducted and recorded at the Usability Lab at the Department of Design Science.

Procedure: As in workshops 1 and 2, the participants were invited to partake in a workshop concerning robot appearance. They all signed a consent form. The participants saw and discussed the two mock-ups (Fig. 3). They were displayed separately at first, one after the other, and then both at the same time. This was done so that the participants could first discuss the mock-ups separately and then compare them. They were given the same sheet of paper as in workshops 1 and 2, showing the different sketches of the robot shapes. They were asked to choose the one they liked best (Fig. 2). They were also asked to write down three wishes they had regarding robot appearance.

Result: An interesting finding in the third workshop was that the favourite choice of shape changed. This time 50% of the participants preferred shape D (Fig. 2). The reason *why* shape D become a favourite, but was only selected once in the other two workshops, may be that one participant argued quite strongly in favour of that shape. The others may have been influenced by his arguments. Another reason may be that when they saw the mock-ups, they realised that the butler shape was bulky. The bright blue colour of the butler was also commented on by many, since it reminded them of the equipment in hospitals and healthcare facilities. Another thing not mentioned before was that the "neck" of the robot was too long. They did not like the feeling that the robot (i.e. the two mock-ups) was looking down at them when they were seated. They repeatedly stated that they wanted to be in charge; they wanted to tell the robot what to do. The butler-shaped mock-up was perceived as too big and bulky while the

Scandinavian shaped mock-up was perceived as more petite and easier to fit into one's home because it was white.

Feedback from the Prospective Older Users: Robot Mock-Up at Home

As a complement to the workshops, we set out to investigate the meaning and values the two robotic assistance mock-ups (Figs. 3 and 4) had for six senior households (five seniors living alone and one couple, making a total of 7). In the mock-up study we wanted to challenge the older peoples' preconceptions and expectations of robots and get them to reflect about robots, their use, and their appearance. They were to do this while living with a robot mock-up for a week in their own homes. Most of the older people involved had probably never thought about having an assistive robot. By letting them have a robot mock-up in their homes for a week, we hoped to encourage them to think carefully about robots as assistants, and give them an opportunity to discuss robots and the mock-ups among themselves and consider them carefully before they were interviewed. The research questions to be addressed were: How do older people perceive a full-scale mock-up? What beliefs, attitudes and needs do older people perceive in regard to robotic assistance? What factors influence their attitudes towards robots?

Fig. 4. The mock-ups in some participants' homes

Participants: Five women and two men, between 72 and 76 years of age (mean age=75 years), participated in the study. They perceived themselves to have moderate vision (6), hearing (3) and mobility impairments (4).They were volunteers who had answered an advertisement.

Procedure: A researcher, who explained that a mock-up of an assistive robot had been developed based on the requirements gathered from participatory design workshops with older people, contacted the participants. The participants agreed to have the mock-up in their homes for one week and thereafter to be interviewed regarding their experience. Four of the participants had the butler mock-up at home and five had the Scandinavian mock-up.

Data Analysis: The interviews were transcribed. They were then read and reread to get a sense of the whole. They were listened to numerous times, as well, to obtain an overall impression. The text regarding the participants' perceptions, attitudes and expectations of assistive robots was extracted and brought together into one text, which constituted the unit of analysis. The text was divided into meaning units that were given a brief description close to the text while still preserving the core (condensed) [8: 106]. The condensed meaning units were grouped together under sub-categories and the various sub-categories were compared based on the differences and similarities, then sorted into categories which constituted the manifest content. The whole process was an iteration of coding the data, identifying the sub-categories and categories and revising the coding scheme until agreement on the final sub-categories, categories, and theme was achieved. As a result, three categories were found: supporting age related changes, supporting social interaction, and robot aesthetics. The latent content, or the underlying meanings of the categories, was formulated into trust, privacy and being in control.

Result: The "supporting age related changes" category held four sub-categories: 1) physical support; 2) cognitive support (i.e. reminders); 3) health promotion; and 4) fear of inactiveness. To increase independent living the participants emphasised that they needed help with daily living tasks that, due to age-related problems, they found difficult or impossible, such as putting on socks, ironing, scrubbing floors, filling and emptying the washing machine, reaching up to get things from a shelf and carrying heavy grocery bags. Regarding decreased cognitive ability, the participants said that reminders to take medication and the robot functioning as a stove safety control would be beneficial. The general opinion among the participants was that an assistive robot was not suitable for people with severe cognitive impairments since the participants thought people with dementia would not be able to learn how to operate it. Fears were also expressed that assistive robots might promote laziness and inactive living: "*Everyone knows that to lose weight you need to eat healthily and exercise but few of us do. Especially if you are in pain! If we have a robot we might be tempted to use it as a servant instead of doing things ourselves, which in the long run will make it even harder to keep fit*". To increase physical activity and promote health, the participants suggested that the robot could remind them about exercise and healthy eating. Although they would not like be told what to do by the robot, they could imagine signing a "contract" with the robot where they agreed to do certain kinds of exercises and gave the robot permission to remind them if they forgot to keep their side of the agreement: "*I see it like a New Year's resolution. You have to make up your mind about going for walks and doing exercises and the robot can help you keep on track with your fitness regime. It has to be a mutual agreement. I would not like the robot to tell me what to do and what to eat if I hadn't programmed it to do so*".

The "supporting social interaction" category held three sub-categories: 1) increased interaction; 2) loss of human contact; and 3) finance. Human contact was mentioned as an important factor in the older people's lives. They could imagine an assistive robot as a complement to home-care service but not as a substitute for human contact: "*I have problems with my back, which makes it difficult to put on my trousers. I have to ask my husband to help me. If I didn't have my husband I would prefer a robot to a stranger to help me with my trousers, but I would also like daily contact with other people*". Some of the participants thought that a robot with a screen could be used short-term to increase social interaction with healthcare professionals when needed. This was especially relevant after a hospital visit when some of them often felt lonely and anxious at home. Compared to the workshops, the mock-up also inspired discussions about finance. Most of the participants perceived a robot as a very expensive tool and they could not see the economic benefits of having one since its functionality was perceived as limited: "*It will take many years until a robot might be really useful. I do not think it will happen during my lifetime*". They also mentioned that they knew of other devices that could do most of the things they perceived an assistive robot might be able to do, such as pick-up grippers (to pick up things from the floor), smartphones and computers (for reminders).

The "**robot aesthetic**" **category** held four sub-categories: 1) non anthropomorphic design; 2) anthropomorphic design; 3) physical space; and 4) indoor adjustments. The attitude towards the design of the mock-ups was that they were bigger than expected (especially the butler), but that they got used to them after a couple of days. Some of the participants (3) could not see the point of the upper part and questioned why it had to have a head; "*Why does it have to have human features like a head? It would be better if it was smaller, without a head*" while others wanted it to have human features (4) "*I thought it was missing a face and that is why I painted a mouth and two eyes on it*". The question concerning appearance and human features addressed the feeling of being under surveillance. Those who preferred a machine-like robot, argued that they might feel under surveillance if it had a face like a human "*If it had eyes I would feel observed all the time. What if I am naked?* ". They could imagine using the "neck" to pull themself up from a chair or the bed. They emphasised that it was important for it to have arms, preferably two, in order to help them. The screen had to be big, vertically and horizontally adjustable to stay free of glare, and should be ergonomically suitable for all. The participants took for granted that in the future they would be able to choose the colour they liked. Regarding physical space, the participants were asked to move the mock-up around in their homes to get a feeling of having it in the bedroom or bathroom. Some of the participants thought the mock-ups were too difficult to move around, while others complained that their bathroom and kitchen were too small to accommodate the mock-up. Concerns were also raised about being watched all the time, but most of them saw this as tolerable if the robot were useful: "Is it going to spy on me? If so, it had better be very useful!" The participants were prepared to make indoor changes to accommodate a robot.

The **latent content analysis** – trust, privacy and being in control: An underlying matter that came up all the time was the question of trust and privacy. Few people seemed willing to place unconditional trust in assistive robots. The trust had to be earned. Assistive robots may earn peoples trust by demonstrating their competence and skills. The findings indicate that an assistive robot has to communicate clearly and carry out requests reliable and effectively. The results suggest that the feeling of being in control is crucial. Bandura's theory of self-efficiency, which refers to the individual's belief in his or her capacity to learn and perform certain behaviour, shows that self-efficiency is an important factor for supporting people to change or adopt a "new" behaviour [1: 117-148]. When applied to assistive robots, if the robots help older people to feel self-efficient in their everyday life, they might feel more in control of their life and be more willing to adopt the robot. Assistive robots might make older people empowered and motivated to manage their everyday life if they feel confident in their ability in operating the assistive robot.

Next Steps: Prototype Testing in the Usability Lab and Prototype Testing in Older People's Homes

New technologies with which people have no experience evoke feelings of uncertainty in how to behave. An explorative Wizard-of-Oz scenario-based method will be used in a lab environment to evaluate the suitability of solutions identified early in the development process [4] (Fig. 5). The suggested framework of robots in older people's practices will be further evaluated and defined when, in the near future, a robot with a similar appearance to the mock-ups will be developed and moved into 18 older people's homes. Their experiences of having the robot in their homes will be studied, since we do not know how assistive robots affect older people and their social connectedness and wellbeing over time. We would also like to determine the benefits of using the approach described above. The question is: How do the requirements captured early on by a mock-up correspond to the ones captured with a "real" prototype?

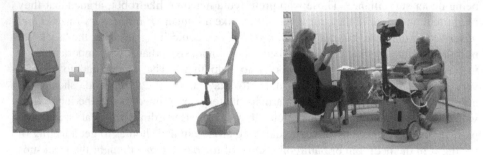

Fig. 5. Wizard-of-Oz scenario in the Usability Lab

2   Conclusions

In this paper we describe a study with a participatory design approach that involves older users, and that explores their acceptance and need of assistive domestic robots. The users' motivation for taking part in the design work was high. The three most common motives for participating in the study included: (a) curiosity about robots and new technological innovations; (b) a desire to add value and share experience and knowledge about what it is like to be old; and (c) a desire to make an impact on the development of robots and prevent the future for older persons becoming too machine-driven, without human contact. The opportunity to participate in the development process of an assistive robot was considered valuable, even though most of the participants believed they would not be alive when assistive robots became an ordinary part of older people's lives. Past contributions from studies involving older people indicate that loyalty to commitment in participation is high among older people [15].Technology is embedded in everyday practice [13]. For assistive robots to be part of older people's practices, they have to be designed to fit into older people's habits and routines. The involvement of older people in the development of an assistive robot is crucial to identify their attitudes, values and intentions. The robot mock-up study has provided valuable experience in how to use mock-ups in older participants' homes in order for them to form rich mental images and an understanding of how they could use assistive robots in their domestic environment. After they imagined themselves actually using an assistive robot, they thought that physical and cognitive support would be crucial. The slogan "form follows function" springs to mind. An assistive robot may constitute an extension of the physical and cognitive effects of ageing in older people's practices. The results of the mock-up pilot study indicate that product mock-ups can be used as tools to broaden the knowledge base of the user's personal goals and device needs in a variety of ways, in this case supporting age-related changes, supporting social interaction and regarding robot aesthetic. In sum, the acceptance of social robots does not depend on a single variable but rather on multiple variables such as; 1) individual variables e.g. personal evaluations of one's disability in relation to a specific social robot, perceived needs and functional status; 2) environmental variables e.g. the social and physical characteristics of the usage context; and 3) the variables of a specific social robot e.g. ease of use, the design of the device and the person-task fit [7].

References

1. Bandura, A.: Perceived self-efficacy in cognitive development and functioning. Educational Psychologist 28(2), 117–148 (1993)
2. Bijker, W.E.: Social Construction of Technology. Wiley Online Library (2009)
3. Buchenau, M., Suri, J.F.: Experience prototyping. In: Proceedings of the 3rd Conference on Designing Interactive Systems: Processes, Practices, Methods, and Techniques, pp. 424–433. ACM (2000)
4. Carroll, J.M.: Making use: scenario-based design of human-computer interactions. MIT Press (2000)
5. Dunne, A., Raby, F.: Design noir: The secret life of electronic objects. Birkhäuser (2001)
6. Frennert, S., Östlund, B., Eftring, H.: Would Granny Let an Assistive Robot into Her Home? In: Ge, S.S., Khatib, O., Cabibihan, J.-J., Simmons, R., Williams, M.-A. (eds.) ICSR 2012. LNCS, vol. 7621, pp. 128–137. Springer, Heidelberg (2012)
7. Gitlin, L.N., Schemm, R.L., Landsberg, L., Burgh, D.: Factors predicting assistive device use in the home by older people following rehabilitation. Journal of Aging and Health 8(4), 554–575 (1996)
8. Graneheim, U.H., Lundman, B.: Qualitative content analysis in nursing research: Concepts, procedures and measures to achieve trustworthiness. Nurse Education Today 24(2), 105–112 (2004)
9. Gunter, B.: Understanding the older consumer: The grey market. Routledge (2012)
10. Koskinen, I., Zimmerman, J., Binder, T., Redstrom, J., Wensveen, S.: Design Research Through Practice: From the Lab, Field, and Showroom. Morgan Kaufmann (2011)

11. Krippendorff, K.: Principles of Design and a Trajectory of Artificiality. Journal of Product Innovation Management 28(3), 411–418 (2011)
12. Kujala, S.: User involvement: a review of the benefits and challenges. Behaviour & Information Technology 22(1), 1–16 (2003)
13. Löwgren, J., Stolterman, E.: Thoughtful interaction design: A design perspective on information technology. MIT Press (2004)
14. McCarthy, J., Wright, P.: Technology as experience. Interactions 11(5), 42–43 (2004)
15. Schön, D.A.: The reflective practitioner: How professionals think in action, vol. 5126. Basic books (1983)
16. Östlund, B.: Images, users, practices: senior citizens entering the IT-society, vol. 1999, p. 9 (1999)

What Older People Expect of Robots:
A Mixed Methods Approach

Susanne Frennert, Håkan Eftring, and Britt Östlund

Department of Design Sciences, Lund University, Sweden
Susanne.Frennert@certec.lth.se

Abstract. This paper focuses on how older people in Sweden imagine the potential role of robots in their lives. The data collection involved mixed methods, including focus groups, a workshop, a questionnaire and interviews. The findings obtained and lessons learnt from one method fed into another. In total, 88 older people were involved. The results indicate that the expectations and preconceptions about robots are multi-dimensional and ambivalent. Ambivalence can been seen in the tension between the benefits of having a robot looking after the older people, helping with or carrying out tasks they no longer are able to do, and the parallel attitudes, resilience and relational inequalities that accompany these benefits. The participants perceived that having a robot might be "good for others but not themselves", "good as a machine not a friend" while their relatives and informal caregivers perceived a robot as "not for my relative but for other older people".

Keywords: Mixed methods, Older people, Expectations, Preconceptions, Robots.

1 Introduction

The growth of an older population is predicted to drive the development of domestic assistive robots that match technological advances to human needs. Older adults as a group are very heterogeneous and age is likely to increase the differentiation within the "group" more than most other "groups" due to life experiences and physical conditions [1]. The diversity of the older population creates challenges for roboticists. Robots may fail because the prospective users do not want or need them [2]. Latour and Woolgar argue that new technologies try to configure the user by setting constraints on what he or she is able to do [3]. The user may be represented either by the developer's personal image of future users (which the developer already has in mind), or by the developer's own preferences, needs and wants [4]. During the development of robots, developers configure the users into the product [3]. Robots are not neutral but come with inscribed and embedded values. However, the developer cannot determine how the robot will be used because the user will find multiple possible usages depending on the context and on the interactive relationship with the robot [5]. How an assistive robot will be used depends on the users' perceptions and interpretations of the robot's mediating role. It is important to understand older users' expectations and preferences for robotic assistance in order to match them with the robots' capabilities.

G. Herrmann et al. (Eds.): ICSR 2013, LNAI 8239, pp. 19–29, 2013.

The paper proceeds as follows: related work is presented to put the research in context. The research process is then described, followed by a discussion of the findings and implications. The findings will feed into the EU-funded HOBBIT Project (www.hobbit-project.eu), which aims to develop a socially assistive robot that helps seniors and old people at home. The project focus is the development of the mutual care concept: building a relationship between the human and the robot in which both take care of each other, similar to building a bond with a pet.

2   Related Work

Several studies have investigated factors that may influence people's acceptance of domestic robots. Salvini et al. [6] claim that the acceptability of a robot is dependent on social acceptance by others, legal issues and the socio-ethical issues (robots are seen as replacements for humans). A key notion in their findings is the social construction of social robots: attitudes of others toward the robot affect the individual's attitude and acceptance of it. Similarly, Forlizzi et al. highlight that the design of assistive robots has to fit into the ecology of older people, support their values, and adapt to all the members of the system who will interact with the robots [7]. The authors claim that robots must be conceived as part of a larger system of exciting products and environments. There is sufficient evidence that social and cultural implications as well as multiple and universal functionality must be considered [7]. In fact, research has suggested that it is important to assess the expectations and needs of a range of stakeholders (older people, family, medical staff) [8]. This is further confirmed in another review by Bemelmans et al., who also claim there is a need for all the stakeholders to understand the benefits of the robot in order for it to be accepted [9]. Kidd et al. state that a supporting socio-material setting is important for the acceptance of a robot in a healthcare context [10]. Heerink et al. present a model of how to assess older adults' acceptance of assistive social technology [11]. Their model has been tested in several studies with older adults. They propose that the use of social agents is determined by social influence, facilitating conditions, perceived usefulness, perceived ease of use, perceived enjoyment, attitude and trust. Likewise, Young et al. confirm in their review that acceptance of robots is dependent on subjective users' perceptions of what robots are, how they work, and what they can and cannot do [12]. They conclude that the factors affecting acceptance are: safety, accessibility, usability, practical benefits, fun, social pressure, status gains, and the social intelligence of the robot. Beer et al. used questionnaires and structured interviews to assess older adults' openness to robotic assistance and to elicit their opinions on using robotic assistance [13]. Their findings indicate that the participants were open to the idea of robotic assistance for cleaning, organising and fetching, but they were concerned about the robot damaging their homes, being unreliable, and being incapable of doing what they were asked. However, one has to keep in mind the limited availability of commercialised robots. The above findings may be due to the novelty of the robots or biased by the participants' willingness to partake in research projects.

3   Methodology

The data collection involved mixed methods including focus groups, a workshop, a questionnaire and interviews. The findings and lessons learnt from one method fed into another. Pre-study focus groups were used to identify preconceptions of assistive robots. A workshop was then organised to verify the findings of the focus groups and to identify major themes in the creation of the questionnaire that was administered to older adults. Based on the workshop and questionnaire results, fourteen qualitative interviews were held with potential users to gain more detailed information about how they imagined the potential role, or lack of one, of robots in their lives. The interviews focused on how the old users imagined they would like certain functions and scenarios to work. Six interviews were also held with relatives/informal caregivers of the elderly people. In total, 88 older people participated in the study, and together the methods provided a large amount of data.

Interpretive phenomenological data analysis was used to gain an understanding of the participants' expectations and preconceptions of robots. The data was gathered from February 2012 to June 2012. Field notes were taken during and after the focus groups, the workshop and interviews. Each interview was recorded. Keywords, phrases and statements regarding expectations and preconceptions of robots were transcribed. The observational notes and the seniors' notes on the attention cards used in the workshop were transcribed. All the answers to the open questions in the questionnaire were read and the ones regarding preconceptions and expectations were separated. All the data were systematically read, and statements in which old adults expressed their preconceptions and expectations of robots were entered into a data file. Condensed meaning units were identified as robot acceptance criteria and labelled with codes. Similar codes were merged into subcategories. The subcategories were sorted into a conceptual scheme. The analysis involved an iterative process, with the researchers constantly moving back and forth between the data sets in order to identify frequency, specificity, emotions, extensiveness, and the big picture [11].

3.1   Research Methods and Results

Focus Groups and Workshop
As a preliminary to the questionnaire and interviews, and to identify major themes that needed to be addressed, two focus groups (8 + 7 participants) were undertaken with older people (age 70+), and a workshop (14 participants with a mean age of 75). Detailed information about the focus groups and the workshop procedure and findings were presented at ICSR2012 [14].

Questionnaire
A questionnaire was developed to validate the findings from the workshop and to identify the themes that needed to be addressed in the subsequent interviews. The themes of the questionnaire were: functionality, safety, operation, mutual care, and appearance. The functionality questions regarded what kind of functionality would be perceived as helpful. The safety questions addressed emergency detection and support

in an emergency, as well as perceived fears. The operation questions referred to how older people wanted to communicate with a robot. The questions addressing mutual care concerned emotional bonding. Finally, the design-related questions addressed the preferences of appearance and size of a prospective robot. The questions were grouped into blocks according to the five main topics. Almost all of the items were formulated as statements. Users were asked to imagine they had a robot at home. They responded to most of the statements on a four-level Likert scale with the alternatives: I agree, I somewhat agree, I rather disagree and I disagree. Some responses simply consisted of a yes/no option. In a final statement on design, users were asked to choose the one picture out of six robot models they liked best. Some statements also contained space for open answers and additional comments by users.

A pilot test of the questionnaire was conducted (one man and two women, ages 73, 76 and 75). Overall, it was well received, but a three statements had to be reframed, as well as the response alternatives.

Participants (estimated to be 100) at a seminar on the HOBBIT Project at the Senior University in Lund, Sweden were asked to complete the questionnaire. Fifteen participants filled out the questionnaire directly after the seminar, while 34 wanted to have it sent to them with a paid response envelope so they could fill it out at home and send it back.

Questionnaire results
In total, 36 older adults, with an average age of 77.6 years, answered the questionnaire. The majority, 72%, were women living on their own, and lived in apartments without any regular home care services or relatives helping out. Most of the respondents used a computer daily (64%).

Table 1. Functionality preferences

FUNCTIONALITY								
I would like my robot to be able to:			I would like my robot to remind me of:			I would like my robot to have the following functions:		
Task	No. of people	%	Reminder	No. of people	%	Function	No. of people	%
Carry small objects for me	21	12	Today's date	25	15	Internet	24	16
Store things (like a drawer)	20	11	TV-Programmes	25	15	Telephone	30	20
Find things I ask for	32	18	Medication intake	31	19	Radio	26	17
Fetch	29	16	Fluid intake	15	9	TV	22	14
Grasp things from the floor	26	15	Appointment/phone calls	26	16	Music	21	14
Get things from shelves	29	16	Meal times	13	8	Games	10	7
Following me	19	11	Body care	9	6	Audio books/news	18	12
			To go shopping	17	10	Other	1	1
			No reminders	2	1			
People may select more than one check box, so percentage may add up to more than 100%								

Table 1 shows the functionality preferred by the respondents. The responses revealed a large diversity among the options. The results indicate there was less interest in the robot offering computer games and reminders about personal hygiene compared to the other functionality options.

Table 2 shows that the majority of the respondents would not find having a robot frightening. They would like the robot to be alert during the night as well as to alert others when there was an emergency.

Table 2. Safety preferences

SAFETY								
My robot can call for help, if something happens to me. That would make me feel safe:			The idea of having a robot at home is frightening:			I would like my robot to be alert during the night:		
Options	No. of people	%	Options	No. of people	%	Options	No. of people	%
I agree	30	86	I agree	3	8	I agree	26	79
I somewhat agree	4	11	I somewhat agree	7	19	I somewhat agree	6	18
I rather disagree	1	3	I rather disagree	0	0	I rather disagree	0	0
I disagree	0	0	I disagree	26	72	I disagree	1	3
The percentage may not add up to 100% if some people decided not to answer the question								

Table 3. Operation preferences

OPERATION								
My robot should only do what I ask it to do:			My robot would make sure its batteries are always charged:			My robot would be allowed to move everywhere in my home:		
Options	No. of people	%	Options	No. of people	%	Options	No. of people	%
I agree	27	75	I agree	36	100	I agree	24	71
I somewhat agree	8	22	I somewhat agree	0	0	I somewhat agree	7	21
I rather disagree	0	0	I rather disagree	0	0	I rather disagree	0	0
I disagree	1	3	I disagree	0	0	I disagree	3	9
How would you like to operate your robot?			How would you like to interact with your robot?					
Options	No. of people	%	Options	No. of people	%			
Keyboard/ buttons	9	10	Touchscreen	24	40			
Touchscreen	19	22	Voice/speech	26	43			
Voice operated system	25	28	Gesture	10	17			
Gestures	13	15						
Remote control	22	25						
People may select more than one check box, so percentage may add up to more than 100%. The percentage may not add up to 100% if some people decided not to answer the question								

Table 3 shows that the majority of the respondents would like the robot to be passive and only do what they tell it to do. It would have to be autonomous and make sure that its battery was always charged. Approximately 70% of the respondents answered that a robot would be allowed to move everywhere in their homes. Touchscreens and voice input and output were the favorite options for communicating with the robot.

Table 4. Mutual care preferences

MUTUAL CARE											
I would like my robot to help me keep in touch with family and friends			I would like my robot to be able to suggest activities that I might do			I would enjoy it if my robot did something together with me			I would like my robot to keep me company		
Options	No. of people	%	Options	No. of people	%	Options	No. of people	%	Options	No. of people	%
I agree	15	43	I agree	14	41	I agree	4	11	I agree	5	14
I somewhat agree	10	29	I somewhat agree	9	26	I somewhat agree	4	11	I somewhat agree	4	11
I rather disagree	2	6	I rather disagree	3	9	I rather disagree	5	14	I rather disagree	6	17
I disagree	8	23	I disagree	8	24	I disagree	22	63	I disagree	20	57
I would give my robot a name			I would help my robot to learn about my environment and habits			I am willing to move some furniture to make it easier for my robot to move around in my home			I think my robot would make me smile		
Options	No. of people	%	Options	No. of people	%	Options	No. of people	%	Options	No. of people	%
I agree	25	76	I agree	26	79	I agree	19	58	I agree	10	30
I somewhat agree	3	9	I somewhat agree	5	15	I somewhat agree	11	33	I somewhat agree	9	27
I rather disagree	0	0	I rather disagree	0	0	I rather disagree	1	3	I rather disagree	5	15
I disagree	5	15	I disagree	2	6	I disagree	2	6	I disagree	9	27
I would show my robot to family and friends			I would take care of my robot			I would reward my robot			I would like my robot to be cuddly		
Options	No. of people	%	Options	No. of people	%	Options	No. of people	%	Options	No. of people	%
I agree	27	82	I agree	28	85	I agree	8	24	I agree	2	6
I somewhat agree	1	3	I somewhat agree	2	6	I somewhat agree	8	24	I somewhat agree	6	19
I rather disagree	1	3	I rather disagree	0	0	I rather disagree	2	6	I rather disagree	4	13
I disagree	4	12	I disagree	3	9	I disagree	15	45	I disagree	20	63
The percentage may not add up to 100% if some people decided not to answer the question											

Table 4 shows that more than half of the respondents would like the robot to help them keep in touch with family and friends. The majority of the respondents believed that they would not enjoy doing things with the robot or for it to keep them company. On the other hand, the majority answered that they were willing to move furniture to accommodate the robot as well as teach it about their habits and environment. They also thought they would show the robot to friends and family.

Table 5. Items regarding design issues

DESIGN								
I would like my robot to have the following height			My robot is:			I would like my robot to have a soft surface		
Options	No. of people	%	Options	No. of people	%	Options	No. of people	%
Knee-high	7	19	Male	1	3	I agree	11	35
Waist-high	15	41	Female	3	9	I somewhat agree	8	26
Chest-high	9	24	Neutral/Genderless	28	88	I rather disagree	4	13
Head-high	4	11				I disagree	8	26
Taller than me	2	5						
I would like to decorate my robot			It is more important to me how the robot look than what it can do			Pictures		
Options	No. of people	%	Options	No. of people	%	Options	No. of people	%
I agree	1	3	I agree	1	3	Care-o-bot	14	44
I somewhat agree	8	25	I somewhat agree	3	9	The yellow HOBBIT	3	9
I rather disagree	6	19	I rather disagree	3	9	Boxie	3	9
I disagree		53	I disagree	25	78	Asimov	5	16
						Paro	3	9
						Geminiod-girl	4	13
People may select more than one check box, so percentage may add up to more than 100%.								

Table 5 shows that 78% of the respondents valued the functionality of the robot more than its appearance. 88% would like the robot to be neutral/genderless and they would give it a name. They could not imagine decorating the robot. Care-O-bot® was the favorite option among the pictures they could choose from.

Interviews

The results of the workshops and questionnaires formed the foundation of the interview guide. The interview guide served to investigate in more detail aspects of the questionnaire findings. Several organisations for seniors and relatives were contacted. The organisations received written information about the EU project, which they were asked to distribute. The seniors and informal caregivers who were interested in being interviewed were asked to contact the first author at the Department of Design Sciences at Lund University to book an interview. Fifteen seniors contacted the researcher and fourteen were interviewed. One was unable to make attend the interview. The participants consisted of three men and 11 women, ranging in age from 65 to 86. Six informal caregivers (three men and three women) were also recruited and interviewed. At the beginning of the interview, a consent form was signed by the participants that informed them about the data collection and procedures of the project. The interviews were conducted either at the participant's home or at the Department. The participant decided where the interview took place. The interviews lasted approximately an hour.

Interview results

The older people interviewed were asked to imagine getting an assistive domestic robot and to describe how they would like it to be presented to them. Most said they would want it to be presented by a knowledgeable representative from the manufacturer or a healthcare professional. They would want someone to demonstrate how to operate the robot. They wanted to know what they could expect of the robot, what the robot could do, how they could make it carry out certain actions, and what to do if something went wrong.

After the initial demonstration, the participants would themselves like to try to operate the robot. Four of them said that they would like to write their own step-by-step guide on how to use the robot at the same time as they were testing it. One lady stated: "I would prefer to make my own user guide since I tend to remember things better if I write them down as I am using the product". The participants emphasised the importance of a reset button. They would want to explore the robot but only if they knew that they could always reset it: "It has to be possible to explore how to use the robot...I like pressing all the buttons and trying out different things if I know that I can always go back to square one," and "I am afraid of breaking it...I would be afraid of using it...what if I did something wrong?"

Most of the participants said they would like a written manual, describing step by step how to make the robot take certain actions like getting keys, or finding a wallet. The participants complained about difficulties in reading and comprehending manuals: "I think the manuals nowadays are automatically translated because the language they use is almost impossible to understand".

The majority of those interviewed mentioned that they would like to be able to phone a helpline if something went wrong or if they needed help managing the robot: "I would like to be able to talk to someone...to describe my problem...and to get someone to explain to me what to do on the phone while I am trying it out to see if it is working or what happens". Some of the participants strongly emphasised the importance of a follow-up visit from healthcare professionals or the manufacturing company representative. Some of the participants complained that they often receive assistive tools that someone initially demonstrates, but after that they hear nothing from the person and they do not know if they are using the tools correctly. During the interviews it became apparent that for the participants to use an assistive robot, they would like proper training. A majority mentioned feeling inhibited about interacting with new technologies since the difficulties they experienced were perceived as due to their inadequacies rather than to the complexity of the system. They insisted that in order for them to use a domestic robot they needed to be trained in how to use it properly; they needed to be able to understand when they made an error and how to recover from the error, or how to get help to fix the problem.

Most of the older people interviewed said they would like to be able to speak to the robot. Some said that they also would like a screen that showed the text at the same time as the robot spoke. One participant mentioned that it is sometimes difficult to hear and interpret what someone is saying, and so they would prefer to be able to read it at the same time. This request may be of cultural origin since in Sweden it is common to have Swedish subtitles in most English-speaking TV series and programmes.

The participants found it difficult to believe that they would form any kind of relationship with the robot. A majority said it is a tool that would make everyday life easier for them, but it would not be a friend. Most wanted a cheerful and happy robot. They emphasised that the robot should be passive in the initial phase and wait for them to give it "orders" but after they were used to having it around, it could be more proactive and come up with suggestions for activities and exercises. One prospective user compared the functionality of the robot to an iPad. She wanted the robot to have basic functionality at the beginning, but after getting used to the robot, she wanted to be able to add functionality as she did when downloading new apps onto her iPad.

All of the informal caregivers thought that their relatives (elderly mum or dad) would not be able to learn how to use a robot. They could see the benefit of having a robot looking after their parent but they did not believe it was feasible since they believed their parents would have difficulties learning new technologies. The majority also thought a robot would scare their elderly parents or relatives.

4   Latent Content of the Result of the Mixed Methods Approach

This paper describes an interpretative research approach consisting of focus groups, a workshop, a questionnaire and interviews with old adults and relatives of older adults. The aim was to find out if Swedish old adults could imagine a robot finding a place in their homes, and if they could imagine what roles the robot might take in their lives.

4.1   "Not for Me But for Other Older People"

An interesting finding was that at first most of the participants in the focus groups were negative towards the idea of robots but after one hour or so, after finding out more about the robots, some of them changed their mind and thought robots might be "good for others but not themselves". Perhaps the unwillingness to imagine having an assistive robot is due to the reluctance to accept the physical and cognitive effects of ageing. However, in the workshop and questionnaire we obtained contradictory results; most of the participants stated that they would gladly show their robot, if they had one, to family and friends.

The reasons for these contradictory results might be pre-knowledge of the research focus – the participants in the focus groups did not know that they were going to talk about robots, while the participants in the workshop, questionnaire and interviews knew they would. Another reason could be that in the workshop, we spent an hour showing different kinds of robots and explained what modern robots are able to do, while the focus groups did not get any information about robots. It has been shown that remaking technology to something close and familiar increases the adoption rate [12]. Perhaps by raising the awareness of what robots can do might decrease the participants' fear of robots and imagining robots in one's life might feel less unfamiliar or strange.

4.2   "Not for My Relative But for Other Older People"

Another interesting finding is that the informal caregivers thought that a robot might be good for other older people but not their elderly relative. There did not seem to be a stigma attached to wanting one's relative to have a robot, but a belief that one's relative would not want a robot or be capable of operating one.

4.3   "As a Machine But Not As a Friend"

Our results suggest that most of the participants would not like to perceive a robot as a social companion for themselves, but for other people who are lonely, fragile and disabled. These findings suggest that there is a stigma attached to having a robot as a "friend". The unwillingness of seeing a robot as a companion may show concerns about the stigma of being dependent on a machine. Having a robot as a social companion seems to be perceived as affecting the individuals' self-image and signals to others that they are lonely and fragile. In fact, other research results indicate that adoption of an innovation is often due to the desire to gain social status [15]. In contrast to having a robot as a "friend" our results indicate that having a robot as a servant, which could do monotonous and challenging chores, was perceived as somewhat acceptable and satisfactory, which confirms to some extent the findings of Dautenhahn et al. [16].

5   Conclusion and Further Work

The results of the combination of methods described above provide an insight into old Swedish adults' expectations and preconceptions of robots. For them to see a robot as being beneficial to their lives, it has to be reliable, adoptable, appealing and easy to use. It is also important that old adults feel that they are in charge of the robot. Their presumptions about robots seem to be ambivalent. Most of the participants could see the benefits of having a robot looking after them and others in that it could help with or carry out tasks they are no longer able to do. On the other hand, they perceived the negative implications of having a robot as being dependent on a machine or being alienated from human contact. However, these expectations and preconceptions toward robots may change if the participants have direct personal experience of robots [13]. During the development of new technologies the developer configures the users into the product [3]. When it comes to robots, roboticists try to guide the users to interact with robots in certain ways when they decide the robot's functionalities and how these are represented. If using Madeline Akrich's terminology, one can say that the roboticists are writing a script that the users need to follow (like an actor using a script in a movie) [17]. Hence, while using robots, the users enter into interactive situations and the possible practices are always ambiguous and multistable [5]. The human-robot relationship will therefore always be two-way, and there will be what Bruno Latour called "symmetry between humans and non-humans" [18]. What happens if the users do not want to follow the script? Will they still get any benefits out of using the assistive robot? Or will they resist using the robot? There is a need to explore the implications of specific robots for older people. Are the impacts of the robot on the older individual's everyday life for better or worse? Is the robot good for some older people and some situations but not for others? An understanding of why people resist using the specific robot is also important – "non-users also matter" [4]. However, not owning the latest technology devices does not equal a lack of interest in new technological innovation, but might indicate a voluntarily selective choice in which technological innovation to adapt. This would be in line with Tornstam's development theory of positive aging, which claims that older people become more selective with age [19]. Since older people have self-knowledge and a lifetime of experience of different technological innovations, they may actually be pickier than younger people when choosing which technological device to buy and adopt. Another explanation of reservations about social robots is that older people may perceive the mental and physical investment needed to use the specific social robot to be greater than the expected benefits of using the robot [20]. In their literature review, Bouma et al. show evidence that indicates that older people, to a higher degree than younger people, compare the immediate cost such as the financial cost or effort for mastering the specific technology, with the future benefit of the promised functionality [21]. Furthermore, Eisma et al. suggests that older people have negative self-efficacy and negative beliefs regarding their abilities to handle new technologies [22]. Older people have a choice to adopt robots or not. The development of robots is still in its infancy, and the shaping of robots and their users is mutual and will continue to be so for a long time. By means of longitudinal studies of the use of robots, we will gain an understanding of how old adults make and remake assistive robots [9] [10].

References

1. Czaja, S.J., Lee, C.C.: The impact of aging on access to technology. Universal Access in the Information Society 5(4), 341–349 (2007)
2. Koskinen, I., et al.: Design Research Through Practice: From the Lab, Field, and Showroom. Morgan Kaufmann (2011)
3. Latour, B., Woolgar, S.: Laboratory life: The construction of scientific facts. Princeton Univ. Pr. (1979)
4. Wyatt, S.: Non-users also matter: The construction of users and non-users of the Internet. In: How Users Matter The Co-construction of Users and Technology, pp. 67–79 (2003)
5. Ihde, D.: Bodies in technology, vol. 5. U. of Minnesota Press (2002)
6. Salvini, P., Laschi, C., Dario, P.: Design for acceptability: improving robots' coexistence in human society. International Journal of Social Robotics 2(4), 451–460 (2010)
7. Forlizzi, J., DiSalvo, C., Gemperle, F.: Assistive robotics and an ecology of elders living independently in their homes. Human–Computer Interaction 19(1-2), 25–59 (2004)
8. Broadbent, E., Stafford, R., MacDonald, B.: Acceptance of healthcare robots for the older population: Review and future directions. International Journal of Social Robotics 1(4), 319–330 (2009)
9. Bemelmans, R., et al.: Socially assistive robots in elderly care: A systematic review into effects and effectiveness. Journal of the American Medical Directors Association 13(2), 114–120 (2012), e1
10. Kidd, C.D., Taggart, W., Turkle, S.: A sociable robot to encourage social interaction among the elderly. In: Proceedings of 2006 IEEE International Conference on Robotics and Automation, ICRA. IEEE (2006)

11. Heerink, M., et al.: Assessing acceptance of assistive social agent technology by older adults: The Almere model. International Journal of Social Robotics 2(4), 361–375 (2010)
12. Young, J.E., et al.: Toward acceptable domestic robots: Applying insights from social psychology. International Journal of Social Robotics 1(1), 95–108 (2009)
13. Beer, J.M., et al.: The domesticated robot: Design guidelines for assisting older adults to age in place. ACM (2012)
14. Frennert, S., Östlund, B., Eftring, H.: Would Granny Let an Assistive Robot into Her Home? In: Ge, S.S., Khatib, O., Cabibihan, J.-J., Simmons, R., Williams, M.-A. (eds.) ICSR 2012. LNCS, vol. 7621, pp. 128–137. Springer, Heidelberg (2012)
15. Rogers, E.M.: Diffusion of innovations. Free Pr. (1995)
16. Dautenhahn, K., et al.: What is a robot companion-friend, assistant or butler? In: 2005 IEEE/RSJ International Conference on Intelligent Robots and Systems (IROS 2005). IEEE (2005)
17. Akrich, M.: The description of technical objects. Shaping Technology/Building Society, 205–224 (1992)
18. Latour, B.: Pandora's hope: Essays on the reality of science studies. Harvard University Press (1999)
19. Tornstam, L.: Gerotranscendence: The contemplative dimension of aging. Journal of Aging Studies 11(2), 143–154 (1997)
20. Melenhorst, A.-S.: Making decisions about future activities: The role of age and health. Gerontechnology 1(3), 153–162 (2002)
21. Bouma, H., et al.: Gerontechnology in perspective. Gerontechnology 6(4), 190–216 (2007)
22. Eisma, R., et al.: Early user involvement in the development of information technology-related products for older people. Universal Access in the Information Society 3(2), 131–140 (2004)

Modelling Human Gameplay at Pool and Countering It with an Anthropomorphic Robot

Konrad Leibrandt[1,2], Tamara Lorenz[2,3], Thomas Nierhoff[2], and Sandra Hirche[2]

[1] Hamlyn Centre for Robotic Surgery,
Imperial College London, London SW7 2AZ, United Kingdom
k.leibrandt12@imperial.ac.uk
[2] Institute for Information-Oriented Control (ITR),
Technische Universität München, 80333 Munich, Germany
{t.lorenz,tn,hirche}@tum.de
[3] General and Experimental Psychology,
Ludwig-Maximilians University, 80802 Munich, Germany

Abstract. Interaction between robotic systems and humans becomes increasingly important in industry, the private and the public sector. A robot which plays pool against a human opponent involves challenges most human robot interaction scenarios have in common: planning in a hybrid state space, numerous uncertainties and a human counterpart with a different visual perception system. In most situations it is important that the robot predicts human decisions to react appropriately. In the following, an approach to model and counter the behavior of human pool players is described. The resulting model allows to predict the stroke a human chooses to perform as well as the outcome of that stroke. This model is combined with a probabilistic search algorithm and implemented to an anthropomorphic robot. By means of this approach the robot is able to defeat a player with better manipulation skills. Furthermore it is outlined how this approach can be applied to other non-deterministic games or to tasks in a continuous state space.

1 Introduction

Robots have been utilized for many years in manufacturing to handle repetitive tasks fast and precisely. Besides in industry, robotic systems are emerging in the private and public sector requiring human robot interaction (HRI). On a high level in HRI one is not only concerted of noticeable human actions but also of underlying intentions [1] in the context of collaborative [2], competitive [3] as well as severe tasks [4]. Regarding HRI these areas have the following challenges in common: (I) a continuous state space, (II) the necessity of being aware of the human. Concerning these two challenges competitive games represent attractive test scenarios because people encounter them in their daily life and furthermore they represent a controllable environment as well as a fixed set of rules. Requiring motor performance and planning capabilities the game of pool is chosen as representative game. Pool has a hybrid state space: the confined space of the table presents a continuous state space whereas the turn-based character of the

G. Herrmann et al. (Eds.): ICSR 2013, LNAI 8239, pp. 30–39, 2013.
© Springer International Publishing Switzerland 2013

game represents a discrete one. In computational pool Monte-Carlo sampling approaches prove to be robust, [5,6]. In contrast, an optimisation approach enables spectacular strokes, ball clusters are broken or multiple balls are sunk with one stroke [7,8]. A fuzzy-logic planner together with a robotic system is considered in [9]. However, the behavior of the human and his level of expertise has thus far not been taken into account. Existing approaches use mathematical models to determine the difficulty of game situations and calculate the best stroke without analyzing the human. But considering the human can improve prediction and planning of the robotic system for two reasons: First, a lateral visual perspective limits the human perception which influences his behavior in contrast to the bird's eye camera perspective of the robot. Second, the conception and evaluation of the game situation of humans is different to mathematical models.

In this paper a new approach is presented that enables the robot to assess the difficulty of game situations from a human perspective. The play of the robot improves by predicting the future actions of its human opponent. In order to do so, the robot needs to be aware of its opponent and requires a model of human pool-playing behavior. Different to all approaches introduced above, this paper concentrates on understanding and modeling human gameplay. It presents the steps to develop such a model, whose core is: (I) the *subjective difficulty* representing how difficult humans perceive a game situation and (II) the *objective difficulty* which is defined as the probability of sinking a ball.

The remainder of this paper is organized as follows: Sec. 2 describes the experimental setup and the important methods and parameters. In Sec. 3 four experiments are presented with the goal of identifying and correlating *subjective* and *objective difficulty*. Finally, in Sec. 4 the results are discussed in context of their application and their transferability to other HRI tasks.

2   Setup, Methods and Parameters

Stroke Parameters in Pool. Pool strokes can be described by five parameters: displacements a,b; slope γ; direction ϑ and stroke impulse p (see Fig. 1). For simplicity a, b=0 and $\gamma = \frac{\pi}{2}$ are chosen. The direction of the cue ball sinking an object ball centrally in the pocket is called ϑ_0 in the following. A detailed description of pool physics can be found in [10,11] and is not included here.

Methods to Measure the Stroke Difficulty. To gauge the component quantifying the stroke difficulty, respective methods from literature are considered. In contrast to the presented method, the following approaches focus on a mathematical optimal stroke and do not include human preferences. In method (I), a lookup table is used to link multiple paradigm game situations with success rates, see [5]. Method II is described by three characteristic parameters: d_1/ d_2: the distances the cue ball/ object has to travel and Θ: the cutting angle between d_1 and d_2, see Fig. 2a. These three parameters are combined in

$$\kappa_{\text{Lit}}^{-1} = \kappa_c = \frac{\cos \Theta}{d_1\, d_2}, \tag{1}$$

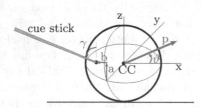 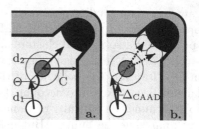

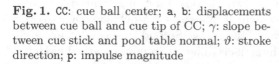

Fig. 1. CC: cue ball center; a, b: displacements between cue ball and cue tip of CC; γ: slope between cue stick and pool table normal; ϑ: stroke direction; p: impulse magnitude

Fig. 2. Parametric description of difficulty methods. a) Geometric parameters. b) Cumulated AAD (Δ_{CAAD}).

to calculate a stroke difficulty measurement [8]. Method III is based on the allowed angular deviation (AAD) which is the maximal angle deviation from ϑ_0 for a successful stroke [12]. Following this, the cumulated allowed angular deviation (Δ_{CAAD}) is defined as the sum of left and right AAD, see Fig. 2b. Since method I and III have a similar approach of using the required precision as a difficulty measure, method III is not dependent on a simulator or the introduction of noise. Thus method II and III are chosen to benchmark the new human difficulty quantification model (DQM) introduced in this paper, see Sec. 3.2.

Pool Setup: Visual System, Pool Table, Robot For the three experiments performed in a real world environment, a pool table with standard 8-ball dimensions ($2.24\,\text{m} \times 1.12\,\text{m}$) was used. A projector, located approximately $2.5\,\text{m}$ above the table, is used to project game situations on the table. Parallel to the projector there is a camera, with a resolution of 1280×960 pixel. The image processing algorithm: (I) locates the pool balls on the table and (II) determines their color, both with a rate of $30\,\text{Hz}$. How ball color and position is extracted from the video is described in [12]. The anthropomorphic robot consists of an omni-directional platform and two seven DOF manipulators with two special end-effectors to perform strokes properly, see [10].

Probabilistic Search Algorithm. To calculate the stroke parameters for the robot the probabilistic search algorithm described in [10] is applied with the following adjustments: (I) optimization is extended to both the ϑ- and p-parameter; (II) a new variable *tactic* is introduced which depends on the game situation: if the robot leads, *tactic* is set to "offensive", while if the robot is behind it is set to "defensive"; (III) a clustering step is introduced after the simulation in order to reduce the search space. For a detailed description and especially for a description of the evaluation process, see [10].

3   Human Difficulty Quantification Model (HDQM)

In the following, four experiments are described which: (I) identify the important parameters humans include into their decision process; (II) identify the decisiveness of the important parameters and determine an equation for the human perceived difficulty (*subjective difficulty*); (III) examination of the correlation between the *subjective difficulty* and the probability for humans to sink a ball (*objective difficulty*) and (IV) comparison of the competitiveness of the robot using (a) the HDQM integrated into a probabilistic search algorithm and (b) performing the easiest stroke according to the Δ_{CAAD}-method.

3.1   Exp. 1: Factors Influencing Subjective Difficulty

According to experts, [13]: Θ, d_1, d_2 are the three parameters to determine the difficulty of a game situation, see Fig. 2a. Since this assumption is not verified by experiments yet, an experiment on a real world pool table is performed.

Participants. In total 25 people (15 male, age: 21-31 years, M = 25) participated in the experiment.

Setup and Procedure. For the experiment 24 scenarios are designed. For each scenario two playable object balls are placed on the pool table. Circles are projected on the cloth to position the balls precisely and to create comparable game situations for all participants. The camera is used to prove correct positioning. In the first 18 scenarios one of the parameters: Θ, d_1, d_2 and Δ_{CAAD} are kept constant for both object balls. In the remaining six scenarios one of the two object balls is easier to play according to its Δ_{CAAD}-value but simultaneously either more difficult to reach or partially blocked by an obstacle ball. For each scenario participants are asked to freely state the easier ball to sink (descision δ_{exp}) and the reasons of their decisions. Every participant judges each scenario once. For data analysis all answers are recorded and equal decision factors were accumulated. Note that sometimes more than one reason is mentioned for choosing one scenario which indicates that the reasons might be interconnected.

Results and Discussion. Results of the first 18 scenarios show that the reason for deciding which is the easier ball to sink is most often a small cutting angle Θ (264 times). Also important for the decision appear to be the distances d_1 (31 times) and d_2 (75 times). Other factors are mentioned fewer times and not consistent over participants and are therefore not considered for further analysis. Thus, in line with the literature (see [13]), the three parameters Θ, d_2 and d_1 are -in this order- most crucial. Regarding the remaining six scenarios in 70% of the cases the ball with the higher Δ_{CAAD} is chosen when the object ball is difficult to reach, while when there is an obstacle ball, in 60% of cases the easier ball is chosen. Therefore it can be assumed that the Δ_{CAAD} has a higher impact on players' decisions than the disturbance by a obstacle. For that reason, both disturbance factors will not be considered in the following any more.

3.2   Exp. 2: Decisiveness of Parameters

In a second experiment the decisiveness of each of the crucial parameters is determined to find an algebraic description.

Participants. In total 23 people (13 male, age: 18-51 years, M = 25) participated in this experiment.

Setup and Procedure. For the experiment, twelve images of pool scenarios, depicting one cue ball and one object ball are created. The scenarios are chosen defining six κ_{Lit}-levels. For each level two parameter combinations are selected: (I) low Θ and high distance (d_1, d_2) values, (II) high Θ and low distance (d_1, d_2) values. To compare every image with every other, six sets of six images are shown to each participant, whereas the images per set are varied systematically. To get an overall ranking the six set-ranking are broken down to multiple pairwise comparisons and then subsequently combined to one overall ranking.

Results and Discussion. For statistical analysis a Friedmann ANOVA is performed which reveals a significant main effect between ranks, $\chi_F^s(11) = 206.31$, $p < 0.001$. Additionally, the rankings are used to calculate a 25% trimmed mean, $\bar{\rho}_{25\%}$ (interquartile mean), representing an average rank for each scenario. These $\bar{\rho}_{25\%}$ are sorted according to their value resulting in an overall ranking R_{exp}. In summary, an order of difficulty (R_{exp}) and an average rank ($\bar{\rho}_{25\%}$) are obtained to ascertain an equation describing the *subjective difficulty*, see following Sec. 3.3 and Fig. 3.

3.3   Equation of Human Subjective Difficulty

The results of Exp. 1 and Exp. 2 are used in a double stage optimization process to determine an equation for human subjective difficulty. Results are: (I) the decision ($\delta_{\text{exp}}(m)$, m $= 1 \ldots 18$) according to Exp. 1, (II) the overall ranking ($R_{\text{exp}}(n)$, n $= 1 \ldots 12$) and (III) the average ranking ($\bar{\rho}_{25\%}(n)$, n $= 1 \ldots 12$) both taken from Exp. 2. To find the most appropriate algebraic description, three template functions are used. The template functions are:

$$\kappa_{\mathrm{I}} = a_1 d_1^{c_1(d_1)} + a_2 d_2^{c_2(d_2)} + a_3 \cos\left(\Theta\right)^{c_3(\Theta)}, \tag{2}$$

$$\kappa_{\mathrm{II}} = \kappa_{\text{Lit}} + a_1 d_1^{c_1(d_1)} + a_2 d_2^{c_2(d_2)} + a_3 \cos\left(\Theta\right)^{c_3(\Theta)}, \tag{3}$$

$$\kappa_{\mathrm{III}} = \frac{d_1^{c_1(d_1)} \, d_2^{c_2(d_2)}}{\cos\left(\Theta\right)^{c_3(\Theta)}}. \tag{4}$$

Equation (2) is linear and (3), (4) are derived from (1), introduced in [8]. The exponents c_1, c_2 and c_3 ($\underline{c} = \{c_1, c_2, c_3\}$) are chosen as polynomials of degree one. The factors a_1, a_2 and a_3 ($\underline{a} = \{a_1, a_2, a_3\}$) are chosen as coefficients. These template functions are then fitted to the experimental data using two cost-functions and the Nelder-Mead algorithm. The underlying approach of the cost-functions is the least-square-method. Applying the difficulty equations (2-4) to the scenario parameters (Θ, d_1, d_2) from Exp. 1 and Exp. 2, the rankings $R_{\text{eq}}(\underline{a}, \underline{c})$, $\hat{\kappa}_{\text{eq}}(\underline{a}, \underline{c})$ and decisions $\delta_{\text{eq}}(\underline{a}, \underline{c})$ are obtained. The two cost functions

$$\Gamma_{\mathrm{I}} = \sum_{n=1}^{12} [R_{\text{eq}}(n, \underline{a}, \underline{c}) - R_{\text{exp}}(n)]^2 + \sum_{m=1}^{18} [\delta_{\text{diff}}(m, \underline{a}, \underline{c})]^2, \tag{5}$$

$$\text{where } \delta_{\text{diff}}(m, \underline{a}, \underline{c}) = \begin{cases} 2, & \text{if } \delta_{\text{eq}}(m, \underline{a}, \underline{c}) \neq \delta_{\text{exp}}(m) \\ 0, & \text{else} \end{cases}$$

Table 1. Best parameters of κ-functions obtained from double stage optimization

EQ.	a_1	a_2	a_3	c_1	c_2	c_3	Γ_{I}	Γ_{II}
κ_{I} (2)	-0.5	1.3	-4.9	-1.08	4.64	$0.2{+}2.2\,\Theta$	24	6E-2
κ_{II} (3)	0.8	-1.4	-4.9	1.94	-0.75	$2.8{+}1.9\,\Theta$	12	8E-2
κ_{III} (4)	–	–	–	0.33	0.38	$4.1{-}2.7\,\Theta$	8	2E-3

$$\text{and } \Gamma_{\mathrm{II}} = \sum_{n=1}^{12} [\hat{\bar{\rho}}_{25\%}(n) - \hat{\kappa}_{\mathrm{eq}}(n,\underline{a},\underline{c})]^2, \tag{6}$$

compare the three experimental results with the rankings and the decision using one of the difficulty equations (2-4). The summation is done over all scenarios in Exp. 1: 18 and Exp. 2: 12. In order to compare the average ranking ($\bar{\rho}_{25\%}$) with the equation value (κ_{eq}), it is necessary to normalize the values (using L^1-norm) which result in $\hat{\bar{\rho}}_{25\%}$ and $\hat{\kappa}_{\mathrm{eq}}$ in (6). The optimization task

$$\min_{\underline{a},\underline{c}}(\Gamma_{\mathrm{I}}), \tag{7}$$

minimizes the value of (5) and ensures that (2-4) best match the rankings and decisions of the participants. However, equation (5) compares the numerical values of R_{eq} and R_{exp} as well as the decisions δ_{eq} and δ_{exp}. Since rankings and decisions in (5) are discrete, parameter intervals for optimal solutions using (7) are found. Hence a second optimization step

$$\min_{\underline{a},\underline{c}}(\Gamma_{\mathrm{II}}), \tag{8}$$

using cost function (6) is performed matching continuous numerical values of the rankings ($\hat{\bar{\rho}}_{25\%}$, $\hat{\kappa}_{\mathrm{eq}}$). In (8), only those $\underline{a}$- and $\underline{c}$-intervals are considered which are obtained from (7). The result of (8) are an optimal fitted equation κ_{I}, κ_{II} or κ_{III} (2-4). Preliminary optimization results show: c_1, c_2 can be chosen as constants and c_3 has to be linear. Final results, show that κ_{III} (4) leads to the best approximation, due to its lowest Γ_{I} (5)- and Γ_{II}-value (6), see Table 1. The *subjective difficulty* is therefore defined as

$$\kappa_{\mathrm{Exp}} = \frac{\cos(\Theta)^{4.1-2.7\,\Theta}}{d_1^{0.33}\, d_2^{0.38}}, \Theta \in [0, 90^\circ]. \tag{9}$$

In line with the impressions from Exp. 1, in (9) the cutting angle of a pool situation is more decisive than the two travel distances, due to the larger exponent. Note regarding (9): Since the exponent of the cos-term is: (I) smaller than the exponent of d_2-term for $\Theta > 79^\circ$ and (II) negative for $\Theta > 87^\circ$, (9) is valid for $\Theta \in [0^\circ, 79^\circ]$. In Fig. 3, the *subjective difficulty* κ_{Exp} is compared with the average ranking of Exp. 2 and with the DQM κ_{Lit} (1) and Δ_{CAAD} (Fig. 2b), which are used in computational and robotic pool. In all scenarios, except number 8, the difficulty perception of participants is described best by κ_{Exp} (9), see Fig. 3. Since it is not guaranteed that a player is able to correctly assess the difficulty of a stroke, a method representing the *objective difficulty* is necessary.

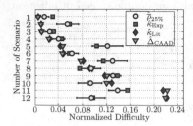 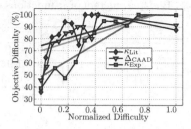

Fig. 3. Comparison of scenario difficulties according to experiment and DQM

Fig. 4. Player P1: Correlation between objective stroke difficulty, and DQM

3.4 Exp. 3: Human Objective Difficulty

The *objective difficulty* of strokes is examined by an experiment in which pool players of similar experience played real pool games against each other.

Participants. Two experienced players (P1, P2) who play weekly and two amateur players (P3, P4) who play yearly took part in the experiment.

Setup and Procedure. The participants are asked to play multiple 8-ball games against each other, to avoid bank-shots and kick-shots and to hit the cue ball centrally. Each participant performs 210-270 strokes. The setup described in Sec. 2 is used to determine the trajectories of all balls and strokes.

Data Analysis. For each performed stroke the parameters Θ, d_1, d_2 are determined and Δ_{CAAD}, κ_{Exp} (9), κ_{Lit} (1) are calculated. The resulting difficulty values (Δ_{CAAD}, κ_{Exp}, κ_{Lit}) of each player are divided into eleven intervals chosen to represent approximately a constant amount of values. A stroke success rate (*objective difficulty*) for each interval is calculated and correlation between the three DQM and the *objective difficulty* is obtained.

Results and Discussion. The resulting correlation coefficients, depicted in Table 2, show the degree of linearity between the *objective difficulty* and each DQM, see Fig. 4. It shows *objective difficulty* and Δ_{CAAD} or κ_{Lit} are less correlated than *objective difficulty* and κ_{Exp}. Regarding κ_{Exp} (9) a high correlation coefficient and a nearly monotonic behavior makes it possible to approximate the correlation with a linear relationship as depicted in Fig. 4. However it is necessary to determine an approximation for every human opponent due to different skill levels. Nevertheless these results show that it is possible to transform the *subjective difficulty* into an *objective difficulty* using a linear approximation. Additionally the *objective difficulty* can be considered as sinking probability and hence can be used in an probabilistic best shot search algorithm.

3.5 Human Decisions, Abilities to Plan Ahead

To predict the stroke a human will perform, the *subjective* and *objective difficulty* of a given situation are not sufficient. In addition, a method to acknowledge the

Table 2. Correlation coefficients between objective difficulty and three DQMs

Player	Strokes	Correlation Coefficient		
		Δ_{CAAD}	κ_{Lit}	κ_{Exp}
One	216	0.37	0.21	0.79
Two	182	0.52	0.42	0.75
Three	164	0.47	0.59	0.85
Four	152	0.33	0.29	0.77

Table 3. Rates of correct predictions using different DQM (max|mean value of CP_R)

Player	Rate of Correct Prediction (%)		
	Δ_{CAAD}	κ_{Lit}	κ_{Exp}
One	71.6|70.8	69.6|69.0	74.9|73.1
Two	74.1|70.6	70.3|68.6	74.3|70.7
Three	70.6|70.1	68.0|66.5	69.7|68.0
Four	67.2|66.4	69.0|67.7	71.7|69.3

planning capabilities is required. Humans make their choice according to the difficulty of a stroke, but especially experienced players plan ahead and consider future game situations. To counter human gameplay, predicting the opponent's next stroke or at least narrowing the choices to few strokes is advantageous. The goal is therefore to increase the rate of correct prediction (CP_R) and to examine the difference in planning between experts and novices. Thus, all possible strokes in a given game situation are simulated and the difficulty value of the current stroke $\kappa_{Exp,0}$ (9) is combined with the difficulty value of the easiest stroke of the simulated game situation $\kappa_{Exp,1}$, using a discounted finite-horizon return function [14]. To combine these values,

$$\kappa_n^{total} = \sum_{i=0}^{n} \delta^i \, \kappa_{Exp,i} , \qquad (10)$$

is used where, δ represents the importance of future situations. It is assumed that (I) the current game situation is most important for a pool player and that (II) an easy game situation is preferred. Hence: $\delta \in [0,1]$. The n-parameter in (10) represents the amount of simulations into the future and is set to $n = 1$. κ_n^{total} (10) calculated for each currently possible stroke is used to determine the human's most probable next stroke. After playing, the performed stroke is compared with the predicted stroke to obtain the rate of correct prediction (CP_R). CP_R is also evaluated based on the two other presented DQMs (Δ_{CAAD} and κ_{Lit}) as shown in Table 3. Here, only the decisions of amateur player three are better described by the Δ_{CAAD} method. It shows that the new developed equation κ_{Exp} (9) predicts the human decision better than Δ_{CAAD} and κ_{Lit} (1), while κ_{Lit} has in general the lowest CP_R. Fig. 5 shows that simulations can increase the CP_R. The CP_R of the experienced player has its maximum at $\delta \approx 0.5$, which means the player considers the actual situation twice as important as the future situation. In comparison, results of the amateur player show a flat, convex CP_R function, thus considering future strokes result in a loss of accuracy of the prediction. These results coincide with the amateurs' statement, only to consider the current game situation. Summarizing, amateurs - in contrast to experts - do not plan ahead.

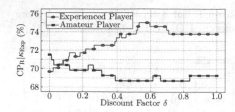

Fig. 5. CP_R using method κ_{Exp}. A maximum rate of 75% shows, actions of human player cannot be exactly predicted.

Fig. 6. Game situation between human pool player and anthropomorphic robot

3.6 Exp. 4: Robot Applying Human Pool Model

In Exp. 4 a robot has to play thirteen 8-ball pool games against a human (Fig. 6), to gauge the competitiveness of the HDQM integrated into the above described search algorithm (see [10]) and the setup from Sec. 2. Two strategies for choosing the robotic stroke are evaluated: (I) human adapted search algorithm strategy (HAS) using the previously described HDQM and probabilistic search algorithm and (II) easiest shot strategy (ES) according to the Δ_{CAAD}. The robot plays seven and six games using the ES and HAS strategy respectively.

Participant. One player (male, age: 24), manipulation skills: $\sigma_p : 0.01\,\text{Ns}$, $\sigma_\vartheta :$ 0.007 rad. The participant is not specifically instructed.

Robot Behavior. For the experiment the same robot as described in Sec. 2 is used ($\sigma_p : 0.03\,\text{Ns}$, $\sigma_\vartheta : 0.012\,\text{rad}$), which has a restricted workspace. In situations in which the cue ball is unplayable for the robot, it is moved towards or away from the cushion. The HDQM is used within the search algorithm whenever it is the human's turn. Furthermore the simulator and the evaluation function consider the pool skills of the human opponent ($\delta = 0.5$) such as the precision of performing a stroke. The algorithm has a search depth of two, meaning it plans two strokes ahead.

Evaluation. Since the number of games is limited the evaluation is based on the strokes' success rate. Using the same robot with both strategies the success rates represents how good the respective strategy prepares future strokes.

Results and Discussion. In games using the HAS strategy, the robot sinks 51% of 67 strokes. In comparison using the ES strategy the robot sinks 28% of 89 strokes. The evaluation shows that the robot plays significantly (two-sample t-test, p=0.0037) better when using the HAS. In addition, the robot adapting to the introduced human pool model while using a probabilistic search algorithm can beat a player of better-than-beginner level.

4 General Discussion and Conclusion

This work presents a human pool player model which determines the *subjective* and *objective difficulty* of game scenarios from a human perspective. Although Smith [15] states that it is not possible to consider the opponent's move, the presented approach is capable of narrowing the amount of resulting situations after a human stroke. Integrated in a probabilistic search algorithm it enables a robot to predict human choices of pool strokes and to approximate their outcome. Although the contribution of the HDQM in the HAS is not examined explicitly it is shown that the complete algorithm improves pool playing abilities of a robot by improving its own stroke preparation and by creating more difficult situations for the opponent. Furthermore, model and algorithm, are not developed for specific pool rules and can therefore easily be transformed to play other pool variants. Varying the parameters in the presented HDQM would change the game behavior and competitiveness of the robot and hence allow the building of a training environment. The method of determining between subjective and objective difficulty can be applied to other non-deterministic games (e.g. carrom, bowls) in which the human behavior should be considered. Furthermore, the approach is transferable to non game related areas. For instance the difficulty of reaching objects (e.g. a cup) in a given situation might be assessable by using geometric parameters such as distances and orientations (e.g. handle orientation) and thus a robot could predict and prevent difficult to reach objects.

References

1. Demiris, Y.: Prediction of intent in robotics and multi-agent systems. Cognitive Processing 8(3), 151–158 (2007)
2. Sebanz, N., Knoblich, G.: Prediction in joint action: What, when, and where. Topics in Cognitive Science 1(2), 353–367 (2009)
3. Wang, Z., Deisenroth, M., Amor, H.B., Vogt, D., Scholkopf, B., Peters, J.: Probabilistic modeling of human movements for intention inference. In: RSS (2012)
4. Howard, N., U. S. D. of Defense: Intention Awareness: In Command, Control, Communications, and Intelligence (C3I). University of Oxford (2000)
5. Smith, M.: Running the table: An AI for computer billiards. In: 21st National Conference on Artificial Intelligence (2006)
6. Archibald, C., Altman, A., Shoham, Y.: Analysis of a winning computational billiards player. In: International Joint Conferences on Artificial Intelligence (2009)
7. Landry, J.F., Dussault, J.P.: AI optimization of a billiard player. Journal of Intelligent & Robotic Systems 50, 399–417 (2007)
8. Dussault, J.-P., Landry, J.-F.: Optimization of a billiard player – tactical play. In: van den Herik, H.J., Ciancarini, P., Donkers, H.H.L.M(J.) (eds.) CG 2006. LNCS, vol. 4630, pp. 256–270. Springer, Heidelberg (2007)
9. Lin, Z., Yang, J., Yang, C.: Grey decision-making for a billiard robot. In: IEEE International Conference on Systems, Man and Cybernetics (2004)
10. Nierhoff, T., Heunisch, K., Hirche, S.: Strategic play for a pool-playing robot. In: IEEE Workshop on Advanced Robotics and its Social Impacts, ARSO (2012)
11. Leckie, W., Greenspan, M.: An event-based pool physics simulator. In: van den Herik, H.J., Hsu, S.-C., Hsu, T.-s., Donkers, H.H.L.M(J.) (eds.) CG 2005. LNCS, vol. 4250, pp. 247–262. Springer, Heidelberg (2006)
12. Nierhoff, T., Kourakos, O., Hirche, S.: Playing pool with a dual-armed robot. In: IEEE International Conference on Robotics and Automation, ICRA (2011)
13. Alian, M., Shouraki, S.: A fuzzy pool player robot with learning ability. WSEAS Transactions on Electronics 1, 422–425 (2004)
14. Sutton, R., Barto, A.: Reinforcement learning: An introduction. Cambridge Univ. Press (1998)
15. Smith, M.: Pickpocket: A computer billiards shark. In: Artificial Intelligence (2007)

Automated Assistance Robot System
for Transferring Model-Free Objects From/To Human
Hand Using Vision/Force Control

Mohamad Bdiwi[1], Alexey Kolker[2], Jozef Suchý[1], and Alexander Winkler[1]

[1] Department of Robotic Systems, Chemnitz University of Technology, Chemnitz, Germany
[2] Automation Department, Novosibirsk State Technical University, Novosibirsk, Russia
mohamad.bdiwi@s2008.tu-chemnitz.de

Abstract. This paper will propose an assistance robot system which is able to transfer model-free objects from/to human hand with the help of visual servoing and force control. The proposed robot system is fully automated, i.e. the handing-over task is performed exclusively by the robot and the human will be considered as the weakest party, e.g. elderly, disabled, blind, etc. The proposed system is supported with different real time vision algorithms to detect, to recognize and to track: 1. Any object located on flat surface or conveyor. 2. Any object carried by human hand. 3. The loadfree human hand. Furthermore, the proposed robot system has integrated vision and force feedback in order to: 1. Perform the handing-over task successfully starting from the free space motion until the full physical human-robot integration. 2. Guarantee the safety of the human and react to the motion of the human hand during the handing-over task. The proposed system has shown a great efficiency during the experiments.

Keywords: Assistance robots, human-robot interaction, visual servoing, force control, handing-over tasks.

1 Introduction

Recently, it has been noticeable that the advanced robot applications have become focused on the tasks which require from the robot to interact with the human physically, e.g. assistance robot for blind, disabled or elderly people for helping in fetching, carrying or transporting objects. In other applications, robot could serve as rescue, service robots or even as member of human-robot teamwork for supporting humans in such applications as space exploration, construction, assembly etc. In such kind of applications, where human hand will interact with robot hand, the initial core is to perform the handing-over task from/to human hand successfully starting from hand tracking in free space motion until to the full physical interaction between the human hand and robot hand.

Numerous papers which have suggested different approaches of service robot systems have discussed the problems of objects transfer between human and robot. With robot HERMES [1] and in the work [2], the transfer tasks are performed exclusively by the human. This means that the robot will bring its hand into a specified pose and

G. Herrmann et al. (Eds.): ICSR 2013, LNAI 8239, pp. 40–53, 2013.
© Springer International Publishing Switzerland 2013

then it will wait until the human places the transported object between the fingers of the gripper. When the robot detects that an object has been placed in its hand, it attempts to grasp the object. In fact, this scenario will not be fit to assist blind, disabled or elderly people or even to support workers concentrating on their work. References [3] and [4] have presented different algorithms of grasp planning during handing-over objects between human and robot. These algorithms depend on human preferences, on the kinematic model of the human and on the availability of the 3D model of the object. Related to that, various papers have focused on analysing and detecting human body movements especially on hand gesture and facial features, e.g. [5], [6] and [7]. The common properties of them are that they detect the human hand in order to direct or to lead the robot for performing of some tasks without any physical interaction with the robot. In addition to that, they have implemented algorithms which are able to handle with only human hand free of load. Hence, what about if the human hand has carried an object and the task needs the robot to interact with the human physically? Moreover, what about if the user is disabled or elderly?

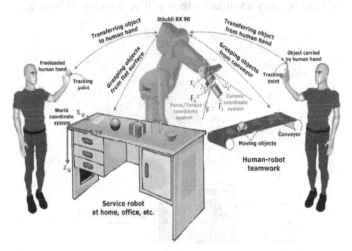

Fig. 1. Handing-over from/to human hand

On the whole, according to our knowledge none of the previous works has proposed a fully automated robot system which combines vision and force control in order to transfer model-free objects from/to human hand. Furthermore, in this work, the proposed robot system will consider the human as the weakest party (negative party) during the handing-over task, i.e. this party could move in a wrong direction (not toward the robot), e.g. elderly or blind or he/she is doing something else at the same time. In this case, the robot system should expect some random motions from human during the task and react accordingly to them. In addition to that, the proposed system has implemented real time vision algorithms for detecting and segmenting loadfree human hand, any objects carried by human hand and any object located on a flat surface. The fusion of vision and force sensors is an optimal solution to guarantee the safety of the user and for performing the handing-over task successfully.

In the next section, the hardware and software equipments will be presented. Section 3 contains general description of the whole algorithms of the proposed system. In section 4, real results will be illustrated for transferring objects from/to human hand. The last section contains conclusion, future work and the benefits of improving the physical interaction between human and robot.

2   System Equipment

As shown in Fig. 2, the overall experimental setup consists of Stäubli RX90 robot with a JR3 multi-axis force/torque sensor together with the eye-in-hand Kinect camera as shown in Fig. 2. The end-effector is installed on the collision protection device. In our application the end-effector is two-finger gripper. JR3 (120M50A) is a six component force/torque sensor and its effective measurment range is ± 100 N for forces F_x, F_y, F_z and ± 10 N.m for torques M_x, M_y, M_z. Kinect camera is RGBD camera which delivers depth and color images with VGA resolution (640x480 pixels). As shown previously in Fig. 1, the active human hand will interact with the robot and it could be free of load or carry an object which will be transferred from it.

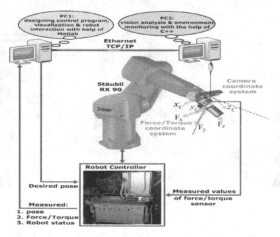

Fig. 2. Hardware/software equipments

The proposed system has used two PCs. PC1 will control the robot to perform the visual and force tracking tasks with the help of MATLAB program and V+ (programming language of Stäubli robot). A special program is designed to integrate Matlab program with V+ language, which means that all V+ instructions, e.g. read/write pose, read force/torque or robot status, could be written directly in Matlab program. PC2 is connected with the Kinect camera, and all the image processing algorithms are performed in PC2 using C++ language (OpenCV and Openkinect libraries). PC2 will send the position of the face and of the object as well as the status of the task to PC1 in every frame using Ethernet TCP/IP protocol as follows:

$$[Obj_{x_C}, Obj_{y_C}, Obj_{z_C}, time\_dif, vision\_status] \tag{1}$$

where $(Obj_{x_C}, Obj_{y_C}, Obj_{z_C})$ is the tracking point of the carried object or of the human hand in the case of loadfree hand, this point will be later the contact point between the object and the gripper. *time_dif* is time difference between two frames. *vision_status* represents the current status of the vision system if only loadfree hand is detected or a carried object is segmented and detected.

3   Description of Proposed Algorithms

This section will present the algorithms of the whole process of the proposed system. As shown in Fig. 3, firstly the user will define the type of the task, either transferring from or to human hand, based on keyboard commands. If the task is to transfer to

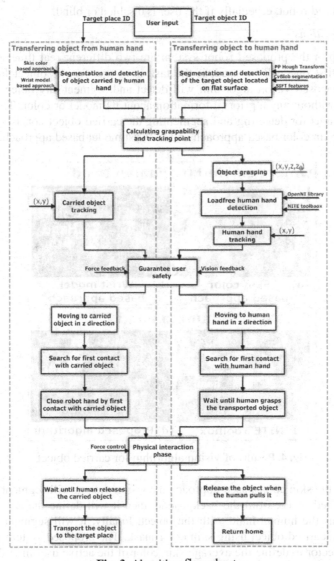

Fig. 3. Algorithm flowchart

human hand, the user should enter either target object ID or name under which its corresponding SIFT features are saved in the database. In other hand, if the task is transferring from human hand, the user needs only to enter the ID of the target place where the object should be transported, e.g. table, workspace, conveyor, etc.

User interface could be improved in the future to be based on voice commands. The currently proposed system is supported only by voice subsystem, i.e. it will announce the current phase (what it is going to do), the status of the operation or whether some errors have occurred. The proposed speaking subsystem will give the human the opportunity to learn and to understand what the robot is doing now and to be prepared if any error has occurred during the task. It will increase the safety factor between human and robot, especially if the user is disabled or blind.

A. Vision algorithms

The next step in the proposed robot system are two different real time vision algorithms. The first one will be used, if the task is to transfer the object from human hand. The proposed vision algorithm will detect and segment any object carried by human hand without any a priori information about its model or color. The proposed vision algorithms for detecting and segmenting the carried object consists of two approaches: 1. Skin color based approach and 2. Wrist model based approach.

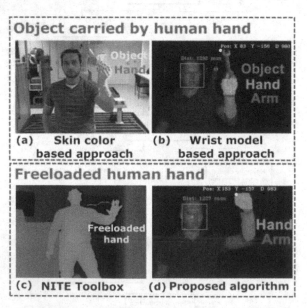

Fig. 4. Results of vision algorithm for carried object

Briefly, in the skin color based approach the vision system will segment the human body and carried object from the background, then it will define the area of interest which contains the human hand with the object. Finally, it will segment the human hand from the carried object using skin color mask. The proposed system will implement face detector to define the average skin color of the active user in HSV space by scanning the HSV values of the detected face. The previous work [8] has defined the general range for human skin color in HSV space as follows (by assuming the range of H, S and V components is 0,...,255):

$$G_{H_{min}} = 0 \ \ and \ G_{H_{max}} = 128$$
$$G_{S_{min}} = 59 \ and \ G_{S_{max}} = 175 \qquad (2)$$
$$G_{V_{min}} = 0 \ \ and \ G_{V_{max}} = 255$$

It is clear that the general range of human skin color in HSV space is very wide, e.g. V values could be spread in whole space. H and S values could be spread in almost the half of the space. Hence, the general ranges could be sufficient to detect skin but they will not be enough to segment the hand from the object in a complicated scene or in different light conditions, especially if the object has almost the same color as the skin such as wood objects. Hence, by scanning the color of the detected face, the color of human skin will be updated every frame if light source, active person and skin reflection are changed. This will help to narrow the ranges of skin color in HSV space and to define exactly the color of the human hand. Fig. 4(a) presents experimental results of the skin color based approach. One limit of this approach is that it cannot be used when the human wears gloves or he/she has Vitiligo disease. Hence, another approach, which depends on wrist model, has been implemented using infrared frames.

Fig. 5 presents the basic principle of the proposed approach. The most important region of the hand is the wrist (the joint between the hand and the arm). By analyzing the cross-section of the human hand parallel with plane xz, it can be concluded that an algebraic equation of the second order can approximately represent all the points of cross section at any value of y in the wrist region as follows:

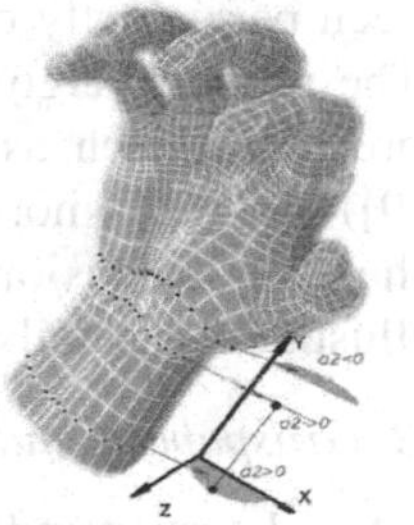

Fig. 5. Human hand profile

$$\bar{Z}_k(x_i) = a_2 \cdot x_i^2 + a_1 \cdot x_i + a_0; \quad i = 1, ..., N \qquad (3)$$

where $\bar{Z}_k(x_i)$ are the approximated depth values. Using least square polynomial approximation, the system can calculate the coefficients a_2, a_1 and a_0. In general, the coefficient a_2 in any second order equation describes the flexion of the curve. Hence, by analysing the human hand profile, it can be concluded the following: If the cross section locates below the wrist joint, the coefficient a_2 will be always positive (as shown in Fig. 5, the region of violet color). The value of coefficient a_2 will become unstable (sometimes negative and sometimes positive) exactly when the cross section crosses the wrist joint upwards. This approach will be useful even if the human wears gloves or has Vilitigo disease and it is able to segment the human hand from the object even if they have the same color starting from complete darkness and ending with different color temperature lamps, as shown in Fig. 4 (b). The frequency of first approach is 5 frames/sec, whereas the frequency of the second approach is 10 frames/sec, i.e. they are fit for fast real time visual servoing.

Furthermore, the proposed vision algorithms are able to detect and segment the loadfree hand as shown in Fig. 4 (d). In contrast, the hand control of NITE toolbox from Kinect camera could be implemented to track only the loadfree hand, as shown in Fig. 4. (c).

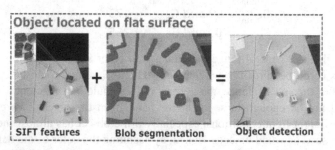

Fig. 6. Result of vision algorithm for object on flat surface

On the other hand, as shown in Fig. 6 the second algorithm for transferring objects to human hand is able to detect and segment any object located on a flat surface. This approach works with SIFT features and it is able to detect the object in real time even if the object has complicated shape, if it has bad contours and if the illumination has been permanently changed and unequally distributed with no need to a priori model. The proposed algorithm in this section has combined different algorithms of image processing such as Canny filter, progressive probabilistic Hough transform (PPHT [9]) and some morphological operations for improving objects contours. In general, the proposed vision algorithms in this work have many contributions which will be illustrated in details in other papers.

B. Graspability and tracking point

After detection and segmentation the target object (even if it is carried by human hand or located on a flat surface), the next phase will include algorithm common to both tasks as shown previously in Fig. 3. In this phase the robot system will calculate the graspability and the position of the tracking point depending on the results of vision processing.

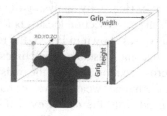

Fig. 7. Object *and robot hand*

As shown in Fig. 7, the graspability will be calculated by comparing the width and the height of the target object (obj_{width}, obj_{height}) with robot hand width $Grip_{width}$ (distance between fingers of the robot hand) and with (height of robot hand) $Grip_{height}$ as follows:

$$if \left(obj_{height} > Grip_{height} \right) \& \left(obj_{width} < Grip_{width} \right) then$$

$$\{Grasp\_ability = 1\} \tag{4}$$

$$else \{Grasp\_ability = 0\}$$

In brief, the object should have grasp area GA(i,j), where its width is less than the gripper width and its height is greater than gripper height in order to activate the graspability. If the robot is able to grasp the object then it will calculate the tracking point of the object which will be later the contact point. The vision system will scan the whole grasp area GA(i,j) and it will define three points of it: 1. Upper point (x_U, y_U, z_U), 2. Left point (x_L, y_L, z_L) and 3. Nearest point (x_N, y_N, z_N). The tracking point (x_O, y_O, z_O) will be calculated as follows:

$$\begin{aligned} x_O &= x_L \\ y_O &= y_U \\ z_O &= z_N \end{aligned} \tag{5}$$

In the task of transferring object to human hand, the next step will be grasping the object from the flat surface. Fig. 8 illustrates the coordinate system of the object. The vision system will control the directions (x, y, z) and the orientation (θz) in order to rive the robot hand to the best grasping position. When the robot arrives at the tracking point of the object, it will move 20 mm in z direction toward the object before establishing any contact with the object. After that the robot will start searching for the first physical contact with the object in x direction. In this way, the robot can guarantee that it will grasp the object successfully by closing the gripper fingers.

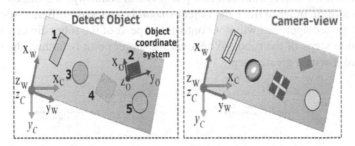

Fig. 8. Grasping object from flat surface

After grasping the object from the flat surface, the vision system will start searching for the loadfree human hand. In this case, the NITE toolbox which supports Kinect camera could be implemented for detecting the human hand and finding the middle point of it, because its frequency is 30 frames/sec.

C. Tracking phase

As shown in Fig. 3, the following phase will be the tracking phase either of the loadfree human hand or of the object carried by human hand depending on the task. The robot will start to track the target object or the hand in two directions, x and y. The tracking algorithm can be easily extended to three directions x, y and z. However, in this phase, a safety distance in z direction between the robot hand and the human hand should be always kept. The robot will not move toward the human hand in z direction to establish the contact until it ensures that the human hand doesn't move anymore. When the human hand is in a stable position and after the robot has already tracked it successfully, the robot system will start moving toward the human hand in z direction.

Hence, the main purpose of the visual tracking in this phase is to preserve the target point of the object or the hand at the middle of camera frame. (Cam_x, Cam_y) represent the position of the middle point of the camera. Hence, the required distance (mm) in order to locate the tracking point of the object in the middle of camera's view (position based visual servoing approach) is as follow:

$$Err_x = Obj_{x_C} - Cam_x$$
$$Err_y = Obj_{y_C} - Cam_y$$

(6)

However, real time tracking of moving object is not an easy task, especially when this object is moving randomly e.g. when tracking the hand of blind user. In our approach, the robot is able track the loadfree hand or the object carried by a human hand smoothly and with sufficient speed in real time in spite of the following difficulties: 1. The position of the object continually changes and it can move in all directions. 2. The speed and acceleration of the object motion are not constant and the motion direction of the object can suddenly change. 3. The motion speed and desired position in the implemented robot (older commercial robot) cannot be changed when the robot is moving, i.e. the robot is not able to receive a new target position unless the previous motion is finished. Hence, the common position controller will not be sufficient, especially when the object direction and position can suddenly be changed by human hand. Therefore, in order to track the object smoothly and with sufficient speed in real time, the proposed approach will not control the speed of the robot directly but it will control motion steps of the robot. Whenever the target point is farther, the robot will move toward it with a greater step.

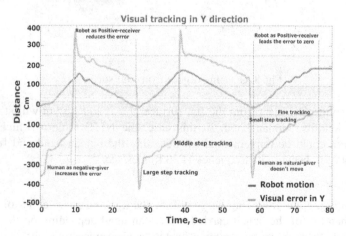

Fig. 9. Visual *tracking in Y direction*

Fig. 9 shows the experimental results of the proposed visual tracking approach in Y direction. The green diagram presents the visual error Err_x (the difference between the tracking point of the object or hand and the camera middle point) in the camera coordinate system, whereas the red diagram represents the motion of robot in the world coordinate system. The different colored zones refer to the different motion step of the robot. From the behaviour of robot motion (red diagram) it is clear that the robot always moves toward the target point as positive party therefore the visual error is diminished to zero. On the contrary, from the behaviour of the human it is clear that the human increases the visual error extremely (as negative party) by moving his/her hand, look e.g. at time t = 10s in Fig. 9).

At time t = 0s, the visual error was near to -350cm which made the robot move upwards to reduce the visual error. Starting in time t = 0s, until t = 2s, the visual error was inside the large step tracking zone and it has diminished. At time t = 2s, the visual error has come inside the middle step tracking zone. However, at time t = 8s the human hand has suddenly been moved and it has changed the visual error from -120cm to 380cm which has led the robot to react rapidly and to change the direction of the motion in order to track the target point. The rapid visual tracking and changing directions will be continued until the human hand acts at least as a natural giver (doesn't move anymore), e.g. at time t = 58s, so the robot will move toward the target point smoothly and the visual error will be diminished until it enters the zone0 (visual error is less than 2cm). It means that the robot has arrived at the target point, after that the visual tracking will be finished and next phase will be started.

D. Safety measures

After tracking phase, the robot will be ready to move toward the human if the safety procedures are taken into account. Numerous papers have been published which have proposed different solutions for improving the safety factor during the human-robot interaction, such as improved mechanical design by reducing the robot mass [10] and motion trajectories related to the human body constraints [11]. However, in our opinion, even if the system would use lightweight robot and predefined trajectories, it is indispensable to integrate vision and force information to guarantee the safety especially when unexpected problems or errors happen during the physical interaction. In this work, three safety measures are implemented: The first one (*SF_body*) is related to the safety of the whole human body depending on the depth map, whereas the second safety (*SF_hand*) is related only to the safety of the fingers during handing-over the object (if the robot is able to grasp the object without touching the human fingers). The third one is (*SF_force*) which monitors the force values, especially when the robot is moving toward the human in z direction. Values of these factors will be set to zero as long as the safety requirements are fulfilled. Otherwise, if any error or dangerous situation of human is recognized or unexpected obstacle is encountered, the safety variables will be immediately activated and the task will be cancelled.

E. Physical interaction phase

If the tracking phase is successfully performed and the robot or the human has grasped the transported object, the physical interaction phase will start. In the first phase, the first party has grasped the transported object but the other party has not released it yet, i.e. the transported object will be like a kind of connection bridge between the human and the robot hand. In this phase, we will assume that the human or the robot may not release the object immediately. Perhaps, the human would like to drive the robot toward another place, so the robot should be able to react to the motion of the human hand, so the robot will comply with the motion of the human hand based on user of the force sensor. Finally, if robot transfers the object to the human hand, it will release the object when the human pulls the object with desired force and after that it will return to home position. Otherwise, by transfer of object from human hand, robot will wait until the human releases the object (no more applied forces) then it will transport the object to the target place.

4   Experiment Results

Some results have been already presented in Fig. 4, Fig. 6 and Fig. 9, whereas, Fig. 10 and Fig. 11 present further experimental results from handing-over task.

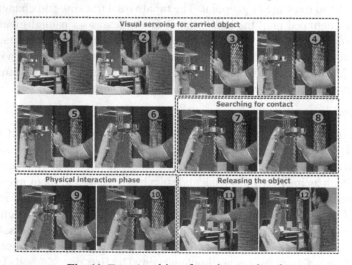

Fig. 10. Transfer object from human hand

In Fig. 10, as shown in the first six pictures, the robot is tracking and moving toward the object carried by human hand in order to grasp it from the human hand with the help of vision system. When robot reaches the tracking point of the object, the robot will start searching for the first contact point with the object using force reading, as shown in pictures 7 and 8. When the force sensor measures that the desired contact with the object has been established, then the robot will close its hand to grasp the object as shown in pictures 9 and 10. Hence, the physical interaction between the human and the robot will start and the robot will react to the motion of the human hand on the basis of force sensor measurements. When the interaction phase is finished, the human will release the object as shown in pictures 11 and 12.

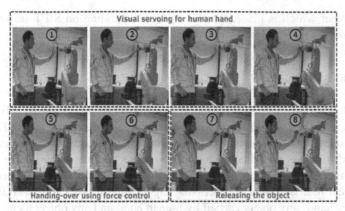

Fig. 11. Transfer *object to human hand*

In Fig. 11, as shown in the first four pictures, robot is moving toward the human hand in order to deliver the cube with the help of vision system. In pictures 5 and 6, the human will start grasping the object by closing his/her hand. In pictures 7 and 8, robot will perceive the contact force. It will release the object when the human starts pulling.

As shown previously in Fig. 10 and Fig. 11, the proposed system will assume that the human as giver/receiver will be the weakest part (blind, elderly etc.) of the task and the robot will play the main role as a positive party to perform the transfer task. In other words, human only needs to open/close his/her hand in order to release/grasp the object.

The proposed algorithms have been repeated for more than 15 different objects which were carried by different users.

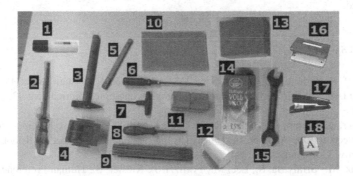

Fig. 12. Dataset *of different transported objects*

Fig. 12 presents the dataset of the transported objects from/to human hand. In all experiments of handing-over objects from human hand, the robot system was able to segment, to track and to grasp the carried object from the human hand successfully. As is shown in the Fig. 12, the implemented objects are textureless, so in the experiments of handing-over objects to human hand, if the robot system was able to detect the target object which is placed on flat surface depending on SIFT features, surely the robot system will be able to deliver that object to the human hand.

5   Conclusion

This paper has just illustrated the general description of the proposed robot system, more details about the vision algorithms, vision/force control structure [12], tracking algorithms, safety procedures and physical interaction phase will be found in other papers. This work has proposed an automated robot system which is able to deliver and receive model-free objects to the human hand automatically. This work has proposed different real time image processing algorithms in order to detect and to segment loadfree hand, any object carried by human hand and any object located on a flat surface. The proposed robot system has implemented visual servoing algorithms for tracking the loadfree hand or any object carried by it in real time. In the proposed system, the transfer object from/to human hand is performed exclusively by robot and the human has been considered as the weakest part in this task (elderly, blind or disabled). Furthermore, for improving the handing-over task the proposed system has combined vision and force control. The fusion of vision and force will ensure the safety of the user during the physical human-robot interaction, it will ensure the fulfilment of the grasping/releasing task and it will make the robot able to react to the motion of human hand during the interaction phase. The proposed robot system could be easily modified in order to fit different scenarios and applications, e.g. in connection with the service, assistance, rescue and mobile robots or even to work simultaneously with the human in industrial human-robot teamwork.

In future work, tactile sensors could be integrated with the system to optimize the handing-over tasks. In addition to that, a voice command subsystem could be implemented.

References

[1] Bischoff, R., Graefe, V.: HERMES - a Versatile Personal Assistant Robot. Proc. IEEE - Special Issue on Human Interactive Robots for Psychological Enrichment 92, 1759–1779 (2004)
[2] Edsinger, A., Kemp, C.: Human-Robot Interaction for Cooperative Manipulation: Handing Objects to One Another. In: 16th IEEE International Conference on Robot & Human Interactive Communication, vol. 2, pp. 1167–1172 (2007)
[3] Cakmak, M., Srinivasa, S., Lee, M., Forlizzi, J., Kiesler, S.: Human Preferences for Robot-Human Hand-over Configurations. In: IEEE/RSJ International Conference on Intelligent Robots and Systems, San Francisco, USA, pp. 1986–1993 (2011)
[4] Kim, J., Park, J., Hwang, Y., Lee, M.: Advanced Grasp Planning for Handover Operation Between Human and Robot: Three Handover Methods in Esteem Etiquettes Using Dual Arms and Hands of Home-Service Robot. In: 2nd International Conference on Autonomous Robots and Agents, New Zealand, pp. 34–39 (2004)

[5] Hussain, A.T., Said, Z., Ahamed, N., Sundaraj, K., Hazry, D.: Real-Time Robot-Human Interaction by Tracking Hand Movement & Orientation Based on Morphology. In: IEEE International Conference on Signal and Image Processing Applications, pp. 283–288 (2011)

[6] Chuang, Y., Chen, L., Zhao, G., Chen, G.: Hand Posture Recognition and Tracking Based on Bag-of-Words for Human Robot Interaction. In: IEEE International Conference on Robotics and Automation, pp. 538–543 (2011)

[7] Yuan, M., Farbiz, F., Mason, C., Yin, T.K.: Robust Hand Tracking Using a Simple Color Classification Technique. The International Journal of Virtual Reality 8(2), 7–12 (2009)

[8] Phung, S., Bouzerdoum, A., Chai, D.: Skin segmentation using color pixel classification: Analysis and comparison. IEEE Transactions on Pattern Analysis and Machine Intelligence 27(1), 148–154 (2005)

[9] Galambos, C., Matas, J., Kittler, J.: Progressive probabilistic Hough transform for line detection. In: IEEE Computer Society Conference on Computer Vision and Pattern Recognition, pp. 554–560 (1999)

[10] Zinn, M., Khatib, O., Roth, B.: A New Actuation Approach for Human Friendly Robot Design. Int. J. of Robotics Research 23, 379–398 (2004)

[11] Mainprice, J., Sisbot, E., Simeon, T., Alami, R.: Planning Safe and Legible Hand-Over Motions for Human-Robot Interaction. In: IARP Workshop on Tech. Challenges for Dependable Robot in Human Environments (2010)

[12] Bdiwi, M., Suchý, J.: Library Automation Using Different Structures of Vision-Force Robot Control and Automatic Decision System. In: IEEE/RSJ International conference on Intelligent Robots and Systems, pp. 1677–1682 (2012)

Robot-Mediated Interviews: Do Robots Possess Advantages over Human Interviewers When Talking to Children with Special Needs?

Luke Jai Wood, Kerstin Dautenhahn, Hagen Lehmann,
Ben Robins, Austen Rainer, and Dag Sverre Syrdal

Adaptive Systems Research Group, University of Hertfordshire, United Kingdom
{l.wood, k.dautenhahn,h.lehmann,b.robins,
a.w.rainer,d.s.syrdal}@herts.ac.uk

Abstract. Children that have a disability are up to four times more likely to be a victim of abuse than typically developing children. However, the number of cases that result in prosecution is relatively low. One of the factors influencing this low prosecution rate is communication difficulties. Our previous research has shown that typically developing children respond to a robotic interviewer very similar compared to a human interviewer. In this paper we conduct a follow up study investigating the possibility of Robot-Mediated Interviews with children that have various special needs. In a case study we investigated how 5 children with special needs aged 9 to 11 responded to the humanoid robot KASPAR compared to a human in an interview scenario. The measures used in this study include duration analysis of responses, detailed analysis of transcribed data, questionnaire responses and data from engagement coding. The main questions in the interviews varied in difficulty and focused on the theme of animals and pets. The results from quantitative data analysis reveal that the children interacted with KASPAR in a very similar manner to how they interacted with the human interviewer, providing both interviewers with similar information and amounts of information regardless of question difficulty. However qualitative analysis suggests that some children may have been more engaged with the robotic interviewer.

Keywords: Humanoid robots, interviews, children, human-robot interaction, disclosure, interaction dynamics, social interaction.

1   Introduction

In recent years research investigating how robots can serve as educational, therapeutic and assistive tools has steadily increased [1, 2]. One area of social robotics investigating social mediation for children with special needs is particularly promising, and indicates that children with special needs such as autism often respond well to robots [3-5]. A new emergent application area of social robotics "Robot-Mediated Interviews", investigates how robots could be used to communicate with children in an interview scenario. Studies suggest that typically developing children

G. Herrmann et al. (Eds.): ICSR 2013, LNAI 8239, pp. 54–63, 2013.

respond well to a robot in an interview scenario [6-9], however, to date there has been very little research investigating how children with special needs would respond to a robot in an interview scenario. This paper endeavours to investigate on a case study basis, how children with a variety of special needs respond to a robot in an interview scenario compared to a human interviewer. Children with special needs such as autism can sometimes be more difficult to communicate with, particularly when communicating with someone unfamiliar to them [10, 11]. These communication difficulties can sometimes be very obstructive when trying to acquire information from a child about sensitive or emotionally provocative events. Establishing if robots could be used to bridge this gap and help facilitate communication may assist local authorities such as police and social services safeguard children that are in potentially unsafe environments. Using robots in this context for children with special needs is a logical step in light of the positive research indicating that children with special needs such as autism do respond well to robots in other contexts [3-5].

To investigate how children with special needs respond to a robot compared to a human interviewer, we interviewed 5 children twice (once by a robot, once by a human) in a counterbalanced study. The primary units of analysis for this study were: word count, information disclosure, time talking, and perceived engagement. These measures allowed us to compare the performance of the robotic interviewer with the human interviewer.

2   Background

The various sub domains of social robotics investigate the wide and diverse applications where social robots could be utilised, from robotic pets and educational aids [1, 2, 12] to therapeutic and assistive tools [4, 13, 14], research in social robotics indicates that both neurotypical children and children with special needs often show enthusiasm towards interacting with social robots [3, 15]. Given that children are often keen to interact with robots, investigating the possibility of robot mediation in an interview setting would be a logical step.

Interviewing children is often a very sensitive and delicate task that is carried out by skilled individuals that have undergone extensive training to perform such a skilled role. One area where the way in which interviews are conducted is particularly well researched and documented is within the police force. The UK police force often refers to a document called Achieving Best Evidence in Criminal Proceedings (ABE) [16]. The ABE was drafted for the UK's Home Office by a team of experts from varying backgrounds including psychology, law and social services. This document provides an effective structured and standardised method for interviewing young children which we adhered to as closely as possible where relevant.

The ABE suggests that interviews should have four phases, an establishing a rapport phase, a free narrative recall phase, a more specific questioning phase, and a closure phase. In the establishing a rapport phase simple non-invasive, topic-neutral questions are asked as well as the ground rules being set. In the free narrative recall phase the child is encouraged to recall as much information as possible without prompting, as this is thought to be the most reliable and accurate information. The question asking phase builds on the free narrative recall phase by asking more specific questions relating to the information gathered in the free narrative recall. It is important that the questions do not lead the interviewee as this could lead to misinformation. In the closing phase the topic returns to a neutral theme before thanking the child for their time and concluding the interview. In our study we used a similar structure but combined the free narrative phase with the questioning phase as we were investigating the relationship between question difficulty and how the child responded to the different interviewers.

When interviewing children it is important to remain neutral so as not to lead them towards a particular response or nonresponse. Maintain ones composure without subtly and unintentionally indicating thoughts and feelings can be difficult, particularly when interviewing children about a stressful or traumatic ordeal. The 2011 ABE states "the interviewer should not display surprise at information as this could be taken as a sign that the information is incorrect" [16]. This level of self-discipline can be quite difficult even for an experienced human interviewer. In addition to this it is also important that the interviewer does not appear to assume that someone is guilty "So far as possible, the interview should be conducted in a 'neutral' atmosphere, with the interviewer taking care not to assume, or appear to assume, the guilt of an individual whose alleged conduct may be the subject of the interview" [16]. More recent research also indicated that gestures can play a significant role in leading a witness [17]. Small gestures such as head nods can affect what someone will say in an interview [17]. Using a robot to interview a child could eliminate any of the subtle unintentional signs in body language that a human interviewer may give away, while the body language of the robot can be fully and precisely controlled by the interviewer.

Jones et al. [18] conducted a systematic review of 17 papers and concluded that children with a disability are up to four times more likely to be a victim of abuse than children without disabilities, however, the number of cases that result in prosecution is relatively low [19, 20]. Interviewing children with special needs can be very difficult, particularly when talking about a sensitive or emotionally provocative topic or event. A Mencap review states that "Those who can't communicate can't tell, and those who can't communicate well won't be believed" [21]. Research has shown that children with special needs such as autism are often very keen and willing to interact with robots [22, 23]. Robots such as KASPAR have been used for therapeutic purposes successfully on many occasions and have encouraged some of these children to be less isolated and more socially interactive [3, 4]. We hypothesis that these positive responses could possibly be replicated in an interview scenario and may prove be advantageous when using a robot in an interview context.

The primary research questions in this study were:

1. Will information disclosure differ between the two conditions?
2. Do the quantitative aspects of the children's responses change when being interviewed by a robot?
3. Are there any differences in the temporal aspects of the verbal children's responses?
4. Will there be a difference in the children's engagement change dependent upon the interview partner?

In light of finding from previous studies we would expect to see a clear preference towards the robotic interview partner, although this may not affect the amount of information they provide in this particular study. We predict that the children will have a particularly clear preference to the robot with regards to how engaged they are with the interview partner and activity.

3   Method

3.1   Participants

This study took place in a UK primary school that specialises in catering for children with special needs. Five (3 male, 2 female) children took part in this study, each of which had different special need that included: Autism, BESD (Behavioural, Emotional and Social Difficulties), MLD (Moderate Learning Difficulties), and SLCN (Speech, Language and Communication Needs). The children were between the ages of 9 and 11, and some had interacted with KASPAR in previous studies in a different context.

3.2   Procedure

The children received a group introduction to both KASPAR the robot (figure 2) and the human interviewer at the school before the interviews commenced. The group introduction was used to familiarise the children with the interviewers and explain the procedure to them. Each child had 2 interviews in total, 1 with the robot and 1 with the human interviewer. The interviews were carried out in a 2 phase counterbalanced structure to minimise any crossover effects, with a 1 day gap between each interview. After each interview the child was asked some questions about the experience, in particular the interviewer. Once all of the interviews had been completed, the children were rewarded with some time to play with KASPAR along with a selection of other robotic toys.

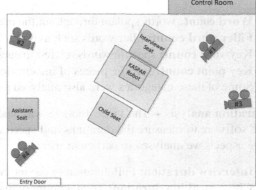

Fig. 1. KASPAR Robot   **Fig. 2.** Room layout

The interviews followed a structure recommended by the ABE [16] of an easy topic neutral followed by a free narrative and questioning phase that focused on the main topic of interest. The interviews then returned to a neutral topic before concluding. The main theme and topic of the interviews focused on animals and pets. This topic was chosen because it offered sufficient scope for various different types and difficulty of questions.

3.3   The Setup

The interviews were conducted in an unused classroom that contained a small lockable cupboard that the children were unable to see into. The cupboard was used as a control room for KASPAR and housed a monitor with a wireless connection to camera #1 to observe the situation and make KASPAR respond appropriately (see Figure 2). The children were unaware that KASPAR was being controlled by a human triggering the correct questions and responses from a pre-recorded list. The interviews were led by the lead investigator either in person or remotely via KASPAR to help maintain consistency between the interviews. The children were escorted to and from the interviews by a second researcher that was unknown to the children. The second researcher remained in the room throughout the interviews, but was as non-reactive as possible. After each interview the children were asked by the second researcher to rate particular aspects of the experience, in particular the interaction partner.

4   Measures

There were four primary units of analysis in this study:

Communicative content – Communicative content refers to the statistical analysis of both the number of words spoken and the amount of qualitative pieces of information gathered from the children.   All interviews were fully transcribed and analysed in detail for:

- **Word count:** words spoken throughout the interview excluding filler words.
- **Filler word count:** filler words such as "err", "errm", "hum".
- **Keyword count:** specific words related to questions e.g. "animal", "feed", "wild".
- **Key point count:** specific pieces of information relating to the questions asked.
  Some of these categories were also analysed proportionately where appropriate.

Duration analysis – The video footage was analysed and coded using the Observer XT software to measure the durations and specific temporal aspects of the interviews. The aspects we analysed in particular were:

- **Interview duration:** Full duration of the interview, start to finish.
- **Child speaking duration:** Total amount of time the child spent speaking.
- **Child pause duration:** Time child pauses for 2 seconds or more whilst speaking.
- **Interviewer speaking duration:** Total amount of time the interviewer speaks.

- **Response time C>I:** Total amount of time between the child speaking and the interviewer responding.
- **Response time I>C:** Total amount of time between the interviewer speaking and the child responding.

Questionnaires – Immediately after each interview the child was asked to rate specific aspects of the experience:

- **Interest:** How interesting the child found the experience.
- **Difficulty:** How difficult the child found the interview.
- **Fun:** How much fun the child had participating.
- **Duration:** How long the child thought the interview took.

Perceived Engagement - In this study we analysed how engaged the children appeared to be in the activity and interaction partner. The engagement was coded by 3 independent blind coders that had no prior knowledge of the study or this line of research. The coders worked individually and were unknown to each other. Each coder viewed all of the videos start to finish before commencing the coding to acquire some familiarity for the children and their behaviour. This was an important phase because accurately assessing a child's engagement relies on having some previous knowledge of their full repertoire of behaviour within that particular context. The videos had been edited so that there was no audio and a portion of the screen had been obscured to hide the interaction partner from the coders. The coders were advised of where an interaction partner was located to order to assist in coding the engagement, but were not provided with any details about the experiment, as this may have adversely altered and biased their coding. The videos were coded with a simple on screen slider that moved with the position of the courser. A smile-meter placed in the top right hand corner of the screen changed according to the position of the slider. If the slider was towards the top this would indicate more engaged and the smile-meter would have a greater smile. If the slider was lower down the screen this would indicate less engaged and the smile-meter would change to more of a frown (Figure 3). This approach of continuous monitoring was originally employed for market research [24], but has also been adopted and adapted for measuring levels of fun and engagement for HRI and HCI applications [25, 26]. The method used in our study takes inspiration from these approaches, but uses a slightly different implementation to suite the requirements of our needs for this particular type of study.

Fig. 3. Engagement coding program

The coders were given brief instructions on what may constitute engagement. However, it was ultimately left to the coder to decide what they personally considered to be engagement. The coders would move the slider up and down on a touch screen according to how engaged they personally felt that the child was in the interaction partner. The coding occurred at the normal video running speed and the program logged where the slider was on the screen in relation to the position of the video. The program logged the position of the slider 10 times every second, and all occurred in one continuous run with no pausing or rewinding.

Typically developing people are often very good at assessing how engaged someone is in an activity or an interaction partner. Rather than trying to code specific behaviours in great detail such as eye contact or body gestures, we tried this more subjective method and combined the results of each coder to establish an overall average for the engagement in each interview.

5   Results

The results from this study in Table 1 are consistent with previous findings in studies with typically developing children [6, 7, 9]. The only significant difference found relating to the perception of the children was how long the children thought the interviews took. The children thought the interviews with the robot were longer which was often correct as the results indicate that the mean duration of the interviews with KASPAR were longer than the human interviewer. Although not significant, it is indicated that on average the children found it more difficult talking to KASPAR. This may be due to the text-to-speech voice of KASPAR being more difficult to

Table 1. Quantitative comparison measures

Measure	KASPAR Mean	Human Mean	Mean Diff.	t(p)	Standard Dev.
1=boring - 5=interesting	3.40	3.40	0.00	0.00 (1.000)	1.428
1=hard - 5=easy	3.20	4.00	-0.80	1.21 (0.242)	1.625
1=no fun - 5=fun	4.60	3.80	0.80	1.21 (0.242)	1.249
1=long time - 5=quick	2.20	5.00	-2.80	2.44 (0.025)*	1.800
Overall key points	18.40	18.80	-0.40	0.09 (0.928)	7.902
All Key words	22.40	18.40	4.00	0.58 (0.568)	10.032
Proportionate all key words	0.17	0.21	-0.04	0.84 (0.412)	0.079
Child word count	155.80	103.20	52.60	0.84 (0.410)	100.019
Proportionate word count	1.06	0.57	0.48	1.01 (0.161)	0.766
Proportionate filler word count	0.07	0.05	0.02	0.85 (0.404)	0.049
Interviewer word count	157.60	180.80	-23.20	2.44 (0.024)*	17.069
Interview duration	328.65	260.78	328.65	1.55 (0.137)	64.817
Child response duration	108.88	79.02	108.88	0.72 (0.480)	50.824
Interviewer response duration	55.25	76.77	55.25	2.94 (0.008)*	12.715
Response time C>I	93.83	34.76	93.83	2.19 (0.040)*	39.454
Response time I>C	36.58	30.47	36.58	0.80 (0.433)	17.070
Child pause duration	29.90	14.51	29.90	1.52 (0.145)	17.038
Perceived average engagement	321.09	278.29	42.81	2.28 (0.084)	72.256

understand. The quantitative results from this study are not significant, however, the qualitative results from looking at the interactions in the 2 conditions would indicate that some of the children seemed to be more engaged with the robot than the human interviewer. From the 5 children that took part in the study it appears that the children with the most severe communication difficulties seem a lot more engaged with the robot than the human interviewer. For example, three of the children's engagement scores were on average 108.99, 47.20 and 44.65 points higher with KASPAR than with the human interviewer. All three of these children had SLCN (Speech, Language and Communication Needs) and two of the children had in addition Autism.

6   Discussion and Conclusion

The majority of the quantitative results from this study were consistent with the results of previous studies. The children interacted with the robot in a similar manner to which they did with a human interviewer. The information the amount and details the children revealed to the robot was similar to what they revealed to the human interviewer. However, the results of the engagement coding seem to suggest that in some cases the children seemed to be more engaged in the interaction with the robot than the human interviewer. Engagement is an important factor in any interaction, and the more engaged a child is in an activity, often the better their participation, which in an interview setting could be very beneficial. From a qualitative perspective the children were all very different and the differences in the interactions where varied. Some of the children had a very obvious preference for the KASPAR, which was reflected in the engagement coding, whilst others responded in a very similar manner in both interviews. It would appear that KASPAR seemed to benefit the children with the more severe conditions the most.

6.1   Limitations and Future Studies

This small case study was with just 5 children, the results may have been clearer had there been more participants. The participants had various different needs and conditions, therefore the results of this study cannot be generalised to a population with particular condition as there is currently insufficient evidence. The children in this study were from one school in one geographical location, covering an age range from 8 to 11. Children from other locations and with different age ranges may respond differently to a robot in this setting as age and cultural differences may have affected the way the children would respond in the interviews. However, the results from multiple studies have consistently shown that children do respond to a robot in an interview context. In futures studies we would like to create a more flexible user friendly system and put it in the hands of a professional interviewer to assess if this approach would be useful in a real world setting. To reach this goal it is important to first conduct studies in a less sensitive context such as the one reported in this article.

Acknowledgements. The authors would like to thank Ben Robins for providing assistance during the experiments and Maria Jesus Hervas Garcia, Rachel Marr-Johnson, Nicholas Shipp and Kinga Grof for help with the engagement and reliability coding. We would also like to thank the staff and children of the school who participated in this work.

References

1. Yoshida, S., Sakamoto, D., Sugiura, Y., Inami, M., Igarashi, T.: RoboJockey: Robotic dance entertainment for all. In: SIGGRAPH Asia 2012 Emerging Technologies. ACM (2012)
2. Tanaka, F., Ghosh, M.: The implementation of care-receiving robot at an English learning school for children. In: HRI 2011: Proceedings of the 6th International Conference on Human-robot Interaction, Lausanne, Switzerland (2011)
3. Wainer, J., Dautenhahn, K., Robins, B., Amirabdollahian, F.: Collaborating with Kaspar: Using an autonomous humanoid robot to foster cooperative dyadic play among children with autism. In: 2010 10th IEEE-RAS International Conference on Humanoid Robots, Humanoids (2010)
4. Robins, B., Dautenhahn, K., Dickerson, P.: From isolation to communication: A case study evaluation of robot assisted play for children with autism with a minimally expressive humanoid robot. In: Proc. the Second International Conferences on Advances in Computer-Human Interactions, ACHI 2009. IEEE Computer Society Press, Cancun (2009)
5. Huskens, B., Verschuur, R., Gillesen, J., Didden, R., Barakova, E.: Promoting question-asking in school-aged children with autism spectrum disorders: Effectiveness of a robot intervention compared to a human-trainer intervention. Developmental Neurorehabilitation 1–12 (2013)
6. Bethel, C.L., Stevenson, M.R., Scassellati, B.: Sharing a Secret: Interactions Between a Child, Robot, and Adult. Presented at the Children with Robots Workshop at the 6th ACM/IEEE International Conference on Human-Robot Interaction, Lausanne, Switzerland (2011)
7. Wood, L.J., Dautenhahn, K., Rainer, A., Robins, B., Lehmann, H., Syrdal, D.S.: Robot-Mediated Interviews-How Effective Is a Humanoid Robot as a Tool for Interviewing Young Children? PLOS One 8(3), e59448 (2013)
8. Bethel, C.L., Eakin, D.K., Anreddy, S., Stuart, J.K., Carruth, D.: Eyewitnesses are misled by human but not robot interviewers. In: Proceedings of the 8th ACM/IEEE International Conference on Human-robot Interaction. IEEE Press (2013)
9. Wood, L.J., Dautenhahn, K., Lehmann, H., Robins, B., Rainer, A., Syrdal, D.S.: Robot-Mediated Interviews: Does a robotic interviewer impact question difficulty and information recovery? In: 12th European AAATE Conference, Vilamoura, Portugal (2013)
10. Wing, L.: The autistic spectrum: A guide for parents and professionals. Constable (1996)
11. Jordan, R.: Autistic spectrum disorders: an introductory handbook for practitioners. David Fulton Publishers, London (1999)
12. Barlett, B., Estivill-Castro, V., Seymon, S.: Dogs or robots—Why do children see them as robotic pets rather than canine machines? In: 5th Australasian User Interface Conference (AUIC 2004), Dunedin (2004)
13. Goris, K., Saldien, J., Lefeber, D.: Probo: a testbed for human robot interaction. In: HRI 2009: Proceedings of the 4th ACM/IEEE International Conference on Human Robot Interaction, La Jolla, California, USA (2009)

14. Kozima, H., Michalowski, M., Nakagawa, C.: Keepon: A Playful Robot for Research, Therapy, and Entertainment. International Journal of Social Robotics 1(1), 3–18 (2009)
15. Michalowski, M.P., Sabanovic, S., Michel, P.: Roillo: Creating a Social Robot for Playrooms. In: The 15th IEEE International Symposium on Robot and Human Interactive Communication, ROMAN (2006)
16. UK Government, Achieving Best Evidence in Criminal Proceedings: Guidance on Interviewing Victims and Witnesses, and Using Special Measures. Home Office Criminal Justice System (2007, 2011)
17. Gurney, D., Pine, K., Wiseman, R.: The gestural misinformation effect: skewing eyewitness testimony through gesture. American Journal of Psychology (2013)
18. Jones, L., Bellis, M.A., Wood, S., Hughes, K., McCoy, E., Eckley, L., Bates, G., Mikton, C., Shakespeare, T., Officer, A.: Prevalence and risk of violence against children with disabilities: a systematic review and meta-analysis of observational studies. The Lancet (2012)
19. Turk, V., Brown, H.: The Sexual Abuse of Adults with Learning Disabilites: Results of a Two Year Incidence Survey. Mental Handicap Research 6(3), 193–216 (1993)
20. McCarthy, M., Thompson, D.: A prevalence study of sexual abuse of adults with intellectual disabilities referred for sex education. Journal of Applied Research in Intellectual Disabilities 10(2), 105–124 (1997)
21. Mencap, Submission to the Sex Offences Review (1999)
22. Robins, B., Dautenhahn, K., Boekhorst, R., Billard, A.: Robotic assistants in therapy and education of children with autism: can a small humanoid robot help encourage social interaction skills? Universal Access in the Information Society 4(2), 105–120 (2005)
23. Robins, B., Dautenhahn, K., Dickerson, P.: Embodiment and cognitive learning – can a humanoid robot help children with autism to learn about tactile social behaviour? In: Ge, S.S., Khatib, O., Cabibihan, J.-J., Simmons, R., Williams, M.-A. (eds.) ICSR 2012. LNCS, vol. 7621, pp. 66–75. Springer, Heidelberg (2012)
24. Fenwick, I., Rice, M.D.: Reliability of continuous measurement copy-testing methods. Journal of Advertising Research (1991)
25. Tanaka, F., Cicourel, A., Movellan, J.R.: Socialization between toddlers and robots at an early childhood education center. Proceedings of the National Academy of Sciences 104(46), 17954–17958 (2007)
26. Read, J., MacFarlane, S., Casey, C.: Endurability, engagement and expectations: Measuring children's fun. In: Interaction Design and Children. Shaker Publishing Eindhoven (2002)

Multidomain Voice Activity Detection during Human-Robot Interaction

Fernando Alonso-Martín, Álvaro Castro-González,
Javier F. Gorostiza, and Miguel A. Salichs

Universidad Carlos III Madrid
{famartin,acgonzal,jgorosti,salichs}@ing.uc3m.es

Abstract. The continuous increase of social robots is leading quickly to the cohabitation of humans and social robots at homes. The main way of interaction in these robots is based on verbal communication. Usually social robots are endowed with microphones to receive the voice signal of the people they interact with. However, due to the principle the microphones are based on, they receive all kind of non verbal signals too. Therefore, it is crucial to differentiate whether the received signal is voice or not.

In this work, we present a Voice Activity Detection (VAD) system to manage this problem. In order to achieve it, the audio signal captured by the robot is analyzed on-line and several characteristics, or *statistics*, are extracted. The statistics belong to three different domains: the time, the frequency, and the time-frequency. The combination of these statistics results in a robust VAD system that, by means of the microphones located in a robot, is able to detect when a person starts to talk and when he ends.

Finally, several experiments are conducted to test the performance of the system. These experiments show a high percentage of success in the classification of different audio signal as voice or unvoice.

1 Introduction

Social Robots are intended for human-robot interaction (HRI) in real environments such as homes. One of the main ways of HRI is by means of oral communication. This implies that the robots are endowed with microphones that work as "artificial ears" for achieving verbal capacities similar to those showed by humans (e.g. to identify if a person is verbalizing, who is talking, his emotional state, his gender, etc.). All these capacities, that the human being is able to perform be means of the auditory canal (exclusively or complemented by other means), are being imitated in social robots.

In this work, we present a Voice Activity Detection (from now on VAD) system intended to be integrated in social robots. This kind of systems classifies audio signals as voice or unvoice. Consequently, a VAD system determines when a human voice starts and ends. They have to avoid the classification of audio signals coming from other sounds (unvoice) as human voice.

G. Herrmann et al. (Eds.): ICSR 2013, LNAI 8239, pp. 64–73, 2013.

VAD systems are very useful in social robots because they can improve the performance of many verbal capacities of the robots such as automatic speech recognition, user localization and identification, user's gender identification, or voice emotion recognition. This is achieved by delimiting the signal to compute, i.e. the voice, to a more precise fragment. Therefore, the VAD system is a key element of the interaction system implemented in our robots, called the Robotics Dialog System (RDS) [Alonso-Martín et al., 2012,Alonso-Martin and Salichs, 2011].

In this paper we present a VAD system that combines in real time statistics (i.e. characteristics extracted from the computation of the audio signal) belonging to three domains: the time, the frequency, and the time-frequency.

There are other techniques to improve the quality of the audio signal during its acquisition, e.g. Acoustic Echo Cancellation (AEC), Active and Passive Noise Cancellation, or Sound Source Separation (SSS). However, these methods are out of the scope of this work and usually they are applied prior to the analysis of the signal with a VAD system. Then, these methods and a VAD system are complementary.

2   Related Work

As already mentioned, a VAD system tries to determine when a verbalization starts and when it ends. Traditionally, this has been achieved considering a certain threshold in the volume: when the volume of the received audio signal exceeds the threshold, this is considered as the beginning of the utterance; after, when the volume goes below the threshold during a certain range of time, it means that the utterance has ended [DesBlache et al., 1987]. This approach fails many times due to its nature: a loud noise, e.g. the noise of an animal or a song, can be easily considered as human voice.

Ghaemmaghami [Ghaemmaghami et al., 2010] presents a VAD system which considers statistics from the time domain. He focuses in two particular statistics: the autocorrelation and the zero-crossing rate. Bachu [Bachu, 2008] also uses statistics from the time domain but, in this case, they are the zero-crossing rate and the energy of the signal.

Other researchers consider statistics in the frequency domain. Moattar [Moattar, 2010] analyzes the magnitude of the peaks in the spectrum that, according to him, they correspond to the vowel sounds. Khoa uses the autocorrelation, similar to Ghaemmaghami, but in the frequency domain. Dou [Dou et al., 2010] considers, in the frequency domain too, the bispectrum or bispectral density which is a statistic addressed to detect and characterize the properties of nonlinear signals. [Nikias and Raghuveer, 1987].

Xiaoling presented a comparison of several techniques of VAD where statistics from both time and frequent domains were used independently [Yang et al., 2010].

The already mentioned domains can be combined in the time-frequency domain. Chen [Chen et al., 2010], by means of the Wavelet Transform, obtains the audio signal in the time-frequency domain. Then, the signal is classified as voice or unvoice using a Support Vector Machine classifier.

Other works combine statistics from several domains. Moattar presents a two-domain VAD system where statistics from the time domain (energy, maximum autocorrelation peak, and autocorrelation peak count) and from the frequency domain (pitch, autocorrelation peak valley ratio, entropy, flatness, peak valley difference, and maximum cepstral peak) [Moattar and Homayounpour, 2011,Moattar and Homayounpour, 2009].   However, to the best of our knowledge, there are not works where statistics from the three aforementioned domains (time, frequency, and time-frequency) are used together. This is the approach followed in this work and it represents the main novelty of this work.

Finally, other researchers apply other AI techniques to achieve a VAD. For example, Aibinu apply artificial neural networks and an autoregressive model for this task [Aibinu et al., 2012]. It is also remarkable the work presented by Cournapeau [Cournapeau, 2010], where he uses Variational Bayes and online Expectation Maximization.

3   The Proposed VAD System

In this section, we present a multi-domain VAD that combines several statistics belonging to three different domains: time, frequency, and time-frequency. The computation of these statistics is performed on-line, analyzing the audio input signal captured in real time. The statistics working in the time domain are directly obtained from the sampled analog signal of the microphone; in the case of the statistics belonging to the frequency domain, the Fast Fourier Transform is applied to the time-domain signal; and the time-frequency domain signal is obtained applying the Discrete Haar Wavelet Transform (Fig. 1).

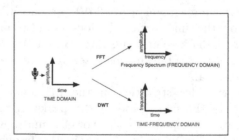

Fig. 1. The three Audio Domains and how they are related

Once the statistics are computed, there is a range of values for each one that is associated to voice. This means, if the value of one statistic lies within the associated voice range, we state that, according to that statistic, the signal can be classified as voice. The combination of several statistics from different domains makes this a robust VAD system.

The extraction of the signal features has been achieved using Chuck [1]. Chuck is an audio programming language specifically designed to analyze sound waves and

---

[1] http://chuck.cs.princeton.edu/

generate non verbal sounds. It has been selected for its capacity for working *on-the-fly*, and for the simplicity of its own audio analyzers. The statistics extracted are computed on-line over time windows.

Following, the statistics used in our VAD system are detailed.

3.1 Pitch

This statistic refers to the frequency of the sound perceived by the human ear. It depends on the number of vibrations produced by the vocal chords. The pitch calculation is not included among the component of Chuck, so it has been implemented by the authors.

In this work we have developed three different Pitch Detection Algorithms (PDA) working in distinct domains:

1. The first method detects the first minimum in the autocorrelation of the input signal in the time domain. That is, the minimum in the cross-relation of the audio signal with itself that corresponds to the pitch or fundamental frequency. This is a simple and well-known method in the literature [McLeod and Wyvill, 2005,Cheveigné and Kawahara, 2002].
2. In the second method, the FFT is applied, so it works in the frequency domain. This algorithm considers the distance among peaks in the spectrum of the signal.
3. The last one works in the time-frequency domain and it uses the Discrete Haar-Wavelet Transform (DWT). It is based in the Larson's work [Larson and Maddox, 2005].

Figure 2 shows an example of these three methods, vertical axis represent the score of the statistic, according its mathematical formula, on consecutive time instants (horizontal axis). The blue line corresponds to the time-domain pitch, the red line represents the frequency-domain pitch, and the green line plots the time-frequency-domain pitch. Focusing on Figure 2(b), it can be observed that, considering a white noise signal, the pitch considerably varies when it is computed using the first two methods (time-domain and frequency domain). Moreover, when it is calculated in the time-frequency domain with the DWT is remains at zero most of the time. On the other side, when the signal corresponds to a human voice (Fig. 2(a)), the pith takes quite similar and homogeneous values. In this case, the peaks in the statistics are due to the gaps between words.

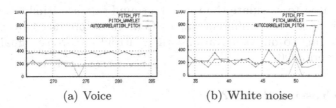

(a) Voice   (b) White noise

Fig. 2. Evolution of the pitch score (hertz) in three domains for voice and white-noise signals

3.2   Rolloff-95

This feature corresponds to the frequency at which the 95% of the signal energy is already contained, considering the window of the spectrum. The signal energy corresponds to the integrate of the signal; therefore, the Rolloff-95 corresponds to the point in the axis of abscissas (frequency) that delimits the 95% of the area of the signal energy defined in the window [Kim and Kim, 2006].

The Rolloff-95 of a voice and white-noise signals are shown in Figure 3. The voice usually takes values of Rolloff-95 around 500 (always below 1000). In contrast, when we analyze a white-noise signal, the Rolloff-95 is around 1000 or higher. Again, the peaks in the voice plot correspond to the gaps between words.

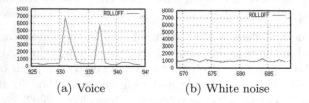

(a) Voice   (b) White noise

Fig. 3. Evolution of RollOff-95 score (hertz) for voice and white-noise signals

3.3   Flux

This statistic is computed in the frequency domain. It is a measure of the spectral rate of change, which is given by the sum across one analysis window of the squared difference between the magnitude spectra corresponding to successive signal frames. In other words, it refers to the variation of the magnitude, in frequency domain, of the signal. When its value is close to 0, it indicates that the magnitude in any frequency is similar. When the value is close to 1, it implies that there are important variations in the amplitude of the signal depending on the frequency; this shows a very sharp spectrum. Flux has been found to be a suitable feature for the separation of music from speech, yielding higher values for music examples [Burred and Lerch, 2004].

3.4   Centroid

The centroid represents the median of the signal spectrum in the frequency domain. That is, the frequency towards the signal approaches to the most. It is frequently used to calculate the tone of a sound or voice (timbre). Our vocal apparatus produces very different values from those produced by a violin or a flute. According to our observations, the centroid of a voice signal oscillates between 850 and 3000.

3.5   Zero Crossing Rate (ZCR)

The ZCR statistic refers to the number of times the signal crosses the axis of abscissas; that is, the number of times the amplitude changes its sign with respect to previous sample. Typically, the ground noise crosses more often the x axis than a verbal sound [Bachu, 2008]. Voiced speech usually shows a low ZCR count, whereas the unvoiced speech shows high ZCR. As it is shown in Figure 4, white-noise has a higher ZCR (around 250 times in each window time) than the voice (between 0 and 50). Once again, the outliers correspond to the silence among words.

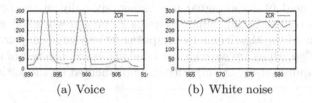

(a) Voice   (b) White noise

Fig. 4. Evolution of the ZCR score for voice and white-noise signals

3.6   Signal to Noise Ratio (SNR)

It relates the current signal volume, or signal energy, to the noise signal volume, or average ground noise energy, in frequency domain. The average ground noise energy is updated only when the VAD system classifies the signal as unvoice. In each iteration the SNR is computed dividing the signal energy by the average ground noise energy. Therefore, this statistic dynamically adapts to environments with different sounds because it takes into account the current ground noise volume.

In this case, a dynamic threshold is set to classify a signal as voice or unvoice. When a voice signal is captured, the SNR passes the threshold everytime a word is uttered (Fig. 5(a)). In the case of a white-noise signal, its SNR is always below the threshold (Fig. 5(b).

The SNR threshold is dynamically updated considering the ground noisy at every moment. Then, if the ground noise is strong (in a noisy environment such as a football stadium), the division between a person's voice and the ground noise (i.e. the SNR) is very low, and then the threshold is reduced. On the contrary, when the ground noise is low (i.e. the environment is calmed), the SNR takes a high value, and the threshold is increased.

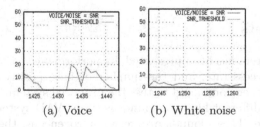

(a) Voice   (b) White noise

Fig. 5. Evolution of SNR score for voice and white-noise signals

3.7   The Algorithm

All the statistics which form our VAD system are summarized in Table 1. This table lists all the signal features, the domain they belong to, and the range of values that corresponds to a voice signal. These voice ranges have been empirically fixed by means of observations of the evolution of statistics in real time.

It is important to emphasize that the statistics applied in this work are well-known in the field of audio signal processing, specially in the area of Music Information Retrieval. In particular, Rebecca Fiebrink[Fiebrink et al., 2008] has developed some of the algorithms to compute certain statistics: *flux*, *rollOff*, and *centroid*. Her algorithms are natively included in Chuck. The rest of the statistics have been computed using the algorithms implemented by the authors.

Table 1. Features analyzed in VAD

Features	Domain	Human voice range fixed
Pitch Autocorrelation	Time	80-600
Pitch FFT	Frequency	80-600
Pitch DWT	Time-Frequency	80-600
Flux	Frequency	> 0.1
Roll-off	Frequency	0-1000
Centroid	Frequency	850-3000
Zero-Crossing Count	Time	0-50
SNR Rate	Frequency	Adaptive

Recalling, the VAD system determines the beginning and the end of an utterance. In order to determine the start of a voice, the audio signal received is analyzed and the statistics are extracted. The values of all the statistics have to be within the corresponding voice range (Table 1). The voice terminates when at least one statistic falls out of its voice range for a predefined amount of time. This time interval has been set to 800 ms. If we consider a shorter time interval, it could happen that the gaps between two words (instants of no-voice) are identified with the end of the voice. In contrast, if this time interval is too much longer, the end of the utterance will be delayed.

This work is focused on the identification of the start and end of a voice robustly. When these events are detected, the VAD system triggers a signal to notify it to the rest of the components in the robot. Actually, these signals are implemented as *ROS topics* [2].

4   Experiments

Once the VAD system has been presented, its performance has to be determined computing the success rate, and the number of false positives. Then, several experiments have been conducted to test the proposed system. The experiments consist in playing different kinds of sounds and our VAD system classifies them as voice or unvoice. These sounds are very heterogeneous, they are originated

---

[2] http://www.ros.org/wiki/

from diverse sources. Of course, one of the sounds is the human voice. Most of the sounds have been acquired directly from the source by the robot's microphone, without any additional technique for controlling the noise (e.g. Echo Cancellation or Sound Source Separation). Some of them have been played using audio files. Each category of sounds has been tested 20 times under different conditions (differents audio volumes, audio capture distances, speakers, etc). The results of the experiments are summarized in Table 2.

Table 2. Features analyzed in VAD

Real sound / Estimation	Voice estimation	Statistics out of voice range
Voice	93%	SNR
White noise	0%	ZCR, SNR, Centroid
Dogs barking	0%	ZCR
Acustic guitar	38.10%	Pitch
Piano-Sintetizador	0%	Pitch, RollOff95, Centroid, ZCR
Knoking	0%	Pitch, RollOff95, ZCR, (sometimes Centroid)
Prerecorded music	0%	Pitch, ZCR, RollOff-95
Engine sounds	0%	Pitch, ZCR, RollOff-95
Pressurized air exhaust	0%	Pitch, rollOff-95, centroid, ZCR

Considering the numbers shown in Table 2, it can be said that our VAD system performs quite well identifying utterance as voice (93% of success). When a voice signal is classified as unvoice, most of the times, it is because a continuous strong noise has appeared just before the voice and it causes an increase of the average ground noise volume. Then, when the voice signal is analyzed, the SNR is very low. This problem usually is solved by the dynamic SNR threshold.

Besides, the system presents few false positives. The worst values are those obtained from the sound of an acoustic guitar. This is because the harmonics of these signals are similar to those in the human voice. When the frequency of a musical note, or a chord, is within the frequency of the human voice, the statistics for this signal are within the predefined voice ranges of our VAD system. Therefore, this is classified as voice. If the musical note is very high-pitched (high frequency), this sound is not classified as voice.

The rest of the unvoice sounds are accurately classified. The last column in Table 2 shows the statistics that are responsible of the classification of these sounds as unvoice; that is, the statistics that lie out of the voice ranges.

Since the VAD system is applied to robots, the required hardware resources are important. The VAD system has been tested in a MacBook White 6.1. It uses between 22% and 30% of the CPU time and 40 MB of RAM memory. The latency for extracting the statistics are variable and it depends on many factors (the spectrum window, the sample rate, graphical plotting, etc.). In our case, this latency has been always lower than 0, 5s.

5   Conclusions

In order to achieve an advanced, robust, voice-based HRI, it is critical to implement a reliable system for detecting the beginning and the end of voice signals. Without it, any sound could be identified with an utterance.

This paper presents a VAD system which is able to classify any sound as voice or unvoice in different environments. It uses the audio signal received by the microphone located inboard our social robots. This signal is analyzed and several statistics belonging to three different domains (time, frequency, and time-frequency) are extracted: *pitch with autocorrelation, pitch with FFT, pitch with DWT, centroid, flux, rollOff-95, autocorrelation, SNR, y ZCR*. For each statistic, a range of human-voice values has been predefined empirically (or dynamically generated in the case of the SNR). The simultaneous combination of all statistics and their voice ranges results in a robust VAD system that determines the start and the end of the utterances.

The main novelty of the VAD system is the combination of on-line computed statistics belonging to three different domains that results in a high success rate.

Several experiments have been run in order to test the performance of the system. It has been observed that the consideration of all the previously listed statistics is necessary to achieve a reliable VAD. For example, if the pitch is not considered, the sound of a acoustic guitar can not be correctly classified. Moreover, the combination of several statistics improves the accuracy of the system; e.g. the classification of engine sounds as unvoice is confirmed by three different statistics.

The system proposed here can be complemented with other techniques, such as echo cancellation, noise cancellation, or sound source separation. Besides, we are already working in a multimodal VAD system which integrates audio and video data to determine who is the speaker in a crowded room.

References

Aibinu et al., 2012. Aibinu, A., Salami, M., Shafie, A.: Artificial neural network based autoregressive modeling technique with application in voice activity detection. In: Engineering Applications of Artificial Intelligence (2012)

Alonso-Martín et al., 2012. Alonso-Martín, F., Gorostiza, J., Malfaz, M., Salichs, M.: User Localization During Human-Robot Interaction. In: Sensors (2012)

Alonso-Martin and Salichs, 2011. Alonso-Martin, F., Salichs, M.: Integration of a voice recognition system in a social robot. Cybernetics and Systems 42(4), 215–245 (2011)

Bachu, 2008. Bachu, R.G.: Separation of Voiced and Unvoiced using Zero crossing rate and Energy of the Speech Signal. In: American Society for Engineering Education (ASEE) Zone Conference Proceedings, pp. 1–7 (2008)

Burred and Lerch, 2004. Burred, J., Lerch, A.: Hierarchical automatic audio signal classification. Journal of the Audio Engineering Society (2004)

Chen et al., 2010. Chen, S.-H., Guido, R.C., Truong, T.-K., Chang, Y.: Improved voice activity detection algorithm using wavelet and support vector machine. Computer Speech & Language 24(3), 531–543 (2010)

Cheveigné and Kawahara, 2002. Cheveigné, A.D., Kawahara, H.: YIN, a fundamental frequency estimator for speech and music. The Journal of the Acoustical Society of America (2002)

Cournapeau, 2010. Cournapeau, D.: Online unsupervised classification with model comparison in the variational bayes framework for voice activity detection. IEEE Journal of Selected Topics in Signal Processing, 1071–1083 (2010)

DesBlache et al., 1987. DesBlache, A., Galand, C., Vermot-Gauchy, R.: Voice activity detection process and means for implementing said process. US Patent 4,672,669 (1987)

Dou et al., 2010. Dou, H., Wu, Z., Feng, Y., Qian, Y.: Voice activity detection based on the bispectrum. In: 2010 IEEE 10th International Conference on (2010)

Fiebrink et al., 2008. Fiebrink, R., Wang, G., Cook, P.: Support for MIR Prototyping and Real-Time Applications in the ChucK Programming Language. In: ISMIR (2008)

Ghaemmaghami et al., 2010. Ghaemmaghami, H., Baker, B.J., Vogt, R.J., Sridharan, S.: Noise robust voice activity detection using features extracted from the time-domain autocorrelation function (2010)

Kim and Kim, 2006. Kim, K., Kim, S.: Quick audio retrieval using multiple feature vectors. IEEE Transactions on Consumer Electronics 52 (2006)

Larson and Maddox, 2005. Larson, E., Maddox, R.: Real-time time-domain pitch tracking using wavelets. In: Proceedings of the University of Illinois at Urbana Champaign Research Experience for Undergraduates Program (2005)

McLeod and Wyvill, 2005. McLeod, P., Wyvill, G.: A smarter way to find pitch. In: Proceedings of International Computer Music Conference, ICMC (2005)

Moattar, 2010. Moattar, M.: A new approach for robust realtime voice activity detection using spectral pattern. In: 2010 IEEE International Conference on Acoustics Speech and Signal Processing (ICASSP), pp. 4478 4481 (2010)

Moattar and Homayounpour, 2009. Moattar, M., Homayounpour, M.: A Simple but efficient real-time voice activity detection algorithm. In: EUSIPCO. EURASIP (2009)

Moattar and Homayounpour, 2011. Moattar, M., Homayounpour, M.: A Weighted Feature Voting Approach for Robust and Real-Time Voice Activity Detection. ETRI J. (2011)

Nikias and Raghuveer, 1987. Nikias, C., Raghuveer, M.: Bispectrum estimation: A digital signal processing framework. Proceedings of the IEEE (1987)

Yang et al., 2010. Yang, X., Tan, B., Ding, J., Zhang, J.: Comparative Study on Voice Activity Detection Algorithm. In: 2010 International Conference on Electrical and Control Engineering (ICECE), Wuhan, pp. 599–602 (2010)

A Low-Cost Classroom-Oriented Educational Robotics System

Mário Saleiro[1], Bruna Carmo[2], Joao M.F. Rodrigues[1], and J.M.H. du Buf[1]

[1] Vision Laboratory, LARSyS, University of the Algarve, 8005-139 Faro, Portugal,
[2] Escola Superior de Educação e Comunicação, University of the Algarve, 8005-139 Faro, Portugal
{masaleiro,jrodrig,dubuf}@ualg.pt, b.santos.carmo@gmail.com
http://w3.ualg.pt/~dubuf/vision.html

Abstract. Over the past few years, there has been a growing interest in using robots in education. The use of these tangible devices in combination with problem-based learning activities results in more motivated students, higher grades and a growing interest in the STEM areas. However, most educational robotics systems still have some restrictions like high cost, long setup time, need of installing software in children's computers, etc. We present a new, low-cost, classroom-oriented educational robotics system that does not require the installation of any software. It can be used with computers, tablets or smartphones. It also supports multiple robots and the system can be setup and is ready to be used in under 5 minutes. The robotics system that will be presented has been successfully used by two classes of 3rd and 4th graders. Besides improving mathematical reasoning, the system can be employed as a motivational tool for any subject.

Keywords: Education, Robotics.

1 Introduction

In the last few years, the use of robots in education has become a very popular way of providing an interdisciplinary, project-based learning activity, with special focus on STEM (science, technology, engineering and mathematics) areas [16]. Robots offer major new benefits to education at all levels [6]. Long-term experiments that involve robots in classrooms have resulted in an increase of students enrolled in such classes, as well as an improvement in general learning, increased motivation and higher performance [5]. Considering the current shortage of student interest in STEM topics, increasing attention has been paid to developing innovative tools for improved teaching of STEM, including robotics [16]. Especially kids from elementary schools (6-10 years) tend to show a big interest in robots, making it an excellent motivational tool for low-grade education [9,15]. Moreover, it is a current belief that getting kids in touch with robotics will spark interest in natural sciences, engineering and computer science [15,19]. The success of educational robotics has attracted so much attention in some countries that robotics can officially be a part of a primary school's curriculum [14].

G. Herrmann et al. (Eds.): ICSR 2013, LNAI 8239, pp. 74–83, 2013.

Much effort in introducing robots in classrooms is focused on introducing robot technologies to the student and underestimate the role of pedagogy. However, robots can be used as an educational tool [17]. They must also be seen as potential vehicles of new ways of thinking about teaching, learning and education at large. In this perspective, teachers can stop functioning as an intellectual "authority" who transfers ready knowledge to students, but rather act as an organizer, coordinator and facilitator of learning for students by raising the questions and problems to be solved and offering the necessary tools for the students to work with creativity, imagination and independence [6].

Using robotics in classrooms is also a good way of teaching programming, since kids are able to view the result of their command sequences in the real world. Success in teaching programming to kids has been reported using Scratch (a virtual programming environment by blocks) with the WeDo kits, enabling children to program simple robotic models [18].

Several robotics systems for education, like the LEGO NXT kits, have been widely used in classrooms all over the world [10,17]. However, they can be quite expensive, they require the installation of specific software and the configuration of the programming connections. This can be difficult for teachers without proper training [16]. When it is not possible to buy ready-to-use robotics kits, some skilled teachers find a way of making their own robot kits [8,13] or buy cheap modules, easily available at online shops, and modify them to better suit their needs [7]. The growing popularity of electronic boards like Arduino [1], Raspberry Pi [3] and tools like 3D printers in the past few years has enabled more and more persons to create robots for purposes like education.

The system that we present here aims at enabling young students, starting at 8 years, to control the behaviour of a tangible model by means of a virtual environment, so that it can be used to learn programming and also as a motivational element when teaching other subjects. With such an approach, kids develop mathematical reasoning and creativity and also become interested in other subjects, as long as knowledge in those areas is embedded in the activity as a requirement for the robot to complete a task. The system that will be described is low cost, it does not require the installation of any specific software, it can be used with computers, tablets or smartphones, it can be setup in under 5 minutes, and it is extremely easy to use. Moreover, young kids need at most 5 minutes to get acquainted with the robotics system. The main difference between the system that will be presented and other commercial systems is that professors and educators can use them without spending too much time in preparing the activities and they do not need to know anything about robotics.

This paper is organised as follows. Section 2 describes the hardware and software of the complete educational robotics system. Section 3 describes two experiments using the system in 3rd and 4th grade classrooms and the educational components involved. Section 4 addresses conclusions and further work.

2   The Educational Robotics System

The system that is going to be presented in this section was designed to be extremely simple to use and fast to set up. When it was designed we had two major criteria in mind: if it cannot be set up in under five minutes, it will not be suitable for teachers, and if it cannot be explained to children in under five minutes, it will not be suitable for kids. Also, the system has been designed to have an extremely low cost. The system is composed of five robots which were named "Infantes" and a single multi-robot controller. This architecture (see Fig. 1) allows the robots to be built with simple and cheap components, while maintaining the usability of a web-based graphical programming interface through a Wi-Fi connection. This architecture also makes the system extremely easy to use at science fairs and exhibitions, allowing visitors to interact with the robots with their own smartphones or tablets.

Fig. 1. Architecture of the educational robotics system, composed of Infante robots, a multi-robot controller, a web-based programming interface that can be accessed by any device with a Wi-Fi connection, and a Javascript-enabled web browser

2.1   The Infante Robots

The robots are very small, each one fitting in a box with size $7.5 \times 7.5 \times 7.5$ cm, which makes them ideal to work on table tops (see Fig. 2). They are differential drive robots, with two motorised wheels and a ball caster. Having the minimisation of cost in mind, we used the cheapest RC servos on the market (SG90 9g servos, ≈1.80€ each) and modified them for continuous rotation, which is a simple process that consists of replacing the internal potentiometer connections with two resistors. The wheels were scavenged from old printers, but cheap wheels can also be bought without a major increase in price. Concerning the electronics, we designed a custom board with a PIC16F88 microcontroller (≈1.80€ per microcontroller), which is also one of the cheapest microcontrollers on the market. The custom board could be replaced by an Arduino for a better documentation support and easier expandability, but this would also increase the cost. For line following and crossings detection we added 4 TCRT5000 infrared sensors (0.15€ each), and for communications with the main controller we used an HC-05 Bluetooth module (4.5€ each) which was also the cheapest that we could find. For the structure of the robot, we used cheap and easy to work materials like expanded PVC and plexiglass. We estimate that taking into account the building materials and some other discrete components, the parts to make each robot cost about 15€, although some work was required to develop them and to put the first one together. However, once the first robot was designed and tested, the others were assembled rather quickly. As power source, the robots require 4 AAA batteries.

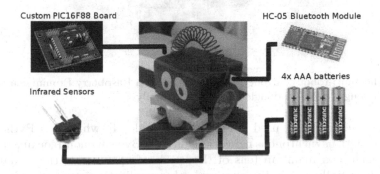

Fig. 2. Infante robot and its electronic parts

As mentioned above, the robots have four infrared sensors which allow them to follow lines and detect line crossings. This is because they were designed to be used on a grid of black lines on a white background, which are considered to be the "roads" for the robot to navigate. The PIC microcontroller has simple firmware that takes care of driving the motors and reading the sensors, so that kids do not have to worry about the low-level part of the robot. The firmware also contains a single communication protocol with only three commands: go forward, rotate 90 degrees to the left, and rotate 90 degrees to the right. Using the programming interface that will be described in Section 2.3, kids are able to build sequences of commands and send the whole sequences to the robot. Every time the robot detects a line crossing, it concludes the current action and it proceeds to execute the action that corresponds to the next command.

2.2   The Multi-robot Controller

The multi-robot controller is an electronic system that acts as a hub for robots and also as a hub for users. It consists of a Raspberry Pi mini computer (35€) running Raspbian Wheezy installed on an SD card, with Bluetooth and Wi-Fi dongles (see Fig. 3). The Raspberry Pi is configured to act as a wireless access point, creating an open Wi-Fi network that users can connect to using their laptops, tablets or smartphones. Besides that, the Raspberry Pi also runs a web server, providing the web-based robot programming interface at a specific address.

Fig. 3. The multi-robot controller, which consists of a Raspberry Pi mini computer, a Bluetooth dongle, a Wi-Fi dongle and a power supply

The web server being used is based on Tornado [4], which is a Python web framework with asynchronous networking library. By using non-blocking network I/O, Tornado can maintain tens of thousands of open connections, making it ideal for long polling, WebSockets, and other applications that require a long-living connection to each user. Since it is a Python framework, it is rather easy to establish a data flow between the web server and the Bluetooth serial connections to the robots. Whenever a user sends a sequence of actions to a specific robot, the multi-robot controller establishes the Bluetooth serial connection to the desired robot, sends the action sequence and exits the connection. This allows to use multiple robots with a single controller. When only a few robots are used, the Bluetooth serial connections to the robots may be kept alive to avoid the delays due to the establishment of the connections.

2.3   Web-Based Robot Graphical Programming Interface

The use of a web-based interface is a big advantage over other existing educational robotics systems, since it allows the users to program the robot using any device with a Wi-Fi connection and a Javascript-enabled web browser, without having to install any additional software.

The programming interface is based on Blockly [2], a google project that allows one to make short programs by dragging and dropping programming blocks to interact with virtual objects on the computer screen. It is similar to Scratch, but works in a web-browser. Since Blockly is an open-source project, we adapted it to contain only three basic operations: go forward, rotate 90° to the left, and rotate 90° to the right (see Fig. 4).

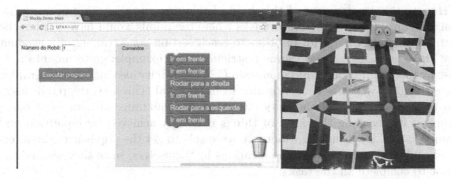

Fig. 4. Example of the programming interface based on blockly and how the instructions translate into real robot actions. The interface is in Portuguese, since it was used with Portuguese children.

This limitation of the instructions was adopted to make it suitable for the young children who would participate in the experiments that we planned to conduct. We also enabled Blockly to interact with real devices, the Infante robots, by using a Python web server based on the Tornado framework, as previously mentioned. Apart from these modifications, we also added a simple input box to the interface in order to select the number of the robot to be programmed.

3 Results

In this section we present the results of two different situations in which robots were used in classrooms. The first case concerns an activity with a class of 4th graders from an elementary school in Faro, Portugal, with the objective of developing mathematical reasoning while teaching geography. The second case features an activity with a class of 3rd graders from another elementary school, also in Faro, in which they were taught about recycling and reusing electronic waste by using the robots.

3.1 Developing Mathematical Reasoning in Geography

Mathematical reasoning may be understood as an activity where a pupil participates while interacting with others in solving a mathematical problem [20]. In this kind of activities, pupils must be guided to explain their reasoning in order to reach a certain conclusion or to justify the way of approaching a specific problem [11]. In order to develop this kind of reasoning, professors must provide activities that allow the pupils to (a) develop methods of mathematical thinking, (b) be encouraged to explore, try and make mistakes, (c) formulate hypotheses, test them and provide arguments about their validity, and (d), question their own and others' reasonings. This kind of activities leads to the development of deductive, inductive and abductive reasoning [12].

Having the development of this kind of thinking and reasoning in mind, we designed an activity with several tasks where the pupils would have to plan the paths of a robot, on a grid, in order to reach certain goals, taking into account that in some tasks there are some restrictions. For example: go to mountain A, pass by mountain B but avoid mountain C. This activity also includes geography, since the goals were always mountains from Portugal. This way, the pupils learn the relative positions of the major Portuguese mountains without even being aware of it. Teaching this kind of things is usually achieved by repetition and memorisation. By using the robots we were able to get the pupils actually interested and motivated to learn the locations by themselves, since they wanted the robot to complete all the tasks.

Fig. 5. 4th graders using the robots. Left: one pupil explains the programming sequences while another uses a computer to program the robot. Right: pupils use a tablet computer to program the robots.

To perform the activity, the class was split into groups of 5 to 6 pupils, with one robot for each group. After a short setup and explanation, the groups were left to work on their own for an hour and a half, during which we observed their reactions, discussions and explanations. As we already expected, getting started with the robots was no problem for them, irrespective of using a laptop or a tablet. We noticed that, at start, most of them could not distinguish left from right. They also forgot to make the robot go forward after a rotation, making it spin around, therefore questioning the programming they had just made in order to find out what was wrong. Another problem that we noticed at start was that the pupils had some difficulty in assuming the point of view of the robot in order to to decide the correct turns. However, after some brief discussions, as one or two of them had already learned how to do it, they taught each other, justifying their way of solving that problem (see Fig. 5). We also noticed that, whenever something did not work as expected, they discussed with each other to try to find the errors, explaining their thoughts. If there were several solutions to complete a task, they would discuss which path to follow, either preferring the shortest path for easier programming, or by choosing a more challenging path, using their creativity. The pupils were so motivated that they finished all tasks sooner than we expected and autonomously. They then used their creativity to invent new tasks themselves and kept playing with the robots. By the end of the activity, all the pupils knew the relative positions of the mountains, they knew how to program the robots, and they could justify their options. Needless to say, but all pupils loved it and did not want the class to end.

3.2 Teaching Pupils about Recycling and Reusing Electronic Waste

In the activity that we described in the previous section, we were amazed by the ease with which 4th graders used the educational robot system. So, we went further to test the system with a class of 3rd graders from another school (see Fig. 6). The activity was similar to the previous one, but was designed to make them aware of recycling and reusing electronic waste. Again, a list of tasks was given to groups of 5 to 6 pupils and some electronic parts like lamps and batteries were placed in the robot's navigation grid. The corresponding recycling bins were also placed in the grid. The tasks were to make the robot reach a specific item, that was then placed on top of the robot, and then carried to the proper bin. We left them again to work on their own and observed their behaviour. As in the previous activity, the pupils learned how to use the system very quickly regardless of the programming device used. We verified that the same initial problems occurred: not making the robot go forward, making it spin around at the same place, confusing left with right, and not being able to assume the point of view of the robot. However, like in the other activity, after a short time they were working as a group, trying to pinpoint their mistakes, and explaining each other why they thought the robot was not doing what they wanted and how they could fix it. Once more, the pupils were extremely motivated. They finished the tasks rapidly and then invented their own tasks and kept on programming the robots until the class ended.

Fig. 6. Third graders using the robots. Left: pupils use a laptop to control the robot in carrying the lamp to the proper recycling bin. Right, a pupil uses a tablet to program the robot while others analyse the solution and say which commands to put in the programming sequence.

4 Conclusions and Further Work

Confirmed by the activities that we realised with 3rd and 4th graders, we conclude that our goal of designing an educational robotics system with an extremely fast setup and which is extremely easy to use has been achieved. The setup time was very short and the pupils learned how to use the robots very quickly. The low-cost robots executed the commands that were sent to them as expected, and the communication between them and the multi-robot controller also worked without any problems. The system proved to be reliable and very easy to use in noisy classroom environments.

As we already mentioned, our objective was also to create a platform that would allow the robots to be used as a tool in teaching any other subject, and not to focus on the process of robot building. Throughout our experiments, we verified that the deployment of robots can be so motivating that pupils actually want to learn things they consider difficult and boring to memorise. Also, pupils perceived the robots as toys and harmoniously played together, confidently discussing how to tackle the challenges and problems they were facing.

We also verified that, regardless of the subject being taught in the activity in which the robots were used, there was a constant development of mathematical reasoning, deductive, abductive as well as inductive. This is extremely important since it enables pupils to formulate hypotheses, to test them, to find out what's wrong, and then formulate new hypotheses that may solve a problem. It also encourages them to work in a group, discussing with each other and communicating their hints to the other group members, who can learn through positive and negative feedback. The development of spatial relations between objects also helps to build mental representations of the world and to use these in reasoning. By deploying the robots as a motivational learning tool in teaching other subjects, the pupils become interested in learning them, since they feel a need to learn to accomplish the tasks that are given to them. By designing activities with robots where there are multiple correct solutions pupils are stimulated to deploy their creativity in pursuit of a solution for a given problem.

As further work we intend to make some improvements of the robots, like changing the power source to Li-ion batteries and adding more features to the communication protocol embedded in the robot's firmware. Adding some distance and other sensors, or a small gripper might also be an option to allow for more creative uses and to increase the user's motivation. We also intend to extend the graphical programming interface by re-adding all the programming blocks that Blockly originally had, such that the system can be more challenging and be used by persons of any age.

Acknowledgements. This work was partially supported by the Portuguese Foundation for Science and Technology (FCT) project PEst-OE/EEI/LA0009/ 2011 and by FCT PhD grant to author SFRH/BD/71831/2010.

References

1. Arduino homepage, http://www.arduino.cc/
2. Blockly: a visual programming editor, http://code.google.com/p/blockly/
3. Raspberry Pi homepage, http://www.raspberrypi.org/
4. Tornado web server, http://www.tornadoweb.org/

5. Alemany, J., Cervera, E.: Appealing robots as a means to increase enrollment rates: a case study. In: Proc. 3rd Int. Conf. on Robotics in Education (RiE 2012), Prague, Czech Republic, pp. 15–19 (2012)
6. Alimisis, D.: Robotics in education & education in robotics: Shifting focus from technology to pedagogy. In: Proc. 3rd Int. Conf. on Robotics in Education (RiE 2012), Prague, Czech Republic, pp. 7–14 (2012)
7. Balogh, R.: Acrob - an educational robotic platform. AT&P Journal Plus Magazine 10(2), 6–9 (2010)
8. Balogh, R.: Educational robotic platform based on Arduino. In: Proc. 1st Int. Conf. on Robotics in Education (RiE 2010), Bratislava, Slovakia, pp. 119–122 (2010)
9. Botelho, S.S.C., Braz, L.G., Rodrigues, R.N.: Exploring creativity and sociability with an accessible educational robotic kit. In: Proc. 3rd Int. Conf. on Robotics in Education (RiE 2012), Prague, Czech Republic, pp. 55–60 (2012)
10. Butterworth, D.T.: Teaching C/C++ programming with lego mindstorms. In: Proc. 3rd Int. Conf. on Robotics in Education (RiE 2012), Prague, Czech Republic, pp. 61–65 (2012)
11. Carmo, B.: Promovendo o raciocínio matemático através da robótica educativa: um exemplo no 4.º ano de escolaridade. In: Proc. II Encontro Nacional TIC e Educação para Alunos do Ensino Básico e Secundário, Lisbon, Portugal, pp. 7–12 (2013)
12. Carmo, B., Guerreiro, A.: Robótica educativa e raciocínio matemático no 1.ºciclo do ensino básico. In: Proc. Encontro Ensinar e Aprender Matemática com Criatividade dos 3 aos 12 anos, Viana do Castelo, Portugal, pp. 89–101 (2013)
13. García-Saura, C., González-Gómez, J.: Lowcost educational platform for robotics, using open-source 3D printers and open-source hardware. In: Proc. 5th Int. Conf. of Education, Research and Innovation (iCERi 2012), pp. 2699–2706 (2012)
14. Ilieva, V.: Robotics in the primary school - How to do it? In: Proc. Int. Conf. on Simulation, Modelling and Programming for Autonomous Robots, Darmstadt, Germany, pp. 596–605 (2010)
15. Jeschke, S., Kato, A., Knipping, L.: The engineers of tomorrow - teaching robotics to primary school children. In: Proc. 36th SEFI Annual Conf., Aalborg, Denmark, pp. 1058–1061 (2008)
16. Matari, M.J., Koenig, N., Feil-Seifer, D.: Materials for enabling hands-on robotics and stem education. In: Proc. AAAI Spring Symposium on Robots and Robot Venues: Resources for AI Education, Palo Alto, California, pp. 99–102 (2007)
17. Miglino, O., Lund, H.H., Cardaci, M.: Robotics as an educational tool. J. Interactive Learning Research 10(1), 25–47 (1999)
18. Olabe, J.C., Olabe, M.A., Basogain, X., Castaño, C.: Programming and robotics with Scratch in primary education. In: Mendez-Vilas, A. (ed.) Education in a Technological World: Communicating current and Emerging Research and Technological Efforts, pp. 356–363. Formatex (2011)
19. Wyffels, F., Hermans, M., Schrauwen, B.: Building robots as a tool to motivate students into an engineering education. In: Proc. 1st Int. Conf. on Robotics in Education (RiE 2010), Bratislava, Slovakia, pp. 113–116 (2010)
20. Yackel, E., Hanna, G.: Reasoning and proof. In: Kilpatrick, J. (ed.) A Research Companion to Principles and Standards for School Mathematics, pp. 227–236. National Council of Teachers of Mathematics (2003)

Social Navigation - Identifying Robot Navigation Patterns in a Path Crossing Scenario*

Christina Lichtenthäler[1], Annika Peters[2], Sascha Griffiths[3], and Alexandra Kirsch[4]

[1] Institute for Advanced Study, Technische Universität München, Garching, Germany
[2] Applied Informatics Group, Bielefeld University, Bielefeld, Germany
[3] Robotics and Embedded Systems, Technische Universität München, Garching, Germany
[4] Department of Computer Science, Tübingen University, Tübingen, Germany

Abstract. The work at hand addresses the question: What kind of navigation behavior do humans expect from a robot in a path crossing scenario? To this end, we developed the "Inverse Oz of Wizard" study design where participants steered a robot in a scenario in which an instructed person is crossing the robot's path. We investigated two aspects of robot behavior: (1) what are the expected actions? and (2) can we determine the expected action by considering the spatial relationship?

The overall navigation strategy, that was performed the most, was driving straight towards the goal and either stop when the person and the robot came close or drive on towards the goal and pass the path of the person. Furthermore, we found that the spatial relationship is significantly correlated with the performed action and we can precisely predict the expected action by using a Support Vector Machine.

Keywords: social navigation, human-robot interaction, spatial relationship.

1 Introduction

Robots will increasingly become part of the habitats and work spaces of humans. Wherever they are located, in the factories as co-workers, in nursing homes or hospitals as care assistants, as guides in a supermarket, or as household-robots, one crucial behavior, which they have all in common, is navigation. A robot has to navigate through spaces where humans live and as Althaus et al. [1] already stated *"The quality of the movements influences strongly the perceived intelligence of the robotic system."*. The way a robot moves affects not only the perceived intelligence, also the perceived safety, comfort and legibility and other factors regarding social acceptance [7,17]. Therefore, one goal in human-robot interaction is to develop methods in order to make robot behavior and in particular navigation socially acceptable [16,8,22].

In the literature several social navigation methods have been proposed. One common approach is to model social conventions (e. g. proxemics [10], keeping to the right side in a corridor,...) and norms by using cost functions or potential fields [14,25,21]. For example Kirby et al. proposed a navigation method which models social conventions and the constraint to pass a person on the right as well as task conventions like time and

---

* This work has partially been founded by the the the Clusters of Excellence "Cognitive Interaction Technology" CITEC (EXC 277) and "Cognition for Technical Systems" CoTeSys (EXC 142). We would like to express special thanks to the Social Motor Learning Lab of the Neurocognition and Action Group (Prof.Dr. Thomas Schack) at Bielefeld University and to William Marshall Land and Bernhard Brüning for assisting the authors with the VICON system.

G. Herrmann et al. (Eds.): ICSR 2013, LNAI 8239, pp. 84–93, 2013.

path length constraints. They use a classical A* path planner equipped with cost functions modeling social and task conventions. Also Tranberg et al. [25] use a proxemics model. Contrary to Kirby et al. they use a potential field instead of cost functions. Another cost function navigation method is proposed by Sisbot et al. [21]. They model not only spacial rules but also other social norms based on their findings in former studies like safety rules, preferred approach directions or visibility constraints.

Another approach is to follow the assumption that robot behavior is socially acceptable if the robot shows similar behavior to humans [26,15]. One of the first methods is proposed by Yoda et al. [26], based on the findings of a human-human experiment they developed an algorithm imitating human passing behavior. Also Kruse et al. [15] developed a Human-Aware Navigation by modeling the findings of a human-human path crossing experiment [3] into a cost function model. However, physical capabilities of robots differ very much from humans' (in particular those of wheeled robots). Therefore, imitating human behavior becomes difficult. Furthermore we showed in former experiments [17] that the legibility and perceived safety of the Human-Aware Navigation method [15] is rather low. In addition, it is not clearly known if humans expect robots to move in a different manner to humans. Is there something like a robot-like behavior that humans expect from robots? A large body of research is dedicated to investigate several aspects of robot navigation [4,20,12,7]. For example Butler et al [4] analyzed the influence of different factors like speed, distance and design on robot motion patterns (frontal approach, passing by, non-interactive navigation). They evaluated how the navigation is perceived by humans regarding the level of comfort. Pacchierotti et al. [20] also tested the conditions speed, and distances in a passing by situation. Proxemics [10] are also widely studied in human-robot interaction [24,19]. In a controlled experiment Dautenhahn et al. [7] identified preferable robot approaching motions by testing different strategies. The aforementioned research presents controlled experiments testing how motion is perceived regarding different conditions like speed, distance and orientation. However, what is only rarely investigated is what humans expect from a robot, especially when the robot has different capabilities than humans. One study towards expected robot motion patterns is presented by Huettenrauch et al. [12]. In order to identify spatial behavior patterns they investigated the spatial behavior (distances according to Hall [10] and orientation according to Kendon [13]) of a human towards a robot during a "Home Tour" study.

In order to find out what kind of behavior humans expect from a robot, and in particular a robot with non-humanlike physical capabilities we let naive participants steer the robot in a path crossing task. The study design of the work at hand is a new kind of the classical "Wizard of Oz" [9], which we call "Inverse Oz of Wizard" following the categorization proposed by Steinfeld et al. [23]. According to the "Oz of Wizard" design described by Steinfeld et al. [23], where the human is simulated, the robot behavior is real, and the robot behavior is evaluated using robot centered metrics we designed our "Inverse Oz of Wizard". We "simulate" the human by instructing a confederate with very strict behavioral rules. The robot behavior is controlled by the participant in order to capture the participants' expectations and the resulting robot behavior is evaluated, by analyzing the captured motions.

In order to perform a structured analysis of the observed robot behavior and particularly with regard to the development of a navigation method we have to formalize the term "robot behavior". Usually behavior is defined as the range of **actions** done by organisms, or artificial entities, which is the **response** of the organism, or artificial entity to various **stimuli**. Or as Arkin shortly states in his book, *"behavior is a reaction to a stimulus"* [2]. Based on this common definition and according to Arkin's behavior-based robotics theory we formalize robot behavior as an **action** a performed by a robot, which is **caused** by a **stimulus** s, i.e. it exist a function f so that $f(s) \mapsto a$. In our navigation scenario possible actions can be driving, stopping or driving a curve and the spatial relationship is a stimulus.

Research Questions. We want to identify expected robot navigation behavior in a human-robot path crossing scenario. Therefore, we want to answer three questions with the study at hand. First we want to identify the preferred actions a, second we want to identify the stimulus s. We expect that the stimulus for an action in a human-robot path crossing situation lies in the spatial relationship of human and robot, third we hypothesize that it is possible to predict the expected action, based on the spatial relationship. i.e. a function f exist so that $f(s) \mapsto a$.

2   Method

In order to answer the aforementioned questions we implemented a within-subject study using our "Inverse Oz of Wizard" study design.

2.1   Participants

We recruited 46 participants with an average age of 28 years - thereof 26 women and 20 men. 89% of the participants had rarely or no contact to robots and 11% had regular contact to robots.

2.2   Technical Setup

Robot. The platform used in this study was the BIRON (BIelefeld Robotic Companion) robot (see Fig. 1(a)). BIRON has an overall size of approximately 0.5m (w) x 0.6m (d) x 1.3m (h). Besides two wheels, BIRON has two rear casters for balance and is constructed with a differential drive (2 degrees of freedom: translation and rotation).

Robot Remote-Control. We used a wireless keyboard to steer the robot. The commands of how to steer the robot were marked on the keyboard with arrows. Five keys corresponded to the five ways of moving the robot: straight forward, rotate around its own axis in a clockwise direction, in an anti-clockwise direction, drive and turn left or right in an arc. The robot only moved by holding down the particular key and the robot stopped by releasing the key. These motions map the actual movement abilities of the robot BIRON. Thus, the actions we investigate in this scenario are drive, stop, curve, and rotate. There was no possibility to accelerate the robot as it was driving at its full speed of 0.7m/s.

Motion Capturing System. To capture the movements of the robot and the interacting person we used a VICON motion capturing system (www.vicon.com). We recorded additional video data with an HD camera.

(a)   (b) Laboratory   (c) Study Setup
BIRON

Fig. 1. Used robot and design of the study

2.3 Study Design

Cover Story. In order to make the scenario realistic all participants were told the same cover story about a grocery store that uses a robot (BIRON) to refill the shelves with goods from the storing place. The participants were asked to navigate the robot from the storing place to a shelf (see Fig. 1(c)) by steering the robot with the wireless keyboard. Furthermore, they were told that the robot might encounter customers in the store.

Setup. According to the grocery store cover story we built up a store scenario with four shelves and a storing place (see Fig. 1(b),1(c)) in a laboratory which measures approximately $133m^2$. Three shelves were placed at the wall on the right side of the storing place with a distance of 2.7 m between them (see Fig. 1(c)). One shelf was placed 7.3 m opposite to the storing place (see Fig. 1(c)). The robot, steered by the participant, had the task to bring items from the storing place to the opposite shelve (see Fig. 1(c)). One experimenter took the role of a *customer*. The *customer* had the task to walk from a fixed point (see Fig. 1(c)) randomly to one of the three shelves at the wall and put an item into his/her basket. In addition to the three randomized aims the *customer* walked randomly in three different walking velocities slow (0.6 to 0.8m/s), normal (1.2 to 1.5m/s), and fast (1.9 to 2.1m/s). The *customer* had to walk straight and maintain the velocity even if the robot would crash into them. To avoid eye contact with the participant the *customers* wore sunglasses. Due to the arrangement of the shelves the robot and the customer coincidentally met each other in 45° and 90° angles (see Fig. 1(c)). Thus, the setup was designed to create completely random and unforeseeable crossing events.

Procedure. First the participants were welcomed and the cover story (see Section 2.3) was explained. In order to familiarize the participants with the setup and with steering the robot BIRON the participants received an introduction to the robot BIRON and an extensive practice of how to steer the robot. Only after the participants managed to drive around obstacles and felt capable of steering the robot, the study began. The participants were told to carry 15 items (only one item at a time) from the storing place to the opposite shelf (see Fig. 1(c)) and then go back to the storing place. Therefore the robot moves 30 times (two times per item) straight through the room. The *customer* crosses the robots path randomly as described in Section 2.3. The movements of the robot and the *customer* were captured by a motion capture system and a video camera (see Section

2.2). Once the participant completed the task he/she were debriefed about the purposes of the study and discussed the study with the experimenters. Demographical data were recorded within the debriefing.

3   Results

We carried out the data analysis in two steps. First we identified robot actions in the video data by observing its motions (drive, stop, curve, rotate). After that we analyzed the spatial relationships of robot and *customer* using the motion capturing data in order to identify the stimulus for a specific action. We only consider crossing situations for our analysis. A crossing situation is defined as a situation where 1) the paths of both, BIRON and the *customer*, will cross and 2) both are located before reaching the crossing point (see Figure 2(b)).

3.1   Video Data

By analyzing the video data we identified four different navigation patterns performed by the robot in crossing situations (see also [18]):

1. stopping before the crossing point (76.7%)
2. driving straightforward and passing behind or in front of the *customer* (18%)
3. driving a curve (3.7%)
4. collision with the *customer* (1.6%) (only shown by two participants)

The action rotate was never performed in a crossing situation and driving a curve was mostly performed by a participant in his/her first trials. The overall navigation strategy, which we can conclude from our observation, was driving straight towards the goal (shelve or storage place) and either stop when both, the *customer* and the robot, came close to the crossing point, or otherwise drive on towards the goal and pass the path of the *customer* without colliding. This strategy was performed by almost all (44 of 46) participants. We assume that the participants anticipate if a collision will happen or not and either stop or drive on. To conclude, from our video analysis we can derive that the actions drive and stop are the preferable actions and we can assume that the stimulus lies in the spatial relationship.

3.2   Motion Data

Spatial Feature Calculation. The raw data from the motion capturing system contains the position of robot r and *customer* c (see Fig. 2(a)) captured with a frame rate of 150Hz. In order to describe the spatial relationship between robot and *customer* we calculated the following spatial features using Matlab (see also Figure 2(b)):

- QTC_c according to Hanheide et al. [11] to determine a crossing situation
- distance between *customer* and robot d
- distance between *customer* and the crossing point d_c
- distance between robot and the crossing point d_r
- angle robot-*customer* α
- velocity *customer* v

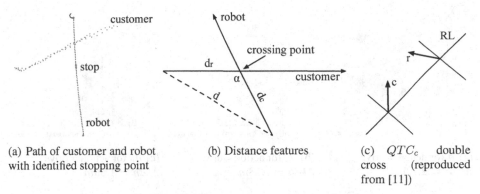

(a) Path of customer and robot with identified stopping point

(b) Distance features

(c) QTC_c double cross (reproduced from [11])

Fig. 2. Spatial features of a crossing situation

Calculating QTC_c Features: QTC_c is a compact representation of spatial relations between two moving objects. It represents the relative motion, with respect to the reference line RL that connects them, as shown in Fig. 2(c) We calculated the QTC_c according to Hanheide et al. [11] as follows:

(1) movement of the robot r with respect to the *customer c*:
   - 1 : r is moving towards c
   +1 : r is moving away from c
(2) movement of c with respect to r
   same as (1), but with r and c swapped
(3) movement of r with respect to RL:
   -1: r is moving to the left-hand side of RL
   0 : r is moving along RL or not moving at all
   +1 : r is moving to the right-hand side of RL
(4) movement of c with respect to RL
   same as (3), but with r substituted by c

Therefore, a crossing situation is given when: (1)= -1, (2)= -1, (3) = -1·(4) or (3)=(4)=0.

Calculating Spatial Features: For the purpose of reducing the amount of data we calculated one feature vector for every 15 frames. Thus, we get 15 feature vectors for one second of recorded data. Additionally, to the aforementioned spatial features we determined the action a (drive, stop, curve, rotate) the robot is performing (see Fig. 2(a)). Thus, we transformed position data points into action related spatial feature vectors containing the action a, distance between customer and robot d, distance between *customer* and the crossing point d_c, distance between robot and the crossing point d_r, angle robot-*customer* α, velocity *customer* v, and the QTC_c values.

$$(a, d, d_c, d_r, \alpha, v, QTC_c)$$

In order to concentrate only on the main strategy we excluded all feature vectors with curve and rotate actions. We identified the crossing situations by using the QTC_c [11] information and excluded all non-crossing situations. We also excluded all feature vectors where the robot is outside of the *customer's* social space ($d > 3.6m$) [10]. Note that we have far more than one feature vector per crossing situation, because different to the video analysis, where we were only counting the reaction for a crossing situation,

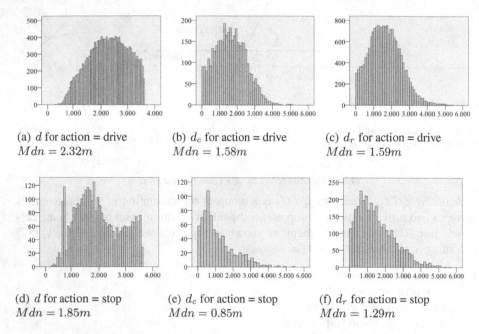

(a) d for action = drive
$Mdn = 2.32m$

(b) d_c for action = drive
$Mdn = 1.58m$

(c) d_r for action = drive
$Mdn = 1.59m$

(d) d for action = stop
$Mdn = 1.85m$

(e) d_c for action = stop
$Mdn = 0.85m$

(f) d_r for action = stop
$Mdn = 1.29m$

Fig. 3. Histograms of the distance values for action = {stop, drive} measured in mm

we now consider every data point of a crossing situation. Furthermore, due to the fact that the robot drives before it stops we have more feature vectors for the action drive than for the action stop.

Statistical Data Analysis. The aim of the statistical data analysis, which was performed with SPSS, was to find support for our hypothesis: "can we predict the action based on the spatial features".

First of all we show the distributions for the distance values (d, d_c, d_r) in Fig. 3.2. The histograms in Figure 3.2 show, that there is a trend for greater distances for the drive action and that most of the participants stop the robot within a distance of approximately 0.7 m to the crossing point whereby the *customer* has a distance of approximately 0.67 m to the crossing point.

By performing inferential statistic tests, we found support for our hypothesis that the spatial features are correlated with the action. Due to the dichotomous action variable we calculated the point-biserial correlation coefficients. The action of the robot was significantly related to the distance between *customer* and robot d, $r_{pb} = .149, p < .01$, to the distance between *customer* and the crossing point d_c, $r_{pb} = .200, p < .01$, to the distance between robot and the crossing point d_r, $r_{pb} = .063, p < .01$, to the angle robot- *customer* α, $r_{pb} = .027, p < .01$, and to the *customer's* velocity v, $r_{pb} = .054$, $p < .01$.

As a next step to support our hypothesis that we can predict the action based on the spatial features, we performed a logistic regression on the normalized spatial feature values. Results are shown in Table 1. Only the angle has no significant influence on the classification model ($B = .06, p = .396$). All other variables have a significant influence and the model is able to predict 86.7 % of the actions correctly. Additionally

Table 1. Results of the logistic regression, performed with SPSS

Value	B (SE)	Lower	Odds Ratio^a	Upper
distance d	-1.023** (.086)	.304	.359	.425
velocity v	-.515** (.037)	.556	.598	.643
angle α	.060 (.71)	.924	1.062	1.221
distance robot d_r	.648** (.081)	1.630	1.912	2.243
distance customer d_c	-.255** (.071)	.674	.775	.892
constant	-2.131 (.059)		.119	

** $p < 0.001$, ^{a}95% CI for Odds Ratio

Model Statistics
$R^2 = .15$ (Cox & Snell), .25 (Nagelkerke)
$\chi^2(8) = 106.097$, $p < 0.001$ (Hosmer & Lemeshow)
accuracy 86.7% (goodness-of-fit)

to the logistic regression we trained a Support Vector Machine SVM with a RBF Kernel [6] in order to show that we can precisely predict the expected action based on the spatial features. The results of a ten-fold-cross-validation, performed with LibSVM [5] are shown in Table 2. The very good prediction results also support our hypothesis.

Table 2. Results of the ten-fold-cross-validation of the SVM model performed with LibSVM [5]

accuracy 99.9527%	f-score 0.999764
precision 100%	recall 99.9527%

4 Discussion

The study at hand was conducted to identify expected robot navigation behavior. In the video analysis we found a prominent robot behavior. Driving straight towards the goal and, when a crossing situation occurs, either stop and wait until the customer passes the robot's path or drive on and pass the path of the person before or behind the person. Thus, the expected actions a are *drive* towards the goal and *stop*. We hypothesized that the stimulus lies in the spatial relationship. Therefore, we used the motion capturing data to identify the stimulus s and calculated several spatial features. We used the QTC_c [11] representation to identify a crossing situation and extended these purely relative representations with distance measures, the crossing angle, and velocity information. We found support for our hypothesis that the spatial relationship is significantly correlated with the action. Furthermore we could show that it is possible to predict the expected action based on our spatial features whereby we found that the distance measures are the most influential values. Additionally, we could show that it is possible to precisely predict the action, by using a Support Vector Machine.

The overall navigation strategy, that we found in our data, is similar to the behavior Basili et al. [3] found in their human-human path crossing experiment. In both studies the participants were going (driving the robot) straight towards the goal and show a collision avoiding strategy by manipulating the speed. In our case by stopping and in their case by decreasing the speed.

However, the findings of the study at hand are significant, but the correlation values are rather low. The reason can be failures in the used equipment, which causes much noise in our data. Sometimes a stop was caused by loosing the connection to the robot or we lost a marker due to occultation. Furthermore, in the study at hand the participants only had a third person view from a fixed point in the room. This fact makes it difficult for the participants to exactly estimate distances, which can also be the reason for the rather low correlation values. Another limitation of the study is the small set of crossing angles and the missing of a frontal approach. Furthermore, the actions are limited. Due to safety reasons we does not provide the option to manipulate the robots' speed.

The aforementioned limitations can be the basis for further investigations of robot navigation behavior. For example, by implementing more controlled experiments, based on our findings, one can avoid the noise and find more precise thresholds for the distance values. Also a first person view by using a camera on the robot can yield more precise results. Furthermore, it could be useful to evaluate the identified robot navigation patterns and test how they are perceived by a human in order to verify our hypothesis that we can find out human expectations about robot behavior by using our "Inverse Oz of Wizard" study design. This can be done by doing it the other way around in a experiment where the behavior of the robot is scripted and the participants are asked to rate the behavior.

5   Conclusion

To sum up, we conducted a study to identify robot behavior patterns in a human robot path crossing scenario. The overall navigation strategy we can conclude from the data is to drive straight towards the goal and only react (stop) to a crossing human when the stimulus based on the spatial relationship predicts to stop, otherwise drive on towards the goal. The expected action can be predicted by using a standard machine learning method like an SVM trained on our dataset. Based on these findings and by using our SVM model, we can develop a social navigation method, that meets human expectations about robot navigation behavior.

References

1. Althaus, P., Ishiguro, H., Kanda, T., Miyashita, T., Christensen, H.I.: Navigation for Human-Robot Interaction Tasks. In: IEEE International Conference on Robotics and Automation (2004)
2. Arkin, R.C.: Behavior-Based Robotics. MIT Press (1998)
3. Basili, P., Sag, M., Kruse, T., Huber, M., Kirsch, A., Glasauer, S.: Strategies of Locomotor Collision Avoidance. Gait & Posture (2012)
4. Butler, J.T., Agah, A.: Psychological Effects of Behavior Patterns of a Mobile Personal Robot. In: Autonomous Robots (2001)
5. Chang, C.-C., Lin, C.-J.: LIBSVM: A Library for Support Vector Machines. ACM Transactions on Intelligent Systems and Technology 2 (2011), Software available at http://www.csie.ntu.edu.tw/~cjlin/libsvm
6. Cristianini, N., Shawe-Taylor, J.: An Introduction to Support Vector Machines and Other Kernel-Based Learning Methods. Cambridge University Press (2000)

7. Dautenhahn, K., Walters, M., Woods, S., Koay, K.L., Nehaniv, C.L., Sisbot, A., Alami, R., Siméon, T.: How I Serve You? A Robot Companion Approaching a Seated Person in a Helping Context. In: ACM SIGCHI/SIGART Conference on Human-Robot Interaction (2006)
8. Dautenhahn, K.: Methodology and Themes of Human-Robot Interaction: A Growing Research Field. International Journal of Advanced Robotic Systems (2007)
9. Green, A., Huttenrauch, H., Severinson Eklundh, K.: Applying the Wizard-of-Oz Framework to Cooperative Service Discovery and Configuration. In: IEEE International Workshop on Robot and Human Interactive Communication (2004)
10. Hall, E.T.: The Hidden Dimension. Anchor Books, New York (1969)
11. Hanheide, M., Peters, A., Bellotto, N.: Analysis of Human-Robot Spatial Behaviour Applying a Qualitative Trajectory Calculus. In: IEEE International Workshop on Robot and Human Interactive Communication (2012)
12. Hüttenrauch, H., Severinson Eklundh, K., Green, A., Topp, E.A.: Investigating Spatial Relationships in Human-Robot Interaction. In: IEEE/RSJ International Conference on Intelligent Robots and Systems (2006)
13. Kendon, A.: Conducting Interaction: Patterns of Behavior in Focused Encounters, vol. 7. CUP Archive (1990)
14. Kirby, R., Simmons, R., Forlizzi, J.: COMPANION: A Constraint-Optimizing Method for Person-Acceptable Navigation. In: IEEE International Symposium on Robot and Human Interactive Communication (2009)
15. Kruse, T., Kirsch, A., Akin Sisbot, E., Alami, R.: Dynamic Generation and Execution of Human Aware Navigation Plans. In: International Conference on Autonomous Agents and Multiagent Systems (2010)
16. Kruse, T., Pandey, A.K., Alami, R., Kirsch, A.: Human-Aware Robot Navigation: A Survey. Robotics and Autonomous Systems (2013)
17. Lichtenthäler, C., Lorenz, T., Kirsch, A.: Influence of Legibility on Perceived Safety in a Virtual Human-Robot Path Crossing Task. In: IEEE International Symposium on Robot and Human Interactive Communication (2012)
18. Lichtenthäler, C., Peters, A., Griffiths, S., Kirsch, A.: Be a Robot! Robot Navigation Patterns in a Path Crossing Scenario. In: Proceedings of the 8th ACM/IEEE International Conference on Human Robot Interaction (2013)
19. Mumm, J., Mutlu, B.: Human-Robot Proxemics: Physical and Psychological Distancing in Human-Robot Interaction. In: ACM/IEEE International Conference on Human-Robot Interaction (2011)
20. Pacchierotti, E., Christensen, H.I., Jensfelt, P.: Human-Robot Embodied Interaction in Hallway Settings: a Pilot User Study. In: IEEE International Workshop on Robot and Human Interactive Communication. IEEE (2005)
21. Sisbot, E.A., Marin-Urias, L.F., Alami, R., Simeon, T.: A Human Aware Mobile Robot Motion Planner. IEEE Transactions on Robotics 23 (2007)
22. Steinfeld, A., Fong, T., Kaber, D., Lewis, M., Scholtz, J., Schultz, A., Goodrich, M.: Common Metrics for Human-Robot Interaction. In: ACM SIGCHI/SIGART Conference on Human-Robot Interaction (2006)
23. Steinfeld, A., Jenkins, O.C., Scassellati, B.: The Oz of Wizard: Simulating the Human for Interaction Research. In: ACM/IEEE International Conference on Human-Robot Interaction (2009)
24. Takayama, L., Pantofaru, C.: Influences on Proxemic Behaviors in Human-Robot Interaction. In: IEEE/RSJ International Conference on Intelligent Robots and Systems (2009)
25. Tranberg Hansen, S., Svenstrup, M., Andersen, H.J., Bak, T.: Adaptive Human Aware Navigation Based on Motion Pattern Analysis. In: IEEE International Symposium on Robot and Human Interactive Communication (2009)
26. Yoda, M., Shiota, Y.: The Mobile Robot which Passes a Man. In: IEEE International Workshop on Robot and Human Communication (1997)

Affordance-Based Activity Placement
in Human-Robot Shared Environments*

Felix Lindner and Carola Eschenbach

Knowledge and Language Processing
Department of Informatics
University of Hamburg
Vogt-Kölln-Straße 30, 22527 Hamburg
{lindner,eschenbach}@informatik.uni-hamburg.de

Abstract. When planning to carry out an activity, a mobile robot has to choose its placement during the activity. Within an environment shared by humans and robots, a social robot should take restrictions deriving from spatial needs of other agents into account. We propose a solution to the problem of obtaining a target placement to perform an activity taking the action possibilities of oneself and others into account. The approach is based on affordance spaces agents can use to perform activities and on socio-spatial reasons that count for or against using such a space.

1 Introduction

As robots increasingly share space with humans, robot behavior needs to take social aspects of space into account. One such social aspect is the awareness of the action possibilities of other agents. In most cases, a socially aware robot should avoid to place itself (or objects it carries around) such that action possibilities of other agents get blocked. Imagine the battery of your robot companion is getting low and thus it seeks a place to recharge. As the recharging process takes some time, the robot chooses a placement that does not interfere with your own action possibilities, e.g., it will not place itself right in a doorway.

In [8] we discuss five types of social spaces and introduce the concept of (social) affordance space motivated by the observation that humans expect others to not block their possibilities for action. We take reasoning about social spaces to be part of robot social intelligence [17], as a robot acting according to social spaces displays the ability to take others' perspectives, and it displays regardfulness by incorporating the needs of others into its own decision making. Empirical research on human reasoning about activity placement under the consideration

* This research is partially supported by the DFG (German Science Foundation) in IRTG 1247 'Cross-modal Interaction in Natural and Artificial Cognitive Systems' (CINACS). Thanks to Jianwei Zhang for providing access to the PR2 and the robot lab. Thanks to Johanna Seibt and Klaus Robering for discussions about affordance spaces in the context of 'Making Space' and 'Friends by Design' and to three anonymous reviewers for their valuable comments.

G. Herrmann et al. (Eds.): ICSR 2013, LNAI 8239, pp. 94–103, 2013.
© Springer International Publishing Switzerland 2013

of the spatial needs of others' action possibilities is rare. Thus, the approach proposed in this article is not based on an existing theory of human social activity placement. Rather, the proposal serves as a technical framework future research results could be integrated to.

More formally, we specify the social activity-placement problem as follows: *Given an activity type ϕ and a potential agent α, a placement π should be determined, such that α can successfully ϕ at π, and π is among the most socially adequate placements for α to ϕ.* The activity-placement problem is a generalization of the robot placement problem (e.g., [14,19]). The potential agent can be the robot itself, but it can also be some other agent. To determine a most socially adequate placement for an activity, a social robot needs to reason about where its own and others' activities can be placed (functional level) and how activities of different agents can spatially interfere (social level).

This article focusses on modelling social affordance spaces. Sect. 2 provides a short review of related work. Sect. 3 introduces a theory of affordances and affordance spaces. This theory serves as a basis for the overall decision process that leads to the most socially adequate placement. To this end, a sampling procedure to obtain an affordance-space map is proposed, which represents the functional level of the placement problem. In Sect. 4, we propose a scheme to evaluate candidate placements with respect to social adequacy. Sect. 5 reports an experiment conducted with a PR2 robot on which the approach was implemented.

2 Related Work

The robot-placement problem has gained much less attention in robotics than other navigation-related problems like path planning or path following. However, selecting a good target pose for a mobile robot is a central problem.

Addressing the functional level of determining a placement, Okada and colleagues [9] introduce the concept of a spot. Spots are virtual poses robots can take up to successfully perform an activity, e.g., a pose that can be taken up to wash the dishes. Action-related places by Stulp and colleagues [14] generalize the notion of a spot. The approach employs a simulation-based learning mechanism that obtains a possibility distribution which relates locations to success probabilities for a specific robot performing a specific type of activity. The probability distribution is centered around the object of the considered action, e.g., the cup to be grasped. Zacharias and colleagues [19] take a complementary approach. They represent the reachability workspace of a robot. A good robot placement can then be found by aligning the reachability space with the objects that shall be manipulated. Keshavdas and colleagues [6] obtain a functional map by simulating the performance of an action at different locations to identify a region that can be used by a robot to perform an activity. The mentioned approaches do not model possible placements for agents with different abilities or for actions outside the robot's inventory.

In the area of social robotics, the social activity-placement problem has been addressed mainly for co-operative activities involving robots and humans. Althaus and colleagues [1] employ a potential field method to control the position of a robot within a formation maintained by people having a conversation. Torta and colleagues [16] propose a framework for dynamically updating a target pose for a robot approaching a human taking obstacle information as well as information about human proxemic preferences into account. Yamaoka and colleagues [18] manually tailor a cost function that captures various proxemic constraints accounted for by a museum tour-guide robot that is to present an exhibit to visitors. Shiotani, Maegawa, and Lee [12] take a similar approach for the case of a robot that projects information onto a wall to be recognized by other people. Sisbot and colleagues [13] propose a planner for generating robot placements for interactions with a human while taking human's and robot's visual views into account, as well as task constraints. Pandey and Alami [10] introduce taskability graphs, which represent at which locations interactions between humans and robots are comfortably possible.

An example of work on shared space of humans and robots is given by Tipaldi and Arras [15]. They obtain an affordance map (mapping locations at which activities take place) from learned data encoding human activity probabilities. The map is used to plan a path of a service robot that maximizes the probability to meet humans, thus enabling interaction.

We observe that most work on activity placement is tailored to the functional level or to specific interaction scenarios between humans and robots. In both cases, the robot needs to find a good solution but does not need to justify its choice. Therefore, the constraints on the candidate placements can be modeled using numeric functions and the decision be based on optimization. However, a robot that is aware of socio-spatial constraints should also be aware of its violating such a constraint. In the best of cases, it should apologize for the violation or even justify its choice. E.g., choosing a power outlet for recharging and thereby partially blocking a whiteboard might be explained by telling that (recharging is really urgent and) the other available choices would lead to blocking a doorway.

In line with [6] and [15], we propose a map-like representation of ability-dependent affordance spaces, i.e., spaces that can be used by agents with fitting abilities. Extending existing work, we take into account that activities have spatial needs beyond agent (base) location. As an example, consider the activity of grabing a cup: besides the location of the robot base also the space between the base and the cup should be clear. Having a robot in mind that is able to explain its choices and to apologize in case of violations, we employ a notion of socio-spatial reason. To decide on the best placement among the available ones, we adapt a qualitative bipolar decision rule that weighs up pros and cons [3].

3   An Affordance-Space Map

The concept of affordance was described in the field of psychology of perception by James J. Gibson [4]. According to Gibson, affordances are dispositional properties of the environment that provide possibilities for action. Dispositional properties are those properties of an entity that can be realized by actual activities. For instance, graspability (affordance) is inherent to a bottle and can be realized by grasping (activity). Gibson's proposal locates affordances in the external world, i.e., in the objects being perceived and acted upon. This view allows for perspective taking as an agent can perceive the affordances it can act upon, as well as the affordances others can act upon (cf., [4], p. 141, for an argument put forward to the social meaning of affordances).

We claim that a general approach to socially aware spatial behavior of robots requires to model the influence of affordances on the functional and social structure of space. To this end we propose a view on affordances that allows for recognizing affordances the environment provides for agents with diverging abilities. For example, humans may be interested in viewing pictures or other kinds of displays. Even though such activities might not be relevant for some robot, acting socially aware in spatial planning involves taking care of such human activities and their spatial requirements.

3.1   Modeling Affordances and Affordance Spaces

We explicate the underlying concepts and relations in more detail, as our approach relies on symbolic representations of the concepts needed to solve the social activity-placement problem. Fig. 1 summarizes the interrelations between the concepts we employ in the description of affordances and affordance spaces.

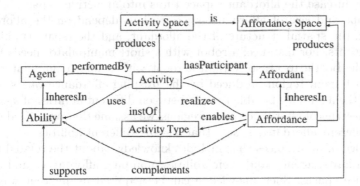

Fig. 1. Concepts and relations in the theory of affordances and affordance spaces

Activities are *instances of activity types, performed by* agents, and can have further *participants*. E.g., a robot that grasps a bottle is agent of and the bottle is participant in the grasping activity. An agent of an activity requires certain *abilities* fitting to the dispositions of the other participant for success. Thus, if a robot grasps a bottle, the grasping *uses* the robot's manipulation abilities and needs to fit the bottles dispositions related to its shape and position. *Affordants* are potential participants of such activities, and *affordances* are the dispositional properties inherent to the affordants, e.g., a graspability (affordance) is inherent to the bottle (affordant). Activities can *realize* affordances but affordances exist independently of an activity taking place. Therefore, we model affordances as primarily related to activity types *enabled* by affordances. If the agent's abilities fit to the affordant's affordances for carrying out the enabled activity type, then we say that the abilities *complement* the affordance relative to the activity type. Hence, if the robot's abilities complement the bottle's affordance with respect to grasping, then we can say that the bottle affords grasping to the robot.

Activities that take place imprint a structure on space. Kendon [5] describes two fundamental regions that come with most interactional activities, viz. an agent region, in which the agents of the activity are located, and the transactional region, which is used beyond the agent region, e.g., the region needed for exchanging objects, for viewing an object from the distance, etc. Activities involving other participants than agents also require a region for locating these participants. To successfully and smoothly perform an activity, an agent needs other agents to not interfere with any of these regions. *Activity spaces* (as described in [8]) are structures *produced by* activities and providing such regions.

Affordance spaces are *produced by* affordances and represent generic spatial constraints for the afforded activity type. The structure of affordance spaces corresponds to the structure of activity spaces. When an activity realizes an affordance, one affordance space turns into the activity space. Thus, an affordance space provides an affordant region (taken up by the affordant), a potential agent region, and a potential transactional region. The attribute "potential" indicates that these regions might become agent region and transactional region respectively in case the affordance space turns into an activity space.

The shape and size of affordance-space regions depend on the afforded activity type, the spatial structure of the affordant, and the (spatial) abilities of the participants. For instance, a robot with a short manipulator needs to move closer to the bottle to grasp it than a robot with a long manipulator. Thus, the potential agent region produced by the grasping affordance varies with the abilities of the agent. To be able to map between different abilities of agents and fitting geometries of affordance-space regions, we assume that one affordance can produce different affordance spaces that *support* different abilities.

Knowledge of affordances thus includes knowledge about the related activity types, affordants, agents, and their abilities, mapping affordances and abilities to affordance spaces. Such knowledge can be acquired in different ways (preprogrammed, learned by simulation, learned by observation, communicated by agents, or even communicated by affordants). In the following we propose a sampling method similar to [14] for mapping affordance spaces based on simulation.

3.2   A Sampling Method for Obtaining an Affordance-Space Map

To construct an affordance-space map, we implemented a procedure that samples affordance spaces for each complementary pair of affordance *aff* and ability *ab*. To this end, it simulates the action of an agent with ability *ab* acting upon *aff* at different poses. These poses are systematically sampled with different distances from and orientations towards the affordant inhering *aff* (Fig. 1). If the simulation is successful, an affordance space is mapped, such that the agent's pose is stored, the region occupied by the agent during simulation is stored as potential agent region, and the region used by the activity beyond agent location is stored as potential transactional region.

The affordance-space map contains geometric representations of affordance spaces, as well as specifications using the concepts and relations introduced in 3.1, i.e., relating to the affordances and the abilities supported by affordance spaces. Fig. 2(b) shows an example affordance-space map obtained from sampling the robot lab taking the bottle affording grasping to humans and robots at the table into account, as well as two doorways affording walking through to humans and robots, the whiteboard on the rightmost wall affording viewing to humans and robot, and two power outlets right to the left door and under the whiteboard affording recharging to the robot.

An affordance-space map supports reasoning about the functional aspect of the social activity-placement problem: Given an activity type ϕ and a potential agent α, those affordance spaces can be obtained that could be used by α to attempt an activity of type ϕ. A *candidate goal affordance space* for the potential agent α to ϕ is an affordance space that supports the ability of α, and is produced by a complementary affordance that enables ϕ.

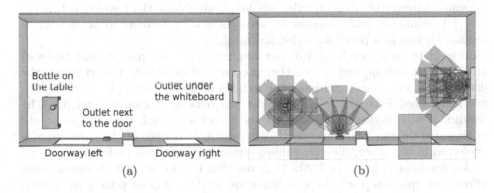

Fig. 2. (a) Map of the robot lab. (b) Affordance-space map of the robot lab: Potential agent regions (grey areas) and potential transactional regions (black-edged areas).

Affordance spaces are bound to the affordants in which the producing affordances are inherent. In dynamic environments, the affordance spaces move with the affordants. To determine affordance-space maps for dynamic environments one has to solve the problem of composing affordance-space maps based on (local) affordance spaces, which is beyond the scope of this article.

4   Social Activity-Placement Based on Reasons

Affordance theory does not solve the problem of choosing among the possible placements. Thus, to rank candidate affordance spaces, the concept of a socio-spatial reason is introduced and put into action using a decision rule. Our reason-based view is inspired by contemporary work in practical philosophy (e.g., [11]).

We refine our earlier work on affordance spaces [8] by differentiating between affordance space with and without social relevance. An affordance space is a *social affordance space*, if it is subject to social considerations deriving from expecting its usage by another agent. Hence, the social considerations we focus on are facts about expected activities.

Facts can induce reasons for or against using a candidate goal affordance space. E.g., assume that it is to be expected that an agent uses affordance space *sp* for walking through a doorway. This fact induces a reason against blocking any affordance-space region that overlaps some region of *sp*. We call such reasons *socio-spatial reasons*. Socio-spatial reasons are modeled as having normative force which takes values in a two-dimensional space spanned by polarity (pro or con) and strength (strong, medium, or weak). To identify relevant facts and determine socio-spatial reasons for (not) using affordance spaces, a knowledge base that tells about which types of agents are expected to inhabit the given environment and their expected behavior at given times is needed (cf., [15]).

Agents can use socio-spatial reasons to evaluate alternatives (and thus to make decisions) and to justify or apologize for their behavior. To evaluate candidate goal affordance spaces, a decision rule is applied that combines and compares reasons. Among the decision rules available, we choose the Levelwise Tallying rule [3]. Bonnefon and collegues [2] show Levelwise Tallying to predict many human choices in a psychological experiment.

Our adaption of Levelwise Tallying determines for each pair of candidate goal affordance spaces sp_1 and sp_2 whether the usage of sp_1 is to be preferred over the usage of sp_2. For both candidates, the number of con reasons of highest strength are subtracted from the number of the pro reasons of highest strength. The candidate with highest score wins. If the scores are equal, then the procedure is repeated on the next lower level of strength. If neither of the candidates is preferred to the other, they are ranked equivalent regarding preference.

An example is given in Table 1: A decision has to be made between using affordance space sp_1, sp_2, or sp_3. Since sp_3 is the only one with a pro reason at the highest level and there is no con reason at the same level, it wins the comparison with sp_1 and sp_2. Affordance space sp_1 is preferred to sp_2 as the reasons at the two highest levels are in balance and sp_2 has more con reasons than sp_1 on the lowest level. Thus, the candidates are ordered as $sp_3 > sp_1 > sp_2$.

Table 1. Example bipolar decision case: Candidate goal affordance spaces sp_1, sp_2, sp_3 and reasons $R_{f_{1-8}}$ having polarity (pro/con) and strength (strong, medium, weak)

	sp_1		sp_2		sp_3	
	Pro	Con	Pro	Con	Pro	Con
Strong					R_{f_6}	
Medium	R_{f_1}		R_{f_3}			
Weak		R_{f_2}		R_{f_4}, R_{f_5}	R_{f_7}	R_{f_8}

After ranking the candidate goal affordance spaces, one of the best candidates is randomly selected (in the example, sp_3 will be selected deterministically). Finally, the goal pose can be extracted from the selected affordance space.

5   Application: A Socially Best Placement for Recharging

To evaluate the approach and to show its feasability, we conducted a lab experiment with the PR2 robot. The knowledge about affordance spaces and socio-spatial reasons was implemented using the web ontology language OWL[1]. The ontology was put to work using ORO [7] as a representation and reasoning framework designed for use in HRI. It uses a RDF triple store to represent knowledge and a description-logics reasoner for inference tasks.

We take the example from the introduction and consider a robot companion whose battery status is getting low. Using its affordance-space map (Fig. 2(b)), the robot first infers the candidate goal affordance spaces, i.e., affordance spaces which i) are produced by an affordance that enables recharging and ii) support the robot's ability (in this case, the limiting factor is the length of the cable).

We assume that the robot has access to a prediction component which recognizes that, during the time it will take to recharge, it is to be expected that some other PR2 might want to grasp the bottle on the table (f_1), that it is likely that a researcher wants to have a look at the whiteboard in the lab (f_2), and that robots and researchers use the doorways (f_3) to enter the lab.

Based on this knowledge, fact f_1 can be used to construct reason R_{f_1} against any candidate whose agent region intersects potential agent regions that can be used by a PR2 to grasp the bottle. This con reason has weak strength, because there are many available possiblities to grasp the bottle, thus blocking one of them does not fully deactivate the grasping possibility. The second fact is used to construct reason R_{f_2}, which is considered stronger, because an activity of a human is involved. Finally, not blocking doorways can even be a safety issue, hence R_{f_3} is a strong reason against candidates that interfere with possible walking-through-doorways activities.

Using the decision making procedure introduced in Sect. 4, the robot obtains an affordance space it has most socio-spatial reasons to use. The result is depicted in Fig. 3(a): The checkerboard region is the potential agent region of the affordance space that was selected. The associated pose can be set as a goal pose to a path planner.[2] Fig. 3(b) shows the robot finally located in the potential agent region of the selected affordance space.

The purpose of this example is to demonstrate the feasibility of the technical solution and to showcase the usefulness of socio-spatial reasons. We are aware that our settings of reasons' polarities and strengths are not experimentally backed. Moreover, rational reasons could be integrated, as well. E.g., if the battery is getting low and there is no time to waste, the urgency of the activity and the distance

---

[1] www.w3.org/TR/owl2-overview/
[2] We employ the standard ROS navigation stack (www.ros.org/wiki/navigation).

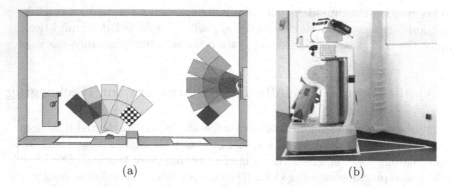

(a)   (b)

Fig. 3. (a) Result of the social activity-placement procedure. Potential agent regions are shaded relative to their social adequacy (the brighter the more adequate). The potential agent region of the selected affordance space is marked by checkerboard pattern. (b) Robot placement in the lab after navigating to the selected affordance space.

to a potential agent region can produce reasons, too. The integration of other kinds of reasons is technically straightforward.

Eventually, imagine a second robot in need of an power outlet. It will have to choose one of the affordance spaces near the outlet close to the whiteboard, thus R_{f_2} is a reason against all of them. As also the second robot is aware of that, it can utter an apology, i.e., that it is aware of the violation (but the battery was getting very low, and all the other possibilities were blocked).

6   Conclusions

In this paper, a general formulation of the social activity-placement problem is proposed. As a solution to the problem, we employ two key concepts: affordance spaces representing the functional aspect of alternatives for action, and facts that yield socio-spatial reasons for or against using an affordance space.

The concept of affordance space allows for taking perspective with respect to the action possibilities of others. Affordance spaces are represented in a data structure called affordance-space map sampled based on knowledge about objects' affordances and agents' abilities. In future, the affordance-space map could also be built from observed data of space use, and be enhanced with knowledge about efforts and success probabilities of performing an activity.

The concept of a (socio-spatial) reason provides a means to evaluate placements and to make choices. How to make use of the explanatory aspect of reasons in human-robot communication is left for future work.

References

1. Althaus, P., Ishiguro, H., Kanda, T., Miyashita, T., Christensen, H.I.: Navigation for human-robot interaction tasks. In: Proc. of ICRA 2004, pp. 368–377 (2004)

2. Bonnefon, J.-F., Dubois, D., Fargier, H., Leblois, S.: Qualitative heuristics for balancing the pros and cons. Theory and Decision 65(1), 71–95 (2008)
3. Dubois, D., Fargier, H., Bonnefon, J.-F.: On the qualitative comparison of decisions having positive and negative features. Journal of Artificial Intelligence Research 32(1), 385–417 (2008)
4. Gibson, J.J.: The Theory of Affordances. In: The Ecological Approach to Visual Perception, pp. 127–143. Lawrence Erlbaum Association Inc. (1986)
5. Kendon, A.: Conducting Interaction: Patterns of Behavior and Focused Encounters. Cambridge University Press, Cambridge (1990)
6. Keshavdas, S., Zender, H., Kruijff, G.J.M., Liu, M., Colas, F.: Functional mapping: Spatial inferencing to aid human-robot rescue efforts in unstructured disaster environments. In: Proc. of the 2012 AAAI Spring Symposium on Designing Intelligent Robots, Stanford University. AAAI Press (2012)
7. Lemaignan, S., Ros, R., Mösenlechner, L., Alami, R., Beetz, M.: Oro, a knowledge management module for cognitive architectures in robotics. In: Proc. of the 2010 IEEE/RSJ Int. Conf. on Intelligent Robots and Systems (2010)
8. Lindner, F., Eschenbach, C.: Towards a formalization of social spaces for socially aware robots. In: Egenhofer, M., Giudice, N., Moratz, R., Worboys, M. (eds.) COSIT 2011. LNCS, vol. 6899, pp. 283–303. Springer, Heidelberg (2011)
9. Okada, K., Kojima, M., Sagawa, Y., Ichino, T., Sato, K., Inaba, M.: Vision based behavior verification system of humanoid robot for daily environment tasks. In: Proc. of the 6th IEEE-RAS Int. Conf. on Humanoid Robots, pp. 7–12. IEEE (2006)
10. Pandey, A.K., Alami, R.: Taskability graph: Towards analyzing effort based agent-agent affordances. In: Proc. of the 2012 IEEE/RSJ Int. Conf. on Intelligent Robots and Systems, pp. 791–796 (2012)
11. Raz, J.: Practical Reason and Norms. Oxford University Press (1999)
12. Shiotani, T., Maegawa, K., Lee, J.H.: A behavior model of autonomous mobile projection robot for the visual information. In: Proc. of the 8th Int. Conf. on Ubiquitous Robots and Ambient Intelligence, pp. 615–620 (2011)
13. Sisbot, E.A., Marin-Urias, L.F., Broquère, X., Sidobre, D., Alami, R.: Synthesizing robot motions adapted to human presence. Int. Journal of Social Robotics 2(3), 329–343 (2010)
14. Stulp, F., Fedrizzi, A., Mösenlechner, L., Beetz, M.: Learning and reasoning with action-related places for robust mobile manipulation. Journal of Artificial Intelligence Research (JAIR) 43, 1–42 (2012)
15. Tipaldi, G.D., Arras, K.O.: I want my coffee hot! Learning to find people under spatio-temporal constraints. In: Proc. of the Int. Conf. on Robotics and Automation, pp. 1217–1222 (2011)
16. Torta, E., Cuijpers, R.H., Juola, J.F., van der Pol, D.: Design of robust robotic proxemic behaviour. In: Mutlu, B., Bartneck, C., Ham, J., Evers, V., Kanda, T. (eds.) ICSR 2011. LNCS (LNAI), vol. 7072, pp. 21–30. Springer, Heidelberg (2011)
17. Williams, M.-A.: Robot social intelligence. In: Ge, S.S., Khatib, O., Cabibihan, J.-J., Simmons, R., Williams, M.-A. (eds.) ICSR 2012. LNCS (LNAI), vol. 7621, pp. 45–55. Springer, Heidelberg (2012)
18. Yamaoka, F., Kanda, T., Ishiguro, H., Hagita, N.: A model of proximity control for information-presenting robots. IEEE Trans. on Robotics 26(1), 187–195 (2010)
19. Zacharias, F., Borst, C., Beetz, M., Hirzinger, G.: Positioning mobile manipulators to perform constrained linear trajectories. In: Proc. of IEEE/RSJ Int. Conf. on Intelligent Robots and Systems (IROS), pp. 2578–2584. IEEE (2008)

Exploring Requirements and Alternative Pet Robots for Robot Assisted Therapy with Older Adults with Dementia

Marcel Heerink[1,2], Jordi Albo-Canals[2], Meritchell Valenti-Soler[3],
Pablo Martinez-Martin[3], Jori Zondag[1], Carolien Smits[1], and Stefanie Anisuzzaman[4]

[1] Windesheim University of Applied Sciences, Almere/Zwolle, The Netherlands
[2] LIFAELS La Salle, Ramon Llull University, Barcelona, Spain
[3] CIEN Foundation and CIBERNED- Alzheimer Center Reina Sofía Foundation, Madrid, Spain
[4] Dignis Lentis, Zuidlaren, The Netherlands
{m.heerink,jori.zondag}@windesheimflevoland.nl,
jalbo@salle.url.edu, {mvalenti,pmartinez}@fundacioncien.es,
chm.smits@windesheim.nl, sp.anisuzzaman@lentis.nl

Abstract. Robot assisted therapy has been applied in care for older adults who suffer from dementia for over ten years. Strong effects like improved interaction and signs of a higher sense of wellbeing have been reported. Still it is unclear which features are needed and which robotic pets would are suitable for this therapy. In this explorative research we interviewed 36 professional caregivers, both experienced and inexperienced in relationship to RAT and compiled a list of requirements. Next, we used this list to compare commercially available robotic pets. We found that many pet robots are usable, although seal robot Paro meets the requirements best, being superior on sustainability, realistic movements and interactivity. Finally, a test with alternative pets showed that different subjects were attracted to different pets and a subsequential questionnaire revealed that some caregivers were not only willing to try alternatives for Paro, but also suggesting that alternative pets could in some cases be more suitable.

1 Introduction

For more than a decade, research has been done on the use of robotic pets for older adults suffering from dementia, suggesting this is a successful form of therapy [1-4]. Although most research has been done in Japan and with the same seal shaped robot called Paro, it is generally assumed that therapeutic use of robotic pets improves mental and physical wellbeing of older adults with dementia and results in a more active interaction of the subjects with their environment [5].

Although there are some alternatives [6-10], Paro is by far the most widely used robotic pet for this purpose. This could be due to the fact that Paro is the only robotic pet that is both especially developed for this purpose and commercially available. However, Paro is quite an investment since it costs close to five thousand dollars [11]. Eldercare professionals that would like to try working with a robotic pet but have a

G. Herrmann et al. (Eds.): ICSR 2013, LNAI 8239, pp. 104–115, 2013.
© Springer International Publishing Switzerland 2013

very limited budget may look for alternatives. These would be pet robots meeting the requirements that make them suitable for robot-assisted therapy in dementia.

In this explorative study we want to elicit and specify these requirements by focusing on professional caregivers working with older adults who suffer from dementia. These caregivers may have experience with similar types of interventions, like using real pet animals, stuffed animals or other techniques that stimulate the senses for which the term 'snoezelen' is used. Snoezelen is also called or Multi-Sensory Stimulation (MSS), and is a widely used and accepted approach to nursing home residents suffering dementia [12].

The caregivers that are subject to our study may or may not be familiar with robot-assisted therapy. If they are not, this could be due to the unfamiliarity of the possibilities of this form of therapy, but also by inaccessibility to practical guidelines: for caregivers who are interested in applying this therapy, there are hardly any practical guidelines available on how to use which type of robot in which state of dementia, how to deal involve family members and how to respond to any negative responses. It could in that case very well be that comprehensible set of guidelines would lead to a wider application of robot-assisted therapy.

Caregivers who are familiar with robot-assisted therapy - and especially the ones who have applied it - may give different responses when asked for the requirements for a suitable robot.

This paper presents the results of an explorative study. The goal was to elicit and specify requirements according to professional caregivers for a pet robot that can be used in therapeutic interventions with older adults suffering from dementia. Moreover, we wanted to establish how familiarity with this form of therapy and the experience of applying it would influence the elicitation and specification of these requirements.

Fig. 1. Paro

In our present study, we wanted to map (a) the familiarity of robot-assisted therapy for professional caregivers in Spain and the Netherlands, (b) the need for guidelines by professional caregivers in Spain and the Netherlands, (c) produce an inventory of requirements for a suitable robot according to these caregivers and (d) produce a comparison of available pet robots based on these requirements (e) describe professionals'reactions to the use of pet robots in a small experiment.

In the following section we will present the project of which this study is a part. Next, we will discuss the used questionnaire and respondents subsequently we will present the results of (a) questions on experience and guidelines and (b) the requirements inventory. After drawing some preliminary conclusions from this, we will compare a few alternative robots guided by these requirements and present a small user study in which we looked for the first response of residents suffering from moderate dementia and caregivers in a care institution.

2   New Friends Framework

The "New friends, old emotions" project is a Dutch-Spanish collaboration which targets the accessibility of robot-assisted therapy for caregivers that work with older adults suffering from dementia. Its first aim is to establish the need for guidelines for robot-assisted therapy by professional and informal caregivers.

Furthermore, the project targets an inventory of (1) experiences that some caregivers already have with robotic pets, (2) available pet robots and their suitability for this form of therapy, and (3) practices by caregivers that can be related to this form of therapy (e.g. using stuffed animals, real pets and activities that otherwise stimulate the senses of the subjects). Moreover, it aims to use the findings of these studies to provide guidelines and to offer supportive workshops for robot-assisted therapy.

The consortium that carries out this project consists of Dutch and Spanish institutions that have technical experience with (pet) robots, experience with field studies concerning older adults, or specific expertise in both studying and working with older adults suffering from dementia. Also a part of the consortium is eldercare institutions in different cities of the Netherlands. The project management is carried out by the Robotics research group of Windesheim Flevoland University of Applied Sciences in Almere, the Netherlands.

3   Developing a Requirements List

To establish our goal, we decided to gather both qualitative and quantitative data from questionnaires completed by caregivers that worked in eldercare institutions in the towns of Almere, Lelystad and Zuidlaren in the Netherlands and in the city of Madrid in Spain. Both in the Netherlands and in Spain, some caregivers had no experience in working with a pet robot, while others had worked with Paro.

The 17 caregivers from the Netherlands were all professionals, aged 19 to 61. They had a lower or higher professional education and they were all female. The 20 caregivers from Spain were aged 21 to 58. They were also female professionals except for one, and their education varied from lower professional to university.

The respondents were asked to fill out the questionnaire individually. This questionnaire (Table 1) consisted of (a) questions on knowledge of and experience with robot-assisted therapy and (b) the need for guidelines and (c) questions on requirements for suitable robots. Four of the questions (in Table 1 these are questions 3 to 6) were actually statements to be replied to on a five point Likert scale, indicating the extent to which they agreed (absolutely agree – agree – neutral - not agree - absolutely not agree). The espondents were aware that the answers on this scale corresponded with an attributed score, varying from 5 (totally agree) to 1 (totally not agree).

Table 1. Questionnaire items

1. Have you ever heard, seen or read about the use of a pet robot for older adults suffering from de mentia?

2. Have you ever used such a robot?

Yes:	2a. Did you use specific directives? Yes: which ones and how did you get t hem? No: Why not? Would you like to have directives?	2b. Did you involve family members? Yes: Did it go well? Did you use directi ves? No: Would you want to? Why would or wouldn't you?
No	2c. Would you like to work with it? 2d. What would hold you back or stimulate you?	

Please indicate how much you agree with the following statements:

3. I believe that activities with pet-like robots may increase the quality of life for people suffering fr om dementia.
4. I (would) like to work with such robots
5. I find it important that there are directives for interventions with such robots
6. These directives should also make it possible for family members to do these interventions

7. What possibillities and properties should suitable pet robot have?
a.   What features and qualities are necessary?
b.   What features and qualities are desirable?
c.   How do you expect that older people respond to these properties?
d.   Which expressions are important? (eg facial expression wagging tail etc)
e.   Why?

8. What possibillities and properties should a suitable pet robot certainly not have?

After the questionnaires were filled out, the respondents elucidated their answers in a conversation with one of the researchers. These were recorded.

4   Questionnaire Results

4.1   Familiarity and Guidelines

Most caregivers were more or less familiar with robot-assisted therapy. Of course, those from Madrid had even applied it, but nine out of eleven from the Netherlands had seen a short television documentary on this subject. Four of them compared it to their own experiences with real pets. In one case this was a dog, but the other three who all worked at the same eldercare institution in the city of Lelystad, reported that they kept a cat on their floor that they made to look like a real street with houses in the seventies. They reported positive effects of cuddling sessions with the cat, but also expressed that a robotic cat would be more beneficial, since it would always be willing to be cuddled.

Four other caregivers reported the use of stuffed animals to be more or less familiar, but even more the practice of "snoezelen", which aims to evoke emotions by stimulating the senses. They expected robot-assisted therapy to be beneficial since it could also evoke emotions.

All caregivers except for one expressed a need for guidelines and stated that robot-assisted therapy would be far more widely applied if these would be commonly available. Some indicated that guidelines were especially needed for dealing with unexpected responses that could also occur with similar activities that evoke emotions. They indicated that occasionally robotic pets could evoke anger, panic or sadness. Moreover several caregivers from the Netherlands reported that some related activities would occasionally evoke resistance, reluctance or even animosity by family members who experienced it as humiliating or insulting to see their fathers or mothers playing with stuffed animals. This could also be expected if it were robotic pets. A set of guidelines should also include directives on how to deal with this. The one caregiver who indicated that no guidelines were needed stated that she expected that this form of therapy would hardly be applied and that developing guidelines would be a waste of time and effort.

As Table 2 shows, the scores on the four Likert scale statements were generally "agree" or "totally agree". For each statement there were only one or two "neutral" scores.

Table 2. Descriptive statistics of s cores on items 3 to 6

	Minimum	Maximum	Mean	Std. Deviation
Item 3	3	5	4,19	,525
Item 4	3	5	4,39	,599
Item 5	4	5	4,64	,487
Item 6	2	5	4,11	,887

Table 3 shows an analysis of the differences between caregivers with and without experience with Paro: none of the questions resulted in significant answers.

Table 3. Difference in experience for items 3 to 6

	Item 3	Item 4	Item 5	Item 6
Mann-Whitney U	154,000	116,000	141,000	143,000
Sig (2-tailed)	,784	,112	,449	,547

Table 4 shows the (Spearman) correlation on the scores for items 3 to 6 plus age of the caregivers. There is significance for the correlation between and between Items 3 and 4, 4 and 5 and 5 and 6. The first correlation is a predictable one: the more care-givers believe in using pet robots, the more they are willing to work with it. The second one is remarkable: the ones who are willing to work with it, generally think they could benefit from good directives. The third indicates that caregivers who think they could benefit from guidelines also think it is good to work with family members.

Moreover, there is a strong correlation between Age and Item 6. This could indi-cate that older caregivers are more willing to involve family members than younger ones.

Table 4. Correlation items 3 to 6 and Age

		Item 3	Item 4	Item 5	Item 6	Age
Item 3	Correlation	1,000	,392*	,255	,321	,188
	Sig. (2-tailed)	.	,017	,127	,053	,265
Item 4	Correlation	,392*	1,000	,328*	,233	-,003
	Sig. (2-tailed)	,017	.	,047	,165	,986
Item 5	Correlation	,255	,328*	1,000	,368*	-,151
	Sig. (2-tailed)	,127	,047	.	,025	,372
Item 6	Correlation	,321	,233	,368*	1,000	,447**
	Sig. (2-tailed)	,053	,165	,025	.	,006

4.2   Requirements

We had asked the caregivers to indicate which requirements were necessary and which ones were desirable. In order to quantify the results some preferred pet charac-teristics were combined by the researchers. For example, some caregivers indicated the skin should be soft, some said it should be furry and some indicated it should be 'pettable like a real animal skin'. All these were categorized under 'soft pettable fur' (listed as requirement 1).

Answers that were given to question 8 were processed in a similar way, since they consistently were the reversed versions of the positive expressions. For example, it was often indicated that the robot should not be noisy which is essentially the same as requirement 3 (mechanical parts are noiseless) and a remark 'It should really not be to breakable' could be categorized under 12 (can withstand rough handling). All these requirement counts that where derived from answers to question 8 where categorized as necessary.

In many cases pet features were mentioned repeatedly, both as necessary and de-sired features and sometimes even again in reversed descriptions answering question 8. In that case, the requirement was only counted once as a necessary feature.

One participant simply stated that the robotic pet should stimulate the user. We did not count this as a requirement, because this is already one of the principle goals of robot-assisted therapy.

Table 5 shows the results of this count, for the caregivers that had worked with Pa-ro (Exp) and the ones with no such experience (Not), followed by the total counts.

Note that each cell contains the counts for necessary (before the slash) and desired (after the slash) requirements.

The 'soft pettable fur' was mentioned in different characterizations by most caregivers of the group with no experience and many of them mentioned appropriate sounds and noiseless mechanical parts. Some mentioned detachable fur (which is actually hardly found for robotic pets).

We may conclude that most caregivers were familiar with robot-assisted therapy. Moreover, they were generally quite willing to apply it if they did not already do. Remarkably they easily linked this form of therapy to familiar activities, like working with real pets, stuffed animals and sensory stimulation. Also, caregivers generally agreed on the need for guidelines.

Table 5. Requirements for caregivers with and without experience with Paro

Requirements	Exp	Not	Total
1. Soft pettable fur	2/-	11/1	13/1
2. Appropriate responses/sounds	4/1	8/7	12/8
3. Looks like a real life pet	5/1	4/1	9/2
4. Mechanical parts are noiseless	-/-	7/2	7/2
5. Young or innocent looking.	4/-	3/1	7/1
6. Nice/not scary	1/-	6/1	7/1
7. Huggable (right size cuddle with)	-/-	6/-	6/-
8. Realistic movements (fluent/natural)	1/-	4/2	5/2
9. Adaptable (shut functions on/off)	1/-	2/2	3/2
10. Autonomous system	-/1	3/-	3/1
11. Mobile (easy to take with you)	2/-	2/-	4/-
12. Can withstand rough handling, solid	1/-	2/-	3/-
13. Easy to use	2/-	3/-	5/-
14. Variety of behaviors and sounds	2/1	1/-	3/1
15. Fur is detachable (to be washed)	1/-	2/-	3/-
16. Cartoonish appearance	-/-	1/-	1/-
17. Flashy/draws attention	-/-	1/1	1/1

Looking at the generated list of requirements we see that a soft pettable fur is mentioned often especially by the caregivers without experience. Remarkable is that the noiselessness of the mechanical part is only mentioned by caregivers without experience.

This list contains 17 items that can be prioritized according to the necessity as indicated by the participants, but also by the frequency of the combined categories. We chose to list them in Table 5 only by the frequency of the necessity.

5   Exploring Alternative Robotic Pets

To explore alternative pets, we selected a few alternative robotic pets and set up a small user study. Subsequently we interviewed the involved caregivers.

5.1   Strategy

We made an inventory of commercially available robotic pets and selected a seal puppy and a cat. Next to realistic looking pet robots, we wanted to use more cartoonish designed pets and selected a baby dinosaur and a bear.

The seal puppy is produced by WowWee, and is an example of the Alive Baby Animals series. Its current price is €35.- and it has the appearance of a Paro seal robot, but is much smaller and lighter. Moreover it is limited in functionality compared to Paro: it can only open and close its mouth and produce baby seal cries. Its mechanical parts are also much noisier.

Fig. 2. Used pet robot – clockwise: Seal puppy, Pleo, Bear and Cat (Cuddlin Kitty)

The cat is a 'FurReal Friends Lulu Cuddlin Kitty', produced by Hasbro. The cost is €60.-. She has a lying position and responds to caressing by shutting her eyes briefly and by making a purring sound. After being petted for a longer time, she lifts her leg and turns on her back so her chest and belly can be petted. When the user stops this, she turns back in her original position. She has multiple sensors in head, back, chest and belly and a microphone. She detects voice and responds to it by meowing. Its mechanical parts are as noisy as the Baby Seal.

The dinosaur is a Pleo robot. It is in fact a baby Camarasaurus, which has just hatched. This means it still has to develop skills and personality when it is received. Its development depends on how it is fed and treated (petting a lot makes it nicer). It features two microphones that are used for voice detection, a camera which is used to localize people and objects and multiple touch sensors on the head and back which make it responsive to petting. Its mechanical parts are much less noisy compared to the previous pets.

The bear is a robot that has been developed by the Robotics Research group of Windesheim Flevoland University of Applied Sciences. It is a regular stuffed bear equiped with a robotic frame made with Arduino, which can easily be transferred to other stuffed animals. This makes it possible to test different embodiments. Moreover, the functionalities (which are still limited at this time) can be turned on and off independently which will enable us in a later stage to establish the importance of each feature. The bear also has WIFI connectivity, so it can be remotely controlled in a wizard-of-oz setup.

We thus had four robots that could all be categorized according to the attributes 'familiar' and 'life like': the seal is not familiar as a pet, but life like; the cat is both life like and familiar; the bear is familiar but not life like and Pleo is neither. However, as Table 6 shows, we have to bear in mind that the available functionalities of the four pets have more differences than these. Nevertheless, they are all more or less comparable to Paro, although Paro fits most requirements and is far superior in weight (it is much heavier – according to some caregivers it is even too heavy) and interactivity to any of the alternative robots.

Table 6. Alternative robots fitting the requierements

Requirements	Seal	Bear	Cat	Pleo	Paro
1. Soft pettable fur	+	+	+	+/-	+
2. Appropriate responses/sounds	+	+/-	+	+	+
3. Mechanical parts are noiseless	+/-	+/-	-	+/-	+/-
4. Young or innocent looking.	+	+	+	+	+
5. Nice/not scary	+	+	+	+/-	+
6. Huggable (right size cuddle with)	+	+	+	+	+/-
7. Realistic movements (fluent/natural)	+	+/-	+/-	+/-	+
8. Looks like a real life pet	+/-	+/-	+	-	+/-
9. Adaptable (shut functions on/off)	-	+	-	-	+
10. Autonomous system	+	+	+	+	+
11. Mobile (easy to take with you)	+	+	+	+	+/-
12. Can withstand rough handling, solid	-	-	-	-	+
13. Easy to use	+	+	+	+	+
14. Variety of behaviors and sounds	-	-	-	+	+
15. Fur is detachable (to be washed)	-	-	-	-	-
16. Cartoonish appearance	-	+	-	+	-
17. Flashy/draws attention	+/-	+/-	+/-	+/-	+/-
+/-	4	6	4	2	8

5.2   Experimental Procedure

We set up a session of one hour with fifteen patients who suffered moderate dementia. They were sitting in a circle as would be usual for group activities, when a caregiver presented the first robotic pet (the cat) to each participant for approximately one minute. The participant could take the robot on his or her lap, touch it and talk to it. When presenting it, the caregiver asked if the participant liked the robot and if he or she thought it was real. After it had been presented to all participants, the next robot was presented (subsequently the seal, the bear and the dinosaur). Researchers were able to observe the responses. They specifically noted smiles caresses, hugs, kisses and talking directed to the robotic pet, and also the response (if any) to the questions of the caregiver.

5.3   Interviews

We interviewed eleven caregivers (nurses and therapists) that were present at the department where we carried out the experiment. They were not only able to see the pet robots we used, but also to pick them up and explore the interaction. All of them had experience with robot-assisted therapy, using Paro.

They were asked to rate the suitability of each of the used pet robots for therapy activities by rating it on a scale from one (absolutely suitable) to five (absolutely not suitable). Subsequently they were asked to elucidate their rating.

5.4   User Study Results

Table 7 shows only a part of the responses we observed. First of all, we felt unable to record the smiles (as has been done in related studies [13]), since we could often not differentiate between a smile caused by the caregiver and a smile caused by the robot and some participants were simply smiling during the entire session.

Also responses to caregivers questions caused some difficulty, since many participants gave no verbal reply to it. For the cat, three participants said it was real and four participants said it was not real. The seal was claimed to be real by one person and not real by four. After the cat and the seal were presented, the caregiver stopped asking this question.

Table 7. Patient responses to the robotic pets

	Pleo	Cat	Seal	Bear
like	6	12	13	6
caress	6	10	11	4
talk	1	7	4	2
hug	0	0	3	6
kiss	4	0	4	0

When analyzing the responses, we noticed that there were clear differences between the participants. Where some responded to the cat and not to the seal, for others the response was the other way around. And some did response more to the bear than to other robotic pets.

Table 7 shows that the seal and the cat scored the highest amount of patients'response, especially on 'likes' and 'caresses'. The cat scored the most 'talks' and the bear the highest amount of hugs. Pleo scored lower on most counts except for kisses, which is in line with the often mentioned requirement of 'looking like a real life pet' in Table 5. We also noted that four participants indicated to be scared of it.

This was something that none of the participants indicated with any of the other pet robots except for one participant with the cat: she indicated that she had always been afraid of cats.

5.5   Caregiver Interview Results

To process the rating scores appropriately so that a higher score would indicate a higher appreciation, we reversed them.

As Table 8 shows, the highest score was for the baby seal robot. The caregivers indicated that they were charmed by its simplicity and softness. Two caregivers even indicated that they liked it more than Paro, because it was lighter, easier to use, more mobile and because they would be less afraid to break it or have it broken. The second highest score was for the cat. Caregivers liked it because it was a realistic representation of an animal that could be referred to as a pet (contrary to all the other pet robots). However, they disliked its movements that were 'too robotic'. The third highest score was for the bear, which was often considered appropriate, but too limited in functionalities and too big. The lowest score was for the Pleo. Many caregivers found it cute, but not familiar enough and 'too reptilious'.

Table 8. Patient responses to the robotic pets

	Total	Minimum	Maximum	Mean	Std. Deviation
cat	49	3	5	4,18	,751
seal	46	3	5	4,45	,688
pleo	39	2	4	3,00	,775
bear	33	3	5	3,82	,751

6   Conclusion, Discussion and Future Research

A first conclusion from the first part of this research is that most caregivers are willing to work with, or at least explore robot-assisted therapy with people suffering from dementia.

A second conclusion could be that furry skin, appropriate response and a silent operating mechanism are the most important requirements according to caregivers. However, much more research can be done on these requirements, for example by focusing on their specification. We could take this list and ask caregivers to attribute a weight to them.

A third finding of this study is that many caregivers spontaneously linked robot-assisted therapy to activities like working with real pets, stuffed animals and evoking emotions by stimulating the senses (snoezelen). When developing guidelines we could indeed learn from caregivers' experiences with these activities and establish if they could be applied to the use robotic pets.

A fourth conclusion is that some older adults in a stage of moderate dementia differ in their response to different types of pet robots. Further research could specify this and establish if there is a predictable pattern (a typology of patients linked to a typology of pet robots) or even that a caregiver should have a collection of different pet robots rather than one specific one.

Finally we conclude that caregivers are open to alternatives to Paro for robot-assisted therapy in dementia and that some of them may even prefer an alternative. This invites us to further explore these alternatives and research the importance of different requirements.

Acknowledgment. We thank the management and caregivers working at De Over-loop and De Archipel in Almere, De Enk in Zuidlaren and at De Hanzeborg in Lelys-tad as well as the staff of the Alzheimer Center Reina Sofia Foundation, Madrid. We are very grateful for their time.

References

1. Wada, K., Shibata, T.: Robot Therapy in a Care House-Its Sociopsychological and Physio-logical Effects on the Residents. In: Proceedings of the 2006 IEEE International Confe-rence on Robotics and Automation, ICRA 2006, pp. 3966–3971 (2006)
2. Wada, K., Shibata, T.: Living With Seal Robots—Its Sociopsychological and Physiologi-cal Influences on the Elderly at a Care House. IEEE Transactions on Robotics [see also IEEE Transactions on Robotics and Automation] 23(5), 972–980 (2007)
3. Inoue, K., Wada, K., Uehara, R.: How Effective Is Robot Therapy?: PARO and People with Dementia. In: Jobbágy, Á. (ed.) 5th European Conference of the International Federa-tion for Medical and Biological Engineering. IFMBE Proceedings, vol. 37, pp. 784–787. Springer, Heidelberg (2011)
4. Bemelmans, R.: Verslag pilot-onderzoek Paro, Hogeschool Zuyd (2012)
5. Broekens, J., Heerink, M., Rosendal, H.: The effectiveness of assistive social robots in el-derly care: a review. Gerontechnology Journal 8(2), 94–103 (2009)
6. Stiehl, W.D., et al.: The Huggable: a therapeutic robotic companion for relational, affective touch. In: Consumer Communications and Networking Conference, Las Vegas, Nevada, USA (2006)
7. Sherwood, T., Mintz, F., Vomela, M.: Project VIRGO: creation of a surrogate companion for the elderly. In: Conference on Human Factors in Computing Systems, pp. 2104–2108 (2005)
8. Kriglstein, S., Wallner, G.: HOMIE: an artificial companion for elderly people. In: Confe-rence on Human Factors in Computing Systems, pp. 2094–2098 (2005)
9. Furuta, Y., et al.: Subjective evaluation of use of Babyloid for doll therapy. IEEE (2012)
10. Boon, B.: Pilot health care scottie (2009)
11. Japantrendshop.com. Japan trend shop Paro (2013),
    http://www.japantrendshop.com/paro-robot-seal-healing-pet-p-144.html
12. Spaull, D., Leach, C., Frampton, I.: An evaluation of the effects of sensory stimulation with people who have dementia. Behavioural and Cognitive Psychotherapy 26(1), 77–86 (1998)
13. Robinson, H., et al.: Suitability of Healthcare Robots for a Dementia Unit and Suggested Improvements. Journal of the American Medical Directors Association 14(1), 34–40 (2013)

Smooth Reaching and Human-Like Compliance in Physical Interactions for Redundant Arms*

Abdelrahem Atawnih and Zoe Doulgeri

Department of Electrical and Computer Engineering
Aristotle University of Thessaloniki
54124 Thessaloniki, Greece
atawnih@ee.auth.gr, doulgeri@eng.auth.gr

Abstract. This work collectively addresses human-like smoothness and compliance to external contact force in reaching tasks of redundant robotic arms, enhancing human safety potential and facilitating physical human-robot interaction. A model based prescribed performance control algorithm is proposed, producing smooth, repeatable reaching movements for the arm and a compliant behavior to an external contact by shaping the reaching target superimposing the position output from a human-like impedance model. Simulation results for a 5dof human-arm like robot demonstrate the performance of the proposed controller.

Keywords: Smooth Reaching Motion, Compliance, Redundant Arms.

1 Introduction

An important factor for the development and acceptance of service robots is their ability to effectively perform tasks with human-like kinematic and compliance characteristics which may unconsciously establish comfort in the coexisting human. Reaching, a basic prerequisite task in service robotics, is related to the problem of addressing the ill-defined inverse kinematics of redundant manipulators yet far more demanding as it requires movements with human-like kinematic and compliance characteristics, with respect to external contact forces. Studies of human unconstrained reaching movements from one point to another reveal the following invariant kinematic characteristics: straight line paths, bell shaped velocities and joint configuration repeatability in repetitive motions [1–3]. In several proposed reaching methods, the target acts as an attractor for the arm but the frame of reference in which motion is coordinated differs; the task space, the joint space [2, 4, 5] or both spaces are used in a multi-referential approach [6] as there is human evidence supporting all the aforementioned approaches [1, 7, 8]. The joint space frame of reference usually involves the solution of the ill-posed inverse kinematics which is resolved either by introducing task priorities like joint limit and constraint avoidance, optimization criteria like distance

* This research is co-financed by the EU-ESF and Greek national funds through the operational program "Education and Lifelong Learning" of the National Strategic Reference Framework (NSRF) - Research Funding Program ARISTEIA I.

G. Herrmann et al. (Eds.): ICSR 2013, LNAI 8239, pp. 116–126, 2013.

minimization from a preferred posture [9, 10] or by reproducing human-like joint configurations regarding the elevation angle [11]. The simple task space feedback control [4] based on the virtual spring-damper hypothesis with the introduction of appropriate joint damping factors successfully imitates the quasi-straight line path of the end-effector but fails to mimic the bell-shaped velocity profile and the joint configuration repeatability in repetitive motions. Partial improvements regarding these issues as well as compliance to external force disturbance are reported in [12] and [13]. However, the authors do not provide any theoretical proof for the stability of the closed loop system when the control law is subjected to the proposed alterations. Compliance protects the human from excessive forces during contact and allows one to override the actions of the robot to a certain extent. It can be achieved passively by using flexible components in the robot's structure or actively by the control method; passive compliance is important for reducing the initial collision force while active compliance is important for improving on post-collision safety and facilitating physical human robot interaction (pHRI) [14]. Strategies for collision detection and reaction and human intent detection and following during physical interaction are similar [15], [16]. A switching strategy is mainly adopted between the current robot task and the reaction law which may negatively affect stability of the overall switched system. Alternatively, within an impedance control scheme the adaptation of force and target impedance when interacting with unknown environments is proposed; the focus is on achieving control targets despite external contact forces rather than complying to them [17]. Last, the prescribed performance control methodology introduced in [18] and detailed in the following sections, was utilized to synthesize the kinematic controller for unconstrained reaching without trajectory planning, outperforming alternative solutions with respect to the smoothness and repeatability attributes [19]. However, the synthesis of a torque controller was left as an open issue as well as the controller's endowing with human-like compliance to external contact forces. This work extends the work of [19] by addressing these open issues via a novel model-based control method.

2 Problem Description

Consider an n-joints manipulator and task coordinates of m dimensions with $m < n$, and let $q \in \Re^n$ be the vector of the generalized joint variables. The general form of manipulator dynamics interacting with the environment can be described by:

$$H(q)\ddot{q} + C(q,\dot{q})\dot{q} + G(q) = u + J^T(q)F \tag{1}$$

where, $H(q) \in \Re^{n \times n}$ is the positive definite robot inertia matrix, $C(q,\dot{q})\dot{q} \in \Re^n$ denotes the centripetal and Coriolis force vector, $G(q) \in \Re^n$ is the gravity vector, $u \in \Re^n$ is the joint input vector, $F \in \Re^m$ is the vector of external contact forces acting on the end-effector and $J(q) \in \Re^{m \times n}$ the manipulator Jacobian. Let $p \in \Re^m$ be the generalized position of the robot end-effector in the task space and $p_d \in \Re^m$ the target position. Let D be the continuous non-linear direct kinematics map associating each q to a unique p, $p = D(q)$. The inverse mapping in the redundancy case is ill-defined in the sense that it associates to each p a set of $q's$ usually containing an infinite numbers of elements. In that respect let q_T be the joint configurations which correspond to the target location, i.e. $q_T = \{q : D(q) = p_d\}$. The joint and task velocities are related by the kinetic relation $\dot{p} = J(q)\dot{q}$ which can be inverted using the Jacobian pseudoinverse:

$$\dot{q} = J^\dagger(q)\dot{p} \tag{2}$$

where $J^\dagger = J^T (JJ^T)^{-1}$ denotes the Moore-Penrose pseudoinverse of matrix $J(q)$ yielding a minimum norm joint velocity solution in case of a full row rank Jacobian. This solution can be augmented by the null space solution $(I_n - J^\dagger J) z$ with $z \in \Re^n$ usually being selected within a task priority or optimization formulation. The aim of this work is to design a torque controller for (1) that brings the arm to the target location p_d with smooth and repeatable movements and human-like compliance under physical contact.

3   Prescribed Performance Preliminaries for Smooth Reaching

The PPC methodology comprises two basic design notions. The first notion refers to the design of a bounded, smooth, strictly positive function of time for each element of the output error $e_i(t)$ defined as the distance between the measured or calculated output value and its reference, called the performance function $\rho_i(t)$ and the setting of a constant $0 \le M_i \le 1$ which together determine the boundaries of a region within which an output error element should ideally evolve as illustrated in Fig.1a and expressed mathematically as follows:

$$-M_i\rho_i(t) < e_i(t) < \rho_i(t) \qquad \forall t \quad \text{in case} \quad e_i(0) \ge 0 \tag{3}$$
$$-\rho_i(t) < e_i(t) < M_i\rho_i(t) \qquad \forall t \quad \text{in case} \quad e_i(0) \le 0 \tag{4}$$

As implied by the above inequalities, either (3) or (4) is employed for each error element, and specifically the one associated with the sign of $e_i(0)$. For reaching, a prescribed performance function that reflects the "minimum jerk trajectory" is proposed producing straight line hand paths and bell-shaped velocity profiles as observed in human subjects performing voluntary unconstrained movements [3]:

$$\rho_i(t) = \begin{cases} \rho_{i0} + (\rho_{i\infty} - \rho_{i0}) \left(10\left(\frac{t}{T}\right)^3 - 15\left(\frac{t}{T}\right)^4 + 6\left(\frac{t}{T}\right)^5\right) & \text{if } t \le T \\ \rho_{i\infty} & \text{if } t \ge T \end{cases} \tag{5}$$

where ρ_{i0}, $\rho_{i\infty}$ and T are positive preset design parameters. The performance boundaries produced by Eq.(5) and a choice of $0 < M_i < 1$ are shown in Fig.1a with red dashed lines. Parameter T can be regarded as the time the output error enters the steady state performance zone expressed by parameter $\rho_{i\infty}$. It is a motion attribute that is imposed a priori by the designer and is independent of the distance to the target and the controller gains. Parameter ρ_{i0} is set greater than $e_i(0)$ so that the error response starts within the performance region.

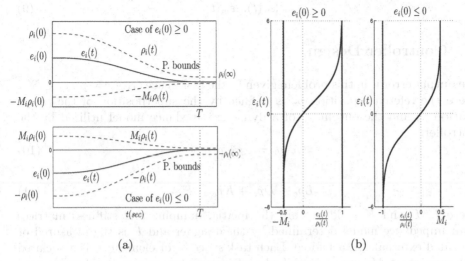

Fig. 1. (a) Prescribed Performance, (b) Natural logarithmic transformation, $T(0) = 0$

The second basic notion of the PPC methodology is the introduction of the transformed errors $\varepsilon_i(t)$ defined by:

$$\varepsilon_i(t) = T_i\left(\frac{e_i(t)}{\rho_i(t)}\right), \quad i = 1,\ldots,m \tag{6}$$

where $T_i(.),\ i = 1,\ldots,m$ is any smooth, strictly increasing function defining a bijective mapping:

$$T_i : \begin{cases} (-M_i, 1) \to (-\infty, \infty), \ e_i(0) \geq 0 \\ (-1, M_i) \to (-\infty, \infty), \ e_i(0) \leq 0 \end{cases} \tag{7}$$

The following transformation function proposed in [20], mapping the origin of the output error space to the origin of the transformed error space, based on the natural logarithmic function, is considered in this work and is illustrated in Fig.1b.

$$T_i\left(\frac{e_i(t)}{\rho_i(t)}\right) = \begin{cases} \ln\left(\dfrac{M_i + \frac{e_i(t)}{\rho_i(t)}}{M_i\left(1 - \frac{e_i(t)}{\rho_i(t)}\right)}\right), \text{ in case } e_i(0) \geq 0 \\[4mm] \ln\left(\dfrac{M_i\left(1 + \frac{e_i(t)}{\rho_i(t)}\right)}{M_i - \frac{e_i(t)}{\rho_i(t)}}\right), \text{ in case } e_i(0) \leq 0 \end{cases} \tag{8}$$

Notice that this choice of transformation function (8) requires the choice of $M_i \neq 0,\ i = 1,\ldots,m$ since otherwise (i.e. $M_i = 0$) $\varepsilon_i(0),\ i = 1,\ldots,m$ becomes infinite.

The PPC methodology's central result borrowed from [21] states that for a generic output error signal $e_i(t)$ and the corresponding transformed error $\varepsilon_i(t)$ defined through (6) that satisfies (7), prescribed performance of $e_i(t)$ in the sense of (3) or (4) is satisfied for all $t \geq 0$, provided $\varepsilon_i(t)$ is bounded. Hence, central to the prescribed performance methodology is to design a controller that guarantees the boundedness of the transformed output error vector given below.

$$\varepsilon = [\varepsilon_1(t)...\varepsilon_m(t)]^T \qquad (9)$$

4   Controller Design

The output error is in this problem given by the task space error $e = p - p_r \in \Re^m$, where the reference position p_r is defined by the superposition of the target location p_d and the output p_e of a dynamic impedance model utilized by the controller:

$$p_r = p_d + p_e \qquad (10)$$

$$L\ddot{p}_e + D\dot{p}_e + Kp_e = \hat{F} \qquad (11)$$

where L, D and $K \in \Re^{m \times m}$ are the inertia, damping and stiffness matrices of an impedance model determined by the designer and $\hat{F}$ is the measured or identified external contact force. Each task space error element e_i is associated with a constant M_i and a prescribed performance function $\rho_i(t)$ (5) embodying human-like smoothness characteristics. Let us further define $a_i(t) \triangleq -\frac{\dot{\rho}_i(t)}{\rho_i(t)} > 0$, $i = 1, \ldots, m$ and $\vartheta T_i \triangleq \frac{\partial T_i}{\partial(e_i/\rho_i)} \frac{1}{\rho_i} > 0$, $i = 1, \ldots, m$ and the positive definite diagonal matrices $a(t) = \text{diag}[a_i(t)]$ and $\vartheta T = \text{diag}[\vartheta T_i]$. Then, assuming that the arm stays away from singular positions, it is possible to define the following intermediate control signal based on [19] :

$$v_{qr} = -J^{\dagger}(q)[a(t)e + K_p \vartheta T \varepsilon - \dot{p}_r] \qquad (12)$$

where K_p is a positive diagonal gain matrix. This control signal can be regarded as an ideal joint velocity. In that respect, we define the joint velocity error $e_s = \dot{q} - v_{qr} \in \Re^n$ and formulate the following model based control law:

$$u = H(q)\dot{v}_{qr} + C(q,\dot{q})v_{qr} + G(q) - K_q e_s - J^T(q)\hat{F} \qquad (13)$$

where K_q is a positive diagonal gain matrix. Assuming measurements of the external contact force using appropriate force/torque or tactile sensors, or identification of the external contact force via a contact detection algorithm, controller (10), (11), (12), (13), applied to the robot dynamic system (1) produces smooth reaching motion and human-like compliant physical contact in redundant robotic arms. In particular, the controller guarantees that the task space error e evolves strictly within the prescribed performance bounds dictated by the designer, while all the other closed loop signals remain bounded. In response to an external contact force, task position references are shaped by displacements produced by a human-like impedance model until force withdrawal; if the reaching target is reached or abandoned, end effector motion is governed by (11), otherwise motion follows the resultant of p_e and the current distance from the target. Depending on the force direction relative to reaching, this may result in smoothly stopping, moving backwards or sideways.

Substituting (13) into (1) yields the following closed loop system :

$$H(q)\dot{e}_s + C(q,\dot{q})e_s + K_q e_s + b(t) = 0 \qquad (14)$$

where $b(t) = J^T(q)\left(\hat{F} - F\right)$ is the bounded measurement or identification error, which satisfies $\|b(t)\| \leq B \ \forall t > 0$.

To proceed with the stability proof we consider the positive definite, radially unbounded function $V = \frac{1}{2}e_s^T H(q)e_s$. By taking its time derivative along the solution of (14) and utilizing the skew symmetry of $\dot{H}(q) - 2C(q, \dot{q})$, we obtain: $\dot{V} = -e_s^T K_q e_s - e_s^T b(t)$. Let k_q denote the minimum entry of K_q; then we can write $\dot{V} \leq -k_q \|e_s\|^2 + \|e_s\|\|b(t)\|$. In case $\hat{F} = F$, then $b(t) = 0$ and $\dot{V}$ is negative definite; hence, e_s is bounded and exponentially convergent to zero. Otherwise, i.e. $\hat{F} \neq F$, we can proceed to prove the uniform ultimately boundedness (uub) of e_s. Completing the squares we get: $\dot{V} \leq -\frac{k_q}{2}\|e_s\|^2 + \frac{1}{2k_q}B^2$. Utilizing the minimum λ_m and maximum λ_M bounds of the arm's inertia matrix we obtain the uub of e_s with respect to the set: $S = \{\|e_s\| \in \Re : \|e_s\| \leq \sqrt{\frac{\lambda_M}{\lambda_m}}\frac{B}{k_q}\}$. The ultimate bound of e_s depends on the magnitude of the measurement error B and can be made smaller by increasing K_q.

Differentiating (9) with respect to time, we obtain the following transformed error dynamics:

$$\dot{\varepsilon} = \left[\frac{\vartheta T_1}{\vartheta\left(\frac{e_1}{\rho_1}\right)}\frac{d}{dt}\left(\frac{e_1}{\rho_1}\right) \cdots \frac{\vartheta T_m}{\vartheta\left(\frac{e_m}{\rho_m}\right)}\frac{d}{dt}\left(\frac{e_m}{\rho_m}\right)\right]^T$$

$$= \left[\frac{1}{\rho_1}\frac{\vartheta T_1}{\vartheta\left(\frac{e_1}{\rho_1}\right)}\left(\dot{e}_1 - e_1\frac{\dot{\rho}_1}{\rho_1}\right) \cdots \frac{1}{\rho_m}\frac{\vartheta T_m}{\vartheta\left(\frac{e_m}{\rho_m}\right)}\left(\dot{e}_m - e_m\frac{\dot{\rho}_m}{\rho_m}\right)\right]^T \quad (15)$$

$$= \vartheta T\left(\dot{e} + a(t)e\right)$$

$$= \vartheta T\left(J(q)\dot{q} - \dot{p}_r + a(t)e\right)$$

Substituting $\dot{q} = v_{qr} + e_s$ into (15) employing v_{qr} from (12) and utilizing $J(q)J^\dagger(q) = I_m$ where I_m is the identity matrix of dimension m, we obtain:

$$\dot{\varepsilon} = -\vartheta T K_p \vartheta T \varepsilon + \vartheta T U \quad (16)$$

where $U = J(q)e_s$. Notice that U is a bounded vector function because $J(q)$ is bounded as it is consisted of bounded trigonometric functions of the joint angles and e_s is proved bounded above; thus, let $\bar{U} = sup_t \|U\|$. Let us further consider the positive definite, radially unbounded function $V = \frac{1}{2}\varepsilon^T \varepsilon$. Taking its time derivative along the solutions of (16) we obtain, $\dot{V} = -\varepsilon^T \vartheta T K_p \vartheta T \varepsilon + \varepsilon^T \vartheta T U$. Let k_p denote the minimum entry of K_p; then, we can write, $\dot{V} \leq -k_p \|\vartheta T \varepsilon\|^2 + \|\vartheta T \varepsilon\|\|U\|$. Completing the squares we get:

$$\dot{V} \leq -\frac{k_p}{2}\|\vartheta T \varepsilon\|^2 + \frac{\bar{U}^2}{2k_p} \quad (17)$$

From (17) we obtain the uub of $\|\vartheta T \varepsilon\|$ with respect to the set $C = \{\|\vartheta T \varepsilon\| \in \Re : \|\vartheta T \varepsilon\| \leq \frac{\bar{U}}{k_p}\}$ and thus its uniform boundedness as well. Let $r > 0$ be an unknown upper bound. Since $\|\vartheta T \varepsilon\| \leq r$, we straightforwardly conclude that:

$$|\vartheta T_i \varepsilon_i| = |\vartheta T_i||\varepsilon_i| \leq r \ \forall \ i = 1, \ldots, m \quad (18)$$

By definition $T_i(\cdot)$ is strictly increasing, hence it is lower bounded away from zero $\frac{\partial T_i}{\partial(e_i/\rho_i)} \geq \bar{c}_i > 0$ $\forall$ $i = 1, \ldots, m$. Moreover, $0 < \rho_i(t) \leq \rho_{0i}$ $\forall$ $i = 1, \ldots, m$ and consequently $\vartheta T_i \geq \frac{\bar{c}_i}{\rho_{0i}} > 0$ $\forall$ $i = 1, \ldots, m$. Thus, from (18) we obtain the uniform boundedness of $\bar{\varepsilon}_i$ $\forall$ $i = 1, \ldots, m$ as follows: $|\varepsilon_i| \leq \frac{r\rho_{0i}}{\bar{c}_i}$ $\forall$ $i = 1, \ldots, m$. The boundedness of ε implies that the task space error will strictly evolve within the prescribed performance boundaries satisfying (3) or (4). Notice that the magnitude of this bound is not in any case involved in the achievement of the prescribed performance but it is related to the control effort bound that may, however, never be reached in practice. The boundedness of ε further implies that all the other signals in the closed loop remain bounded and in particular, ϑT, v_{qr} as implied from (12) and hence $\dot{q}$ as implied by the boundedness of e_s.

5   Simulation Results

For simulations, a 5dof arm with geometry and parameter values corresponding to the average male adult, was borrowed from [4]. It is consisted of the upper arm with double axes of rotation, the shoulder joint, the forearm, the palm and the index finger. The last four axes allow motion on a plane while the first axis allows up and down movements. The arms's length is $74.5cm$ at full extension. Its kinematic and dynamic parameters are shown in [4]. Initially, the arm is stationary at joint position $q(0) = [10 \ 22 \ 43 \ 40 \ 45]^T deg$, corresponding to the end-effector position $p(0) = [0.3408 \ 0.4218 \ 0.1652]^T m$. The target location is at $p_d = [0.0436 \ 0.6426 \ 0.0775]^T m$ corresponding to a distance of $(38.0504cm)$. The prescribed performance parameters for the task space error are preset to the values, $T = 1.5s$ reflecting the desired motion duration, $\rho_{i\infty} = 0.001m$ reflecting a desired accuracy of less than $1mm$, $M_i = 1$ while $\rho_{i0} = 2max(|e_i(0)|)$ $\forall i = 1, \ldots, 3$. Control gain values are set to $K_p = 1 \times 10^{-4}I_3$, $K_q = 25I_5$, and desired impedance matrices $L = 0.15I_3$, $D = 9.5I_3$ and $K = 148I_3$, where I_3 and I_5 are the identity matrix of dimension 3 and 5 respectively.

Simulation scenarios include the unconstrained forward movement to the target and the return to the initial position as well as the influence of an external contact force F applied to the end-effector in the perpendicular to the motion direction during the motion, i.e. at time $t = 0.5sec$ (Fig.2a). The measured force shown in Fig.2b is produced by adding white gaussian noise to F with a signal to noise ratio equal to 15. It is clear that the proposed controller is exhibiting human-like motion characteristics in unconstrained movements as shown by the straight line paths in the forward and return movements (Fig.3a), the bell-shaped Cartesian velocity profile (Fig.3b), the smooth joint velocities (Fig.4a), as well as its return close to its initial joint configuration (Fig.5). Task space errors remain within the performance boundaries (Fig.6a) and velocity errors converge to zero (Fig.7a) confirming theoretical results. Notice how the error evolution is shaped similar to the performance boundary. Results produced under the influence of an external contact force exerted during the motion (Fig.2) show that the proposed method allows a smooth compliant reaction, returning to the target after the force is removed as depicted in Fig.3a and Fig.3b. Joint velocities are

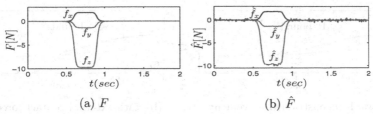

(a) F (b) $\hat{F}$

Fig. 2. The actual and measured external contact force applied to the end-effector

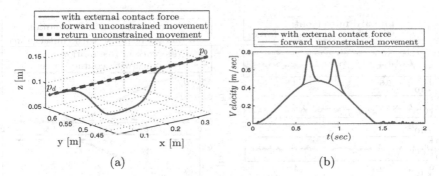

(a)   (b)

Fig. 3. (a) End-effector path, (b) Velocity profile $\|\dot{p}\|$

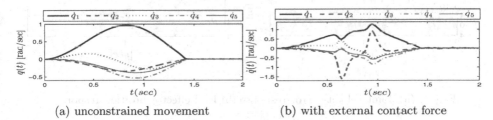

(a) unconstrained movement   (b) with external contact force

Fig. 4. Joint velocities

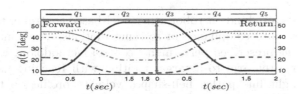

Fig. 5. Joint trajectories during a return trip

in this case shown in Fig.4b and task space error evolution in Fig.6b; the latter is similar to the unconstrained case. Joint velocity error responses depicted in Fig.7a reflect to an extent the noisy force measurements. Reference and actual end effector trajectories are shown in Fig.7b for both cases of unconstrained movement and under the external force. Notice how the existence of the external force shapes the reference position trajectory for the end effector.

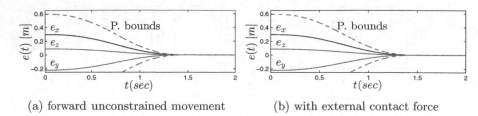

(a) forward unconstrained movement   (b) with external contact force

Fig. 6. Task space error coordinate response

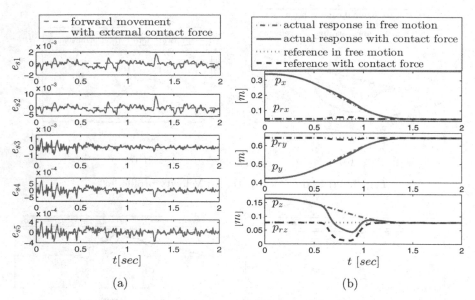

(a)   (b)

Fig. 7. (a) Joint velocity error response (b) End effector position response

6   Conclusions

This work proposes a redundant robot arm model-based controller that achieves human-like motion and compliance in reaching motions. The controller is designed following the prescribed performance control methodology and is proved to practically drive the arm to a target position with reaching motions characterized by straight line paths, bell-shaped velocity profiles, path and joint configuration consistency in return movements as well as compliance to external contact forces. Simulations of a 5dof arm using the main structure and data from a human arm demonstrate human-like reaching motions being consistent with theoretical findings.

References

[1] Morasso, P.: Spatial control of arm movements. Experimental Brain Research 42(2), 223–227 (1981)

[2] Bizzi, E., Accornero, N., Chapple, W., Hogan, N.: Posture control and trajectory formation during arm movement. The Journal of Neuroscience 4(11), 2738–2744 (1984)

[3] Flash, T., Hogan, N.: The coordination of arm movements: An experimentally confirmed mathematical model. Journal of Neuroscience 5(7), 1688–1703 (1985)

[4] Arimoto, S., Sekimoto, M.: Human-like movements of robotic arms with redundant DOF: virtual spring-damper hypothesis to tackle the bernstein problem. In: Proc. IEEE International Conference in Robotics and Automation, pp. 1860–1866 (2006)

[5] Sciavicco, L., Siciliano, B.: A solution algorithm to the inverse kinematic problem for redundant manipulatotrs. IEEE Journal of Robotics and Automation 4(4), 403–410 (1988)

[6] Hersch, M., Billard, A.: Reaching with multi-referential dynamical systems. Auton. Robot. 25(1-2), 71–83 (2008)

[7] Lacquaniti, F., Soechting, J., Terzuolo, S.: Path constraints on point to point arm movements in three dimensional space. Neuroscience 17(2), 313–324 (1986)

[8] Desmurget, M., Pelisson, D., Rosseti, Y., Prablanc, C.: From eye to hand: Planning goal-directed movements. Neuroscience and Biobehavioural Reviews 22(6), 761–788 (1998)

[9] Nakamura, Y., Hanafusa, H., Yoshikawa, T.: Task-priority based redundancy control of robot manipulators. The International Journal of Robotics Research 6(2), 3–15 (1987)

[10] Nakanishi, J., Cory, R., Mistry, M., Peters, J., Schaal, S.: Comparative experiments on task space control with redundancy resolution. In: IEEE International Conference on Intelligent Robots and Systems, pp. 3901–3908. IEEE (2005)

[11] Zanchettin, A., Rocco, P., Bascetta, L., Symeonidis, I., Peldschus, S.: Kinematic analysis and synthesis of the human arm motion during a manipulation task. In: IEEE International Conference on Robotics and Automation (ICRA), pp. 2692–2697 (2011)

[12] Sekimoto, M., Arimoto, S.: Experimental study on reaching movements of robot arms with redundant dofs based upon virtual springdamper hypothesis. In: Proceedings of the 2006 IEEE/RSJ International Conference on Intelligent Robots and Systems, pp. 562–567 (2006)

[13] Seto, F., Sugihara, T.: Motion control with slow and rapid adaptation for smooth reaching movement under external force disturbance. In: Proceedings of 2010 IEEE/RSJ International Conference on Intelligent Robots and Systems, pp. 1650–1655 (2010)

[14] Heinzmann, J., Zelinsky, A.: Quantitative safety guarantees for physical human-robot interaction. Int. J. of Robotics Research 22(7/8), 479–504 (2003)

[15] De Luca, A., Albu-Schaffer, A., Haddadin, S., Hirzinger, G.: Collision detection and safe reaction with DLF-III lightweight manipulator arm. In: Proc. of IEEE/RSJ International Conference on Intelligent Robots and Systems (IROS), pp. 1623–1630 (2006)

[16] Erden, M.S., Tomiyama, T.: Human-intent detection and physically interactive control of a robot without force sensors. IEEE Transactions on Robotics 26(2), 370–382 (2010)

[17] Yang, C., Ganesh, G., Haddadin, S., Parusel, S., Albu-Schaeffer, A., Burdet, E.: Human-like adaptation of force and impedance in stable and unstable interactions. IEEE Transactions on Robotics 27(5), 918–930 (2011)

[18] Bechlioulis, C., Rovithakis, G.: Robust adaptive control of feedback linearizable MIMO nonlinear systems with prescribed performance. IEEE Transactions on Automatic Control 53(9), 2090–2099 (2008)

[19] Doulgeri, Z., Karalis, P.: A prescribed performance referential control for human-like reaching movement of redundant arms. In: Proc. 10th Intern. IFAC Symposium on Robot Control, pp. 295–300 (2012)

[20] Karayiannidis, Y., Doulgeri, Z.: Model-free robot joint position regulation and tracking with prescribed performance guarantees. Robotics and Autonomous Systems 60(2), 214–226 (2012)

[21] Bechlioulis, C., Rovithakis, G.: Prescribed performance adaptive control for multi-input multi-output affine in the control nonlinear systems. IEEE Transactions on Automatic Control 55(5), 1220–1226 (2010)

Recognition and Representation of Robot Skills in Real Time: A Theoretical Analysis

Wei Wang, Benjamin Johnston, and Mary-Anne Williams

Centre for Quantum Computation & Intelligent Systems,
University of Technology, Sydney, Australia
Wei.Wang-10@student.uts.edu.au,
{benjamin.johnston,Mary-Anne.Williams}@uts.edu.au

Abstract. Sharing reusable knowledge among robots has the potential to sustainably develop robot skills. The bottlenecks to sharing robot skills across a network are how to recognise and represent reusable robot skills in real-time and how to define reusable robot skills in a way that facilitates the recognition and representation challenge. In this paper, we first analyse the considerations to categorise reusable robot skills that manipulate objects derived from R.C. Schank's script representation of human basic motion, and define three types of reusable robot skills on the basis of the analysis. Then, we propose a method with potential to identify robot skills in real-time. We present a theoretical process of skills recognition during task performance. Finally, we characterise reusable robot skill based on new definitions and explain how the new proposed representation of robot skill is potentially advantageous over current state-of-the-art work.

Keywords: Human-Robot Interaction, Robot-Robot Interaction, Social networking, Internet of Things, Robots.

1   Introduction

Social robots are desired to interact with people and assist people to perform diverse tasks with little oversight in the future [1]. This requires social robots to be equipped with a certain level of autonomy and be capable of developing themselves during lifelong time [1] [2].

In order to help robots achieve autonomy, researchers have been trying to find better approaches to developing robot skills than traditional means like hand coding and human demonstration. A number of attempts have been made in terms of defining primitive actions and their reuse to benefit new task performance. In RoboEarth, reusable knowledge is created as an action recipe in which all parameters are processed manually to regenerate a new action recipe according to the semantics defined [3]. ESS (Experimental State Splitting) is an algorithm defined on an action schema from which agents are able to learn and transfer knowledge from one situation to another [4]. CST (Constructing Skill Trees) provides a way to regenerate new skills in new but similar tasks by reusing predefined primitive skills [5]. QLAP (Qualitative Learner of Actions and Perception) applies qualitative representations for primitive actions and uses a dynamic Bayesian network to train the optimal course of actions in different situations

G. Herrmann et al. (Eds.): ICSR 2013, LNAI 8239, pp. 127–137, 2013.

[6]. Task Transfer generates new tasks by using numerical primitive actions [7]. Among these approaches, ESS is built on Piaget's Cognitive Theory [8], and CST, QLAP and Task Transfer are all developed based on Autonomous Mental Development theory [2].

Inspired by online social networking in human society and taking advantage of the Internet, we are attempting to create a social network for robots that will allow them to share and reuse skills on their own in lifelong learning development. Since existing methods have not provided clear answers to the questions like what kind of skills can be reused and how to represent these skills can facilitate better skill acquisition and transfer. Hence, a challenge here is to recognise reusable robot skills in real-time autonomously [5]. This challenge firstly requires a resolution of *identifying reusable robot skills in the way that suits for real-time recognition and sharing across the network by robots, and finding out a way to allow robot to understand the starting points to apply these skills and the terminations to finish.*

The emphasis in this research is placed on sharing robot skills to manipulate physical objects motivated by R.C.Schank's script framework [9], and this paper will answer the following three questions:

- What are reusable robot skills?
- How can these reusable skills be recognised in real-time?
- How can these skills be characterised for future reuse?

The rest of this paper is organised as follows. We first introduce our understanding of reusable robot skills in Section 2 in order to define reusable robot skills with three types and the features of robot skills in this research in Section 3. Then an explicit demonstration of the process to recognise reusable robot skills in real-time is provided in Section 4. After that, in Section 5, we introduce the representations for each type of reusable robot skills and highlight its edges over CST, the state-of-the-art work regarding autonomous skill acquisition via primitive skill abstraction reuse. Finally, we summarise and point out future work.

2   Understanding Reusable Robot Skills

To understand reusable robot skills, we first review Schank's framework, in which it provides a consideration of what kind of skills is of high demand for a robot to develop as they are important for human. In this framework, 11 basic human motions are selected and considered as fundamental blocks to build more complex human motions, namely (1)PROPEL, (2)GRASP, (3)MOVE, (4)ATRANS, (5)PTRANS, (6)ATTEND, (7)INGEST, (8)EXPEL, (9)SPEAK, (10)MTRANS, (11)MBUILD [9]. An important category of these motions is to interact with physical object (the first five basic motions). Therefore, we choose robot skills to interact with physical objects as our focus.

A robot skill is considered to be a set of actions consisting of a series of motion primitives [12]. For example, a robot skill "open a door" consists of a few sub motions including navigation, approaching, pushing the handle and passing. These sub motions are usually repeatedly applied in other tasks with specific adjustments. For instance, navigation can be not only applied in opening a door, but also applied in the case when a robot tries to find lift; approaching can be reused in the situations where a robot needs to get closer to the object it interacts with and so forth. We call such skills are *reusable robot skills*. However, it is unrealistic to enumerate all reusable skills for robots to learn. A better way is to extract those reusable skills into skill abstractions so that skill instances suitable for a particular situation can be customised [10] [11].

2.1   The Changing Relationship between Agent and Target

Since the robot skills we focus on occur during the interaction between the robot and the physical object, understanding this changing relationship between agent and object will help figure out the reason and consequences of a particular robot skill and in turn find a better way to characterise these skills (See Figure 1).

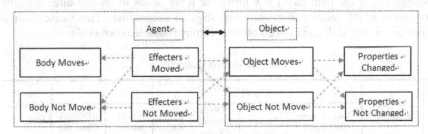

Fig. 1. The relationship between agent and object during interaction

In Figure 1, **arrows** point to the consequences that result from the part where arrows originally come. For instance, *"body not move"* is the consequence because of either *"effecters move"* or *"effecters not move"*; agent's *"body moves"* will result in either *"object moves"* or *"object not move"*, and in turn probably result in the changes of the *object's properties*.   The meaning of each item appeared in Figure 1 are briefly explained below:

- *Body Moves/Not Move*:   refers to whether or not the robot body changes position (corresponding to translation) or direction (corresponding to rotation).
- *Object Moves/Not Move*:   refers to whether or not the physical object changes position (corresponding to translation) or direction (corresponding to rotation).
- *Properties Changed/Not Changed:*   The properties of a physical object refer to its unique features including color, texture, shape and mass and so on.
- *Effecters Moved/Not Move*:   effecters defined in each robot are usually a set of joints forming a part of limb of the robot body such as left arm or right leg. For example, a humanoid robot often has the following effecters: Head, Left Arm, Right Arm, Left Leg, Right Leg and the Torso.   "Effecters Moved" refers to some outstanding changes happened with respect to position (corresponding to translation) or direction (corresponding to rotation) of the joints belonging to an effecter. "Effecters Not Move" refers to no outstanding translation or rotation movements happened on any effecters of the robot body.

2.2   Considerations to Categorise Reusable Robot Skills

In order to find out a reasonable way to define reusable robot skills, we surveyed 55 human motion that robot can do and pick up 18 representatives to analyse the role that object plays in terms of identifying reusable robot skills (see Table 1).

We mainly concern two situations: (1) *if the motion affects object or not,* and (2) *if the motion requires object or not.* Since it is agent-object interaction, there must be at least one object engaged in the task. Nevertheless, sometime the object is only a reference, skills applied do not directly work on the object, like navigation and waving hand to people; some robot skills require touching the object and exerting influence on the object, such as grasp, open the door and cutting.

Another issue we concern by digging into this table is that, although these motions correspond to atomic human motion defined in Schank's framework, most of them are not atomic robot skills as they usually require more than one stage to complete. Typical examples include "open the door" and "get object from fridge". Even simpler motions like Navigation, if the path has a few turns, the robot needs to change direction along with the turns which require more than one stage to accomplish. These cases indicate that a reusable robot skill can contain other even tiny reusable robot skills.

Table 1. Motion Examples

| | Motion Example | Object | | | | Agent | | Type in Schank's Framework [9] |
		Require/work on Object	Object Moves /Stays	Object Properties Changed	Affect Object	Body Moves	Effectors Changed	
1	Navigation	Yes/No	No	No	No	Yes	Yes	3,5
2	Walking	Yes/No	No	No	No	Yes	Yes	3,5
3	Raising Arm	Yes/No	No	No	No	No	Yes	3
4	Wave hand	Yes/No	No	No	No	No	Yes	3
5	Dancing	No/No	No	No	No	Yes/No	Yes	3,5
6	Shake hand	Yes/Yes	Yes	No	Yes	No	Yes	2,3
7	give an object to someone	Yes/Yes	Yes	No	Yes	Yes/No	Yes	4
8	cutting	Yes/Yes	Yes	Yes	Yes	Yes/No	Yes	1,2,3
9	shake object	Yes/Yes	Yes	No	Yes	No	Yes	2,3
10	driving	Yes/Yes	Yes	No	Yes	No	Yes	2,3
11	open the door	Yes/Yes	Yes	No	Yes	Yes	Yes	1,2,3,5
12	get object from fridge	Yes/Yes	Yes	No	Yes	Yes	Yes	2,3,5
13	pull	Yes/Yes	Yes	No	Yes	Yes	Yes	1
14	grasp	Yes/Yes	Yes	No	Yes	Yes	Yes	2
15	carry object	Yes/Yes	Yes	No	Yes	Yes	Yes	5
16	Painting wall	Yes/Yes	Yes/No	Yes/No	Yes	Yes	Yes	1,2,3,5
17	looking around	Yes/No	No	No	No	No	Yes	3
18	turn over	Yes/No	No	No	No	Yes	Yes	3

3   Reusable Robot Skills Definition

Based on the findings above, in this section we define the target an agent operates on, three types of reusable robot skills, and the robot skills this research focuses on.

Definition 0: Target.   *A target is a physical existence in the environment that the agent is going to interact with or operate on. It could be alive like an animal or people, or not alive like a physical object.*

Definition 1: Innate Skill.   *Innate skills are those abilities each robot originally has that are 6 fundamental motions of transition and rotation. Innate skill is the atomic block of any robot skills of any complex level.*

Definition 2: Basic Skill.   *Basic skill is a functional block consisting of a set of innate skills with the ability to exert certain effect on or respond to the working environment. It is the fundamentally functional component of a task.*

According to the effects that the agent exerts on the target, basic skill is categorised into two types:

Definition 2.1: Non target-oriented Basic Skill (NTBS).   *This type of basic skill does not operate on the target with the purpose of not causing any change of the target, such as navigation, raise arm, wave hand to people, or pointing.*

Definition 2.2: Target-oriented Basic Skill (TBS).   *This type of basic skill directly operates on the target with the purpose of causing status or property changes of the target, such as grasp, push, lift.*

The differences between these two types of basic skills is that target-oriented basic skill could consist of more than one non target-oriented basic skills but non-target-oriented basic skill could not contain any target-oriented basic skills. In CST, basic skill defined here is called macro-action, skill abstraction or option [5] [11].

Definition 3: Task-level Skill.   *A task is a process to reach the goal by applying more than one basic skill to realize.* For example, to open a door, the robot needs to apply basic skills including navigation, approaching, and pushing to accomplish.

To redefine robot skills into innate skill, NTBS and TBS basic skill and task-level skill has threefold consideration. The first one is to facilitate agent to realize the start and the change points during task performance without too much predefinition [5] [11]. The second is the consideration of granularity of skills in order to benefit for skill discovery, selection, decomposition and reconstruction [5] [12]. The third one is to avoid using too much human verbs to define every step of robot task performance as the only way to make robots reach to real autonomy is not through symbolic method but more robot-understandable methods [7] [13]. The usage of these definitions will be provided in Section 4 via a real example.

On top of three reusable robot skills definitions, we now scope the robot skills we are going to focus on in this research by Definition 4 below.

Definition 4: Robot Skill. Robot skills are performed to achieve position or status change of a physical object. Such skills are assumed to satisfy following features.

Assumption 4.1: Working Environments. *The skills are performed in high-dimensional and continuous domains rather than low-level and discrete domain. "High-dimensional" means robot will be required to take more than one time-step to finish, and "continuous" means any state in a task will not appear twice* [11].

Assumption 4.2: Tasks. *In any given task, there must be at least one physical object engaged in the environment. The robot will perform skills on only one physical object at a time. The physical object starts from steady status.*

Assumption 4.3: Type of Engaged Robot Skills. *Robot skills in this research will be recognised in the form of innate skills, basic skills and task-level skills.*

Assumption 4.4: Motivation to Trigger Skills. *Robot skill is triggered by visual features of the object rather than verbs or any symbolic forms.*

Assumption 4.5: Reusability. *A skill applied in one task can be applied in different tasks occurring in different contexts due to the similarity.*

Assumption 4.6:   Trajectory. *Any skill has a routine way to realize.*

Assumption 4.7: Gravity. *All skills occurred are assumed to be completed without agent's body falling down.*

Based on these definitions, we are ready to introduce how to recognise major stages and the applied robot skills in real-time in the next section.

4   Robot Skill Recognition

4.1   Recognising Major Stages of a Task in Real Time

According to Figure 1, we obtain four parameters as the identifications to recognise basic skills and they are "Effectors Move", "Body Moves", "Object Moves" and "Property Changed". Possible situations during the interaction between agent and object can be summarized using Table 2, in which, 0 stands for no movement or changes happened, 1 stands for either movement or changes happened.

One point required to be noticed is the recognition of the change of basic skills (NTBS and TBS). Since different basic skills have dissimilar sets of effecters and joints. Therefore, if the set of effecters (sometime we also consider the most engaged joints belonging to them if the joints are more than one) do not change, then we consider the current active basic skills keep the same. To recognise the most engaged effecters and joints are realised by making use of Temporal Theory [14].

In order to allow agents to recognise main steps during performing a task, three fundamental rules need to be defined:

- The object must be in the field of view of the robot first [11].
- Before interacting with the object, robot body has to face towards the object. In that case, the heading and orientation of robot need to adapt as required [11].

Table 2. Possible Situations during the Interaction between Agent and Object

	Effectors Move	Body Moves	Object Moves	Property Changes	Type of Skills
1	0	0	0	0	No skills happening /All skills finished.
2	0	0	0	1	TBS finished
3	0	0	1	0	TBS finished
4	0	0	1	1	TBS finished
5	1	0	0	0	NTBS happening
6	1	0	0	1	TBS happening
7	1	0	1	0	TBS happening
8	1	0	1	1	TBS happening
9	1	1	0	0	NTBS happening
10	1	1	0	1	TBS happening
11	1	1	1	0	TBS happening
12	1	1	1	1	TBS happening

- In order to find the optimal touch point to operate on the object, the distance between the endpoint of the robot's effecter and the centre of the object must be zero. This will be for finding optimal policy during learning.

4.2 A Case Study

We now take PR2 robot pushing down a colour box as an example to theoretically illustrate how a robot can recognise major steps in real-time performance using our method. The scenario is designed as follows: a PR2 robot locates half meter far to a table on which a color box with yellow front and red bottom stands. In order to reach to the box, PR2 has to firstly find the box and then approach to it, finally exert force on it to push it down on the table (red bottom is seen). In this case, we identify major stages of the performance by integrating temporal theory and the changing relationship between robot and object. These two functional modules will run simultaneously during the task performance.

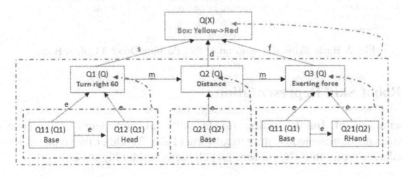

Fig. 2. The Recognition of Effects and Major Effecters during This Process

We apply temporal theory to figure out the most engaged effecters in order to capture change points of the performance. We start by monitoring joint-level changes that are innate skills, and analysing the most engaged effecters and joints and their orders of occurrence, as normally the change of the most engaged effecters means the whole performance moves forward into the next step. For example, in Figure 2, we first capture the most engaged effecters that are Base with four wheels and its Head (see Q11 and Q12). Shortly, the effecter "Head" stops changing and only "Base" is moving (see Q21), which identifies the performance is moving to the second stage. The analysis will stop when the performance satisfies the requirements defined by the task. In this way, we obtain all sub processes Q1, Q2 and Q3 according to the temporal order of their occurrences [14]. Then the parameter Q of each sub process as the condition to trigger Q1, Q2 and Q3 to happen will be calculated and marked. At the end, the final effect will be assigned to parameter X of Q as the condition to trigger Q (the whole task) to happen.

The role of monitoring the changing relationship between the agent and the object is to identify the type of skill applied. As we can see in Figure 3, there are three major stages which correspond to Qi, i=1, 2, 3, recognised in Figure 2. According to Table1, we can see Q1 is recognised as a NTBS as there is neither object movement nor change of object's properties. Q2 is same to Q1, but Q3 is different from Q1 and Q2 as the object moved and changed afterwards. Thereby, it is recognised as a TBS.

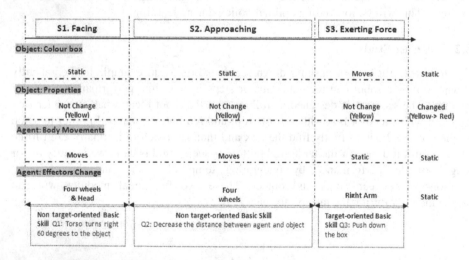

Fig. 3. Basic Skills Recognition of PR2 Pushing Down a Colour Box

5   Robot Skill Representation

Representation of Innate Skills. Innate skills are considered as commonsense knowledge and represented by numerical form (Table 3). Other two types of reusable robot skills are represented by options and characterised by Figure 4.

Table 3. Innate Skills Representations

Category		Representation	Meaning
Translation	Moving up and down	X, Y, Z' = Zi + ΔZ, p, yaw, r	ΔZ > 0, up, Z' > Zi ΔZ< 0, Down, Z' < Zi
	Moving left and right	X, Y' = Yi + ΔY, Z, p, yaw, r	ΔY > 0, right, Y' > Yi ΔY < 0, left, Y' < Yi
	Moving forward and backward	X' = Xi + ΔX, Y, Z, p, yaw, r	ΔX > 0, forward, X' > Xi ΔX < 0, backward, X'< Xi
Rotation	Tilting forward and backward (pitching)	X, Y, Z, P' = Pi +ΔP, yaw, r	ΔP > 0, up, P' > Pi ΔP <0, down, P' < Pi
	Turning left and right (yawing)	X, Y, Z, P, yaw' = yaw + Δyaw, r	Δyaw >0, right, yaw' > yaw Δyaw <0, left, yaw' < yaw
	Tilting side to side (rolling)	X, Y, Z, P, yaw, r' = ri +Δr	Δr > 0, inside, r' > ri Δr< 0, outside, r' <ri

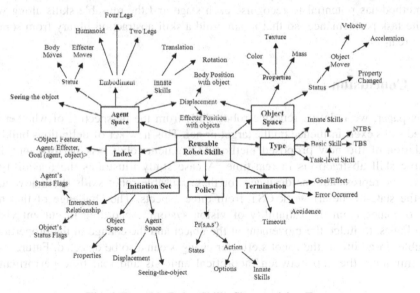

Fig. 4. Reusable Robot Skills Characterising Tree

Representation of Basic Skills and Task-level Skills. These two types of reusable skills are represented by options in the form of *(initiation sets, policy, termination)* [12]. In this case, option framework will be applied in order to construct selection strategy [11][12]. In order to represent customised option instance for searching purpose, we create an index for each generated option instance using a tuple *<object.feature, agent. effecters, goal(agent, object)>*. Basic-level options with same object.feature/goal will be chained into a task-level option instance.

Finally, we compare with CST to show the potential **advantages** of our method:

(1) **Minimizing the Usage of Human Language.** In CST, primitive skill is defined as skill abstraction which is represented by a pair of robot's effecter and the name

of the object such as <body-button> [11]. Since all elements in the proposed representation can be real-time detected and mostly represented using numerical parameters facilitating both robots and people to understand, thus it can reduce the usage of human words to define skills.

(2) **Reducing Ambiguity.** Solving-problem based Skill abstractions in CST usually have same representations but different meanings [11]. For example, one skill abstraction named <body-button> has two descriptions, one is "align with the button" and the other is "Turn and approach the button" [11]. Our method identifies skills using three distinct perspectives which are not prone to have same representations. For instance, same feature might have dissimilar goals and same goals might require different effecters to realise and vice versa. Thus it can reduce the chance of ambiguity.

(3) **Constructing Skill Abstraction Library from Scratch [11].** CST creates skill abstraction library by analysing a demonstrated task from the end to the beginning in order to prepare a skill abstraction library for reuse [11]. Our method has potential to recognise each stage and the reusable skills along with the task performance, so that it can build a skill abstraction library from scratch in real-time.

6   Conclusion

In this paper, we redefine reusable robot skills from the perspective of whether the applied skills exert influence on the target or not. This new set of definitions build the foundation of the new proposed method to segment robot task performance into reusable skill abstractions in real-time. A case study illustrates the feasibility in theory. The representations designed for these reusable robot skills show advantage over the state-of-the-art work CST from three aspects. The challenge of this new method comes from the reliability of vision system. According to current vision technologies, to judge the movement of the object and the change of the properties is available. In addition, the robot's effecter changes can also be detected. Future work is to minimize the gap between theoretical analysis and real time performance.

References

1. Castro-González, Á., Malfaz, M., Salichs, M.A.: Learning the Selection of Actions for an Autonomous Social Robot by Reinforcement Learning Based on Motivations. International Journal of Social Robotics 3(4), 427–441 (2011)
2. Weng, J., McClelland, J., Pentland, A., Sporns, O., Stockman, I., Sur, M., Thelen, E.: Autonomous Mental Development by Robots and Animals. Science 291, 599–600 (2001)
3. Tenorth, M., Perzylo, A., Lafrenz, R., Beetz, M.: The RoboEarth Language: Representing and Exchanging Knowledge about Action, Objects and Environments. In: ICRA, pp. 1284–1289 (2012)
4. Cohen, P.R., Chang, Y.-H., Morrison, C.T.: Learning and Transferring Action Schemas. In: IJCAI, pp. 720–725 (2007)

5. Konidaris, G.D., Kuindersma, S.R., Grupen, R.A., Barto, A.G.: Robot Learning from Demonstration by Constructing Skill Trees. International Journal of Robotics Research 31(3), 360–375 (2012)
6. Mugan, J., Kuipers, B.: Autonomous Learning of High-Level States and Actions in Continuous Environments. IEEE Transactions on Autonomous Mental Development 4(1), 70–86 (2012)
7. Zhang, Y.L., Weng, J.Y.: Task Transfer by a Developmental Robot. IEEE Transactions on Evolutionary Computation 11(2), 226–248 (2007)
8. Piaget, J.: The Construction of Reality in the Child, New York (1954)
9. Schank, R., Abelson, R.P.: Scripts, Plans, Goals and Understanding: An Inquiry into Human Knowledge Structures. Erlbaum (1977)
10. Wiering, M., Otterlo, M.V.: Transfer in Reinforcement Learning Reinforcement Learning State-of-the-Art., ch. 5. Springer, Berlin (2012)
11. Konidaris, G.D., Kuindersma, S.R., Grupen, R.A., Barto, A.G.: Autonomous Skill Acquisition on a Mobile Manipulator. In: AAAI, pp. 1468–1473 (2011)
12. Sutton, R.S., Precup, D., Singh, S.: Between MDPs and semi-MDPs: A Framework for Temporal Abstraction in Reinforcement Learning. Artificial Intelligence 112(1-2), 181–211 (1999)
13. Nicolescu, M.N., Matari, M.J.: A Hierarchical Architecture for Behavior-Based Robots. In: Proceedings of the First International Joint Conference on Autonomous Agents and Multi-agent Systems: Part 1. ACM, Bologna (2002)
14. Allen, J.F.: Maintaining Knowledge about Temporal Intervals. Communications of the ACM 26(11), 832–843 (1983)

Robots in Time:
How User Experience in Human-Robot Interaction Changes over Time

Roland Buchner, Daniela Wurhofer, Astrid Weiss, and Manfred Tscheligi

HCI& Usability Unit, ICT&S Center, University of Salzburg, Austria
firstname.lastname@sbg.ac.at

Abstract. This paper describes a User Experience (UX) study on industrial robots in the context of a semiconductor factory cleanroom. We accompanied the deployment of a new robotic arm, without a safety fence, over one and a half years. Within our study, we explored if there is a UX difference between robots which have been used for more than 10 years within a safety fence (type A robot) and a newly deployed robot without fence (type B robot). Further, we investigated if the UX ratings change over time. The departments of interest were the oven (type A robots), the etching (type B robot), and the implantation department (type B robot). To observe experience changes over time, a UX questionnaire was developed and distributed to the operators at three defined points in time within these departments. The first survey was conducted one week after the deployment of robot B (n=23), the second survey was deployed six months later (n=21), and the third survey was distributed one and a half years later (n=23). Our results show an increasing positive UX towards the newly deployed robots with progressing time, which partly aligns with the UX ratings of the robots in safety fences. However, this effect seems to fade after one year. We further found that the UX ratings for all scales for the established robots were stable at all three points in time.

Keywords: Industrial Robots, Measurement, Semiconductor Factory, User Experience.

1 Introduction

For effective and highly productive industrial manufacturing, robots have already shown their usefulness in many sectors of production. With that kind of automation, a vast, cheap, and fast production has become reality. However, most of these systems are placed within a safety fence. During production, no human is allowed to enter the working space of the robot and therefore restricting access, any interaction, and/or cooperation with the robot. However, there are claims that more powerful human-robot interaction with the human and the robot working as a team is needed in order to be highly competitive [1]. That means it is necessary to break the general known paradigm of strictly separating

G. Herrmann et al. (Eds.): ICSR 2013, LNAI 8239, pp. 138–147, 2013.

the workplace into a human and a robotic area by creating a shared work space, where the human and the robot can benefit from each others strengths. A shared workspace between operator and robot strongly influences the social structure in the factory and thus makes social robotics a key issue in the factory. As soon as interaction between human and robot comes into play, e.g. turn-taking tasks, User Experience (UX) becomes an important issue [2]. Alben [3] particularly points out the relevance of UX research in all areas in which humans interact with a system. This also includes the context of a semiconductor factory [4]. Karapanos et al. [5] report in their work on prolonged interaction with mobile phones that UX factors are not static, but change over time. Our aim is to extend research in this area regarding other contexts and the effects of time on specific UX factors. In particular, our research objective was to *investigate the adoption process of robots in the factory. We therefore monitored the temporal change of an operator's experience regarding cooperation, perceived safety, perceived usability, stress, and general UX (i.e., the overall positive/negative experience of the robotic system) when interacting with a robot.* Research in the context of a semiconductor factory is a difficult endeavor. Researchers are facing several challenges (e.g. meeting cleanroom standards, different shift cycles, ambient noise etc.) whereas entering the cleanroom for studies at pre-defined points in time is one of the biggest (as it heavily depends on the order situation).

Within this paper, we point out how we studied long-time experience of human-robot interactions in a semiconductor factory cleanroom.

We first describe the context in which the study was carried out in more detail. Then, we present the state of the art on UX measuring over time in general and UX research in IIRI specifically. Next, we describe how the questionnaire was compiled, the study procedure, and its results. In the closing section we discuss our insights and point out relevant future work on UX over time in human-robot interaction.

2   Research Context Factory

Researching the context of a semiconductor factory is very challenging. Integrated circuits on silicon wafers have to be highly reliable and manufactured fast, and accurately. Wafers are put into so-called lot boxes which contain up to 50 wafers. The production cycle in a semiconductor factory is 24/7/365 meaning that it is produced throughout the whole year. Therefore, the work is organized in three shifts, each eight hours (morning shift 06:00-14:00; afternoon shift: 14:00-22:00; night shift: 22:00-06:00). The integral part of the wafer production is the cleanroom. Almost all dust needs to be eliminated as wafers are an extremely sensitive and expensive good which can be easily damaged through the smallest dust particle. These lot boxes are usually transported by operators from equipment X to equipment Y to be processed in their working area.

Within our study, we addressed two types of robots. Type A robot (Fig. 1A) is placed within working cells, not allowing any direct interaction with operators. This kind of robot is established in the oven department for more than 10 years.

Type B robot (Fig. 1B) acts on rails and allows direct interaction with operators if needed (e.g., if robot obstructs an equipment item, the operator can move the robot using its touch interface). This type of robot has been introduced recently in the etching and in the implantation department.

The operators deliver lot boxes to both types of robots for further processing of the wafers. Both types of robots are pre-programmed and act autonomously. In the oven department, robot A takes out single wafers to add a new layer on them. This is done by coating a light-sensitive liquid on the wafer and then heatening it in the oven to harden the new layer. In the etching department, robot B takes out the wafers from the lot box and puts them into the equipment to remove (etch) unwanted material from the wafer. In the implantation department, robot B takes out the wafers from the lot box and puts it in the equipment to change the electrical characteristics of a wafer by firing energized ions into special areas of it to create electro-conductive and non-conductive layers. After the processing of the wafers is finished, the robots put the processed wafers back in the lot boxes. In contrast to the task of robot B, which was formerly performed by operators, the task of robot A cannot be performed by human operators as it needs highly accurate and precise movements.

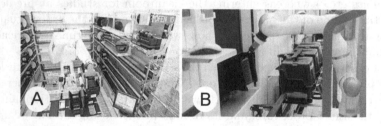

Fig. 1. Robot A is operated within a safety fence (oven department) whereas robot B is operated without (etching and implantation department)

3   Related Work

Assuming that humans and robots will work increasingly closer together in a factory context, it is necessary to research how the interaction is experienced by workers and how these interactions are influenced by temporal aspects [1]. Recent research in the field of Human-Computer Interaction (HCI) has already shown that it is important to know how UX changes over time.

Karapanos et al. [5] found that when users interact with mobile phones over a longer period of time, their experience regarding the mobile phone changes with progressing time. Their study resulted in a UX framework explicitly taking time into account. The framework consists of three phases: the orientation phase, the incorporation phase, and the identification phase. Karapanos [6] tried to explain the differences between initial and prolonged experiences in terms of the way users form overall evaluative judgments about a product. Another temporal approach on UX was developed by Pohlmeyer et al. [7]. They proposed their ContinUX framework which spans the whole UX lifecycle. The model takes the users' memories into account as long as the UX continues and the memories last. With prolonged interaction, a relationship between the user and the system will be created, emphasizing more on the personal experience over time [8] and the factor usability is becoming more important [9].

Regarding human-robot interaction, we are not aware of much research focusing on temporal aspects. As claimed by Dautenhahn [10], there is a need to investigate this aspect in more detail. When studying HRI over a longer period of time in the field, the working context offers one important advantage: certified robotic products that can be safely used on an everyday basis.

Even investigations of service robots in the field over a longer period became more frequent, e.g., in homes [11], schools [12], and health care settings [13] [14]. In such studies, it was reported that the users' attitude towards the robot changed over time, e.g., became more accepted over time [13] [11]. Leea and Forlizzi [15] showed that personalized services which are based on the users' service usage and interactions with the robot could improve rapport, cooperation, and engagement with the robot. It was found that new social norms and practices were developed over time by study participants when interacting with the robot [16]; participants became used to the robot and let it come closer than at the beginning of the study [12]. It is important to keep the users' interest [17], to match the functionalities of the robot to its appearance [18], and to give feedback at the right time [19]. As Karapanos et al. [5] pointed out, usability becomes more important over time; studies in HRI have also shown that it is crucial to consider usability aspects [14] for long-term human-robot interaction and that the interaction with the robot should not increase the general workload of the user [20]. Longitudinal studies can help to gain a better understanding of how certain experience factors, such as user acceptance and the feeling of the robot as social actor, change due to the interaction. However, the previously mentioned studies did not explicitly aim to measure single UX factors and were mostly carried out with social robots; such factors which were identified may also be relevant for the factory context.

Several questionnaires already exist to measure the UX of an interactive system, such as the Attrakdiff [21] and the UX with robots [22] [23]. These questionnaires use a semantic differential to measure how the system is perceived between two contrasting items (e.g., pleasant - unpleasant). We argue that for measuring specific UX factors, items which have to be rated on a Likert-scale are more promising as this provides more detailed information on the item addressed. Similarly, several tools in HCI already exist to measure how UX changes over time (e.g. DrawUX [24], iScale [25], UX Curve [26]). However, we have not identified any applications to HRI scenarios. In our research context, an item-based survey rated on a Likert-scale was considered as the easiest and least obtrusive way to gather data on UX changes over time in the cleanroom setting. We did not want to disturb the production cycle in the factory and to distract the operators to much from their routine.

4   Method and Procedure

Based on an existing UX questionnaire designed for measuring specific UX factors in HRI [27], we developed a survey adapted for measuring the UX over time in the factory context. The questionnaire is intended to cover different aspects of UX and consists of five scales (consisting of x-y items each). *Cooperation* addresses differences in the collaboration between the already established and the new robot. *Perceived safety* addresses workers' concerns regarding safety related robots in working cells versus new robots without safety fence. *Perceived usability* relates to how effective, efficient, and satisfying workers perceive the interaction with the robot. *Stress* indicates how much workload the interaction with the robot requires from the workers. *General UX* collects the overall positive/negative experience of the robotic system.

The final questionnaire included demographic data and 25 items (5-point Likert scale: 1-totally agree; 5-totally disagree). The internal consistency (Cronbach's Alpha) of the scales was computed for each of the three times the questionnaire was handed out, revealing an $\alpha \geq .7$ for all scales except perceived safety (α=.626). As for studies with small sample sizes a Cronbach's Alpha around 0.6 can still be considered as sufficient [28], we will also report on perceived safety.

The first survey was conducted one week after the type B robot was deployed in the etching department. During that time, the robot was still in its testing phase (six weeks in total), meaning it ran only for eight hours a day, under supervision. During the test phase, the operators could also ask the supervisor next to the robot for help if something was not working properly. After this 6-week period the robot had been running 24/7. Two additional robots of the same type were deployed in the same area of the etching department, also with a testing phase of four to six weeks. Three months after the deployment of the first robot, all three robots were running 24/7 without supervision.

For the deployment of type B robots in the implantation department (18 months later), the company followed the same procedure as before. In order to research the temporal change of UX, the questionnaires were filled in by operators in the cleanroom at three different points in time. The operators were either working in the oven (robot A), the etching (robot B), or in the implantation (robot B) departments. The sample group was randomly chosen among eight shift groups. In the first round of questionnaires t_1 (testing phase etching, end of February 2011), a researcher randomly distributed the questionnaires to the operators in the etching and oven departments. Six months later t_2 (August 2011), a second round was conducted and in September 2012 (testing phase implantation, t_3), the third round of questionnaires was conducted. At that time of measurement, we also handed out questionnaires to the operators working in the implantation department with type B robots for the first time.

5   Results

At t_1, 23 questionnaires were filled in (5f, 18m). A total of 13 respondents indicated to currently work with robot B, 10 with robot A. The mean age of respondents was 31 (SD=10). At t_2, 21 questionnaires were filled in (10f, 11m). A total of 14 respondents indicated to currently work with robot B, 7 with robot A. The mean age of respondents was 35 (SD=12). At t_3, a total of 23 questionnaire were filled in (8f, 15m); respondents had a mean age of 32 years (SD=11). Thereby, 5 respondents indicated to work with robot A, 8 with robot B in the etching department, 8 in the implantation department (2 missing). For participants in the oven department, the average cleanroom working experience is 4 years (SD=5.1). In the etching department, the average cleanroom working experience is 8 years (SD=9.4) and in the implantation department it is 12 years (SD=10.6).

At t_1, an independent samples t-test[1] was conducted to compare UX ratings for type A and B robots. We found a significant difference in all scales of UX, with a more positive UX regarding type A robots (*cooperation* M_A=4.06, SD_A=0.71, M_B=2.82, SD_B=0.70, t(21)=-4.78, p=.000; *perceived safety* M_A=4.30, SD_A=0.43, M_B=3.61, SD_B=0.94, t(17)=-2.33, p=.032; *perceived usability* M_A=4.34, SD_A=0.14, M_B=3.05, SD_B=0.75, t(13)=-6.09, p=.000; *stress* M_A=1.88, SD_A=0.29, M_B=3.07, SD_B=0.52, t(21)=6.37, p=.000; *general UX* M_A=4.20, SD_A=0.35, M_B-2.77, SD_B=0.52, t(17)=-5.40, p=.000).

Compared to newly introduced robots, the longer established robots are experienced more positively in general by the operators. They are more positively perceived regarding collaboration, evoke less concerns regarding safety issues, indicate less stress, and are perceived more effective, efficient and satisfying. This is in line with our assumption that the well known robots in the oven department within their safety fences are better experienced in the beginning (due to the fences, but also because operators are already used to them). Additionally, the ratings of the established robots in terms of UX did not change over all points of measurement as expected, as an "experience saturation" for these robots was already achieved among the employees. Therefore, we used the ratings from t_1 of robot A as a benchmark to be compared with robot B from the other two departments for all three points of measurement. Results for type B robots in the etching department at t_2 (*cooperation* M_B=3.31, S_B=0.54; *perceived safety* M_B=3.3.92, S_B=0.19; *perceived usability* M_B=3.67, S_B=0.70; *stress* M_B=2.81, S_B=0.45; *general UX* M_B=3.46, S_B=0.69) and t_3 (*cooperation* M_B=3.10, S_B=0.78; *perceived safety* M_B=4.92, S_B=0.21; *perceived usability* M_B=3.66, S_B=0.39; *stress* M_B=2.72, S_B=0.72; *general UX* M_B=3.06, S_B=0.77) show similar results as mentioned above, where all scales but *perceived safety* were rated significantly better for type A robots at t_1 (see Fig 2).

To get insights on the development of UX over time regarding type B robots in the etching department, a one-way between subjects ANOVA was conducted. After the first six months of introduction (t_1 vs t_2), we found a significantly better rating of perceived usability F(2,32)=3.59, p=.039 and a noteworthy result regarding general UX, even if it did not achieve the significant level of .05. Post hoc comparison using the LSD test indicate the mean score for t_2 (M_B=3.68, SD_B=0.70) was significantly higher than for t_1 (M_B=3.04, SD_B=0.75).

---

[1] All data were normally distributed.

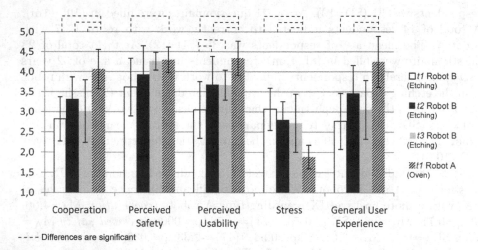

Fig. 2. All scales at t_1, t_2, and t_3 from the etching department were significantly lower rated, with the exception of the factor perceived safety at t_2 and t_3 in comparison to t_1 of the oven department

The mean values for the UX scales stress (M_B=2.82, SD_B=0.45) and perceived safety (M_B=3.92, SD_B=0.71) also improved, however, not on a statistically significant level. After one and a half year of introduction (t_1 vs t_3) the only scale which was rated significantly better was perceived usability. A post hoc comparison using the LSD test indicate the mean score for t_3 (M_B=3.66, SD_B=0.39) was significantly higher than for t_1 (M_B=3.04, SD_B=0.75). Figure 2 shows the UX ratings for both types of robots in the oven and etching departments over time.

Regarding the adoption of new robots, our data shows an increasing positive UX in the first six months after introduction in general. The robot is better rated in terms of perceived effectiveness, efficiency and satisfaction (i.e., perceived usability) when compared to the beginning.

To get insights on how technology was adopted in other departments, we also handed out the same questionnaires at the implantation department, where the company introduced type B robots at t_3. To compare the first deployment of the robot B in the etching department (t_1) with the first deployment of the same robot in the implantation department (t_3), we ran an independent samples t-test. We found no significant difference regarding operators initial experience with regard to perceived usability, stress, perceived safety, cooperation and general UX. This indicates that the workers in both departments had a similar initial experiences with type B robots. Consequently, there is a significant better rating of the established robots in the oven department compared to the newly introduced robots in the implantation department (*cooperation* M_A=4.06, SD_A=0.46, M_B=2.93, S_B=0.33, t(16)=-5.76, p=.000; *perceived safety* M_A=4.30, SD_A=0.43. M_B=3.39, S_B=0.51, t(16)=-4.05, p=.001; *perceived usability* M_A=4.34, SD_A=0.14, M_B=3.54, S_B=0.30; t(9)=-6.85, p=.000;

stress M_A=1.88, SD_A=0.29, M_B=2.93, S_B=0.34, t(16)=6.98, p=.000; *general UX* M_A=4.20, SD_A=0.35, M_B=2.81, S_B=0.45, t(16)=-7.50, p=.000).

Regarding influences from the user context, we observed that age plays a role when it comes to UX changes over time. An independent samples t-test of operators working with type B robots showed significant differences regarding the age. Younger operators (age ≤30) rated the factors cooperation (M_B=3.35, SD_B=0.59; t(41)=2.34, p=.024) and perceived usability (M_B=3.73, SD_B=0.42; t(41)=2.06, p=.45) significantly better than older operators (age>30) (cooperation M_B=2.90, SD_B=0.60, perceived usability M_B=3.31, SD_B=0.71). Further we found a moderate negative correlation between the age of the operators working with robot B and cooperation (Pearson's r=-.342, n=43, p=.025) as well as perceived usability (Pearsons's r=-.346, n=43, p=.023). By means of an independent samples t-test we observed that stress was significantly lower rated by younger operators (age≤30) (M_B=2.59, SD_B=0.56; t(41)=-3.08, p=.004) when compared to older operators (age>30) (M_B=3.06, SD_B=0.41). Our results show a moderate correlation between stress and age (Pearson's r=.425, n=43, p=.000).

Regarding type B robots, we further identified a moderate negative correlation between working experience and cooperation (Pearson's r=-.358, n=41, p=.022), as well as a moderate positive correlation between working experience and stress (Pearson's r=.327, n=41, p=.037). This means the younger the operators are, the less usage problems occur, the better collaboration is perceived, and the less stressful the interaction with robot B is perceived. These findings suggest a close relation between age and user experience.

Regarding robot A, we could identify a moderate negative correlation of perceived usability and age (Pearson's r=-.489, n=21, p=.024), and of general UX and age (Pearson's r=-.489, n=21, p=.024). This indicates a better usability for younger operators and a better overall positive experience for them.

6   Conclusions and Future Work

The presented study demonstrated that time can be a crucial factor in human-robot interaction, as it impacts how users experience the system. This was shown by investigating the UX of a newly introduced robotic arm at three points of time: one week after its introduction, six months after and 18 months later. A robotic arm (within a safety fence) which has already been used over 10 years served as a control group, since the rating on UX for that kind of robot were stable at all three times of measurement. We found that the newly introduced robotic arm received low ratings with regard to the UX factors cooperation, perceived safety, perceived usability, stress, and general UX shortly after its introduction when compared to the already established ones. After six months, the ratings of the newly introduced robotic arm significantly improved regarding perceived usability and showed a noteworthy result for general UX. These results suggest that initial experiences positively change with prolonged interaction. However, at round three after 18 months, the UX ratings regarding the new robot did not significantly increase nor decrease when compared to the second round 12 months before. We could also receive initial insights of how technology is adopted in other departments for first time interactions. The robots which were deployed first in the implantation department just before the third survey round showed similar ratings on all UX scales to the robots in the etching department when deployed for the first time. Differences in the ratings of the two robot types could be rooted in the fact that the newly deployed robots substitute activities formerly performed by human operators, whereas the established robots perform activities which could not be done by humans (as they require too precise and accurate movements). Our results imply a higher overall acceptance of robots which perform tasks which cannot be accomplished by humans in contrast to robots replacing human working routines.

For future work we plan to further explore this issue. To deepen the understanding of how specific UX aspects change over time, we will focus on first time and longer time interactions at more different departments. To enrich our quantitative results, we will conduct semi-structured interviews with operators the oven, etching, and implantation department. We plan to support the interviews with retrospective methods, e.g. the UX Curve [26].

Acknowledgments. We greatly acknowledge the financial support by the Federal Ministry of Economy, Family and Youth and the National Foundation for Research, Technology and Development (Christian Doppler Laboratory for "Contextual Interfaces").

References

1. Weiss, A., Buchner, R., Fischer, H., Tscheligi, M.: Exploring human-robot cooperation possibilities for semiconductor manufacturing. In: Workshop on Collaborative Robots and Human Robot Interaction (2011)
2. Kose-Bagci, H., Dautenhahn, K., Nehaniv, C.L.: Emergent dynamics of turn-taking interaction in drumming games with a humanoid robot. In: RO-MAN 2008, pp. 346–353. IEEE (2008)
3. Alben, L.: Quality of experience: defining the criteria for effective interaction design. Interactions 3(3), 11–15 (1996)
4. Obrist, M., Reitberger, W., Wurhofer, D., Förster, F., Tscheligi, M.: User experience research in the semiconductor factory: A contradiction? In: Campos, P., Graham, N., Jorge, J., Nunes, N., Palanque, P., Winckler, M. (eds.) INTERACT 2011, Part IV. LNCS, vol. 6949, pp. 144–151. Springer, Heidelberg (2011)
5. Karapanos, E., Zimmerman, J., Forlizzi, J., Martens, J.B.: User experience over time: an initial framework. In: Proc. of CHI 2009, pp. 729–738 (2009)
6. Karapanos, E.: User experience over time. In: Karapanos, E. (ed.) Modeling Users' Experiences with Interactive Systems. SCI, vol. 436, pp. 61–88. Springer, Heidelberg (2013)
7. Pohlmeyer, A.E.: Identifying Attribute Importance in Early Product Development. Exemplified by Interactive Technologies and Age. PhD thesis (2011)
8. Kujala, S., Vogel, M., Pohlmeyer, A.E., Obrist, M.: Lost in time: the meaning of temporal aspects in user experience. In: CHI 2013 Extended Abstracts on Human Factors in Computing Systems, CHI EA 2013, pp. 559–564. ACM, New York (2013)
9. Kujala, S., Miron-Shatz, T.: Emotions, experiences and usability in real-life mobile phone use. In: Proc. of the SIGCHI Conf. on Human Factors in Computing Systems, CHI 2013, pp. 1061–1070 (2013)

10. Dautenhahn, K.: Methodology & Themes of Human-Robot Interaction: A Growing Research Field. Int. Journal of Advanced Robotic Systems 4(1), 103–108 (2007)
11. Sung, J.Y., Christensen, H.I., Grinter, R.E.: Robots in the wild: understanding long-term use. In: HRI, pp. 45–52 (2009)
12. Koay, K.L., Syrdal, D.S., Walters, M.L., Dautenhahn, K.: Living with robots: Investigating the habituation effect in participants' preferences during a longitudinal human-robot interaction study. In: RO-MAN 2007, pp. 564–569. IEEE (2007)
13. Castellano, G., Aylett, R., Dautenhahn, K., Paiva, A., McOwan, P.W., Ho, S.: Long-Term Affect Sensitive and Socially Interactive Companions. In: Proc. of the 4th Int. Workshop on Human-Computer Conversation (2008)
14. Coradeschi, S., Kristoffersson, A., Loutfi, A., Von Rump, S., Cesta, A., Cortellessa, G., Gonzalez, J.: Towards a methodology for longitudinal evaluation of social robotic telepresence for elderly. In: 1st Workshop on Social Robotic Telepresence at HRI 2011 (2011)
15. Lee, M.K., Forlizzi, J.: Designing adaptive robotic services. In: Proc. of IASDR 2009 (2009)
16. Lee, M.K., Forlizzi, J., Kiesler, S., Rybski, P., Antanitis, J., Savetsila, S.: Personalization in hri: A longitudinal field experiment. In: HRI 2012, pp. 319–326. IEEE (2012)
17. Lee, M.K., Forlizzi, J., Rybski, P.E., Crabbe, F., Chung, W., Finkle, J., Glaser, E., Kiesler, S.: The snackbot: documenting the design of a robot for long-term human-robot interaction. In: Human-Robot Interaction (HRI), pp. 7–14 (2009)
18. Fernaeus, Y., Håkansson, M., Jacobsson, M., Ljungblad, S.: How do you play with a robotic toy animal?: A long-term study of pleo. In: Proc. of the 9th Int. Conf. on Interaction Design and Children, pp. 39–48 (2010)
19. Kidd, C.D.: Designing for Long-Term Human-Robot Interaction and Application to Weight Loss. PhD thesis, Massachusetts Institute of Technology. MIT (2008)
20. Mutlu, B., Forlizzi, J.: Robots in organizations: the role of workflow, social, and environmental factors in human-robot interaction. In: Human-Robot Interaction (HRI), pp. 287–294 (2008)
21. Hassenzahl, M., Burmester, M., Koller, F.: Attrakdiff: Ein fragebogen zur messung wahrgenommener hedonischer und pragmatischer qualität. Mensch & Computer 2003: Interaktion in Bewegung (2003)
22. Bartneck, C., Kulić, D., Croft, E., Zoghbi, S.: Measurement instruments for the anthropomorphism, animacy, likeability, perceived intelligence, and perceived safety of robots. Int. Journal of Social Robotics 1(1), 71–81 (2009)
23. Lohse, M.: Bridging the gap between users' expectations and system evaluations. In: RO-MAN 2011, pp. 485–490. IEEE (2011)
24. Varsaluoma, J., Kentta, V.: Drawux: web-based research tool for long-term user experience evaluation. In: Proc. of the 7th NordiCHI 2012, pp. 769–770 (2012)
25. Karapanos, E., Zimmerman, J., Forlizzi, J., Martens, J.B.: Measuring the dynamics of remembered experience over time. Interact. Comput. 22(5), 328–335 (2010)
26. Kujala, S., Roto, V., Väänänen-Vainio-Mattila, K., Sinnelä, A.: Identifying hedonic factors in long-term user experience. In: Proc. of the 2011 Conf. on Designing Pleasurable Products and Interfaces, DPPI 2011, pp. 17:1–17:8 (2011)
27. Weiss, A., Bernhaupt, R., Lankes, M., Tscheligi, M.: The usus evaluation framework for human-robot interaction. In: AISB 2009: Proceedings of the Symposium on New Frontiers in Human-Robot Interaction, pp. 158–165 (2009)
28. Pedhazur, E., Schmelkin, L.: Measurement, Design, and Analysis: An Integrated Approach. Erlbaum (1991)

Interpreting Robot Pointing Behavior

Mary-Anne Williams[1], Shaukat Abidi[1], Peter Gärdenfors[2],
Xun Wang[1], Benjamin Kuipers[3], and Benjamin Johnston[1]

[1] Social Robotics Studio, University of Technology, Sydney, Australia
mary-anne.williams@uts.edu.au
[2] University of Lund, Sweden
peter.gardenfors@lucs.lu.se
[3] University of Michigan, USA
kuipers@umich.edu

Abstract. The ability to draw other agents' attention to objects and events is an important skill on the critical path to effective human-robot collaboration. People use the act of pointing to draw other people's attention to objects and events for a wide range of purposes. While there is significant work that aims to understand people's pointing behavior, there is little work analyzing how people interpret robot pointing. Since robots have a wide range of physical bodies and cognitive architectures, interpreting pointing will be determined by a specific robot's morphology and behavior. Humanoids and robots whose heads, torso and arms resemble humans that point may be easier for people to interpret, however if such robots have different perceptual capabilities to people then misinterpretation may occur. In this paper we investigate how ordinary people interpret the pointing behavior of a leading state-of-the-art service robot that has been designed to work closely with people. We tested three hypotheses about how robot pointing is interpreted. The most surprising finding was that the direction and pitch of the robot's head was important in some conditions.

Keywords: Human-robot interaction, human-robot collaboration, socio-cognitive skills, attention, joint attention, pointing.

1 Introduction

Service robots that communicate and collaborate with people can undertake tasks that stand-alone robots cannot. In order to realize their full value in enhancing productivity, service robots will need to interact and work with people fluently. Designing intelligent collaborative robots for deployment in open, complex and changing environments where they are expected to communicate, interact and work with people presents enormous scientific challenges.

There is a wealth of research on the role attention plays in communication among people [1, 3, 7, 8, 9, 10, 12, 14, 15, 22, 28], and the need to establish shared and joint attention when agents cooperate and undertake collaborative actions [2, 5, 17, 19, 20, 25, 25]. Pointing is a vital attention directing behavior employed by people to draw other people's attention to relevant and important objects and events. It is a basic skill in human communication. Young children master the art of pointing early. The ability

G. Herrmann et al. (Eds.): ICSR 2013, LNAI 8239, pp. 148–159, 2013.

to draw people's attention to relevant objects and events within perceptual vicinity will be a critically important skill for service robots.

There is considerable and seminal research on understanding and interpreting human pointing and other gestures [17, 18, 26] that robot designers can use. This paper investigates how untrained people interpret robot pointing and how they determine the target of a robot's pointing behaviour. Wnuczko and Kennedy [26] (Experiment 4) found that people typically point to align their eyes, with their index finger and the object (taking an egocentric perspective). In contrast, most onlookers interpret the pointing along the direction of the arm. Hence, there is a conflict in how the person pointing and the attendant onlooker interpret the pointing gesture (and presumably identify the target). These two pointing strategies are the basis for our hypotheses. In one of the experimental conditions we found that the direction of the robot's head is also important for human interpreters. Thus, a strategy of following head pitch and yaw directions was added as a third hypothesis about how humans interpret robot pointing.

2   Attention in People and Robots

Before we turn to the role of pointing in human-robot collaboration, we must prepare by discussing the role of attention in people and robots. Attention plays a crucial role in facilitating collaboration and is typically understood to be the cognitive process of selectively concentrating on a specific aspect of perceptions, normally the most relevant aspect, while ignoring other aspects at the same time. Kopp and Gärdenfors [15] argue that being able to attend is a minimal criterion for an agent to be intentional.

According to Andersen [1] attention is best conceived as the allocation of processing resources and this interpretation is helpful when analyzing and designing robot systems that can pay attention and interact with humans. Novianto *et al.* [19] used this idea as the basis for the innovative robot attention based architecture ASMO. It seems obvious that in order to collaborate robots need to be able to pay attention to important aspects of the job at hand, but it turns out that designing robots that can do that on the fly in social and dynamic settings presents major difficulties.

What is required of a robot if it is to be able to attend to something? First of all it should be noted that there are several different kinds of attention. Brinck [5] distinguishes three particular kinds of attention: scanning, attraction and focusing. Scanning involves the continuous surveying of the environment and is directed at discovering opportunities to act. Attention attraction occurs when something happens that is at odds with the expectations of an agent. Attention focusing, finally, is the intentional form of attention where the agent itself deliberatively choses what to attend to. In this case the agent may even have its attention directed to something that does not exist or not present in the current environment. In other words, what is attended to is a representation rather than a physical object in the real world. For example, if one is looking for chanterelles in the forest, one's attention is focused on yellow mushrooms, even if there is not a single chanterelle in the environment. This kind of attention thus satisfies the criterion proposed by Kopp and Gärdenfors [15] that an agent is intentional with respect to some goal, only if the agent has a representation of the goal.

In this paper we are concerned with intention-based attention as Tomasello [23, 24] argues intention is a fundamental prerequisite for pointing and for understanding pointing. Building on the work of Kopp and Gärdenfors [15] the following capabilities are required for a robot to point and these requirements are assumed to be the case when people interpret pointing of other people and robots: (i) search mechanism for discovering relevant objects and events, (ii) identification mechanism for such objects/events, (iii) selection mechanism to attend to one of the identified objects/events, (iv) directing mechanism to direct sensors towards the selected object/event; and (v) focusing mechanism to maintain attention on the selected object/event.

These capabilities demand representations of objects/events, which is a radical departure from behavioristic principles. What is a "relevant" object for capability (i), above, can be determined by the goal of the agent. For example, locations and places form a special class of objects that are in focus when the agent has navigational goals. No set of stimulus-response couplings is sufficient for identifying and selecting an object/event in a complex dynamic environment. For this the robot needs a way of internally marking a set of features as being characteristic of a particular object/event. This set is often used to constitute the representation of an object/event.

This kind of representation is an inevitable control variable in the attention mechanism of a robot. The representation "binds together" the stimulus-response pairs of a reactive system and allows them to cooperate in new ways. In other words, the perception-action interactions are enhanced by perception-representation-action combinations that are made possible by adding, for example, object/event representations that can act as a "hidden variable" in the control mechanisms of the system.

A special case of capability (v) is that the robot should be able to track an object, that is, focus on the object even if it moves. Since we see attention focusing as a minimal form of intentionality, it should be noted that the five criteria proposed here align with Cohen and Levesque's [10] proposal that *intention is choice with commitment*.

3   Pointing as Directing Attention

People use many gestures to direct attention to objects or happenings; they also use gestures to control and influence other people's behavior by directing attention. Young children can point, and apes can be taught to point and often do so to signal to adults to give them food that is out of reach. Pointing can have many meanings; it can be used to inform, request, identify, show something to share, and indicate preferences. There is an important distinction between imperative and declarative pointing. Robots will need to direct people to help them achieve goals. Imperative pointing is the most elementary and one in which a robot would treat a human as causal agent that can be influenced by drawing their attention to an object or event. In principle, imperative pointing has been interpreted as an extension of power, rather than a form of communication. Robots can enact imperative pointing using representations of their goals as well as physical pointing skills.

Declarative pointing, on the other hand, would involve a robot directing the attention of a person towards a specific object/event. This type of pointing is significantly more challenging to design and develop in robots because it requires a robot to follow a person's gaze to see whether they are looking at the desired object/event. In contradistinction to imperative pointing, declarative pointing always involves intentional communication since the pointer wants to affect the state of the mind of the attendant. The crucial difference with imperative pointing is that the pointer need not desire to obtain the object pointed at, but rather to achieve joint attention to the object with the attendant. The pointer thus takes the mental state of the attendant into account [31].

4   Experiment: Interpreting Robot Pointing

4.1   Experiment Objective

The experiment was designed to enhance our understanding of how ordinary people interpret robot pointing behavior. In the experiment we wanted to find out how people determine the target object/event that a service robot, the PR2, is attempting to draw their attention to by pointing.

4.2   The PR2 Robot

The PR2, Personal Robot 2, is a service robot designed to work with people and its features are depicted in Figure 1. The PR2 is designed as a personal robot to undertake applications that enable people to be more productive at home, at work and at play. The PR2 combines the mobility to navigate human environments, and the dexterity to grasp and manipulate objects in those environments. The robot arms are back-drivable and current controlled so that the robot is able to manipulate objects in unstructured environments; the arms have a passive spring counterbalance system that ensures they are weightless. The PR2's wrist has two continuous degrees of freedom and sufficient torque for the robot to manipulate everyday objects. The two-pronged gripper can grasp all kinds of everyday objects. The PR2's head and neck area, illustrated in Figure 2(a) above, has numerous sensors including multiple cameras and a Microsoft Kinect sensor. The PR2's head can pan 350° and tilt 115°.

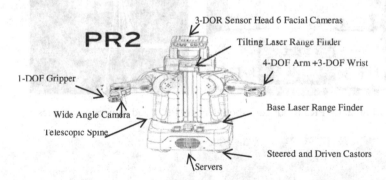

Fig. 1. Personal Robot, PR2, specifications (Source: Willow Garage)

The PR2 is capable of pointing at an object or event that it has perceived through its sensors, i.e. six cameras, a kinect and a laser range finder on its head - see Figure 1 and 2. The trajectories from all six cameras on the PR2 head are plotted in Figure 2(b). The initial point is the camera while the second point of trajectory is the top of the right-gripper. To quantify pointing, these trajectories are extrapolated to the ground to highlight the variations in target locations. Seven trajectories are illustrated

- one from each of the six head cameras and one from the shoulder of the right hand through the end of the gripper. Figure 2(b) highlights the variation of targets the PR2 robot might be considered pointing at, when viewed from the front. Each trajectory is designated a different color: from Camera 1 light cyan, Camera 2 pink, Camera 3 yellow, Camera 4 blue, Camera 5 cyan, Camera 6 red, and from the top of the shoulder the trajectory is silver.

4.3   Experimental Hypotheses

Following Wnuczko and Kennedy's [26] experiment, we wanted to compare two different ways for people to point to an object in front of them. One method is to align one's line of sight with the end of the pointing finger and the object at focus. Another method is to align the angle of your arm (or pointing device) in the direction of the object. These two basic pointing methods are illustrated in Figure 3(a).

Pointing straight downward leads to essentially the same target for people, however, the further away the object/event of interest, the larger the distance between the targeted points determined by the facial camera-gripper trajectory and the angle of the arm trajectory. The discrepancy between the targets illustrated in Figure 3 show it can be significant.

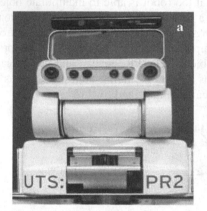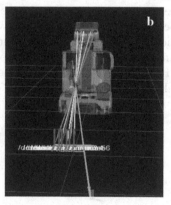

Fig. 2. (a) PR2 head   and shoulders; (b) PR2 head camera trajectories

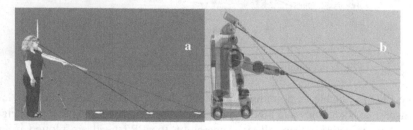

Fig. 3. Two basic pointing trajectories a longitudinal section for (a) a person and three for (b) a PR2.   Figure 3(a) (reproduced from Wnuczko and Kennedy, 2011, p. 1486).

During the second condition of our experiment it became clear, however, that the direction of the robot's head was important for how our subjects interpreted what the robot was pointing to (see Fig 3(b)).

We therefore formulate three hypotheses for how humans interpret robot pointing when the target is in the direction of an outstretched arm:

1. Humans follow the direction of the robot's arm.
2. Humans follow the line of sight from the robot's head and end of the gripper.
3. Humans follow the direction of the robot's head.

4.4   Experiment Design

In order to determine how people interpret what a PR2 robot is pointing at, the robot's arm was placed in a pointing configuration and subjects without background in robotics were asked to identify what the robot was pointing at.

The PR2 arms are not identical to human arms and there are limitations to how a PR2 arm can be configured to point. For example the PR2's arm cannot always stretch out in straight line because the elbow joint cannot extend to 180 degrees depending on the angle the upper arm makes with the shoulder unlike a normal human arm. During the experiment the PR2 arm was not entirely a straight line, i.e. the inside angle made by the upper and lower arms at the joint was less than 180 degrees. Figure 4 shows how the robot arm was configured during the pointing experiment. The PR2 is able to maintain 180 degrees between the upper and lower arm only in some configurations.

In the experiment we collected data from 196 participants; 122 male and 74 female. The experiment consisted of two conditions. During both Condition 1 and Condition 2 of the experiment the robot's arm was static; pointing in a single direction. The robot arm did not follow moving objects; the robot was mute, in a fixed pose and did not interact directly with participants.

The participants' view of the robot is depicted in Figure 5. Participants were asked "What is the robot pointing at?". Participants faced the robot front-on during both conditions and they indicated verbally what the robot was pointing at. They were not given any information about what the robot was trying to do, nor did the participants have access to any visualizations of the robot's perception.

The experiment took place in the atrium of Building 10 at the University of Technology, Sydney campus. The robot was located at the western end of the building and its right arm was raised and placed in a pointing posture in the direction of the eastern entrance to the building. The robot was pointing upwards towards the bridge area – see Figure 4. We placed three beanbags (green, pink and purple) on the bridge and two banners (red and blue) with UTS in large letters were hung from the ceiling. These objects were salient things for the robot to point at, but participants were not restricted to indicate these objects in their answers. Table 2 summarizes the answers of the participants.

During Condition 1 of the experiment the PR2 was configured so that its right arm pointed upwards to its right and its head faced the front direction. In this configuration the participants were able to see the robot's pointing arm clearly. We collected responses from 32 participants during Condition 1 when the PR2's head faced the participant directly rather than in the direction of its pointing arm (thus this condition had more similarities to that of declarative pointing). The robot's outstretched arm was raised to the right in a pointing gesture.

During Condition 2 of the experiment the robot arm remained in the same configuration but its head was turned to face the direction of its pointing arm as seen in Figure 4. The attenuating angle of the head was deliberately made slightly higher than the attenuating angle of the arm. In this way the predictions from hypotheses 1 and 3 could be separated. Participants faced the robot and could clearly see the relative trajectories. Responses were collected from 164 participants during Condition 2.

Fig. 4. (a & b) The PR2 points in the atrium of Building 10 in Condition 2. (c) Robot head facing direction of the pointing arm.

The time each participant took to answer the question "what is the robot pointing at" was measured. Participant answers were timed and noted as *immediate*, not immediate but *up to 3 seconds*, and *more than 3 seconds* during both conditions of the experiment. The reason for recording response times was to see if the task of identifying the target of the robot's pointing was easier for people when the robot was facing away from its pointing arm or facing the same direction as the pointing arm.

After the data collection, participants were invited to describe how they determined what the robot was pointing at and to provide general comments on the task and their experience in determining the target of the robot's pointing.

5   Results

Figure 4 shows the potential target objects in the robot's and participants' field of view, e.g. the bridge, banners, and beanbags. Participants were asked to nominate what the robot was pointing at, and they identified the following objects: the bridge, the railing, lights on the bridge, the banners, "U" on red banner, "U" on blue banner, green beanbag, pink beanbag, purple beanbag, sky through the glass ceiling, and the far building on the other side of the street which was visible by looking under the bridge through the glass wall at the end of Building 10.

Table 1. All identified target objects classified by robot trajectories

Trajectory hypothesis	Described objects
H1. Lower Arm Trajectory	Banners, red banner, blue banner, "U" on red banners "U" on blue banner, above beanbags, lights on the bridge,
H2. Line of sight from facial cameras to gripper	Far building
H3. Head trajectory	Bean bags, wooden beam of bridge
Unknown	Bridge, bridge wall, roof, sky

All the target objects identified by participants were grouped according to the trajectory that they were on using the robots head, lower arm and head-gripper line of sight – see Table 1. In some cases the trajectory was unknown because the target object indicated covered a large nonspecific area and so the trajectory was too ambiguous to classify.

Table 2 presents the results of the PR2 pointing experiment. It lists the number of participants who identified objects on the trajectory of the robot's head, lower arm and line of sight from facial cameras to gripper.

6   Discussion

Wnuczko and Kennedy [26] (Experiment 4) found that people in general point as to align with their eyes, index finger and object from an egocentric perspective, while most onlookers interpret the pointing along the direction of the arm. Hence, there is a conflict in how the person pointing and the attendant onlooker interpret the pointing trajectory/target. Conforming to their results, there was only one participant in Condition 1 and one in Condition 2 who lined up the robots camera with the end of the gripper to identify a target.

The most surprising result from the experiment is that people do not interpret the pointing behavior of robots in the same way they interpret the direction other people point. The results from Condition 2 show that over half (51%) of the participants used the direction of the robot's head in determining the target, while only 38% used the direction of the arm.

Furthermore, 50% of participants indicated general targets, e.g. the bridge, sky, roof, in Condition 1, but in Condition 2 general targets only accounted for 10% of the responses. Again this indicated that people found it was much easier to determine specific pointing targets when the robot was facing the direction of its arm. Participant comments confirmed this interpretation of the evidence.

Table 2. Robot Pointing Targets during Condition 1 and Condition 2

Trajectory hypothesis	Condition 1: Facing away from gripper	Condition 2: Facing gripper
H1. Lower arm	14 (44%)	63 (38%)
H2. Line of sight from facial cameras to gripper	1 (3%)	1 (1%)
H3. Head	2 (6%)	84 (51%)
Unknown	15 (47%)	16 (10%)
Total	**32**	**164**

The explanation for the discrepancy between the two conditions lies in the difference between the human visual system and that of the PR2. Human eyes/gaze can move while the head is stationary, however the angle of the PR2 head determines the direction of its fixed facial cameras. Furthermore, an onlooker can easily determine the angle of the PR2 head's trajectory using the angle and direction of the head. It is more difficult to translate the angle of a human head to the gaze direction in human subjects. As shown in Figure 5, below, it is obvious that the angle of the robot's head is readily observable and the results suggest that almost half the participants used the direction of the robot's head to determine what the robot was pointing at.

One of the most striking patterns in the response data was that almost half (46%) the participants took more than 3 seconds to determine the target during Condition 1 when the robot was not facing the direction of its pointing arm. In contrast, only 30% took more than 3 seconds when the robot was facing the direction it was pointing. This is evidence that it was more difficult for people to determine what the robot was pointing at when its head/camera was not facing the direction of the pointing arm.

Table 3. Participant response times during Condition 1 and Condition 2

Robot head direction	Response time of participants		
	Immediate	< 3 seconds	> 3 seconds
Condition 1: Right angle to direction of arm	10 (32%)	7 (22%)	15 (46%)
Condition 2: Facing direction of arm but aiming lower than arm	29 (18%)	86 (52%)	49 (30%)

It is noteworthy that participants only identified physical objects during the experiment; even though events took place in the field of view of the robot. During the data collection period for Condition 1 and Condition 2 people walked across the footbridge, sat in the beanbags, and waved to people below. Participants were not asked to exclude events, but none of the 196 participants indicated that the robot was pointing at an event.

7   Conclusion

Using the non-humanoid PR2 robot, this paper reports an investigation of the way people interpret robot-pointing behavior. We designed an experiment in which 196 human participants provided judgments of the target of interest that a stationary PR2 robot was pointing towards.

The first key experimental result demonstrates that some people interpret robot-pointing behavior in a similar way to interpreting other people's pointing, i.e. via the direction of the arm. However, when the robot's head faced the same direction as the pointing arm more than half of the participants used the camera trajectory based on the pitch angle of the robot's head to determine the target, rather than the trajectory of the robot's arm. We have postulated that because the angle of the camera can be estimated from the side view of the PR2, people tend to take the head trajectory into consideration when they determine the robot's pointing target. In addition, since the cameras are fixed, the head angle provides a close approximation to what the robot is "looking" at.

The second key experimental result is that it took much longer for participants to make their judgment when the robot was not looking in the direction of its extended pointing arm. Therefore we can say that it is easier for people to determine what a robot is pointing at if the onlooker can determine what the robot might perceive while it is pointing. This also highlights the dominance of visual sensors in the way people interpret what has the robots attention. This can be compared with the fact that for many animals, their elongated head makes it comparatively easy for an onlooker to determine where it is looking. Figures 3 and 4 illustrates the ease with which the PR2 head pitch can be discerned, due to its elongated and rectangular nature.

The experiment is a first attempt to investigate how humans interpret robot pointing. In a future study, it would be interesting to compare our result to a corresponding study with humanoid robots that have a human-like head. The unhuman-likeness of the head of PR2 and its particular way of indicating a direction contributed to the unexpected result concerning the importance of the head. Furthermore, when the robot is looking in the direction of its arm, the most natural interpretation for a human is that it is a case of imperative pointing. However, when the robot is looking at the human while pointing in another direction, the human interpretation is perhaps that it is a case of declarative pointing. However, declarative pointing builds on joint attention but the corresponding behavior is not exhibited by PR2.

Several implications for robot design arise from this work. A first concerns designing robots with intentions, a second concerns designing robot pointing behaviors, and a third concerns the need for more experiments in more complex social settings. The experiment described herein has provided a first answer to the central question posed in the beginning of this paper, namely, how do people interpret robot-pointing behavior. Furthermore, it generates clues for how to design important future experiments that will investigate how people interpret robots pointing to events, the role of robot intention in robot attention seeking behavior, and the impact of displaying sensor data so experiment participants "see" what the robot perceives.

Another experiment by Wnuczko and Kennedy [26] (Experiment 1) showed that if the pointer was blindfolded, she pointed by aligning the direction of her arm with the object pointed at. A blindfolded person behaved as if they were reaching out to grasp the object. The discrepancy between our experimental results and those from Wnuczko and Kennedy's Experiment 4 suggest that a future experiment in which the robot is visibly blindfolded could yield interesting results as human onlookers would perceive the robot was unable to "see". Since robots, including the PR2 have multiple sensors, this kind of experiment would yield new insights.

References

1. Anderson, R.: Cognitive psychology and its implications, 6th edn. Worth Publishers (2004)
2. Axelrod, R.: The Evolution of Cooperation. Basic Books, NY (1984)
3. Wright, R.D., Ward, L.M.: Orienting of Attention. Oxford University Press (2008)
4. Wolfe, J.M.: Guided search 2.0: a revised model of visual search. Psychonomic Bulletin Review 1(2), 202–238 (1994)
5. Brinck, I.: Attention and the evolution of intentional communication. Pragmatics & Cognition 9(2), 255–272 (2001)
6. Boyd, R., Richerson, P.J.: Culture and Evolutionary Process, Chicago (1985)
7. Castiello, U., Umilta, C.: Size of the attentional focus and efficiency of processing. Acta Psychologica 73(3), 195–209 (1990), doi:10.1016/0001-6918(90)90022-8; Corballis, M.C.: From Hand to Mouth. The Origins of Language, Princeton, NJ (2002)
8. Deutsch, J.A., Deutsch, D.: Attention: some theoretical considerations. Psychological Review 70, 80–90 (1963)
9. Donald, M.: Origins of the Modern Mind. Three Stages in the Evolution of Culture and Cognition, Cambridge, MA (1991)
10. Cohen, P.R., Leveque, H.: Intention Is Choice with Commitment. Artificial Intelligence 42, 213–261 (1990)
11. Eriksen, C., St James, J.: Visual attention within and around the field of focal attention: A zoom lens model. Perception & Psychophysics 40(4), 225–240 (1986); Eriksen, C.W., Hoffman, J.E.: The extent of processing of noise elements during selective encoding from visual displays. Perception & Psychophysics 14(1), 155–160 (1973)
12. Givon, T., Malle, B. (eds.): The Evolution of Language out of Pre-Language, Philadelphia (2002)
13. Hauser, M.: The Evolution of Communication, Cambridge (1996)
14. Jonides, J.: Further towards a model of the mind's eye's movement. Bulletin of the Psychonomic Society 21(4), 247–250 (1983)
15. Kopp, L., Gärdenfors, P.: Attention as a minimal criterion of intentionality in robots. Cognitive Science Quarterly 2, 302–319

16. Lavie, N., Hirst, A., de Fockert, J.W., Viding, E.: Load theory of selective attention and cognitive control. J. of Experimental Psychology 133(3), 339–354 (2004)
17. McNeill, D.: Hand and Mind. What Gestures Reveal About Thought, Chicago (1992)
18. McNeill, D.: Gesture & Thought. U. of Chicago Press (2005)
19. Novianto, R., Johnston, B., Williams, M.-A.: Attention in the ASMO Cognitive Architecture. In: International Conference on Biologically Inspired Cognitive Architectures (2010)
20. Rizzolatti, G., Arbib, M.A.: Language within Our Grasp. Trends in Neuroscience 21, 188–194 (1998)
21. Schilbach, L., Wilms, M., Eickhoff, S.B., Romanzetti, S., Tepest, R., Bente, G., Shah, N.J., Fink, G.R., Vogeley, K.: Minds made for sharing: initiating joint attention recruits reward-related neurocircuitry. Journal of Cognitive Neuroscience 22(12), 2702–2715 (2010)
22. Tomasello, M.: Why don't apes point? In: Enfield, N.J., Levinson, S.C. (eds.) Roots of Human Sociality: Culture, Cognition and Interaction, Berg, pp. 506–524 (2006)
23. Tomasello, M.: Origins of Human Communication. Cambridge University Press (2009)
24. Treisman, A., Gelade, G.: A feature-integration theory of attention. Cognitive Psychology 12(1), 97–136 (1980)
25. Williams, M.-A.: Robot Social Intelligence. In: Ge, S.S., Khatib, O., Cabibihan, J.-J., Simmons, R., Williams, M.-A. (eds.) ICSR 2012. LNCS (LNAI), vol. 7621, pp. 45–55. Springer, Heidelberg (2012)
26. Wnuczko, M., Kennedy, J.M.: Pivots for pointing: Visually-monitored pointing has higher arm elevations than pointing blindfolded. Journal of Experimental Psychology: Human Perception and Performance 37, 1485–1491 (2011)
27. Camaioni, L., Perucchini, P., Bellagamba, F., Colonnesi, C.: The Role of Declarative Pointing in Developing a Theory of Mind. Infancy 5(3), 291–308 (2004), doi:10.1207/s15327078in0503_3
28. Carpenter, M., Nagell, K., Tomasello, M.: Social cognition, joint attention, and communicative competence from 9 to 15 months of age. Monographs of the Society for Research in Child Development 63(4, Serial No. 255) (1998)
29. Sodian, B., Thoermer, C.: Infants' under-standing of looking, pointing, and reaching as cues to goal-directed action. J. of Cognition and Development 5, 289–316 (2004)
30. Gärdenfors, P.: How Homo Became Sapiens. Oxford University Press, Oxford (2003)
31. Gärdenfors, P., Warglien, P.: The development of semantic space for pointing and verbal communication. To Appear in Conceptual Spaces and the Construal of Spatial Meaning: Empirical Evidence from Human Communication. Cambridge University Press, Cambridge

Coping with Stress Using Social Robots as Emotion-Oriented Tool: Potential Factors Discovered from Stress Game Experiment

Thi-Hai-Ha Dang and Adriana Tapus

Robotics and Computer Vision Lab, ENSTA-Paristech, France

Abstract. It is known from psychology that humans cope with stress by either changing stress-induced situations (which is called problem-oriented coping strategy) or by changing his/her internal perception about the stress-induced situations (which is called emotion-oriented coping strategy). Social robots, as emerging tools that have abilities to socially interact with humans, can be of great use to help people coping with stress. The paper discusses some recent studies about ways of dealing with stress in stressful situations by using social robots. Moreover, we focus on presenting our experimental design and the discovered factors that can allow social robots to assist humans in dealing with stress in an emotion-oriented manner.

1 Introduction

1.1 Mental Stress and Stress Coping in Humans

Mental stress in humans is quantified as his/her response to an environmental condition or stimulus, generally considered as humans way to react to challenges. These stimulus are called stressors, and can be external (from the outside world) or internal (surged from thoughts). When a stressor occurs, human's sympathetic nervous system generates a reaction of fight-or-flight in term of physiological (such as heart rate variation, skin conductance change) and behavioural (e.g. facial expression, gestural change) responses, as depicted in Figure 1. In psychology literature, these responses are part of coping strategies, which can be classified in two categories: problem-oriented coping and emotion-oriented coping, as suggested by Lazarus in [9]. The problem-oriented coping style addresses the stress by managing the stressors, while emotion-oriented coping style aims at tackling the person's affective responses towards the stressors. These coping responses may affect positively or negatively the situation in order to optimize the person's performance and/or to maintain his/her ideal mental state.

Yerkes and Dodsons law [17] explains the link between the arousal level and the performance (see Figure 2). The individual's optimal performance is generally attained when he/she is moderately aroused. In order to optimize task performance, human subject can use different coping strategies to moderate his/her arousal level. When an individual is under-aroused, a coping strategy

G. Herrmann et al. (Eds.): ICSR 2013, LNAI 8239, pp. 160–169, 2013.

to positively affect the stressors so as to increase the arousal level is employed; and when an individual experiences stress (i.e., being over-aroused), a coping strategy to negatively affect the stressors in order to decrease the arousal level is used.

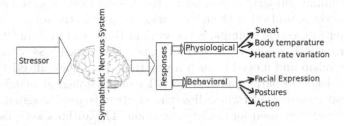

Fig. 1. Mechanism of mental stress

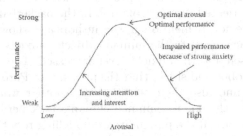

Fig. 2. Hebbian version of the Yerkes Dodson curve

Researches in Human-Machine Interaction have long been focusing on humans' stress experience in order to improve task performance, either by detecting stress and/or by helping humans to deal with stress. The next section presents an overview on recent works in stress detection and stress coping in human-machine interaction context.

1.2   Study about Stress in Human-Robot Interaction Research Domain

Stress Detection: Stress Detection via body movements includes facial unit movements [2] [3] and behavioural features [6]. To assess the accuracy of stress detection via body movements, physiological signals are used as ground truth. Physiological signals that are commonly used in most of research works on stress detection are, for example, heart rate variability (HRV), galvanic skin response (GSR), and finger temperature (FT). An overview of the state of the art on Stress Detection from physiological signals can be seen in [12].

Given the strategic role of dealing with stress during task performance, researchers in human-robot interaction have also studied several aspects of stress during the interaction, including how robots can induce stress during interaction with humans [1][8][10]. For example, the authors in [8] examined the ability of a robotic arm in motion to induce specific affective states in humans. In their experiment, human subjects seated facing the robotic arm, observed its various predefined motions, and rated their impression related to the robotic arm's motions. Another interesting example is the work of Rani et al. [10]. In [10], human subjects were asked to play a basketball game. The basket was attached to a robotic mechanism and moved in such a way that could elicit affective reaction in human subjects, in particular anxiety. A lot of physiological signals (such as electro-dermal response, electrocardiogram, electromyographic signal, and finger temperature) were used for anxiety detection. The authors were also able to demonstrate the possibility of the robot to elicit affective response in humans, and the possibility of detecting anxiety from physiological signals.

Stress Coping in Human-Robot Interaction: The literature shows that social robots are being used either as problem-oriented coping tools (as in [10], [18]) or as emotion-oriented coping tools [15]. In the basketball game experiment of Rani et al. [10] previously mentioned, the authors also showed that by altering the robotic basket according to the human subject's anxiety level, the human's performance increased and the anxiety level decreased. The study developed in [18] uses Paro seal robot and shows that the presence of Paro impacts positively elderly's social life and also their stress level. Another work that focuses on using robot's social behavior to enhance human's task performance is that of [15]. The authors described the possibility of modelling robot's personality so as to improve task performance in a scenario designed for post-stroke rehabilitation. Their findings suggest that by matching the personality of the robot to that of the human user, human's interest in continuing the post-stroke exercise is higher and their experience with the robot is more favourable.

In this work, we study potential factors to be used in social robots so as to help people deal with stress. We present in the next section our experiment (called Stress Game) and its results concerning the emotion-oriented coping assistance that we designed for the humanoid robot NAO.

2   Our Study in Stress Coping Strategies in Human-Robot Interaction

2.1   Data Collection: Stress Game Experiment

The goal of our "Stress Game" is to evaluate the robot's capability of increasing human's task performance and interest to the game. The game is intended to elicit stress and thus frustration in the player. The robot continuously monitors the performance level of the player (and therefore his/her state of frustration) and acts accordingly in order to help to lower the player's frustration level and to enhance his/her task performance.

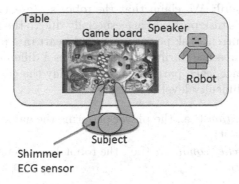

Fig. 3. Experiment set-up

Stress Game Description: A sketch of the experimental design is depicted in Figure 3. The game board and the robot are placed on a table. The player sits in front of the game board.

Before the experiment, the participants filled-up a general questionnaire (e.g., age, gender, background) and a Big Five Personality Test [14] to determine their personality traits (e.g., introverted/extroverted). During the game, the participant has to pick-up as many objects as possible during a given period of time. An annoying sound is played when the participant touches the edge of an opening. Each touch is counted as a mistake. The game has two levels of difficulty:

- normal: There are no false alarms.
- stressful: There are false alarms (i.e., annoying sounds are played when no mistake occurs), which can make the player more stressed and/or confused, and can decrease his/her task performance.

Robot Behaviour: The robot's behaviour depends on the player's personality (collected before the game) and the robot's strategy (defined before the game). In our design, the robot's behavior refers to spoken phrases accompanied by appropriate arms' gestures. The behavior of the robot during the game can be to encourage the player, to challenge the player, and/or show empathy, depending on the player's on-going performance. The strategies of action of the robot are listed in Table 1.

Table 1. Robot Behavior Strategy in terms of Player's Personality

Player's Personality	Robot strategy
Introverted	Empathetic
Average Introverted	Encouraging
Extroverted	Challenging

Experiment Protocol: We claim that the robot's presence during the game will increase the user's task performance especially due to the fact that the robot reacts to the user's current task performance and heart rate parameter. In order to see if the robot's presence during the game makes a difference in terms of the user's task performance, each participant had to play the game in two different conditions for each difficulty level of the game:

- *(a) no robot condition* (i.e., the player performs the game without the intervention of the robot);
- *(b) robot stress relief condition* (i.e., the robot has an encouraging and motivating behavior).

The conditions are presented to the players in a random order. After each condition, the players are asked to fill-up a post-condition questionnaire that allows us to evaluate the robot's behavior and its influence on the user's task performance. During the robot condition, the Nao robot introduces the game, as follows: "Hello! My name is Nao and I'm a robot. Today we will play a game. The game is called "Operation". Here are the instructions of the game: There are 13 openings on the board. In each opening there is an object. With the help of the tweezers you have to extract all the objects. Each time you do an error a noise will be played. As soon as you extract an object please press the button to validate. You have 1 minute to extract as many objects as possible. I'm here next to you so as to encourage you and monitor your errors. When you are ready please press the button to start! Good luck! "

Online Questionnaire: During the onsite experiment, participants were limited to interact with only one personality type of the robot (user-robot personality matching). In order to better understand participants preference about the robots personality, we designed an online questionnaire where participants watched videos of a person playing the game with robots having different personalities. We then invited participants to fill-up the online questionnaire in order to evaluate how they perceive the robots personality and which one they prefer.

Equipments: For the retrieval of heart rate data of the player, we use Shimmer ECG sensor to acquire the player data in real-time and transfer this data via Bluetooth communication to the computer for further processing. ECG sensor is strapped on the player's body and connected to the four electrodes as described in the Shimmer ECG User Guide[1]. The robot used in this Stress Game was the humanoid robot NAO. The pair of tweezers to be used by the user is connected to the game board and help to produce annoying sounds when user makes mistakes during the game.

---

[1] http://www.shimmer-research.com

2.2 Results and Discussion about the Robot's Stress Coping Assistance

The on-site experiment was tested with 17 individuals (16 male and 1 female), the age of participants was between 23 to 36 years old and they were all with a technical sciences background. For the online questionnaire, we had 13 participants' answers (they were all previously involved in the on-site experiment).

In the scope of this paper, we will analyze the efficacy of our robot's behavior in helping people to cope with stress, and discuss solutions to enhance the acceptability of robot's assistance in stressful situations.

Efficacy of Stress Coping for Human Partner. Eysenck in his work on personality [4] stated that introverted individuals have higher arousal level than extroverted individuals. In our experiment, we are able to observe a correlation between the stress level (i.e., arousal level) and the personality of the human player. Among the 17 participants of our experiment, there are 9 introverted, 2 average introverted, and 6 extroverted (i.e., these personality traits are determined from their Big5 Scores as follows: Introverted: $\leq$35; Average Introverted: in-between 35-70; Extraverted: $\geq$70).

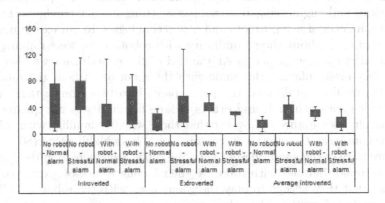

Fig. 4. Statistic about heart rate peaks in terms of personality

As shown in Figure 4, the average number of heart rate peaks (those that surpass the baseline heart rate level) during the game of introverted participants is significantly higher than those of extroverted participants. Besides, our results reveals that the number of heart rate peaks of average introverted participants is lower than those who are introverted or extroverted. However, we only have two average introverted participants among the 17 participants of the stress game experiment, which makes our observation about average introverted people not sufficient for any conclusions.

During robot's intervention, the number of heart rate peaks of introverted participants decreases in the difficult level of the game (see Figure 4). For extroverted participants, robot's intervention increases people's heart rate in easy game level and decreases it in the difficult level. This can be considered as positive impact of the robot's behavior because according to the theory presented in [17] (which is graphically summarized in Figure 2), it is preferable to keep humans in a moderate stress level. In our case, the robot was able to decrease the high stress level of introverted people and moderate the stress level of extroverted people in different difficulty levels. Our design of robot's behavior in the context of Stress Game experiment is thus appropriate to assist people in coping with stress.

While examining the heart rate data to construct the stress detection algorithm, we noticed an interesting phenomenon: players' heart rate seemed to accelerate when the end of the game approached. To test if this co-occurrence actually happens, we checked the occurrence of heart rate peak at the end of the game. This occurrence exists at the rate of 98.15 % (and if only heart rate peaks higher than the baseline are considered the occurrence happened at the rate of 62.69 %). This can be an interesting phenomenon to be considered in human-robot interaction as it suggests that human reacts also at the beginning and at the end of a long event (such as the case of our game). An assistant robot can use this information to choose an appropriate action in order to better assist the user.

Familiarity with Robotics and Acceptance of Robots: Peoples familiarity with robotics and engineering research has a strong impact on the acceptance and the willingness of using robots and new technologies. In our experiment, we asked participants about their familiarity with robotics (answers ranking from 1 (not at all) to 7 (very much)). At the end of the experiment, we also asked whether they prefer playing the game with the robot or without the robot. To analyze the results, we divided the participants into two groups: those whose familiarity is greater than 4, and others (whose familiarity is not greater than 4). The results (see Figure 5) suggest that those who are familiar with robotics (i.e., their familiarity to robotics is greater than 4) have highly rated (80 %) their acceptance to play the game further with the robot. Meanwhile, in the group of those who are not familiar with robotics (their familiarity is not greater than 4), the chance that they accept to play the game along with the robot is 43 %.

This aspect on familiarity has also been studied several times in technological research domain (see [14] for an overview). User's familiarity with technological systems normally correlates with user's understanding on how the system works thus facilitates users effort to make use of the system, and generally affects user's acceptance rate. This applies in any robotic application that involves human-robot interaction, from assistance to cooperation. For example, in [5] where a relation between social-demography and user acceptance towards assistive technology is depicted, the authors have shown that user acceptance towards assistive systems is superior when the user has higher education and/or technological experience. In [7], where the authors discuss about Human-Robot Team performance under stress, it was noticed that when human operator doesn't know what, how, and why the robot is doing what it is doing, he/she will not accept to cooperate with the robot and the Human-Robot collaboration will just fail. Moreover, Tessier et al. [16] discuss about the main challenges in human-robot

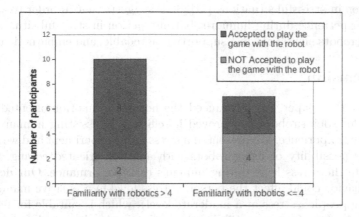

Fig. 5. Acceptability in terms of user's familiarity to robotics

cooperation. They also insisted that situations where the operator and the decision algorithm "do not understand each other" should be avoided in order to have better cooperation between human operator and the involved intelligent systems.

Participant's Preference towards Robot's Assistance Strategy: From the result of the online questionnaire, we were able to evaluate people's preference about robot's personality in the context of the stress game experiment. Among the 13 people who answered the online questionnaire, 9 chose the Encouraging robot, 3 chose the Empathetic robot, and only 1 chose the Challenging robot as shown in Figure 6 (No correlation is found between participant's personality and his/her preference about the robot's personality).

This means that people prefer the robot to encourage and motivate them during stressful task performance, and do not want the robot to challenge them or criticize them when they make errors in such a situation. This tendency of preferring supportive robots to challenging robots is originated from the human's need

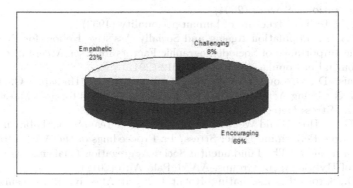

Fig. 6. Participant's preference about robot's assistance strategy in stress game

for affiliation in stressful situations [11]. This suggests that in order to optimize the user's experience during human-robot interaction in stressful situations the designated robots should act in a supportive, agreeable, and empathetic manner.

3   Conclusion

Throughout this paper, we introduced the notion of emotion-oriented coping assistance for social robots and proved its efficacy in assisting humans during stress-related experience. We presented a stress game experiment and we demonstrated the possibility of using robots with emotion-oriented coping strategy to moderate the stress level during human's task performance. Our design on robot's behavior could affect the person's stress level in a positive manner (i.e., maintaining people's stress at a moderate level), which is suitable for people to attain their best performance [17]. Furthermore, we also discussed people's preferences about the robot's social behavior in stress-induced situations. We were also able to study the acceptability of robots in stressful situations. People's acceptance rate towards our robot correlates positively with their familiarity level with robotics. This reveals the importance of user's knowledge about any kind of intelligent systems with which humans have to interact/cooperate to accomplish a common goal. The better the human operator knows about the system, the better the human-system cooperation will be.

References

1. Wendt, C., Popp, M., Faerber, B.: Emotion induction during human-robot interaction. In: Proceedings of the 4th ACM/IEEE International Conference on Human Robot Interaction (HRI 2009), pp. 213–214. ACM, New York (2009)
2. Dinges, D.F., Rider, R.L., Dorrian, J., McGlinchey, E.L., Rogers, N.L., Cizman, Z., Goldenstein, S.K., Vogler, C., Venkataraman, S., Metaxas, D.N.: Optical computer recognition of facial expressions associated with stress induced by performance demands. Aviation, Space, and Environmental Medicine 76(6), sec. II (2005)
3. Metaxas, D.N., Venkataraman, S., Vogler, C.: Image-Based Stress Recognition Using a Model-Based Dynamic Face Tracking System. In: International Conference on Computational Science (2004)
4. Eysenck, H.J.: The structure of human personality (1953)
5. Flandorfer, P.: Population Ageing and Socially Assistive Robots for Elderly Persons: The Importance of Sociodemographic Factors for User Acceptance. International Journal of Population Research 2012 (2012)
6. Giakoumis, D., Drosou, A., Cipresso, P., Tzovaras, D., Hassapis, G., Gaggioli, A., Riva, G.: Using Activity-Related Behavioural Features towards More Effective Automatic Stress Detection. PLoS ONE 7(9) (2012)
7. Kruijff, G.-J.: How Could We Model Cohesiveness in Team Social Fabric in Human-Robot Teams Performing Under Stress? In: Proceedings of the AAAI 2012 Spring Symposium on AI, The Fundamental Social Aggregation Challenge, and the Autonomy of Hybrid Agent Groups. AAAI, Palo Alto (2012)
8. Kulic, D., Croft, E.: Estimating Robot Induced Affective State using Hidden Markov Models. In: The 15th IEEE International Symposium on Robot and Human Interactive Communication (ROMAN 2006), pp. 257–262 (2006)

9. Lazarus, R.S.: From psychological stress to the emotions: A history of changing outlooks. Annual Review of Psychology 44, 1–24 (1993)
10. Rani, P., Liu, C., Sarkar, N.: Interaction between Human and Robot - an Affect-inspired Approach. Interaction Studies 9(2), 230–257 (2008)
11. Rofe, Y.: Stress and affiliation: A utility theory. Psychological Review 91, 235–250 (1984)
12. de Santos Sierra, A., Avila, C.S., Guerra Casanova, J., Bailador del Pozo, G.: Real-Time Stress Detection by Means of Physiological Signals. In: Yang, J. (ed.) Advanced Biometric Technologies (2011)
13. Schneider, S.: Exploring Social Feedback in Human-Robot Interaction During Cognitive Stress. Master thesis, Bielefeld University (2011)
14. Sun, H., Zhang, P.: The role of moderating factors in user technology acceptance. International Journal of Human-Computer Studies 64(2), 53–78 (2006)
15. Tapus, A., Tapus, C., Matarić, M.J.: User-Robot Personality Matching and Robot Behavior Adaptation for Post-Stroke Rehabilitation Therapy, Intelligent Service Robotics. Special Issue on Multidisciplinary Collaboration for Socially Assistive Robotics 38(2) (April 2008)
16. Tessier, C., Dehais, F.: Authority management and conflict solving in human-machine systems. AerospaceLab, The Onera Journal 4 (2012)
17. Yerkes, R.M., Dodson, J.D.: The relation of strength of stimulus to rapidity of habit-formation. Journal of Comparative Neurology and Psychology 18 (1908)
18. Wada, K., Shibata, T.: Social and physiological influences of living with seal robots in an elderly care house for two months. Gerontechnology 7(2), 235 (2008)

Study of a Social Robot's Appearance Using Interviews and a Mobile Eye-Tracking Device

Michał Dziergwa[1], Mirela Frontkiewicz[1], Paweł Kaczmarek[1], Jan Kędzierski[1], and Marta Zagdańska[2]

[1] Authors are with Institute of Computer Engineering, Control and Robotics, Wrocław University of Technology, W. Wyspiańskiego 27, 50-370 Wrocław, Poland
[2] M. Zagdańska is with Realizacja Sp. z o.o., Kolbaczewska 6a, 02-879 Warszawa, Poland

Abstract. During interaction with a social robot, its appearance is a key factor. This paper presents the use of methods more widely utilized in market research in order to verify the design of social robot FLASH during a short-term interaction. The proposed approach relies on in-depth interviews with the participants as well as data from a mobile device which allows the tracking of one's gaze.

Keywords: social robot, eye-tracking, human-robot interaction, perception, in-depth interviews.

1 Introduction

Over the last decade a significant increase of interest in social robots can be seen, both among scientists and potential customers. The introduction of robots into the human environment is not only associated with the development of technology including control systems, mechatronic or sensory solutions. The influence of appearance of the robot on quality of interaction is also an important factor. This issue is now widely explored and its impact on how humans develop attachment to a machine is still an open research problem.

We stipulate that the techniques used in the creation of commercial products should play a supporting role in the process of designing the appearance of social robots. Market research utilizes methods that have been developed for years on the basis of social studies in order to verify the usability, ergonomics and attractiveness of products to their potential buyers [1]. In addition to traditional data collection methods such as interviews or questionnaires, techniques drawn from neuropsychology, such as mobile and stationary eye-tracking, are used increasingly often [2].

Recent research shows that humans tend to locate faces in visual stimuli and fixate on them with extreme precision and quickness [3], [4]. Inside the facial area our gaze tends to focus on specific elements with emphasis on eyes, nose and mouth [5]. Even though head seems to play the most important role during interaction, body posture is also assessed, especially in the case of threatening

G. Herrmann et al. (Eds.): ICSR 2013, LNAI 8239, pp. 170–179, 2013.

behaviors [6]. It is natural to assume that interaction with a social robot should follow these same basic guidelines.

This paper presents findings obtained from an extensive experiment carried out with the use of robot FLASH. It's aims were to study various elements of interaction with a social robot using the aforementioned techniques. Since the amount of acquired data is too extensive to present in this paper, we will only address selected issues regarding the mechanical design of the robot and the psychological aspects that are associated with it. These problems have been summarized by the following hypotheses:

– Gaze patterns of people interacting with FLASH follow rules observed in interaction between humans.
– Anthropomorphic elements of FLASH's design attract attention the most and cause the robot to be perceived as humanlike.

2   Methodology

The main experiment was preceded by a pilot study on a group of 19 people. In experiments, as research tools, we used a mobile eye-tracker, which allows to achieve qualitative (heat and focus maps, gaze path, etc.) and quantitative (statistical quality indicators called KPIs) results. Interviews with the participants have also been carried out. This chapter provides general information about the main experiment, a short description of FLASH (whose appearance has been the subject of the experiment), the information on the eye-tracker, the course of the experiment and data analysis methods.

2.1   General Information

The experiment was carried out at the Main Railway Station in Wrocław. The choice of this location was motivated by the need to acquire a large number of participants with diverse demographic and social background. Test subjects were between 15 and 82 years of age and 26,9 was the average. 90 people (42 females and 48 males) participated in the experiment with the eye-tracking device and have filled out questionnaires afterwards. Additionally, 66 people took part in an in-depth interview.

2.2   Robot Used during Experiments

FLASH (Flexible LIREC Autonomous Social Helper) [7], [8] is a social robot developed at the Wroclaw University of Technology within the EU FP7 LIREC project. He is an anthropomorphic robot based on a balancing platform equipped with head called EMYS [9]. The robot's appearance and dimensions are presented in Fig. 1. The employment of a balancing platform, functionally similar to that of Segway, resulted in obtaining natural, fine and smooth robot movements, fully acceptable by humans. The remaining robot's components, i.e. the

Fig. 1. FLASH - the robot used during experiments

head and the arms fixed to the torso, are tasked with expressing emotions by means of facial expressions and gesticulation. These capabilities increase the acceptance degree of the robot by humans and lays foundations for the establishment of long term human-robot relationship.

2.3   Eye-Tracking Glasses

The mobile eye-tracking device from SensoMotoric Instruments [10], provided by Simply User [11], is a high-end piece of equipment allowing the recording of a person's point of view and the path that their sight follows. It is completely unobtrusive to the user and consists of a sleek pair of glasses connected to a laptop that the test subject carries with them in a backpack. This kind of technology allows various eye-tracking experiments, such as shelf-testing, driving research or even analysis of training methods in sports, to be conducted out in the field.

2.4   Indiviudal In-Depth Interviews

To obtain more accurate and valid data, social research methods such as individual in-depth interview (IDI) and paper-and-pencil questionnaire [12] have been implemented in the described experiment. This paper presents only the qualitative part, i.e. the IDI's. Individual in-depth interviews are a widely used, semi-structured technique of gathering information from individual subjects [13]. This method is applied, if the deepest level of attitudes stemming from individual approach, way of thinking or life history of the respondents is to be examined. Participants are encouraged to express their feelings and opinion freely. The whole meeting lasts from 30 minutes up to 1,5 hour. It is usually recorded and transcribed before analysis.

2.5   The Course of Experiment

Experiment participants were recruited from among passengers waiting at the railway station. They were told only that their help was needed in a scientific experiment involving a robot. After being recruited the participants entered a secluded part of the station, where they were equipped with the eye-tracking glasses. After that, they were led into the room with FLASH. At this point, they knew only that they would meet the robot and play with him for a while. Upon entering the experiment room the participant had to stand next to a container filled with various colorful toys. When the person was detected, FLASH would "wake up" and start interacting with them.

During the whole experiment FLASH operated autonomously and delivered all the instructions to the participants himself. The scenario implemented for this experiment was divided into 2 parts:

Introduction - the robot greeted the person and held out his hand for a moment to allow the person to shake it. Next, FLASH said a few words about himself (where, when and for what purpose he was created) and demonstrated his gesticulative abilities. As the last part of introduction the robot complimented the color of the participant's clothes. All of these actions were undertaken to familiarize the person with the machine and create a bond between them.

Game - the main part of the experiment consisted of a simple game. The participant had to show toys to the robot in any order they wished. FLASH then reacted depending on the choice of a toy by pronouncing an emotional comment and expressing his emotions with facial expression and gesticulation. Robot's commentary added context to the emotions that he expressed, e.g. when expressing sadness he would inform the participant that *"They never let me play with this toy!"* or when surprised he would say *"I was sure that I had lost this toy!"*. This was done to increase the recognizability of the emotions. Four emotions were implemented for the experiment: anger, happiness, sadness and surprise. The game ended when the participant showed six toys after which the robot "fell asleep".

After the main part of the experiment, paper-and-pencil and individual in-depth interviews were conducted.

2.6 Data Analysis

Processing of the eye-tracker data has been done through BeGaze software from SMI. The footage obtained from each person has been divided into segments corresponding to parts of experiment (such as introduction, compliment, game, etc.). For each segment a Reference View, i.e. a still image of the robot, has been defined. The next step was to determine the set of Areas of Interest (AOIs), such as head, torso, arms, etc. BeGaze gives the researcher a side-by-side view of a particular Reference View and the footage taken during the experiment allowing them to mark the gaze path of the test subject on a still image.

As mentioned before, all 66 individual in-depth interviews were transcribed. Afterwards, the semantic layer of the respondents' statements was analyzed and grouped into categories by an experienced market research specialist. Occurrences of certain opinions were counted and the resulting frequencies were used to obtain an intuitive visual presentation of the results.

3   Results

3.1   Eye-Tracking Data

BeGaze software allows the user to present data in various forms such as focus and heat maps, scan paths, Key Performance Indicators or graphs presenting the most important statistics for every AOI.

For the purpose of this paper, a focus map obtained from the eye-tracking device, shown in Fig. 2, was chosen to present the qualitative results. The whole image is darkened and only the parts that participants gazed at are revealed.

Robot's body was divided into 9 AOIs shown in Fig. 2. Statistical data are generated by BeGaze in the form of Key Performance Indicators (KPIs). They are summarized in Table 1. The order in which the various AOIs are listed in the table corresponds to the order in which participants were looking at them during the experiment. KPIs used in this analysis are as follows:

- **Entry time** – average duration that elapsed before the first fixation into the AOI
- **Dwell time** – the sum of all fixations and saccades within an AOI for all subjects divided by the number of subjects
- **Avg. fix. (average fixation)** – the average length of a single fixation in an AOI
- **First fix. (first fixation)** – the average length of the first fixation in an AOI
- **Fix. count (fixation count)** – the average number of fixations in an AOI

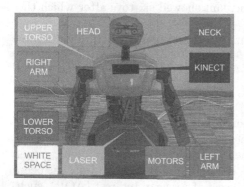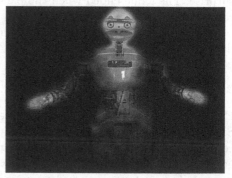

Fig. 2. On the left: Areas of Interest. On the right: Focus map.

Immediately, some regularities come to light. The head of the robot is drawing the most attention. This effect is reflected by the focus map as well as the performance indicators. On the focus map the head is clearly visible which suggests that subjects gazed at it intensely. Statistical data give a more complete view of the situation. On average, during the experiment the subjects looked at the head for 32.4% of the time, and there were approximately 147 fixations in

Table 1. Key Performance Indicators

AOI	Entry time [ms]	Dwell time	Avg. fix. [ms]	First fix. [ms]	Fix. count
Head	2507	52820 (32,4%)	225	215	147
Right Arm	4893	3892 (2,4%)	166	179	16,6
Upper torso	5103	14245 (8,7%)	173	183	56,2
Left Arm	12711	2903 (1,8%)	166	154	12,2
Neck	13334	6112 (3,7%)	206	180	23,2
Kinect	14234	3405 (2,1%)	170	168	15,3
Motors	14745	2724 (1,7%)	154	171	12,3
Lower Torso	16403	2487 (1,5%)	141	135	10,7
Laser	47164	146 (0%)	49	50	0,8

this AOI. This means that participants were looking almost 3 times more often and nearly four times longer at the head than at any other AOI. In addition, statistically, the face was the first area that the subjects fixated on upon entering the experiment area (on average 2386 ms earlier than on any other AOI). This is compliant with the findings of Cerf, Frady and Koch [14]. It is worth noting that the average length of a single fixation is greatest for the head (30% more than the average length of fixation for any other AOI, with the exception of the neck). During the analysis it was also noted that most of fixations were limited to both eyeballs and the "nose" (opening for a camera) of the robot. A similar effect was observed by researchers studying how we perceive human faces [15], [16]. However, since the individual elements of the face were not included in the analysis as separate AOIs, there are no specific statistical data to support this statement.

In the second place in terms of the gaze dwell time is the upper torso. On average the subjects looked at this area for 8.7% of the experiment. Fixations were located mostly in the center and to the top of this region, and did not descend below the opening for the Kinect sensor. Interestingly, although the focus map may suggest differently, the Kinect opening does not seem to attract a lot of attention. On average, throughout the experiment, people were looking at this area for 2.1% of the time, which is a value comparable to the time spent looking at the arms of the robot.

The third area that has attracted considerable attention was the neck. It is in the third place in terms of average total fixation time with the value of 3.7% of experiment duration time. This is less than half of the value of the same indicator for the upper torso but still noticeably greater than the other areas of interest. This effect has not been described in any of the publications known by the authors of this paper. It is possible that the observed attractiveness of the neck is caused only by its location between the head and the upper torso. What seems to disprove this argument is the average duration of a single fixation in this area. It is nearly as long as that recorded for the head of the robot (206 ms - 19% longer than for any other part of the robot, except the head).

When analyzing entry times for different AOIs, the right hand is worth discussing. The fact that it is gazed at in the beginning of the experiment is in no way surprising since, as mentioned in section 2.5, the scenario that FLASH follows began with a handshake. The left arm of the robot was gazed at less intensely. If we look at the focus map, it is clear that fixations are not distributed evenly in this region. Most people glanced at the hands, which was to be expected as the robot gesticulated lively with his fingers in the early parts of the experiment. Interestingly, there were many fixations on FLASH's forearms, while the area from elbow to shoulder hardly attracted any attention at all. This may be due to the robot's mechanical structure. His forearms are unusually massive (if we look at them in human terms) and contain electronic components and drives responsible for moving the fingers.

3.2 Interview Data

Data gathered from in-depth interviews have been analyzed and summarized in the form of four graphs, shown in Fig. 3 and 4. They contain answers to the most important questions posed during the interviews.

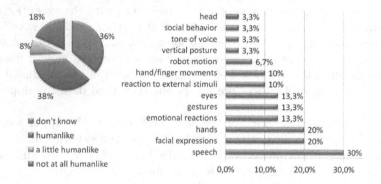

Fig. 3. On the left: Is FLASH humanlike? On the right: What makes FLASH humanlike?

Only 18,2% of respondents declare that FLASH does not remind a human being. More than one third of survey participants were convinced that robot is a humanlike creature. Beside the ability to speak, express emotions and gesticulate, appearance was a crucial factor in perceiving FLASH as a living creature. According to interviewed subjects, the following parts of body play an important role in creating an impression that the robot is humanlike: facial expressions (20%), hands (20%), eyes (13,3%), hand/finger movements (10%), head (3,3%).

According to respondents' declaration hands and fingers (51%) together with head and face (48,5%) attract attention the most. More than one third of test subject indicated eyes as the most noticeable element of the robot's body. For some respondents electronic parts and cables were also very visible.

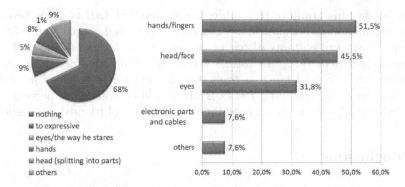

Fig. 4. On the left: What in FLASH's behavior was disturbing or unpleasant? On the right: What elements of FLASH attract your attention the most?

According to 68% of survey participants, there was nothing disturbing or unpleasant in FLASH's behavior. There were some respondents who did not like the robot's hands (i.e. *"I did not want to shake hands with him, because of the fear that he might crush my hand"*, *"His hands were Terminator-like. Macabre"*, *"Hands like those of a skeleton"*, *"I associate his hands with uncovered veins. That's disgusting"*), eyes (i.e. *"I felt uneasy when he looked at me this way"*, *"The eyes were bulging"*). One person mentioned that they did not like the head splitting into parts.

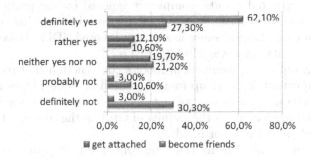

Fig. 5. Could you get attached to FLASH? Could FLASH become your friend?

Most of the respondents claim that they could get attached to FLASH (74,2%). Popular reasons given by participants include: *"People are able to get attached to different things nowadays. That is why they will get attached to FLASH as well."*, *"Since he can communicate I believe I could get attached pretty easily"*, *"It will take time, but it is possible"*, *"Yes. Especially, if FLASH would have some more sophisticated options and functions"*. While getting attached does not seem to be difficult, developing friendship with the robot raises more objections. One third of respondents is convinced that people (including the respondents themselves) will definitely not become friends with FLASH (30,3%). Main obstacle pointed out by this group of test subjects was that *"FLASH is a machine with no real emotions and consciousness"*. Some of interviewed people emphasize the fact that *"Their expectations of friend cannot be met by a machine"*, *"A robot will always be schematic"*, *"All robot's reactions are programmed, so it cannot be spontaneous"*, *"Friendship is a term that should only be reserved for human beings or some living creatures, i.e. dogs"*. It should be noted that percentage of undecided participants is stable over the attachment and friendship question.

4   Conclusions

The experiment proved to be instructive and managed to verify the proposed research methodology and the possibility of long-term autonomous operation of the robot. The results suggest the validity of our hypotheses.

The face of the robot draws the most attention which lets us suspect that just as in human relationships, the respondents maintain eye contact with the robot. FLASH's torso as well as his hands and forearms were also drawing participants' gaze. This kind of behavior was anticipated based on psychological research [4]. Eye-tracking data led to some unexpected conclusions as well. Firstly, neck effectively diverts attention from the face of the robot. It is likely that this effect is caused by the fact that the robot's neck consists mainly of an uncovered servomotor. It has a rough, rectangular shape that stands out against other elements of FLASH. In addition, the motor was constantly moving during the experiment. Secondly, motors located in the exposed parts of the body are, contrary to the expected results, completely ignored by the participants, even though their size is substantial. It is also interesting that openings in the casing for Kinect sensor and laser scanner are mostly ignored. IDI's showed that what attracted respondents' gaze was blinking LEDs.

Results point out to the elements of robot's construction that require modification. It is important to cover up motors, signaling diodes, wires and tendons since they can attract gaze or trigger negative associations. This applies only to the elements that are located in the vicinity of vital (in the context of interaction) body parts, such as the head or hands.

Based on data gathered using IDI's we conclude that people could become attached to FLASH. However this is not a result of specific characteristics of the robot. It is rather caused by the same mechanism that underlies accustoming to objects of everyday use (i.e. mobile phones, cars).

This experiment has shed light on important methodological issues that should be addressed with special care during further experiments regarding social robots appearance. Proper calibration of the eye-tracking device is essential in order to obtain reliable results and should be double-checked. The respondent's eyes need to be clearly visible to the camera in the glasses at all times (obstacles include improper positioning of the glasses, eyelash extensions or swollen eyes).

Future analysis of gathered data will concern the effects that particular traits of the test subjects (such as age, gender, education, technical knowledge) affect interaction with the robot. Also worth exploring is the matter of ascribing intentions to the robot [17], and whether it affects the participants' perception of FLASH.

Acknowledgement. The research work reported here was supported by a statutory grant from Wrocław University of Technology as well as a grant from the National Science Centre of Poland. The authors are indebted to anonymous reviewers whose comments significantly influenced the final form of this paper.

References

[1] Grover, R., Vriens, M.: The Handbook of Marketing Research: Uses, Misuses, and Future Advances. SAGE, Los Angeles (2006)

[2] Zurawicki, L.: Neuromarketing: Exploring the Brain of the Consumer. Springer, Heidelberg (2010)

[3] Nather, F.C., et al.: Time estimation and eye-tracking movements in human body static images. Fechner Day 25, 399–404 (2009)

[4] Cerf, M., et al.: Faces and text attract gaze independent of the task: Experimental data and computer model. Journal of Vision 9, 1–15 (2009)

[5] Chelnokova, O., Laeng, B.: Three-dimensional information in face recognition: An eye-tracking study. Journal of Vision 11(13), 1–15 (2011)

[6] Bannerman, R.L., et al.: Orienting to threat: faster localization of fearful facial expressions and body postures revealed by saccadic eye movements. Proceedings of the Royal Society B 276, 1635–1641 (2009)

[7] FLASH: Flexible LIREC Autonomous Social Helper (2013), http://flash.lirec.ict.pwr.wroc.pl

[8] Kędzierski, J., Janiak, M.: Budowa robota spolecznego FLASH. Prace Naukowe - Politechnika Warszawska (2012) (in Polish)

[9] Kędzierski, J., et al.: EMYS - emotive head of a social robot. International Journal of Social Robotics 5(2), 237–249 (2013)

[10] SMI: Mobile eye-tracker, SensoMotoric Instruments (2013), http://www.smivision.com/

[11] Simply User: Homepage (2013), http://www.simplyuser.pl

[12] Groves, R.M., et al.: Survey Methodology. John Wiley & Sons, New Jersey (2009)

[13] Hesse-Biber, S.N., Leavy, P.: The Practice of Qualitative Research. SAGE, Los Angeles (2008)

[14] Cerf, M., et al.: Faces and text attract gaze independent of the task: Experimental data and computer model. Journal of Vision 9(12), 1–15 (2009)

[15] Snow, J., et al.: Impaired visual scanning and memory for faces in high-functioning autism spectrum disorders: It's not just the eyes. Journal of the International Neuropsychological Society 17(6), 1021–1029 (2011)

[16] Perlman, S.B., et al.: Individual differences in personality predict how people look at faces. PLoS ONE 4(6), 1–6 (2009)

[17] Wiese, E., Wykowska, A., Zwickel, J., Müller, H.J.: I See What You Mean: How Attentional Selection Is Shaped by Ascribing Intentions to Others. PLoS ONE 7, 1–9 (2012)

Playful Interaction
with Voice Sensing Modular Robots

Bjarke Heesche, Ewen MacDonald, Rune Fogh,
Moises Pacheco, and David Johan Christensen

Department of Electrical Engineering
Technical University of Denmark
Kgs. Lyngby, Denmark
bjarke.heesche@gmail.com, {emcd,rufo,mpac,djchr}@elektro.dtu.dk

Abstract. This paper describes a voice sensor, suitable for modular robotic systems, which estimates the energy and fundamental frequency, F_0, of the user's voice. Through a number of example applications and tests with children, we observe how the voice sensor facilitates playful interaction between children and two different robot configurations. In future work, we will investigate if such a system can motivate children to improve voice control and explore how to extend the sensor to detect emotions in the user's voice.

1 Introduction

We are developing the Fable modular robotic system to enable and motivate non-expert users to assemble and program their own robots from user-friendly building blocks [9]. A user assembles a robot by connecting several robotic modules together where each module provides functionality to perform a specific task, such as identifying users, characterizing the environment, moving, or manipulating its surroundings. In this paper we describe the development of a voice sensor for the Fable system that is able to measure different features (energy and F_0) of the user's voice (in particular children's). Our objectives with developing this sensor include the following:

1. Facilitate playful interaction between robots and users.
2. Enable users to create their own voice enabled robot applications.
3. Motivate users to improve the control of their voice (e.g., pitch and voice level).

In this paper, we focus on the first objective by demonstrating four applications that use the voice sensor to create games and playful interaction with the Fable system (see Fig. 1). In future work we will address the other two objectives by integrating the voice sensor and our programming tool-kit to enable the users to also program their own applications. As well, we plan to extend the sensor functionality to detect emotion in children's voices.

G. Herrmann et al. (Eds.): ICSR 2013, LNAI 8239, pp. 180–189, 2013.

Fig. 1. User testing with two Fable robots controlled using the voice sensor (voice energy and F_0 sensing). The humanoid has controllable arms and the quadruped has controllable locomotion.

2   Related Work

As the name implies, modular robotic systems consist of open-ended modules that can be combined in different ways to form a variety of different configurations [18, 19]. Our approach to designing modular playware [7] combines modular robotics with the concept of construction toys (such as LEGO) to facilitate playful creativity and learning. Examples of modular robotic playware include the Topobo system which enables the user to record and playback motor sequences using programming-by-demonstration [11] and the roBlocks (now Cubelets) which utilizes a programming-by-building strategy where the behavior emerges from the interaction of the modules with each other and their environment [15]. Similarly, LEGO Mindstorms, LEGO WeDo, and PicoCricket [13] are robotic construction kits that enable direct-user-programming of actuated and sensing models. In some ways, our Fable system is similar to these examples and is also influenced by Papert's learning theory of constructionism [10]. Our aim is to enable everyone to become a robot designer and encourage them to imagine, create, play, share and reflect (i.e., the kindergarten approach to learning [12]).

We hypothesize that by enabling the creation of socially interactive robots, we can motivate a wider range of users to participate in this activity. In order to create these robots, we must develop sensors that can reliably interpret human communication. While, much research has been dedicated to machine recognition of human communication (e.g., [6]) and its applications to social robots (e.g., [2,3, 14,17]), our approach is to simplify the sensor design by limiting the detection to simple units of communication (e.g, gestures). Previous work along these lines to track and detect a user's posture using the Fable system has been promising [8]. In this paper, we limit the sensor to the detection of a single acoustic variable, the fundamental frequency or F_0, and then examine if playful interaction can be sustained using input only from this detector.

3   The Voice Sensor

While the main focus of the sensor is on the estimation and tracking of the user's F_0 over time, we must first detect if the user is vocalizing. Thus, the voice sensor consists of two elements: an energy detector and and F_0 estimator.

3.1   Energy Detector

The simplest approach to detecting if a user is vocalizing is to measure and track in real-time the acoustic energy picked up by a microphone. A simple approach is to calculate the root mean square (RMS) of the signal over some window. However, to reduce the number of computations, the signal is squared and low-pass filtered using an exponential moving average (EMA). Thus, the energy E, for a given signal $x[i]$ is estimated as follows:

$$y[i] = \alpha \times y[i-1] + (1-\alpha) \times x[i]^2 \tag{1}$$

where α is a weighting factor. For our application, with a sampling frequency, F_s of 8 kHz, we set $\alpha = 0.990$. Thus, the time constant of the EMA was 15 ms and resulted in a relatively stable measure of E during vocalizations. The output is energy in dB relative to $E_{ref} = 1$ is as follows:

$$E_{dB}[i] = 10 \times log_{10}(y[i]) \tag{2}$$

If a signal is detected with an energy level, $E[i] > \beta$, where $\beta = 0.05$, it is assumed to be a sound produced by the user.

3.2 F_0 Sensor

In speech, F_0 refers to the fundamental frequency, which corresponds to the frequency at which the vocal folds open and close. However, not all speech is voiced (i.e., uses the vocal folds to produce a periodic acoustic excitation). If the vocal tract is sufficiently constricted, then the air flow will become turbulent, which produces sound. As well, for plosive sounds, airflow can be stopped and the pressure can be allowed to build significantly before the air is released in a burst. Thus, speech can be separated into voiced (periodic) or un-voiced (aperiodic) sounds. However, when estimating F_0, we currently assume that the signal is always voiced. In the future, we plan to add a harmonicity detector to determine if the vocalizations produced are voiced or unvoiced.

As with the energy detector, the sensor is designed to estimate F_0 in near-real-time (i.e., low latency) to facilitate interaction. Further, as the goal is to

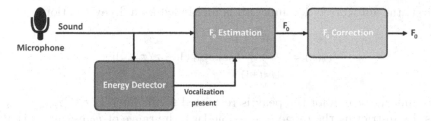

Fig. 2. Elements in the F_0 sensor

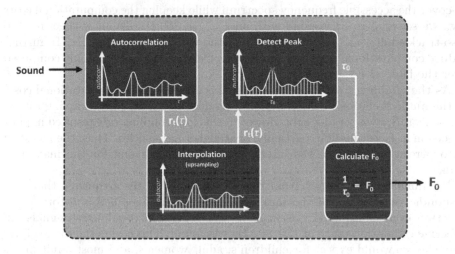

Fig. 3. Elements in F_0 estimation

develop inexpensive modular sensors, the algorithm used should be feasible to implement using current embedded processors for autonomous control.

The F_0 sensor consists of three major elements (see Fig. 2). The energy detector, described previously, is used to determine if the user is vocalizing. If speech is detected, the second element estimates F_0 of the microphone signal. The final element, F_0 correction, monitors the F_0 estimates to detect and removes outliers caused by noise or unvoiced sounds.

F_0 **Estimation:** The F_0 estimator uses an autocorrelation method based on [4], which is simple and computationally efficient (see Fig. 3). The autocorrelation function $r_t(\tau)$ is computed for short time windows of length W_{auto}. If the signal is periodic then the autocorrelation function will have peaks at lags τ corresponding to the period of the signal. Thus, the fundamental frequency is estimated by finding the time lag corresponding to the first peak in the autocorrelation function. To improve performance in picking the correct peak (i.e., the one corresponding to the fundamental frequency), two modifications are made.

First, the autocorrelation function is multiplied by a decay function.

$$r_t(\tau) = \sum_{(j=t+1)}^{(t+W_{auto})} [x_j x_{j+\tau}(1 - \tau/\tau_{max})] \tag{3}$$

Second, the search for the peak is restricted to the interval $\tau \in [\tau_{min} : \tau_{max}]$. Thus, by restricting the range of possible lags, the range of F_0 estimates is also restricted. While the average fundamental frequency for a child is 300 Hz for speech [5], F_0 can be higher for non-speech vocalizations such as humming. Thus, to cover the accessible frequency spectrum while keeping the computational costs low, we set τ_{min} to correspond with an F_{max} of 800 Hz. To allow the sensor to also track adults, τ_{max} was set to correspond with an F_{min} of 44 Hz. To further reduce computational costs, the modified correlation function was only computed over the interval $\tau \in [\tau_{min} : \tau_{max}]$.

As the sampling frequency of the signal is increased, the computational costs of the autocorrelation function increase significantly. Thus, the sampling rate of the sensor, F_s, was set at 8 kHz. However, this low sampling rate resulted in poor precision in F_0 estimation, particularly at higher frequencies. Thus, the modified autocorrelation function was interpolated before searching for the maximum peak.

In summary, the output from the F_0 estimator is the frequency that corresponds with the lag of the maximum peak of the modified autocorrelation function over the interval $\tau \in [\tau_{min} : \tau_{max}]$. Using sinusoidal test signals, an effective estimation range of 120–800 Hz was found. This covers the range of frequencies we would expect for children's, adult women's, and most adult men's vocalizations.

F_0 Correction: While the F_0 estimator works well, some artefacts are observed in running speech due to unvoiced speech sounds or when the sensor is placed in a noisy environment. Thus, an F_0 correction stage was developed.

For outlier detection, the F_0 estimates are monitored and the interquartile range (IQR; i.e., range between the first and third quartile) is estimated. Using this, upper and lower boundaries are set to $Q_3 + 1.5 \times IQR$ and $Q_1 - 1.5 \times IQR$ respectively. Estimates outside this range are considered to be outliers. Before discarding the outlier, an octave correction is tested. The outlier is shifted up or down by an octave towards the frequency range where most F_0 estimates are present. If the difference between two neighbour estimates is less than 10 Hz, the correction is considered to be successful. Otherwise, the outlier is discarded.

Performance: An example of F_0 estimation is shown in Fig. 4. To evaluate the performance of the system, 1150 children's utterances [16] were tested. Each utterance was processed by both the voice sensor and Praat [1] and the F_0 estimates were compared. As there were occasions where Praat detected voiced speech but the sensor did not (and vice versa), the comparison was restricted to time points where both Praat and the voice sensor produced F_0 estimates.

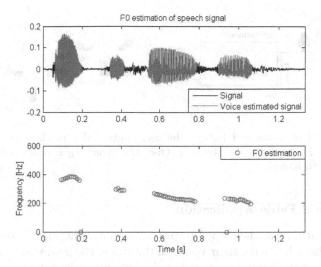

Fig. 4. Example of voice and F0 estimation for a speech signal

Overall, when F_0 correction was not used in the voice sensor, the mean error was 14 Hz. When F_0 correction was employed, the mean error was reduced to 4 Hz. In summary, the system is sufficiently accurate for our applications, but we can improve its performance in future work.

4   Test of Applications

4.1   Test Setup

A total of four different applications were tested, across two configurations of the Fable system (one using a humanoid and three using a quadruped; see Fig. 1). Children's vocalizations were recorded using headsets connected to a computer. The voice sensor processing was conducted in Matlab running on a standard PC, which sent commands over a wireless serial connection to control the Fable actuators.

Four girls and four boys ranging in age from 7-10 years participated in the testing. The children were divide into four pairs according to their age: younger girls (7-8 years old), older girls (10 years old), younger boys (8-9 years old) and older boys (10 years old). As the average F_0 can vary across age and gender the individual parameters $F_{0,min}$ and $F_{0,max}$ were measured for each child prior to the test.

The aim of the tests was to demonstrate if the Fable system and the voice sensor can be successfully used by children to create playful interactions. The tests were designed as small games in order to motivate the users to engage and play with the system.

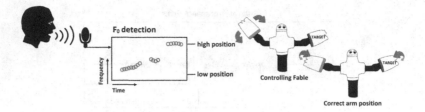

Fig. 5. F_0 control of humanoid Fable. The user controlled arm changes its position according to the F_0 produced by the user. Once the target F_0 is produced, the robot starts waving both hands.

4.2   Humanoid Fable Application

The first application was developed to provide an introduction to how the users could control the robot with their voice. In the game, the goal was to angle both of the Fable arms the same way (see Fig. 5). One arm, the target, was initially angled by the experimenter using a gamepad. The other arm was controlled by the F_0 of the user's voice. The user hummed or sung a tone into the microphone and the arm would raise or lower based on the F_0 that was produced. Once the controlled arm matched the target, the Fable started waving its arms. This waving continued for as long as the user could hold the tone. This task was selected as a good introduction because the children received clear visual feedback from the robot (i.e., the angle of its arm) that corresponded to the pitch they produced. Three different random target values were tried for each of the eight children.

The task was easily understood and all of the children were able to complete it. However the older children were slightly better at the task. Initially, two of the younger children had difficulties because they were changing their pitch too fast and/or in large steps. As well, the younger girls required more instruction before understanding the relationship between their voice and Fable's arm position. Regardless, the most common strategy for completing the task was to start at a random pitch and adjust it up or down seeking the target position. Even though the task was developed as an introduction to the concept of controlling a robot by voice, the children were very engaged in the activity and appeared to be enjoying themselves.

4.3   Quadruped Fable Applications

In the following applications the objective was to move the quadruped robot through a track with four gates. The F_0 of the users' voices was used to select between four different movements (forward, backward, left-turn, right-turn). In addition to the robot's movement, visual feedback was also provided on a computer screen. The F_0 being produced was plotted as a large dot that moved over a background that illustrated the target pitch corresponding to each motion. Three different applications were tested:

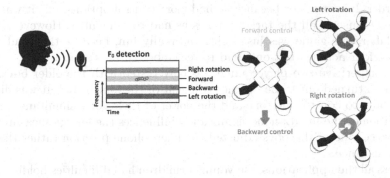

Fig. 6. F_0 control of quadruped Fable. By producing an F_0 that corresponds to one of the targets, the user can select different movements and control the robot.

Single Player: In this game, there were two different levels of difficulties. At level 1, there were only two pitch targets corresponding to forward and left-turn movements. At level 2, there were four pitch targets, each corresponding to one of the four movements (see Fig. 6).

Two Player - Collaborative: In this game, two players collaborated to control the quadruped using two pitch targets. One user controlled forward and backward movement while the other controlled turning left and right.

Two Player - Competitive: In this game, two players competed to gain control over the quadruped and move it to a specific area to win the game. One user controlled forward and left-turn and the other user controlled backward and right-turn. The user that was loudest, while maintaining a target pitch, controlled the robot. However, to avoid the users screaming as loud as possible, a maximum value was set for the voice energy. If the user exceeded this level, control was lost until the voice level was reduced below the maximum threshold.

All eight children participated in the single player game at level 1. The four oldest quickly understood the task and learned how to shift between tones in order to get the robot moving through the track. In contrast, only one of the younger boys performed well. The others needed many attempts before completing the task. The two older boys tried level 2 and were able to control the robot successfully. While we anticipated that level 2 would be more difficult, the boys were able to complete the route faster than at level 1.

As the younger children had difficulty with the singleplayer game it was decided that they would participate in the collaborative two-player game with the older children. The collaborative game added an additional complexity as the users needed to communicate with each other without triggering the robot by mistake. This challenge engaged the children to communicate with gestures or mute the microphone, and they were surprisingly good at completing the task. One of the boys commented that the single player game on level 1 was a bit hard, because he could only turn left. He also felt that level 2 was both easier and harder. It was easier because he had more control options, but it was more difficult because of all the target pitches he had to remember. However, he felt the collaborative game was just right, and really fun, because they had access to all the four actions, but only had to remember two notes at a time.

The competitive two-player game was only tested with the older boys. This application turned out to be the most difficult of the four tested. It was difficult for the boys to control the tone and the power of their voice simultaneously. As the position of the microphone significantly influences the energy measurements of the voice, one of the boys adjusted the microphone position rather than the level of his voice.

Across all the applications, the younger children had difficulties holding a specific pitch for a long period of time. Instead, they would hum a small melody. This approach might have been easier for them had they been instructed to change between two melodies instead of two tones. As a result, the older children seemed to have more fun when playing with the system. The two older boys continuously asked to play the collaborative game again after finishing. If the interactions are not changed, it is suggested that future tests focus on the older age group (i.e., 9 years and older). In summary, these example applications confirmed that we can use the voice sensor to facilitate playful interaction between the children and Fable robots. In addition, we believe that the children were also learning to better control their voices but further work is needed to test this.

5   Conclusion and Future Work

This paper presented a voice sensor for the Fable modular robotic system. The sensor measures the energy in the user's voice and estimates F_0, which is related to the pitch of the voice, with very low latency. The performance of the F_0 estimation was tested and found sufficient for the purpose. Four different applications utilizing the voice sensor were tested with eight children and we found that the system facilitated interaction that was challenging, playful and motivating.

In future work, more voice sensor functionality will be added and the possibility of extending the current processing for use in emotion detection will be explored. In addition, we will integrate the voice sensor with our existing software tool-kit to enable users to program their own applications.

Acknowledgements. This work was performed as part of the "Modular Playware Technology" project funded by the Danish National Advanced Technology Foundation.

References

1. Boersma, P., Weenink, D.: Praat [computer software]. University of Amsterdam, Amsterdam (2006), http://www.fon.hum.uva.nl/praat/
2. Breazeal, C., Aryananda, L.: Recognition of affective communicative intent in robot-directed speech. Autonomous Robots 12(1), 83–104 (2002)

3. Breazeal, C., Scassellati, B.: A context-dependent attention system for a social robot. rn 255, 3 (1999)
4. De Cheveigné, A., Kawahara, H.: Yin, a fundamental frequency estimator for speech and music. The Journal of the Acoustical Society of America 111, 1917 (2002)
5. Jacobsen, F., Poulsen, T., Rindel, J.H., Christian Gade, A., Ohlrich, M.: Fundamentals of acoustics and noise control, Note no 31200 (2010)
6. Lippmann, R.P.: Speech recognition by machines and humans. Speech Communication 22(1), 1–15 (1997)
7. Lund, H.H.: Modular interactive tiles for rehabilitation: evidence and effect. In: Proceedings of the 10th WSEAS International Conference on Applied Computer Science, ACS 2010, Stevens Point, Wisconsin, USA, pp. 520–525. World Scientific and Engineering Academy and Society (WSEAS) (2010)
8. Magnsson, A., Pacheco, M., Moghadam, M., Lund, H.H., Christensen, D.J.: Fable: Socially interactive modular robot. In: Proceedings of 18th International Symposium on Artificial Life and Robotics, Daejeon, Korea (January 2013)
9. Pacheco, M., Moghadam, M., Magnusson, A., Silverman, B., Lund, H.H., Christensen, D.J.: Fable: Design of a modular robotic playware platform. In: Proceedings of the IEEE Int. Conference on Robotics and Automation (ICRA), Karlsruhe, Germany (May 2013)
10. Papert, S.: Mindstorms: Children, computers, and powerful ideas. Basic Books, Inc. (1980)
11. Raffle, H.S., Parkes, A., Ishii, H.: Topobo: a constructive assembly system with kinetic memory. In: Proceedings of the SIGCHI Conference on Human Factors in Computing Systems, CHI 2004, pp. 647–654. ACM, New York (2004)
12. Resnick, M.: All i really need to know (about creative thinking) i learned (by studying how children learn) in kindergarten. In: Proceedings of the 6th ACM SIGCHI Conference on Creativity & Cognition, pp. 1–6. ACM (2007)
13. Resnick, M., Silverman, B.: Some reflections on designing construction kits for kids. In: Proceedings of the 2005 Conference on Interaction Design and Children, pp. 117–122. ACM (2005)
14. Saldien, J., Goris, K., Yilmazyildiz, S., Verhelst, W., Lefeber, D.: On the design of the huggable robot probo. Journal of Physical Agents 2(2), 3–12 (2008)
15. Schweikardt, E., Gross, M.D.: roblocks: a robotic construction kit for mathematics and science education. In: Proceedings of the 8th International Conference on Multimodal Interfaces, ICMI 2006, pp. 72–75. ACM, New York (2006)
16. Steidl, S.: Automatic Classification of Emotion. Related User States in Spontaneous Children's Speech. PhD thesis, University of Erlangen-Nuremberg (2009)
17. Stiehl, W.D., Lee, J.K., Breazeal, C., Nalin, M., Morandi, A., Sanna, A.: The huggable: a platform for research in robotic companions for pediatric care. In: Proceedings of the 8th International Conference on Interaction Design and Children, IDC 2009, pp. 317–320. ACM, New York (2009)
18. Stoy, K., Brandt, D., Christensen, D.J.: Self-Reconfigurable Robots: An Introduction. Intelligent Robotics and Autonomous Agents Series. The MIT Press (2010)
19. Yim, M., Shirmohammadi, B., Sastra, J., Park, M., Dugan, M., Taylor, C.J.: Towards robotic self-reassembly after explosion. In: Video Proceedings of the IEEE/RSJ Intl. Conf. on Intelligent Robots and Systems (IROS), San Diego, CA (2007)

Social Comparison between the Self and a Humanoid

Self-Evaluation Maintenance Model in HRI and Psychological Safety

Hiroko Kamide[1], Koji Kawabe[2], Satoshi Shigemi[2], and Tatsuo Arai[3]

[1, 3] Department of Engineering Science, Osaka University
Machkaneyamacho 1-3, Toyonaka, Osaka 560-8531, Japan
[2] Honda R&D Co., Ltd.
Honcho 8-1, Wako, Saitama 351-0188, Japan
kamide@arai-lab.sys.es.osaka-u.ac.jp

Abstract. We investigate whether the SEM model (Self-evaluation maintenance model) can be applied in HRI in relation to psychological safety of a robot.
The SEM model deals with social comparisons, and predicts the cognitive mechanism that works to enhance or maintain the relative goodness of the self.
The results obtained from 139 participants show that the higher self-relevance
of a task is related to a lower evaluation of the robot regardless of actual level
of performance. Simultaneously, a higher evaluation of performance relates to
higher safety. This study replicates the prediction of the SEM model. In this paper, we discuss the generality of these results.

Keywords: Social Comparison, Self, Humanoid, Psychological Safety, Self-
evaluation Maintenance Model.

1 Introduction

With the ongoing development in the field of robotics, the harmonious coexistence of
robots and humans is expected to be realized. Previous studies have discovered important robot parameters in real-time interaction between humans and robots [1-4]. At
the same time, focusing on social relationships, not only on immediate interactions,
should give significant suggestion to understand how coexistence can be possible. We
aim to clarify the cognitive mechanism of social comparisons between humans and a
robot in relation to psychological safety of a robot. This study is the first to adopt a
social psychological theory which predicts a cognitive mechanism to maintain self-
evaluation in interpersonal relationships into HRI to understand social relationships
between humans and robots.

2 Self, Others and Social Comparison

2.1 The Self and a Social Comparison with Others

The consciousness of the self can be discussed in two aspects: the self as the knower
(I), and the self as known (me) [5]. Humans (I) have a basic motivation to evaluate

G. Herrmann et al. (Eds.): ICSR 2013, LNAI 8239, pp. 190–198, 2013.

the self (me) to work on the social environment efficiently [6]. That is, if humans can obtain an objectively correct self-evaluation, they can react to their social environment in a more effective manner. To assess the self correctly, it is necessary to compare the self with others in terms of opinions or abilities. This is called a social comparison [6]. This process emphasizes the social effects on the self. Through social comparisons, humans are getting to know the aspects of the self and are composing the self-concept. While humans need to assess the self correctly, at the same time they want to evaluate the self positively [7].

2.2 Self-Evaluation Maintenance Model

The self-evaluation maintenance (SEM) model predicts cognitive dynamism to maintain or enhance a positive self-evaluation through social comparisons in social relationships [8, 9], while depending on three basic factors: (1) the performance level of the target (i.e., whether the target as compared with the self is successful), (2) the importance of the comparison contents to the self (i.e., whether the task or ability is self-relevant), and (3) the closeness of the comparison target to the self (i.e., whether the target is similar to the self in age, sex, social status, and so on) [10-12]. For example, when a target shares their social status with the self (i.e., a friend who is a student at the same school), the target's performance should be perceived more positively for certain tasks or abilities of self-irrelevance (to the self), and less positively for certain tasks or abilities of high self-relevance. If the target is psychologically distant (i.e., a stranger), this tendency should be attenuated.

The self-evaluation is defined as the relative goodness that people recognize in the self, or a meta-cognition of the evaluation of others. Unlike self-esteem, which is defined as a stable trait, a self-evaluation is affected by the relationships with the targets, as well as the targets' behaviors, and is changeable at different times. To maintain or enhance their self-evaluation, humans regulate three basic factors ((1) the performance level of the target, (2) the importance of the comparison contents to the self, and (3) the closeness of the comparison target) depending on the situation in terms of cognition, behavior, or both. The SEM model deals with daily human life in relation to social relationships with others and an adaptation of the self. Studies have verified the SEM model not only in experimental laboratories [10-12], but also in daily interpersonal relationships with friends [13]. [13] revealed that the SEM model covers the prediction of mental health in addition to a positive self-evaluation[10-12].

3   Sem Model in HRI

3.1   Apply SEM Model in HRI

Robots can support people but they may worry about replacing humans with robots and show a negative repulsion toward robots [14]. Such anxieties occur exactly through a social comparison between the self and robots. Therefore, it is necessary to consider how humans compare robots to the self and develop comfortable relationships. This study challenges to apply the SEM model in relationships between humans and a robot. If a prediction from the SEM model can be applied to HRI, it will be revealed why people resist/accept the activities of robots.

As the first step, we deal with two factors of the SEM model: the performance level of the target and the importance of the tasks to the self. As for the task to compare, this study uses a presentation task as one of most feasible task. If the SEM model fits HRI, when the presentation task is relevant to oneself, one's evaluation of the performance level of the robot's presentation will decrease. At the same time, when the task is self-irrelevant, such evaluation will not decrease [8-12]..

3.2   SEM Model and Psychological Safety

We investigate the SEM model in relation to perceived comfort, especially psychological safety with robots which is an important evaluation for robots [15]. [16] revealed that there are three core factors such as psychological safety of robots: *Acceptance*, *Performance*, and *No-glitch*. *Acceptance* indicates how comfortable and at ease people are with robots. *Performance* is the high-level cognitive and behavioral abilities of robots. *No-glitch* indicates that a robot does not break down or fall accidentally. As [16] showed, robots that are higher in performance will be perceived as safer. Therefore, if robots demonstrate excellent performances, people will feel comfortable with them. However, it is necessary to consider this relationship by taking self-relevance into account, and assessing the SEM model and psychological safety comprehensively.

Additionally, we measure the amylase level by buccal secretion which can be used to determine one's level of stress [17]. While the main focus is psychological safety, a physiological test such as saliva amylase will also be explored additionally. We measured the amylase of the subjects three times: the first before observing the robot, the second immediately after the observation, and the third after all experiments were completed.

3.3   An Actual Level of Performance by a Robot

We manipulate the performance level objectively to confirm that cognitive mechanism of SEM model would work regardless of the actual excellence of performance. We define an excellence of presentation as audience's correct understanding of main messages of the presentation. To check the manipulation, we set up a surprise test after observing a presentation. We manipulated the robot's behaviors based on the behaviors of teachers toward their students in a classroom [18, 19] and prepare four conditions which differ in the level of excellence.

Excellent teachers keep attentions of students and emphasis the important points of the classes. These two activities are crucial for the learning of the students. Open postures and eye-contact toward the students are effective to keep their attention. Pointing behaviors on the screen in front of the students mean the emphasis of importance of the pointed words or contents. Thus, we set the four patterns of behavior of the robot. First includes both behaviors of keeping attention and emphasis of important words of the screen. Second and third is assigned respectively to only keeping attention or only emphasis of words. Fourth doesn't include either and the robot shows simple motions to wave arms and legs repeatedly.

4   Experiment on a Robot Presentation

4.1   Participants and Procedure

A total of 139 people ranging in age from 10 into their 60s (mean age, 39.57; SD, 15.85; 67 males and 72 females) participated in this experiment. We used ASIMO as an example of a robot. The presentation was about activities by Honda. The contents are how to clean the beach by a car automatically, why Honda wants to do this, and the details of the place to be cleaned and so on. The overall length of the presentation was about 10 min. The robot speaks the same explanation in all conditions. There were no trouble through all experiments, therefore participants had sit and watched the presentation of the robot which made no mistake at all during 10 minutes..

One experiment included six participants, each sitting in a box separated by partitions (Figure 1). The participants first signed a letter of consent, after which, they gave a sample of their buccal secretion using a special tip. This process took about one minute. They completed the first questionnaire about self-relevance. They then observed the robot's presentation and provided another buccal secretion. The participants took a surprise test which was limited to 4 minutes. After the test, they completed a questionnaire, which included psychological safety, and an evaluation of the robot's performance compared to the participants' self perception. Finally, the participants gave a third buccal secretion sample.

Fig. 1. Observing place and presentation by the robot

4.2   Measurement

The participants completed the following items related to self-relevance (three items), i.e., "It is important to me to give an accurate presentation to the audience," "It is important to me to give an attractive presentation to the audience," and "It is important to me to give an excellent presentation to the audience." We combined these three items together to make a single variable as self-relevance.

After observing the presentation, the participants were given a surprise test, which included 11 questions about main points of the presentation. The scores were summed and the correct rate computed. Finally, the participants completed a questionnaire related to psychological safety [16]. The scale for psychological safety contained 33 items ranked on a seven-point Likert scale with responses that ranged from 1 (strongly disagree) to 7 (strongly agree). The scale has five factors which compose the psychological safety and we used main three factors; *Acceptance*, *Performance*, and *No-glitch*. *Acceptance* reflects personal acceptance of the robots in people's life. *Performance* encompasses perceived high technology of the robots. *No-glitch* means the degree which the robots don't seem to be broke down easily. We used three items for each factor which show high loading to the corresponding factor in the result of [16]. Examples of each factor are as following, "The robot looks as if it could comfort me. (*Acceptance*)", "The robot looks as if it can correctly understand human speech.(*Performance*)", and "The robot looks rather fragile. (*No-glitch*)"

As for objective excellence of presentation, [18, 19] showed that both approaches toward the audience and the presentation content are important to giving a communicative presentation. We manipulated both behaviors as high and low level, therefore four conditions were prepared. The level of performance was checked through the surprise test. In particular, for the serial position effect [20, 21], we used the correct rate of questions of the middle three minutes of the presentation.

5   Results and Discussion

5.1   Manipulation Check of the Performance Level

Table 1 shows the descriptive statistics of the variables such as SR (self-relevance), SC (social comparison). The amylase scores have missing data owing to problems with the monitor. We used the available data, all of which include all three measurements (N=119), for our analysis of the participants' level of amylase.

We conducted an ANOVA (for the four conditions of behaviors) for the correct rate for the middle of the presentation. The results showed that there is a significant difference between the four conditions ($F (3, 132) = 3.19$, $p < .05$). That is, the different behaviors of the robot had significantly an effect on the correct rate and the objective excellence of the performance is determined different among the conditions.

Table 1. Descriptive Statistics

	N	Min	Max	M	SD
SR(Self-relevance)	139	1.00	7.00	5.57	1.42
SC (Social comparison of the performance between the robot and the self)	139	1.00	7.00	5.12	1.36
Test score (correct rate)	139	0.14	1.00	0.50	0.17
Psychological safety					
Performance	139	2.00	7.00	4.54	1.06
Acceptance	139	1.00	7.00	4.24	1.33
No-glitch	139	2.33	7.00	4.81	1.14
Acceptance of robot's activity					
Social acceptance rate	139	25.00	200.00	98.51	21.57
Personal acceptance rate	139	25.00	300.00	100.33	36.51
Amylase T1: before observation, T2: after observation, T3: after experiment					
Amylase T1	119	3.00	283.00	42.52	38.17
Amylase T2	124	3.00	186.00	41.96	35.71
Amylase T3	123	3.00	187.00	34.15	26.69

5.2   Structural Equation Modeling; Main Analyses

We conducted structural equation modeling by AMOS 18 [22]. Initially, the entire relationship among the SR, SC, and psychological safety, which is the main issue of this study, is analyzed. The model was modified based on the modification indices. We show the significant path and no error variables in the figures.

The overall fit of the proposed model was examined using the goodness-of-fit index (GFI) [23], the adjusted GFI index (AGFI) [24], and the comparative fit index (CFI) [25]. Values for the GFI, AGFI, and CFI generally range from 0 to 1 with larger values indicating a better fit of the model to the data. [26] recommend a cutoff of .90 for the GFI. For the AGFI, [27] suggested that a value of less than .80 can be regarded as inadequate. [25] recommended that values of the CFI should not be less than .90. The χ^2 statistic can be used to evaluate the goodness-of-fit of a model, with a significant χ^2 suggesting a poor fit. In total, the following model showed a good fit: GFI = .99, AGFI = .93, CFI = .96, and χ^2 = 5.3 (df = 3, p = .15).

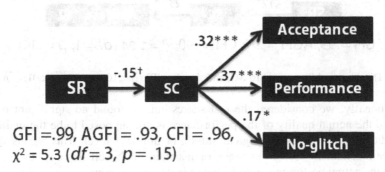

Fig. 2. The relationship among SR, SC, and psychological safety ($^\dagger p<.10$, $^* p<.05$, $^{**} p<.01$, $^{***} p<.001$)

The arrows represent causal relationships between variables. If the score is positive (+), the relationship between the scores is positive. If the score is negative (-), the scores are related with each other negatively. As shown in Figure 2, the SR relates negatively to the SC and SC relates positively to the three factors of psychological safety. A higher level of self-relevance led to a lower evaluation of the robot. When people view a task of presentation as more important to the self, admitting the ability of the robot may be very difficult. Thus, people may try to recognize the performance as being worse than their selves. This relationship fits the prediction of the SEM model, although the path is a significant tendency.

However, the psychological safety is predicted by a higher evaluation of the robot's performance. The factor of *Performance* seems easy to connect the evaluation of the performance because the items of *Performance* ask the degree of high technology of the robot; however, the results show a significant relationship between SC and *Acceptance* and *No-glitch*, too. Considering comprehensively, if the task is self-relevant, a better performance may decrease psychological safety of the robot based on a social comparison.

As for amylase, as indicated in Figure 3, the SC is significantly related to Amylase T2, while Amylase 1 and 3 do not show a significant relationship. It is surprising that the relationship between the SC and Amylase T2 is negative. This means that a higher evaluation of the robot's performance is related to a higher level of stress. This result indicates the opposite meaning as the results shown in Figure 3.

The participants interpreted the information during the presentation through cognitive processing to answer the items of safety. In contrast, the amylase data was physically taken immediately after watching the robot. The timing to take the data and quality of the data between safety and amylase are quite different. The amylase data may reflect a physical arousal from watching an unfamiliar robot. However, in this study we emphasis the result of psychological safety. To realize coexistence with humans and robots, it is the first to consider people's interpretations toward robots such as subjectively perceived safety. A comprehensive understanding of the data for both cognition and physiology should be discussed in future research using more efficient data.

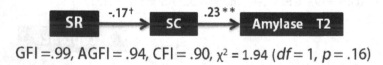

$$\text{SR} \xrightarrow{-.17^{\dagger}} \text{SC} \xrightarrow{.23^{**}} \text{Amylase   T2}$$

$$\text{GFI} = .99, \text{AGFI} = .94, \text{CFI} = .90, \chi^2 = 1.94 \ (df = 1, p = .16)$$

Fig. 3. The relationship among SR, SC, and Amylase T2 ($^{\dagger}p<.10$, $^{*}p<.05$, $^{**}p<.01$, $^{***}p<.001$)

Additionally, we considered the test scores but we found no significant relationships and the actual quality of the performance is not supposed to be the main factor for predicting the SC, psychological safety, or amylase level. This means that the cognitive mechanism used to enhance or maintain one's self has a robustness compared to the actual performance of the robot in this experiment.

6   Conclusion

We challenged to apply the SEM model in HRI regarding psychological safety. We found that a similar cognitive process occurred in HRI with a social comparison of interpersonal relationships. When the task is relevant to the self, people are likely to give a worse evaluation of the performance of the robot. It seems that even if the target is a robot, it is possible to decrease the self-evaluation through a social comparison with the robot. At the same time, a higher performance of the robot positively enhances the psychological safety. In sum, to feel more comfortable with a robot, it is necessary for the robot to perform "better" on the task which people see "self-irrelevant".

This result suggests that if the service the robot provides has significant meaning to the self of the user, the user will be threatened by the robot. However, if the service is unrelated to the self, then an excellent job by the robot will be admitted as safer.

The amylase shows the opposite prediction and this result is important in a sense that subjective data and physiological data are not always congruent. However, it is necessary to add more physiological data to conclude the meaning of physiological reactions. We did not manipulate the closeness of the robot. It is difficult to manipulate the closeness with the robot but it is necessary to discuss how to manipulate closeness with robots, and to investigate this factor using the SEM model in HRI. We have to confirm this result in other tasks, too.

Acknowledgement. This work was supported by JSPS KAKENHI Grant Number 23730579. This research was supported by Global COE Program "Center of Human-Friendly Robotics Based on Cognitive Neuroscience" of the Ministry of Education, Culture, Sports, Science and Technology, Japan.

References

1. Kanda, T., Miyashita, T., Osada, T., Haikawa, Y., Ishiguro, H.: Analysis of Humanoid Appearances in Human-robot Interaction. IEEE Trans. on Robotics 24, 725–735 (2008)
2. Broadbent, E., MacDonald, B., Jago, L., Juergens, M., Mazharullah, O.: Human reactions to good and bad robots. In: Proc. of the IEEE/RSJ International Conference on Intelligent Robots and Systems, San Diego, pp. 3703–3708 (2007)
3. Muto, Y., Takasugi, S., Yamamoto, T., Miyake, Y.: Timing Control of Utterance and Gesture in Interaction between Human and Humanoid robot. In: Proc. of the IEEE International Symposium on Robot and Human Interactive Communication, Toyama, pp. 1022–1028 (2009)
4. Kanda, T., Ishiguro, H., Imai, M., Ono, T.: Development and Evaluation of Interactive Humanoid Robots. Proceedings of the IEEE (Special Issue on Human Interactive Robot for Psychological Enrichment) 92, 1839–1850 (2004)
5. James, W.: The Principles of Psychology. Holt, New York (1890)
6. Festinger, L.: A Theory of Social Comparison Processes. Human Relations 7, 117–140 (1954)
7. Thornton, D., Arrowood, A.J.: Self-evaluation, self-enhancement, and the locus of social comparison. Journal of Experimental Social Psychology 2 (suppl. 1), 40–48 (1966)
8. Tesser, A.: Toward a self-evaluation maintenance model of social behavior. Advances in Experimental Social Psychology 21, 181–227 (1988)

9. Tesser, A., Millar, M., Moore, J.: Some affective consequences of social comparison and reflection processes: The pain and pleasure of being close. Journal of Personality and Social Psychology 54, 49–61 (1988)

10. Tesser, A., Campbell, J.: Self-definition: The impact of the relative performance and similarity of others. Social Psychology Quarterly 43, 341–347 (1980)

11. Pleban, R., Tesser, A.: The effects of relevance and quality of another's performance on interpersonal closeness. Social Psychology Quarterly 44, 278–285 (1981)

12. Tesser, A., Campbell, J.: Self-evaluation maintenance and the perception of friends and strangers. Journal of Personality 59, 261–279 (1982)

13. Kamide, H., Daibo, I.: Application of a Self-Evaluation Maintenance Model to Psychological Health in Interpersonal Context. The Journal of Positive Psychology 4, 557–565 (2009)

14. Kamide, H., Kawabe, K., Shigemi, S., Arai, T.: A Psychological Scale for General Impressions of Humanoids. In: Proc. of 2012 IEEE International Conference on Robotics and Automation, Saint Paul (2012)

15. Bartneck, C., Kulic, D., Croft, E., Zoghbi, S.: Measurement instruments for the anthropomorphism, animacy, likeability, perceived intelligence, and perceived safety of robots. International Journal of Social Robots 1(1), 71–81 (2009)

16. Kamide, H., Kawabe, K., Shigemi, S., Arai, T.: New Measurement of Psychological Safety for Humanoid. In: Proc. of the 7th ACM/IEEE International Conference on Human-Robot Interaction 2012, pp. 49–56 (2012)

17. Nater, U.M., Rohleder, N., Gaab, J., Berger, S., Jud, A., Kirschbaum, C., Ehlert, U.: Human salivary alpha-amylase reactivity in a psychosocial stress paradigm. International Journal of Psychophysiology 55(3), 333–342 (2005)

18. Richmond, V.P.: Nonverbal communication in the classroom, 2nd edn. Tapestry Press, Acton (1997)

19. Ambady, N., Rosenthal, R.: Half a minute: Predicting teacher evaluations from thin slices of nonverbal behavior and physical attractiveness. Journal of Personality and Social Psychology 64(3), 431–441 (1993)

20. Hartley, J.: Note-taking: A critical review. Programmed Learning and Educational Technology 15, 207–224 (1978)

21. Atkinson, R.C., Shiffrin, R.M.: Human memory: A proposed system and its control processes. In: Spence, K.W., Spence, J.T. (eds.) The Psychology of Learning and Motivation: Advances in Research and Theory, vol. 2, pp. 89–195. Academic Press, San Diego (1968)

22. Cunninham, E.G., Wang, W.C.: Using Amos Graphics to Enhance the Understanding and Communication of Multiple Regression. IASE/ISI Satellite Publications (2005), http://www.stat.auckland.ac.nz/~iase/publications/14/cunning ham.pdf

23. Bentler, P.M.: Multivariate analysis with latent variables: Causal modeling. Annual Review of Psychology 31, 419–456 (1980)

24. Joreskog, K.G., Sorbom, D.: Lisrel 7: A guide to the program and applications. SPSS, Chicago (1989)

25. Bentler, P.M.: Comparative fit indexes in structural models. Psychological Bulletin 107, 238–246 (1990)

26. Joreskog, K.G., Sorbom, D.: Lisrel V: Analysis of linear structural relationships by the method of maximum likelihood. National Educational Resources, Chicago (1981)

27. Cuttance, P.: Issues and problems in the application of structural equation models. In: Cuttance, P., Ecob, R. (eds.) Structural Modeling by Example: Applications in Educational, Sociological, and Behavioral Research, pp. 241–279. Cambridge Univ. Press (1987)

Psychological Anthropomorphism of Robots

Measuring Mind Perception and Humanity in Japanese Context

Hiroko Kamide[1], Friederike Eyssel[2], and Tatsuo Arai[3]

[1,3] Department of Engineering Science, Osaka University
Machkaneyamacho 1-3, Toyonaka, Osaka 560-8531, Japan
[2] Department of Psychology, Center of Excellence in Cognitive
Interaction Technology, University of Bielefeld
P.O. Box 100131, 33501 Bielefeld, Germany
kamide@arai-lab.sys.es.osaka-u.ac.jp

Abstract. Using a representative sample, we explored the validity of measures of psychological anthropomorphism in Japanese context. We did so by having participants evaluate both robots and human targets regarding "mind perception" (Gray et al., 2007) and "human essence" (Haslam, 2006)", respectively. Data from 1,200 Japanese participants confirmed the factor structure of the measures and their overall good psychometric quality. Moreover, the findings emphasize the important role of valence for humanity attribution to both people and robots. Clearly, the proposed self-report measures enlarge the existing repertoire of scales to assess psychological anthropomorphism of robots in Japanese context.

Keywords: Anthropomorphism, Robots, Japanese, Mind Perception, Human Essence.

1 Introduction

The vision that robot companions facilitate everyday life is not far-fetched, specifically not in Japanese society.. For one, this is due to demographic changes in society and a decrease in human work force. Second, robots are an important part of Japanese popular culture as evident in movies or manga. Previous research indicates that Japanese people hold a different attitude toward technology and robots compared to people from Western countries [1-3]. Especially, Japanese shows strong preference to robot-like robots but not to highly human-like robots compared to Americans [4]. That is, humanness of robots may be perceived differently by Japanese from Western people. Epley and colleagues [5] state that, "imbuing the imagined or real behavior of nonhuman agents with humanlike characteristics, motivations, intentions, and emotions is the essence of anthropomorphism" (pp. 864-865). Accordingly, psychological anthropomorphism goes beyond the mere attribution of "lifelikeness," "naturalness," "humanlikeness," [5] or "animism" [6, 7]. Existing research by Eyssel and colleagues

G. Herrmann et al. (Eds.): ICSR 2013, LNAI 8239, pp. 199–208, 2013.

has already focused on mind perception and human essence attribution to nonhuman entities [8-12], however, this has not been studied yet in Japanese context. Therefore, we extended the literature by validating these prominent meausures of anthropomorphism   validated Japanese versions of scales measuring psychological and sought to explore culture-specific effects.   This is the first step to understand anthropomorphisation of robots in Japan which may lead to deeper understanding of anthropomorphism in relation to collectivistic cultural background.

2   Related Works in Anthropomorphism

Gray, Gray and Wegner [13] have proposed two dimensions of "mind perception". *Agency* refers to the capacity to act, plan, and exert self-control. *Experience*, on the other hand, encompasses the capacity to feel pain, pleasure, and other emotional states. Moreover, Haslam [14] has suggested two distinct senses of humanness at the trait-level, namely "uniquely human" (*UH*) and "human nature" (*HN*) traits. *UH* traits imply higher cognition, civility, and refinement, and individuals who lack this sense of humanness are implicitly likened to animals. *HN* traits, however, reflect emotionality, warmth, desire, and openness. In this sense, Gray et al. [15]   have emphasized that two dimensions of *agency* and *experience* parallel *UH* and *HN* traits, respectively.

To date, mind attribution and human essence attribution have been studied largely in the human interpersonal or intergroup context and only recently, Eyssel and colleagues have adapted these measures to assess psychological anthropomorphism in various robot prototypes [8-12]. The scale by [16] is also widely used in social robotics. However, in this case, participants are asked to report the extent to which they perceive a robot as "fake", "machinelike", "unconscious" or the like. Obviously, thus, the instrument focuses on the machine's human-likeness – a notion that is clearly distinguished from the process of psychological anthropomorphism as framed by Epley and colleagues [5].

Thus far, mind attribution and HN and HU traits have not been introduced as measures of psychological anthropomorphism to the large community of Japanese researchers who work in the domains of   human-robot interaction and social robotics.

Therefore, this step is taken in the present research that sought to validate the respective constructs in Japanese context.

3   Evaluation of Robots in Japan

It is no news that robots are an important part of contemporary Japanese culture. Nevertheless, evidence shows that Japanese people exhibit relatively negative reactions towards robots, especially when they appear highly humanlike [4].   To date, there is no research that explores mind perception and the attribution to typically human traits [13, 14] to a variety of robot and human targets in a representative Japanese sample. Thus, the present research sheds light on the research question whether Japanese participants differentially attribute mind, human nature and uniquely human traits to robots that vary in humanlike appearance.

Furthermore, in a collectivist society such as Japan, adherence to the ingroup's norms is of essential social value. Japanese, just like Western people distinguish quickly and automatically between ingroup and outgroup members. Clearly, social categorization of others in "us" versus "them" (i.e., positive versus negative) serves to strengthen ingroup trust. Collectivist societies in particular foster the notion that one should trust ingroup members only. To illustrate, Japanese people are more likely to give positive feedback to ingroup members than to out-roup members [17] because it is part of the collectivist norm that ingroupers are to be treated preferentially [18, 19].

It is therefore plausible that Japanese participants make predominantly valence-based jugments to distinguish socially acceptable characteristics from unacceptable ones when evaluating new robot targets.

We deliberately chose a variety of human and robot targets that differed in human-like appearance. We did so to obtain a relative large variance in ratings of mind perception and human essence attribution and to be able to generalize across a wide-range of stimuli.

4   Method

4.1   Targets for Evaluation

Eight targets (Fig. 1) were used to assess mind perception and human essence, covering various humanlike stimuli ranging from wakamaru [20], HRP-2 [21], HRP-4C [22], to Geminoid F [23] and Geminoid HI [24] . These robots clearly differ in perceived humanlikeness of appearance [25]. Additionally, we selected ASTERISK [26] and two humans (Models for Geminoid F and Geminoid H1-4, respectively) as control stimuli. Figure 1 shows photos that were used for the investigation with the expected levels of humanness presented.

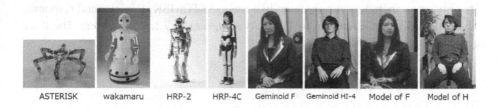

ASTERISK   wakamaru   HRP-2   HRP-4C   Geminoid F   Geminoid HI-4   Model of F   Model of H

Fig. 1. Target stimuli

4.2   Participants and Procedure

A representative sample of 1200 Japanese participants (50% female; mean age = 38.37, SD = 12.03) took part in our online study. In a between-subject design, 150 people per cell of the design were randomly assigned to rate one of the eight targets depicted in Figure 1. None of the participants worked in the engineering field or was a robot expert.

4.3   Measures

Anthropomorphism

To assess psychological anthropomorphism of the 8 targets, we used the items proposed by [13, 14]. The scales have already been tested and validated in the context of social robotics and therefore represent suitable measures of psychological anthropomorphism [8-12]. The self-report measures were originally developed in English language and were thus translated to Japanese, and back-translated into English to confirm the appropriateness of the translation. The first measure assesses mind perception [13] and comprises 18 items (Table1) of *Agency* and *Experience*. For both dimension of mind perception, we asked participants to rate how much the target was capable of each item on a 5-point scale from 1 (*not capable at all*) to 5 (*very capable*).

To measure human essence attribution [14], participants were presented with a list of 20 personality traits reflecting *UH* and *HN*. Each subscale contained 5 positive and 5 negative traits (Table 2). Participants rated how well each term describes the target on a scale from 1 (*not at all descriptive*) to 6 (*very descriptive*).

Humanlikeness as General Impressions of Robots

Even though the main goal of the present research was to explore the psychometric quality of the scales related to mind perception and human essence, we also assessed perceptions of humanlikeness by asking participants to complete three items taken from   a Psychological Scale for General Impressions of Humanoid ([25]). The scale of general impressions measures general impressions of humanoids. For   the purpose of the present research, we only used three items: "I could easily mistake the robot for a real person", "I am amazed at the progress of technology when I look at the robot" and "The robot looks like a human." Due to the fact that the item content does not apply to human targets, we utilized the scale of general impressions on the robot targets only, even including the machine-like robot ASTERISK. Participants' responses were collected using a scale from 1 (*totally disagree*) to 7 (*totally agree*)..The three items were highly reliable, $\alpha = .82$.

5   Results and Discussion

5.1   Perceived Humanlikeness of General Impression

We conducted an ANOVA on perceived humanlikeness of the six robots. The main effect was significant, $F (5, 892) = 83.16$, $p < .001$. Figure 2 shows results from multiple comparisons (Bonferroni method) and illustrates differences in perceived humanlikeness across the robot prototypes.

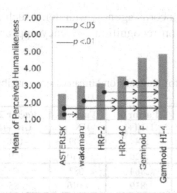

Fig. 2. Perceived humanlikeness as a function target

5.2 Factor Analyses on the Scales of Mind Perception and Human Essence

An exploratory factor analysis with promax rotation was conducted on mind perception and human essence items (Table 1, 2).

Table 1. Results of Factor Analysis on the scale of Mind Perception

	Factor 1 Experience (+)	Factor3 Experience(-)	Factor 2 Agency
Hunger	.244	**.568**	-.050
Fear	-.075	**.930**	-.019
Pain	-.077	**.972**	-.010
Pleasure	**.466**	-.049	.275
Rage	.398	**.532**	-.134
Desire	**.726**	.160	-.043
Personality	**.745**	-.022	.060
Consciousness	**.822**	-.008	.072
Pride	**.727**	.134	-.018
Embarrassment	.028	**.507**	.259
Joy	.378	-.024	.394
Self-control	-.135	.191	**.738**
Morality	-.101	.160	**.755**
Memory	-.052	-.189	**.831**
Emotion recognition	.253	.042	**.527**
Planning	.063	.000	**.694**
Communication	.115	-.156	**.698**
Thought	.145	.063	**.632**
α	**.875**	**.879**	**.894**

Table 1 shows that mind perception is composed of three subfactors, whereas, as predicted, the "experience" factor reflects positive and negative valence. The "joy" item had low and redundant loadings, but all other items corresponded to the factor structure observed previously [8], except that the experience factor was differentiated by item valence. Factor 2 concerns agency, with items such as "self-control," "morality," "memory," and "emotion recognition." Agency appears to be a unidimensional, univalent construct.

Table 2. Factor analyses on human essence attribution items

	Factor 1 HN(+)	Factor 3 HN(-)	Factor 2 UH(+)	Factor 4 UH(-)
Curious	**.784**	.209	-.058	-.082
Friendly	**.820**	-.110	.149	.080
Fun-loving	**.864**	.087	-.026	-.048
Sociable	**.859**	-.016	.085	.074
Trusting	**.406**	.046	.391	.013
Aggressive	.158	**.747**	-.177	-.035
Distractible	.252	**.485**	-.126	.205
Impatient	.195	**.880**	-.062	-.100
Jealous	-.072	**.703**	.081	.062
Nervous	-.232	**.722**	.253	-.005
Broadminded	.504	-.112	.367	.099
Humble	.155	-.170	**.712**	.020
Organized	.129	.120	**.739**	-.123
Polite	.090	-.136	**.848**	-.067
Thorough	-.033	.126	**.825**	-.096
Cold	-.228	.356	.239	.274
Conservative	-.138	.162	**.545**	.239
Hard-hearted	-.085	.228	.042	**.604**
Rude	.083	.154	-.174	**.789**
Shallow	.049	.025	-.057	**.847**
α	**.90**	**.845**	**.86**	**.87**

Table 2 reveals that both *UH* and *HN* differentiate into positive (+) and negative (-) factors; however, "broadminded" and "cold" overlapped across factors. Factor 1 (*HN+*) includes positive human nature- related items, such as "curious," "friendly". Factor 2 comprises positive traits, e.g., "humble," "organized". One might argue that "conservative"   represents a negative characteristic -   however, from a collectivistic perspective, being conservative does not always have a negative meaning. Moreover, conservative people do not bother others or disturb situations. In this sense, being conservative is a positive trait within a collectivistic culture. All items in Factor 2 originally belonged to the *UH* factor; therefore, this factor was named *UH(+)*. Factor 3 included "aggressive," and so on. These items correspond to the negative components of HN. Factor 3 was thus termed *HN(-)*. Factor 4 comprised "hard-hearted," "rude", all reflecting the *UH* factor. This factor was labeled *UH(-)*.

Figures 3-5 show mind perception as a function of target type – evidently, this scale clearly distinguishes between the different robot prototypes, although mind attribution occurs at relatively low level in the present research study [27]. Moreover, of the subscales of mind perception and human essence show high internal consistencies. In sum total, the developed scales are usable for evaluation of anthropomorphism in Japan.

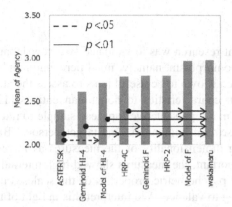

Fig. 3. Mean attribution of agency as a function of target type

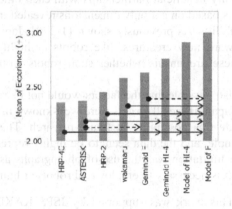

Fig. 4. Mean attribution of experience (+) as a function of target type

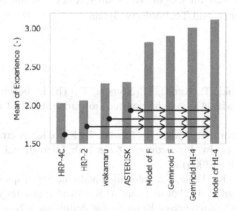

Fig. 5. Mean attribution of experience (-) as a function of target type

6   Conclusion

The goal of the present research was to validated two new Japanese measures of   of psychological anthropomorphism, namely, mind perception and human essence attribution. These have been shown to be useful tools to assess the attributions of typically and essentially human characteristics to nonhuman entities [12-15]. We did so by asking participants from a representative Japanese sample to rate a variety of targets - a machinelike robot, several humanoids, and target persons. Broadly, we replicated the findings regarding dimensionality of the measures as in research from Western countries. Equally important, the measures showed high internal consistencies which, too, confirm the good psychometric properties of the scales. Furthermore, the scales differentiated according to valence . We interpret this in light of the fact that in collectivist societies. social categorization may be particularly crucial [17, 18]. The collectivists trust in-group members. According to collectivist social norms, ingroupers should maintain mutually beneficial partnerships with each other. Therefore, anthropomorphism in Japan is based on not only dimensions revealed in previous studies but also on social acceptability. As previously shown [1-2, 4], Japanese citizens display negative reactions toward new creatures, like robots or highly human-like robots, suggesting that Japanese are unsure whether such robots would be acceptable in-group members.

In this sense, it is also possible that the factors would not be separated into positive and negative even in Japan, if the all targets were well-known in-group members. It is necessary to investigate this possibility in future research. The current study had a satisfactorily large sample, and the data seem to be highly representative of Japanese people. However, one limitation was the use of photographs as stimuli and not real robots or humans and it is necessary to employ real robots or humans as stimuli.

Acknowledgement. This work was supported by JSPS KAKENHI Grant Number 23730579. This research was supported by Global COE Program "Center of Human-Friendly Robotics Based on Cognitive Neuroscience" of the Ministry of Education, Culture, Sports, Science and Technology, Japan.

References

1. Bartneck, C., Suzuki, T., Kanda, T., Nomura, T.: The influence of people's culture and prior experiences with Aibo on their attitude towards robots. AI and Society 21, 217–230 (2007)
2. Bartneck, C., Nomura, T., Kanda, T., Suzuki, T., Kato, K.: A cross-cultural study on attitude towards robots. In: Proceedings of the 11th International Conference on Human-Computer Interaction (2005)
3. Shibata, T., Wada, K., Tanie, K.: Tabulation and analysis of questionnaire results of subjective evaluation of seal robot in Japan, U.K., Sweden and Italy. In: Proceedings of the IEEE International Conference on Robotics and Automation, New Orleans (2004)

4. Bartneck, C.: Who like androids more: Japanese or US Americans? In: Proceedings of the 17th IEEE International Symposium on Robot and Human Interactive Communication, RO-MAN 2008, Munchen, pp. 553–557 (2008)
5. Epley, N., Waytz, A., Cacioppo, J.T.: On Seeing Human: A threefactor theory of anthropomorphism. Psychological Review 114, 864–886 (2007)
6. Waytz, A., Cacioppo, J.T., Epley, N.: Who sees human? The stability and importance of individual differences in anthropomorphism. Perspectives on Psychological Science 5, 219–232 (2010)
7. Scholl, B.J., Tremoulet, P.D.: Perceptual causality and animacy. Trends in Cognitive Science 4, 200–209 (2000)
8. Gray, H.M., Gray, K., Wegner, D.M.: Dimensions of mind perception. Science 315, 619 (2007)
9. Haslam, N.: Dehumanization: An integrative review. Personality and Social Psychology Review 10, 252–264 (2006)
10. Gray, K., Knobe, J., Sheskin, M., Bloom, P., Barrett, L.F.: More than a body: Mind perception and the surprising nature of objectification. Journal of Personality and Social Psychology 101(6), 1207–1220 (2011)
11. Bartneck, C., Croft, E., Kulic, D., Zoghbi, S.: Measurement instruments for the anthropomorphism, animacy, likeability, perceived intelligence, and perceived safety of robots. International Journal of Social Robotics 1, 71–81 (2009)
12. Eyssel, F., Kuchenbrandt, D., Hegel, F., de Ruiter, L.: Activating elicited agent knowledge: How robot and user features shape the perception of social robots. In: Proceedings of the 21st IEEE International Symposium in Robot and Human Interactive Communication (RO-MAN 2012), pp. 851–857 (2012)
13. Eyssel, F., Reich, N.: Loneliness makes my heart grow fonder (of robots)? On the effects of loneliness on psychological anthropomorphism. In: Proceedings of the 8th ACM/IEEE Conference on Human-Robot Interaction (HRI 2013), pp. 121–122 (2013)
14. Eyssel, F., Kuchenbrandt, D., Bobinger, S.: Effects of anticipated human-robot interaction and predictability of robot behavior on perceptions of anthropomorphism. In: Proceedings of the 6th ACM/IEEE Conference on Human-Robot Interaction (HRI 2011), pp. 61–67 (2011)
15. Eyssel, F., Kuchenbrandt, D., Bobinger, S., de Ruiter, L., Hegel, F.: 'If you sound like me, you must be more human': On the interplay of robot and user features on human-robot acceptance and anthropomorphism. In: Proceedings of the 7th ACM/IEEE Conference on Human-Robot Interaction (HRI 2012), pp. 125–126 (2012)
16. Jin, N., Yamagishi, T., Kiyonari, T.: Bilateral Dependency and the Minimal Group Paradigm. Japanese Journal of Psychology 67, 77–85 (1996) (in Japanese with an English abstract)
17. Yamagishi, T., Cook, K.S., Watabe, M.: Uncertainty, trust, and commitment formation in the United States and Japan. American Journal of Sociology 104, 165–194 (1998)
18. Yamagishi, T., Kikuchi, M., Kosugi, M.: Trust, gullibility, and social intelligence. Asian Journal of Social Psychology 2, 145–161 (1999)
19. Kawachi, N., Koyuu, Y., Nagashima, K., Ohnishi, K., Hiura, R.: Home-use Robot wakamaru. Mitsubishi Juko Giho 40(5), 270–273 (2003) (in Japanese)
20. Kaneko, K., Kanehiro, F., Kajita, S., Hirukawa, H., Kawasaki, T., Hirata, M., Akachi, K., Isozumi, T.: Humanoid Robot HRP-2. In: Proc. of the 2004 IEEE International Conference on Robotics & Automation, New Orleans, pp. 1083–1090 (2004)

21. Nakaoka, S., Kanehiro, F., Miura, K., Morisawa, M., Fujiwara, K., Kaneko, K., Kajita, S., Hirukawa, H.: Creating facial motions of Cybernetic Human HRP-4C. In: Proc. IEEE-RAS International Conference on Humanoid Robots, Paris (2009)
22. http://www.kokoro-dreams.co.jp/actroid_f/index.html (in Japanese)
23. Ogawa, K., Bartneck, C., Sakamoto, D., Kanda, T., Ono, T., Ishiguro, H.: Can An Android Persuade You? In: Proc. 18th International Symposium on Robot and Human Interactive Communication, pp. 553–557 (2009)
24. Theeravithayangkura, C., Takubo, T., Ohara, K., Mae, Y., Arai, T.: Dynamic Rolling-Walk Motion by the Limb Mechanism Robot ASTERISK. Advanced Robotics 25(1-2), 75–91 (2011)
25. Kamide, H., Kawabe, K., Shigemi, S., Arai, T.: Development of a Psychological Scale for General Impressions of Humanoid. Advanced Robotics 17(1), 3–17 (2013)
26. Eyssel, F., Kuchenbrandt, D.: Social categorization of social robots: Anthropomorphism as a function of robot group membership. British Journal of Social Psychology 51, 724–731 (2012)

The Ultimatum Game as Measurement Tool for Anthropomorphism in Human–Robot Interaction

Elena Torta, Elisabeth van Dijk, Peter A. M. Ruijten,
and Raymond H. Cuijpers

Eindhoven University of Technology,
Den Dolech 16, 5600MB Eindhoven, NL
{e.torta,p.a.m.ruijten,r.h.cuijpers}@tue.nl,
e.t.v.dijk@student.tue.nl

Abstract. Anthropomorphism is the tendency to attribute human characteristics to non–human entities. This paper presents exploratory work to evaluate how human responses during the ultimatum game vary according to the level of anthropomorphism of the opponent, which was either a human, a humanoid robot or a computer. Results from an online user study (N=138) show that rejection scores are higher in the case of a computer opponent than in the case of a human or robotic opponent. Participants also took significantly longer to reply to the offer of the computer rather than to the robot. This indicates that players might use similar ways to decide whether to accept or reject offers made by robotic or human opponents which are different in the case of a computer opponent.

Keywords: Anthropomorphism, Ultimatum Game, Human–Robot Interaction, Human–Computer Interaction, Social–Robotics, Evaluation Metrics, Implicit Measurements.

1 Introduction

People tend to attribute human characteristics to non-human entities such as computers and robots [3,5]. This tendency is generally referred to as 'anthropomorphism' (e.g.,[6,3,4]). For instance, we may say that a pet 'remembers' something, that a computer 'thinks' or that the hammer that fell on our foot did so 'on purpose'. This term can also be used to describe some aspects of a system that are human-like and therefore likely to elicit anthropomorphic responses (e.g., [11,10]). Anthropomorphism of an object may result in responses on different levels, affecting affective and cognitive states as well as behavior and therefore influences the quality of interaction between human and technology.

Many of these responses are unconscious (e.g., [6,12]). For example, people can experience empathic feelings towards a sad or a happy robot. However, when asked explicitly, people would state that they do not believe the robot can experience such emotions.

G. Herrmann et al. (Eds.): ICSR 2013, LNAI 8239, pp. 209–217, 2013.
© Springer International Publishing Switzerland 2013

Most studies concerning attributions of moral accountability and anthropo-morphism to robots measure users' views with surveys. Kiesler and Goetz [7], for instance, developed surveys to assess the complexity of users' mental models of robots. Others have set out to identify the different aspects of anthropomorphism in order to create standardized questionnaires ([4,15,1]). The greatest advantage of surveys is that they are easy to administer. However if anthropomorphism is a complex unconscious process, surveys might not be the most suitable mea-surement tool and cannot, by themselves, give the full picture. Thus, employing additional measurements of (aspects of) anthropomorphism can help us gaining a broader view of the subject.

In the current study, we examine the possibility of using the ultimatum game as a measurement tool for anthropomorphism by looking at the decisions peo-ple take during the game when facing non–human opponents (humanoid robots and computers). Although previous studies showed that people exhibit different rejection behaviours if they play against computers rather than humans [13,9], it is still unclear how people react when the opponent is a robot.

1.1 Decision Making in the Ultimatum Game

Game Theory studies general strategic decision-making. One game from Game Theory is the so-called 'Ultimatum Game' (UG). In this game there are two players and some amount of goods or money to be distributed. Player 1 (the 'opponent') can propose a division of the goods between himself and player 2 (the 'receiver'). The receiver can subsequently decide to accept or reject the proposed division. If the receiver accepts, both get the proposed amount. If the receiver rejects, both get nothing. An overview of the possible outcomes of the UG is visible in Figure 1. If this game is played with two rational agents,

Proposer action / Receiver action	Give MIN (MIN = minimal amount)	Give x (x > MIN)	Proposer's outcome / Receiver's outcome
Accept	p - MIN	p - x	
	MIN	x	
Reject	0	0	
	0	0	

Fig. 1. The outcomes of the Ultimatum Game for the different actions of the players. p is the amount initially given to the proposer. The receiver's outcome is always lower when rejecting: $0 < \text{MIN}$ and $0 < x$. Assuming the proposer knows this, he is always best off giving as little as possible: $p - \text{MIN} > p - x$.

the outcome is clear: the receiver will accept any non-zero offer, as the profit in accepting is always higher than rejecting. The proposer, knowing this, will always offer the lowest possible non-zero amount to the receiver, knowing that a rational receiver will always accept. Studies have shown that human players deviate from rationally optimal behavior when playing the UG. Sanfey et al [14] found that human receivers reject more than 50% of low offers (< 20% of the amount of goods). The deviation from rationally optimal behaviour is due to the social settings of the game and its consequences (accepting a low offer will cause great advantage to the proposer). 'Social' here is not used to describe the decision-makers' motives, but rather the situation the decision-makers are in. The settings or 'game' used in Game Theory are social in the sense that the outcome for a given agent not only depends on that agent's decisions and behavior, but also on the other agents' decisions and behavior.

1.2   The Current Study

The purpose of our study is to get insight into the behavior of human players towards non–human opponents (in particular robots) in the UG. Previous research on this topic has shown that, when offers are made by a computer rather than a human, players' rejection rates are much lower [14,9]. This might indicate that the social context of the UG influences the decision making behaviour of human receivers. In particular, the different levels of anthropomorphism of the opponents (human or computer) might be responsible for the observed differences in decision outcomes.

Although there is some information on how computers are treated in the UG, less is known about the responses robotic players evoke. In the current study we want to investigate how humans respond to robotic opponents when playing the ultimatum game and how the response patterns differ from those observed with human and computer opponents. We also want to investigate how the responses differ from explicit measures of anthropomorphism.

2   Method

2.1   Participants

A total of 138 participants took part in an online experiment. The age of participants ranged from 18 to 69 years (M = 25.68, SD = 11.45). Out of the 138 participants, 80 were male, 58 female. 78 Participants indicated they had some personal experience with robots. 22 Participants indicated they had heard of or played the UG before. As previous experience with both robots and the game proved to have no significant effect on participants' responses in the UG, data from these 22 participants was kept and included in the analyses.

2.2   Experiment Design

The experiment had a 3 (Opponent Type: human vs. robot vs. computer) × 4 (Offer: 50% vs. 30% vs. 20% vs. 10%) mixed design. The Opponent Type factor represents the nature of the opponent: human, robot or computer. This factor was varied between subjects. Eight different instances of the same opponent were used during the experiment. The Offer represents the percentage of the initial

sum proposed to the participant and it comprises the same four levels as used by
Sanfey et al. [14]. In each round we named the initial sum as 100 Money Unites
(MU) and the offer to the receiver could be 50 MU, 30 MU, 20 MU or 10 MU.
The Offer was varied within participants and each offer was presented twice, so
that each participant played a total of eight rounds. Each offer was presented by
a different instance of the opponent. The order of the offers was counterbalanced
using a balanced Latin square [2].

2.3   Materials and Measures

Opponents were introduced to participants with pictures. For the human opponents, eight pictures of Caucasian males with neutral expressions, showing the
person from the front, looking straight into the camera were selected from the
Radboud Faces Database [8]. An example is depicted in Figure 2(a).

The robot pictures were selected based on a pre-test in which participants
(N=27) were asked to rate 20 pictures of different humanoid robots in terms of
six scales: attractiveness, valence, naturalness, human likeness, and socialness.
The eight pictures whose ratings on these scales were closest to the mean were
selected. An example is depicted in Figure 2(b).

For the computer opponents, a set of eight different pictures of desktop computers were taken from the internet. Each picture showed a desktop computer, a
black screen and a keyboard and mouse on a white background see Figure 2(c).
During the experiment, the responses of participants in the UG rounds were

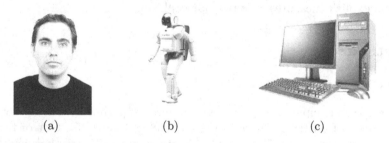

(a)   (b)   (c)

Fig. 2. Examples of the pictures used as human (a), robot (b) or computer (c) opponents in the experiment.

recorded. In addition, the time in milliseconds it took for participants to respond
to each offer was recorded. A set of five questions related to anthropomorphism
suggested by Epley et al.[4] ($\alpha = 0.938$) was administered to subjects at the end
of the experiment and used for comparison with the UG results.

2.4   Procedure

Participants performed the experiment online. Upon arrival at the experiment
website, the participant was welcomed and explained that the experiment would

consist of some simple games and a number of questions. Additionally, the introductory text notified the participant of the relevant conditions concerning participation and privacy, and the incentive arrangement of the experiment. Then, the Ultimatum Game was explained to the participant. After reading the instruction, the participants started the first of eight game rounds. In each round, the first page showed a picture of the opponent. Participants were instructed to take a good look at their opponent and then continue with the game by clicking on a button. On the second page, participants were instructed to click a box that subsequently displayed the offer the opponent made for that round. At the bottom of the page, the participant could then indicate his/her choice by clicking at the accept or reject button. The third page showed a simple confirmation that the round had been finished and what the result was (offer accepted or rejected, MU gained by participant and opponent). After filling out all the questions, participants were redirected to a final page, where they were thanked for their participation and where they could download a document detailing the nature and purpose of the study.

3 Results

3.1 Response Scores

In order to investigate the effect of our manipulations, offer and type of opponent, on response scores we computed a mixed model univariate ANOVA with response as dependent variable and offer type and opponent type as factors. Four outliers were removed from the data set as the reaction time was longer than the third quartile plus three times the interquartile range. On average participants displayed significantly different rejection behaviours according to the type of offer they received ($F(3, 134) = 234.9$, $p < 0.001$). As expected, the lower the offer the lower the acceptance ratio (see Figure 3(a)). Participants accepted offers made by computer less than those made by the robot or the human (see Figure 3(b)) although this effect is only marginally significant ($F(2, 134) = 3.0$, $p = 0.051$). The observed differences between opponents are dependent on the offer that was proposed to participants as indicated by the significant interaction effect ($F(6, 134) = 2.1$, $p = 0.049$). In particular, see Figure 3(c), we see no difference between opponents at the highest offer (50 MU). A higher acceptance rate for the human opponent than for the robot or computer opponent at 30 MU. And a higher acceptance rate for the robot than for the human or computer at low offers, 20 MU and 10 MU.

3.2 Decision Time

Decision times provide an indication of how our experimental manipulations (the type of opponent and the type of offer) influence the difficulty of the decision task. In order to investigate the effect of these factors on the decision time, a univariate ANOVA with response time as dependent variable and offer and opponent type as factors was computed. Decision time was significantly dependent

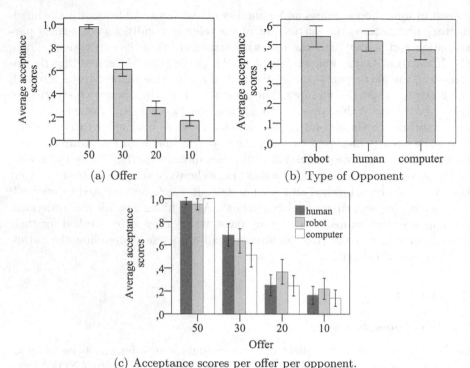

(a) Offer

(b) Type of Opponent

(c) Acceptance scores per offer per opponent.

Fig. 3. Mean acceptance ratio per offer (a) and opponent (b). Mean acceptance ratio per offer clustered by opponent (c). Error bars display 95% confidence interval.

on the type of offer that participants received ($F(3, 134) = 12.3$, $p < 0.001$). As one could expect, it took the shortest time to respond to the highest offer (50 MU), see Figure 4(a). The longest time was reached when responding to offers of 30 MU, which indicates a difficulty participants had when accepting or rejecting that offer. Decision times were also significantly affected by opponent type ($F(2, 134) = 5.44$, $p = 0.004$). In particular, it took longer to reply to offers made by a computer than to offers made by a human or a robot (see Figure 4(b)). The pattern of decision times across opponent type (depicted in Figure 4(b)) is consistent with the pattern of acceptance scores across opponent type (see Figure 3(b)). Participants rejected offers made by a computer more often than offers made by a human or a robot. We can observe the same pattern in decision times since it took longer to reply to an offer made by a computer than an offer made by a robot or a human. An independent sample t-test reveals a significant difference only between the robot and computer condition ($t(702) = 2.75$, $p = 0.006$).

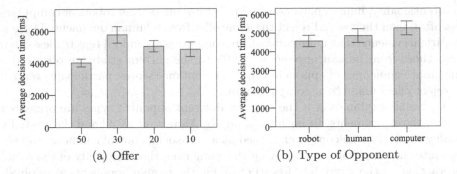

(a) Offer   (b) Type of Opponent

Fig. 4. Mean values of decision times in ms per offer (a) and opponent (b). Error bars display 95% confidence interval.

3.3 Epley Questionnaire

In order to check whether participants recognize differences in anthropomorphism between the different actors, the results of the Epley questionnaire on anthropomorphism were analyzed. A Mixed model ANOVA with the mean of the answers given to the Epley questionnaire as dependent variable and opponent type as independent variable shows a significant effect of agent type ($F(2, 134) = 584.9$, $p < 0.001$). The mean scores for the three types of agents are reported in Figure 5. A higher score indicates a higher level of anthropomorphism. As expected, on a conscious level, participants recognize human opponents as most anthropomorphic followed by robots and computers.

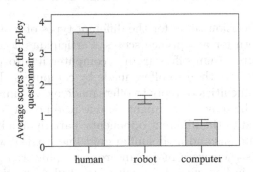

Fig. 5. Average scores of the Epley questionnaire per opponent type. Error bars display 95% confidence interval.

4 Discussion

Previous research on the UG showed that human players tend to reject unfair offers less when the opponent is perceived as lacking intentionality. One would therefore expect that if robots or computers are seen as 'less human' with respect to intentionality, human players would reject low offers from robots or computers less often than they would reject the same offer from a human opponent. There is empirical evidence that unfair offers from computers are indeed rejected less often than those from human opponents [14,9]. In the current study we observed a marginal significance of opponent type on acceptance scores; participants tended to reject offers made by a computer more.

A possible explanation is the way the different opponent types are presented during the experiments. In both the study by Moretti et al. [9] and the study by Sanfey et al. [14] the computer is used as a non-social control condition to help separate the simple action of playing the game from the complexity of the social interaction. In the study by Moretti et al. [9], the human opponents were physically present at the beginning of the round and left the room when participants played the game. Although they were hidden from view by an opaque partition, they were still within earshot of the subject. In the rounds where subjects played against a computer, subjects were instructed explicitly that the offers from the computer were randomly generated. In the experiments by Sanfey et al. [14], subjects were introduced face-to-face to all their human opponents before each game round. The computer opponent was not introduced beforehand.

For the purpose of having an unbiased comparison of human and computer, however, the methods of Moretti et al. [9] and Sanfey et al. [14] are less suitable. In the current study, the differences between the opponent types are given much less emphasis. The exact same instructions were used, regardless of which type of opponent the subject would face. For all three opponent types, only pictures were used to introduce the opponent. In addition, the between-subjects manipulation of the Opponent Type factor gave subjects less opportunity to compare the opponent types than the within-subject design used by Moretti et al. [9] and Sanfey et al. [14].

The pattern on decision times for the different types of opponents are consistent with the pattern for acceptance scores. Participants accepted offers made by a robot more often than offers from a computer it also took less to reply to offers made by robots than to offers made by computers. This suggests that people have more difficulties to reply to offers made by a computer than to offers made by robots or humans. This might be due to different levels of anthropomorphism associated to the different opponents: participants knew how to reply to offers made by humans and used similar guidelines to deal with offer from the humanoid robots as they were more 'anthropomorphic' than computers. This explanations is also in line with the result of the Epley questionnaire.

Indeed, the Epley questionnaire results show a very clear pattern across the different opponent types: humans are anthropomorphized the most, computers the least and robots are somewhere in between. This is the kind of pattern one would expect to find based on rational consideration: a human is obviously very much like a human, a computer is more of a machine and a humanoid robot is like a computer, but usually has physical features and interaction modalities similar to a human's and is therefore somewhere in the middle.

In conclusion, our data show a difference between opponents in the implicit measurements of the UG as well as in the explicit measurements of the Epley questionnaire. Contrary to previous research, we did not observe an increase in acceptance scores for the computer opponent, but we observed longer decision times for the computer than for the robot or the human. This might be due to difference in anthropomorphism attributed to the different opponents.

Acknowledgments. The authors gratefully acknowledge the contribution to this paper given by our colleague Andreas Spahn.

References

1. Bartneck, C., Kuliç, D., Croft, E., Zoghbi, S.: Measurement instruments for the anthropomorphism, animacy, likeability, perceived intelligence, and perceived safety of robots. International Journal of Social Robotics 1(1), 71–81 (2009)
2. Bradley, J.V.: Complete counterbalancing of immediate sequential effects in a Latin square design. Journal of the American Statistical Association 53(282), 525–528 (1958), doi:10.1080/01621459.1958.10501456
3. Duffy, B.R.: Anthropomorphism and the social robot. Special Issue on Socially Interactive Robots, Robotics and Autonomous Systems 42, 170–190 (2003)
4. Epley, N., Waytz, A., Cacioppo, J.T.: On seeing human: a three-factor theory of anthropomorphism. Psychological Review 114, 864–886 (2007)
5. Fong, T., Nourbakhsh, I., Dautenhahn, K.: A survey of socially interactive robots. Robotics and Autonomous Systems 42(3), 143–166 (2003)
6. Kennedy, J.S.: The new anthropomorphism. Cambridge University Press (1992)
7. Kiesler, S., Goetz, J.: Mental models and cooperation with robotic assistants. In: Proceedings of Conference on Human Factors in Computing Systems, pp. 576–577. ACM Press (2002)
8. Langner, O., Dotsch, R., Bijlstra, G., Wigboldus, D.H.J., Hawk, S.T., van Knippenberg, A.: Presentation and validation of the radboud faces database. Cognition & Emotion 24, 1377–1388 (2010)
9. Moretti, L., di Pellegrino, G.: Disgust selectivity modulates reciprocal fainess in economic interactions. Emotion 10, 169–180 (2010)
10. Nowak, K.L.: The influence of anthropomorphism and agency on social judgment in virtual environments. Journal of Computer-Mediated Communication 9(2) (2004)
11. Nowak, K.L., Biocca, F.: The effect of the agency and anthropomorphism on users' sense of telepresence, copresence, and social presence in virtual environments. Presence: Teleoperators & Virtual Environments 12(5), 481–494 (2003)
12. Ruijten, P.A.M., Midden, C.J.H., Ham, J.: I didn't know that virtual agent was angry at me: Investigating effects of gaze direction on emotion recognition and evaluation. In: Berkovsky, S., Freyne, J. (eds.) PERSUASIVE 2013. LNCS, vol. 7822, pp. 192–197. Springer, Heidelberg (2013)
13. Sanfey, A.G.: Expectations and social decision-mkaing: biasing effects of prior knowledge on ultimatum responses. Mind and Society 8, 93–107 (2009)
14. Sanfey, A.G., Riling, J.K., Aronson, J.A., Nystrom, L.E., Cohen, J.D.: The neural basis of economic decision-making in the ultimatum game. Science 13, 1755–1758 (2003)
15. Zhang, T., Zhu, B., Lee, L., Kaber, D.: Service robot anthropomorphism and interface design for emotion in human-robot interaction. In: IEEE International Conference on Automation Science and Engineering, CASE 2008, pp. 674–679 (August 2008)

Human-Robot Upper Body Gesture Imitation Analysis for Autism Spectrum Disorders

Isura Ranatunga[1], Monica Beltran[1], Nahum A. Torres[1], Nicoleta Bugnariu[2],
Rita M. Patterson[2], Carolyn Garver[3], and Dan O. Popa[1]

[1] Dept. of Electrical Engineering, University of Texas at Arlington,
416 Yates Street, Arlington, TX 76011 USA
isura.ranatunga@mavs.uta.edu, popa@uta.edu
[2] University of North Texas Health Science Center,
3500 Camp Bowie Blvd., Fort Worth, TX 76107, USA
{nicoleta.bugnariu,rita.patterson}@unthsc.edu
[3] Autism Treatment Center,
10503 Metric Drive, Dallas, TX 75243, USA
cgarver@atcoftexas.org

Abstract. In this paper we combine robot control and data analysis techniques into a system aimed at early detection and treatment of autism. A humanoid robot - Zeno is used to perform interactive upper body gestures which the human subject can imitate or initiate. The result of interaction is recorded using a motion capture system, and the similarity of gestures performed by human and robot is measured using the Dynamic Time Warping algorithm. This measurement is proposed as a quantitative similarity measure to objectively analyze the quality of the imitation interaction between the human and the robot. In turn, the clinical hypothesis is that this will serve as a consistent quantitative measurement, and can be used to obtain information about the condition and possible improvement of children with autism spectrum disorders. Experimental results with a small set of child subjects are presented to illustrate our approach.

1 Introduction

Autism Spectrum Disorder (ASD) is a developmental disorder characterized by deficiencies in social interaction, speech, cognition, motor coordination and imitation [4]. According to the Autism and Developmental Disabilities Monitoring (ADDM) Network of the Centers for Disease Control and Prevention (CDC) an average of 1 in 88 children in the US is diagnosed with an ASD [2]. Although the cognitive capacity of individuals in the Autism spectrum vary greatly, most of the individuals have sensorimotor abnormalities. Although Autism was recognized as early as 1943 by Kanner and accepted to have a biomedical origin by the 1980s, there is a lack of quantitative diagnosis tools. Currently ASD diagnosis mainly focuses on qualitative behavior observation which results in imprecise and sometimes arbitrary categorization of individuals in the Autism spectrum [4,14].

G. Herrmann et al. (Eds.): ICSR 2013, LNAI 8239, pp. 218–228, 2013.
© Springer International Publishing Switzerland 2013

Robotic systems have been developed for use in the therapy of individuals in the autism spectrum such as FACE, AuRoRa, Kaspar, Nao, and Keepon [14,15], but many of them do not engage in dynamic gestural interaction in a truly autonomous, interactive manner. Studies show that the appearance of the robot plays an important role in how children relate to and interact with such robots [11,10,12] and suggest that imitation and turn-taking are types of interactions useful in motivating and engaging children with ASD. In these types of projects, the interaction capacity of the robot is restricted due to lack of objective criteria to rate imitative gestural Human-Robot Interaction (HRI). A robot called Bandit was used to guide older adults to perform imitative exercises [3]. This project had a robot perform upper body gestures that the subject imitated, performance criteria were related to achievement of target poses. Projects involving robots interacting with individuals, specifically children with ASD tend to be interdisciplinary projects that have a small number of participants, which makes a good clinical conclusion difficult. These projects and others involving the use of robots for assisting humans in a social, collaborative setting can be considered part of a relatively new field called Socially Assistive Robotics (SAR) [14].

Our recent multi-institutional project aims to create the technological and clinical basis for a novel Human-Robot Interaction System - RoDiCA, which will be used in the future as an early diagnostic and treatment tool for children with ASD. The clinical hypothesis is that motor developmental problems including imitation are present and can be used in early identification and diagnosis of ASD. The technological hypothesis is that a humanoid robot can motivate children with ASD to engage in motor activities and these interactions can be analyzed for diagnosis and therapy.

Fig. 1. Zeno and Child during an Experiment - Photo credit: Fort Worth Star-Telegram/Max Faulkner

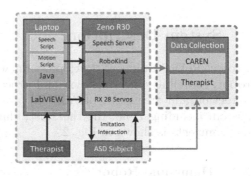

Fig. 2. System diagram

This project builds on our HRI work in [7, 8, 16, 9]. We have integrated the functionalities developed by the NGS Lab at UT Arlington, Hanson RoboKind, and the Osteopathic Heritage Foundation Physical Medicine Core Research Facility (OHFPMCRF) at UNT Health Sciences in a HRI framework that enables us to study the interaction of human subjects with the Zeno robot.

According to Baer et al. [1] any behavior whose structure follows that of a model functionally and is close to the model temporally can be considered imitation. In this respect they are specific about what constitutes imitative behavior; the imitator should be following cues from the model. From this definition it is clear that imitation does not generate the exact behavior that is exhibited by a model. One of the points being that the structure or topography of the action is given more importance than the temporal similarity of the behavior. This means that we can assess the quality of imitation by ignoring some temporal differences in the observed behavior.

The contribution of this paper, is the development of a robotic system which uses Dynamic Time Warping (DTW) as a tool for the diagnosis and treatment of Autism. DTW [5, 13] was used extensively in the speech processing community, and is a dynamic programming algorithm, which gives a distance measure that is locally temporally invariant. It has also been used extensively by the data mining community recently [6]. Here, we use DTW as a similarity measure for comparing arm motions initiated by robot Zeno and imitated by children. Results show that the DTW similarity measure can serve as both a meaningful and objective measure for evaluating the HRI quality. In the future, more trials will test the clinical hypothesis of using DTW as a tool to identify a motor marker for ASD.

This paper is organized as follows: section 2 describes the experimental platform, data collection and provides details about the subjects of the study; section 3 describes the method of data analysis and the development of DTW as a tool to identify a motor marker for ASD; section 4 shows preliminary results from the experiment and the initial reactions from ASD children; section 5 concludes the paper and describes our future work.

2   System Description

Project RoDiCA aims at developing a new motor cortex for the Zeno humanoid robot, as well as data collection and processing components to enable the early detection and treatment of ASD [9]. An overall system diagram of the HRI system including Zeno, the therapist, child and the associated data collection environments is shown in Fig. 2.

2.1   Humanoid Robot

Zeno is a 2 foot tall articulated humanoid robot with an expressive human like face shown in Fig. 1. It has 9 degrees of freedom (DOF) in the upper body and arms, an expressive face with 8 DOF, and a rigid lower body [16, 9]. The robot is capable of moving the upper body using a waist joint, and four joints each on the arms implemented using Dynamixel RX-28 servos. It has a 1.6 GHz Intel Atom Z530 processor onboard and is controlled by an external Dell XPS quad core laptop running LabVIEW [9].

Two different modes of interaction are implemented on Zeno. The first is called 'Dynamic Interaction,' which uses a Kinect sensor and LabVIEW to allow full teleoperation control of the arms and waist DOFs. This allows a therapist to interact with the child through the robot, or allows the child to control the robot directly. The second mode called 'Scripted Interaction,' is based on the Zeno RoboKind software which allows preprogrammed motions and conversations using text to speech software. In this paper we look at the data captured during both scripted and dynamic interaction modes where the child imitates Zeno's motion.

2.2 Data Collection and Testing of Human Subjects

The data collection is performed using a motion capture system from Motion Analysis Corp, Santa Ana, CA, and CAREN, a virtual reality system from Motek Medical BV described in detail in previous work [9]. This system consists of cameras that capture motion at 120 Hz. In this work the human subject and the robot Zeno are both instrumented using 40-50 reflective markers.

For the present pilot study interaction experiments were conducted with Zeno facing the subject as seen in Fig. 1. A scripted routine of gestures was run on Zeno, which the child was prompted to imitate using Zeno's speech functionality. The following 42 second scripted routine from [9] was run on Zeno:

1 Wave hello with right arm [8.3s to 15.8s]
2 "Tummy rub" with right arm [16.8s to 25.3s]
3 Fist bump right with arm [25.7s to 30.3s]
4 Fist bump left with arm [31.0s to 34.5s]
5 Wave goodbye with left arm [34.5s to 42.0s]

For this phase of the analysis, we only used the time series of the joint angles of Zeno's and the subject's motion during a right handed wave. Similarity in motion between child and robot is measured using the joint angles, since Cartesian positions are more difficult to compare due to dimensional and pose differences e.g. robot may be much smaller than child, and rotated in space to face each other.

3 Method

In this section, we describe how the data is processed, describe the DTW algorithm and propose the development of DTW as a tool to identify a motor marker for ASD.

3.1 Data Representation and Inverse Kinematics

In this paper, we record the imitative behavior using the four joint angles of the arm as depicted in Fig. 3. The angles are α and β in the shoulder joint and γ and θ in the elbow joint of Zeno. The data from the motion capture system is in the form of Cartesian joint positions of the shoulder $P_s = (x_s, y_s, z_s)$, elbow $P_e = (x_e, y_e, z_e)$ and the hand $P_h = (x_h, y_h, z_h)$ respectively. The joint angles are then calculated from these positions using the trigonometric equations below.

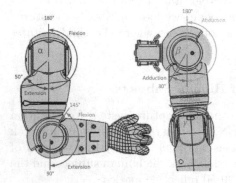

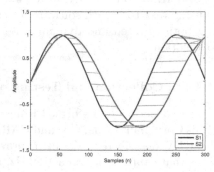

Fig. 3. Zeno's Arm Angles

Fig. 4. Example DTW match between signals S_1 and S_2

We define n_c the surface normal to the plane containing P_s, P_e and P_h, $n_c = V_{se} \times V_{eh}$, V_{se} is the vector from P_s to P_e, V_{eh} the vector from P_e to P_h, V_{eh} the vector from P_e to P_h and V_{es} the vector from P_e to P_s.

The following equations are used to obtain the joint angles:

$$\alpha = tan^{-1} \left(\frac{V_{se}(3)}{V_{se}(2)} \right), \tag{1}$$

$$\beta = cos^{-1} \left(\frac{V_{se}(1)}{|V_{se}|} \right), \tag{2}$$

$$\gamma = cos^{-1} \left(\frac{n_i \cdot n_c}{|n_i||n_c|} \right) \mathrm{sgn}([1\ 0\ 0]\ (n_i \times n_c)), \tag{3}$$

$$\theta = cos^{-1} \left(\frac{V_{eh} \cdot V_{es}}{|V_{eh}||V_{es}|} \right), \tag{4}$$

Where $V_{se}(1)$, $V_{se}(2)$ and $V_{se}(3)$ are the x, y and z components of V_{se} respectively, and n_i is defined as:

$$n_i = R_{\alpha\beta} \begin{bmatrix} 0 \\ 0 \\ 1 \end{bmatrix} \tag{5}$$

$$R_{\alpha\beta} = \begin{bmatrix} cos(\beta) & -sin(\beta) & 0 \\ cos(\alpha)sin(\beta) & cos(\alpha)cos(\beta) & -sin(\alpha) \\ sin(\alpha)sin(\beta) & sin(\alpha)cos(\beta) & cos(\alpha) \end{bmatrix} \tag{6}$$

Since the imitation between Zeno and the human is mirrored, Zenos right arm is compared to each subjects left arm angle trajectories. This data is then pre-processed by z-normalization, suggested by various researchers as an important step to be performed before running any tests [6]. The z-normalization removes offsets and scaling issues in the data, see following equation:

$$s_z = \frac{s - \mu_s}{\sigma_s} \tag{7}$$

where μ_s and σ_s are the mean and standard deviation of the signal s.

3.2   Motion Analysis Using Dynamic Time Warping

There is a need in the Autism research community to obtain quantitative measurements of imitation quality. We propose using the DTW algorithm to obtain a similarity measure between time series joint angle signals. DTW is an established signal processing method that offers a distance measure between signals similar to the Euclidean distance. However, time-warping is applied to signals to align them optimally, prior to taking the difference. Optimal alignment in this context, is the alignment of the signal time samples that makes the total distance between the signals as small as possible. This alignment induces a non-linear mapping between the two signals, e.g. warping of the signals. A good description of the DTW algorithm is given by Keogh et al. [6]. The strength of DTW is in its ability to compare the similarity between signals by ignoring time-delays and uneven time sampling. This situation is very relevant in the context of our problem, since the motion of child and robot experiences both these effects.

Given two signals that are time dependent $X = \{x_1, x_2...x_n\}$ and $Y = \{y_1, y_2...y_n\}$ where $n \in \mathbb{N}$. DTW finds the optimal distance between these two signals using a local distance measure d, we use the euclidean distance as the distance measure $d(X, Y) = \sqrt{\sum_{i=1}^{n}(x_i - y_i)^2}$.

The DTW cost $\mathfrak{D}(n, n)$ can be calculated using the following recursion:

$$\mathfrak{D}(i, j) = d(s_i, r_j) + min\{\mathfrak{D}(i - 1, j - 1), \\ \mathfrak{D}(i - 1, j), \\ \mathfrak{D}(i, j - 1)\} \tag{8}$$

DTW is applied to each set of angle trajectories generating a DTW value for each of the four angles. A range of motion is then obtained in order to weigh each angle using equation 9, where X represents a column vector of unnormalized joint angles:

$$W = max(X) - min(X). \tag{9}$$

The combined DTW distance for all four angles is calculated by a weighted average as shown in 10.

$$A_w = \frac{W_\alpha D_\alpha + W_\beta D_\beta + W_\gamma D_\gamma + W_\theta D_\theta}{W_\alpha + W_\beta + W_\gamma + W_\theta},\tag{10}$$

where,

W_* - weight per joint

D_* - calculated DTW distance for each joint

The pseudo-code used for implementing our version of DTW is shown in Algorithm 1.

Algorithm 1. DTW Cost
H

1: **procedure** DTWCOST(S,R)
2:     $m \leftarrow$ row(S)
3:     $n \leftarrow$ row(R)
4:     $\mathfrak{D} \leftarrow$ zeros($n + 1, m + 1$)
5:     $\mathfrak{C} \leftarrow 0$
6:     $\mathfrak{D}(1,1) \leftarrow 0$
7:     **for** $i \leftarrow 2, m$ **do**
8:         $\mathfrak{D}(i,1) \leftarrow \mathfrak{D}(i$-$1,1) + $ d(S(i),R(1))
9:     **end for**
10:     **for** $j \leftarrow 2, n$ **do**
11:         $\mathfrak{D}(1,j) \leftarrow \mathfrak{D}(1,j$-$1) + $ d(S(1),R(j))
12:     **end for**
13:     **for** $i \leftarrow 2, n$ **do**
14:         **for** $j \leftarrow 2, n$ **do**
15:             $\mathfrak{C} \leftarrow$ d(S(i),R(j))
16:             $\delta \leftarrow$ min($\mathfrak{D}(i$-$1,j),\mathfrak{D}(i,j$-$1),\mathfrak{D}(i$-$1,j$-$1)$)
17:             $\mathfrak{D}(i,j) \leftarrow \mathfrak{C} + \delta$
18:         **end for**
19:     **end for**
20:     **return** $\mathfrak{D}(m,n)$
21: **end procedure**

Fig. 4 shows a typical result when using the DTW algorithm from 1, the gray lines depict the nonlinear map between the signals. It can be seen from the right side of Fig. 4, where many points from S_1 are mapped to a single point of S_2, this is because DTW matches every point of the signals together.

4   Results

A set of clinical experiments with several children were performed to test the robot, capture the motion, and to compare the motion data using DTW. During experiments, children were directed by the robot to follow along performing several hand gestures, such as wave with both hands, tummy rub, fist bump, etc. Each motion was performed three times. The best one was picked for analysis.

Table 1. DTW distance per joint

Subject	Age	Gender	Type	α	β	γ	θ	$\beta\ \theta$ Combined
3	6	Male	Control	0.4421	0.6730	0.2775	0.1145	0.3075
2	6	Male	ASD	1.2128	1.2128	1.1359	0.7232	0.8913
12	9	Male	Control	0.6436	0.0792	0.5253	0.3073	0.2247
4	9	Male	ASD	0.3387	0.6738	0.6899	0.9251	0.7331
13	11	Male	Control	0.6164	0.2267	0.6409	0.2854	0.2666
6	11	Male	ASD	0.4473	0.1881	0.4319	0.3228	0.2757
11	12	Male	Control	1.1141	0.2583	0.5157	0.1159	0.1636
7	12	Male	ASD	0.2350	0.6023	0.4192	0.2091	0.4102

Wave gestures were compared for four different age groups (6, 9, 11, 12) as seen in Table 1, they consist of pairs of control and ASD children recruited from the Dallas Autism Treatment Center. Each age group consists of a control subject and an ASD subject. Representative joint angle trajectories for each DOF are shown in Fig. 5. The angle trajectories of subject 12 gives insight into the significance of each joint in performing each action. Here the elbow angle θ has the best match during the imitation, this is because the elbow performs a periodic motion when executing a wave, so it is easy to script and imitate. β which is the other main angle used in a hand wave is also fairly similar. The difference seen in Fig. 5 for the normalized angles varies more in the human trajectory because it is difficult for humans to keep their shoulder in the exact place over a period of time like the robot. The computation of the DTW for the different subjects performing the wave motion is shown in Table 1 and summarized in Fig. 6. The weights for α and γ in Equation 10 are set to zero to calculate a weighted average using only β and θ, this is because β and θ angles are more representative for the motions performed in this study.

By looking at this data we can see that the combined average for the control subjects are all lower than for the ASD subjects. This shows that DTW in combination with weighting based on range of motion for the most important angles, β and θ, for a hand wave shows a promising method of comparing human to robot imitation. Also, notice by looking at Table 1 the DTW values for β and θ also are consistent for each age group with the exception of β for the group of eleven year olds. This discrepancy is simple due to a better imitation by the ASD subject for β, but as mentioned before β will always have variation due

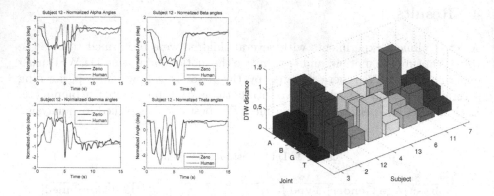

Fig. 5. Normalized Joint Angles   **Fig. 6.** DTW values

to human nature. However, this inconsistency helps show that the value of the overall performance is not affected by one angle, but instead provides a consistent value based on both angles used. This demonstrates that DTW is a good method to use in motion comparison analysis.

These preliminary results show that the DTW similarity measure can serve as both a meaningful and objective measure for evaluating the quality of imitation behavior. With age all children improve their imitation behavior, but the children with ASD consistently perform worse than their age-match controls. The combined weighted joint average for the wave motion has a higher value in all ASD children, indicating that their imitation of Zenos waving motion is less accurate.

5   Conclusions and Future Work

The contribution of this paper is the development of a robotic system which uses Dynamic Time Warping (DTW) as a tool for the diagnosis and treatment of Autism. This paper describes the development of a robotic system capable of measuring the quality of imitation interaction between a humanoid robot and a human subject. A humanoid robot called Zeno, is used to perform gestures which the human subject imitates. The similarity of the gestures performed is measured using DTW, and used to objectively analyze the quality of the imitation interaction between the human and the robot. The hope is that this type of system will enable consistent objective measurement of the quality of imitation, and can be used to obtain information about the condition and possible improvement of children with ASD.

From the graphs of the wave motion it is clear that the robot motion should be derived from nominal human motion trajectories. The initial data analysis shows that DTW can be a good tool for comparing imitation interaction since it allows the comparison of temporally inexact imitative motion. However, it is not clear what robot motions to select, and what DOFs to compare using DTW in order to best affect ASD treatment.

In the future we will comprehensively evaluate the DTW algorithm in a larger cross-longitudinal study. The motions generated by the Zeno robot will be made more natural and human like. Data collection, natural human motion generation and analysis will be contained in the newly developed motor cortex of the Zeno robot. The imitation script will be modified to increase its ability to engage the children with autism. The analysis of imitation will be extended to facial gestures.

Acknowledgment. This work was supported by the TxMRC consortium grant: Human-Robot Interaction System for Early Diagnosis and Treatment of Childhood Autism Spectrum Disorders (RoDiCA) and by the US National Science Foundation Grant CPS 1035913.

The authors wish to thank David Hanson, Richard Margolin, Matt Stevenson and Joshua Jach of Hanson RoboKind and Intelligent Bots LLC for their help with Zeno R-30. The authors thank Robert Longnecker, Janet Trammell, David Cummings and Carolyn Carr from UNTHSC for their help with the experimental data collection presented in this paper. The authors also thank Abhishek Thakurdesai from the NGS group for help with Zeno programming.

References

1. Baer, D., Peterson, R., Sherman, J.: The development of imitation by reinforcing behavioral similarity to a model. Journal of the Experimental Analysis of Behavior 10(5), 405 (1967)
2. Centers for Disease Control and Prevention: Autism and developmental disabilities monitoring (addm) network (Sep 2012),
   http://www.cdc.gov/ncbddd/autism/addm.html
3. Fasola, J., Mataric, M.: Using socially assistive human-robot interaction to motivate physical exercise for older adults. Proceedings of the IEEE 100(8), 2512–2526 (2012)
4. Geschwind, D.H.: Advances in autism. Annual Review of Medicine 60(1), 367–380 (2009)
5. Itakura, F.: Minimum prediction residual principle applied to speech recognition. IEEE Transactions on Acoustics, Speech and Signal Processing 23(1), 67–72 (1975)
6. Keogh, E., Ratanamahatana, C.A.: Exact indexing of dynamic time warping. Knowledge and Information Systems 7, 358–386 (2005), doi:10.1007/s10115-004-0154-9
7. Rajruangrabin, J., Popa, D.: Robot head motion control with an emphasis on realism of neck-eye coordination during object tracking. Journal of Intelligent & Robotic Systems 63(2), 163–190 (2011), doi:10.1007/s10846-010-9468-x
8. Ranatunga, I., Rajruangrabin, J., Popa, D.O., Makedon, F.: Enhanced therapeutic interactivity using social robot zeno. In: Proceedings of the 4th International Conference on PErvasive Technologies Related to Assistive Environments, PETRA 2011, pp. 57:1–57:6. ACM, New York (2011)
9. Ranatunga, I., Torres, N.A., Patterson, R., Bugnariu, N., Stevenson, M., Popa, D.: Rodica: a human-robot interaction system for treatment of childhood autism spectrum disorders. In: Proceedings of the 5th International Conference on PErvasive Technologies Related to Assistive Environments, PETRA 2012. ACM, New York (2012)

10. Ricks, D.J., Colton, M.B.: Trends and considerations in robot-assisted autism therapy. In: 2010 IEEE International Conference on Robotics and Automation (ICRA), pp. 4354–4359 (2010); iD: 1
11. Rinehart, N., Bradshaw, J., Brereton, A., Tonge, B.: Movement preparation in high-functioning autism and asperger disorder: A serial choice reaction time task involving motor reprogramming (2001)
12. Robins, B., Dautenhahn, K., Dubowski, J.: Does appearance matter in the interaction of children with autism with a humanoid robot? Interaction Studies 7(3), 509–542 (2006)
13. Sakoe, H., Chiba, S.: Dynamic programming algorithm optimization for spoken word recognition. IEEE Transactions on Acoustics, Speech and Signal Processing 26(1), 43–49 (1978)
14. Scassellati, B., Admoni, H., Mataric, M.: Robots for use in autism research. Annual Review of Biomedical Engineering 14(1), 275–294 (2012), doi:10.1146/annurev-bioeng-071811-150036
15. Tapus, A., Peca, A., Aly, A., Pop, C., Jisa, L., Pintea, S., Rusu, A.S., David, D.O.: Children with autism social engagement in interaction with nao, an imitative robot. Interaction Studies 13(3), 315–347 (2012)
16. Torres, N.A., Clark, N., Ranatunga, I., Popa, D.: Implementation of interactive armplayback behaviors of social robot zenofor autism spectrum disorder therapy. In: Proceedings of the 5th International Conference on PErvasive Technologies Related to Assistive Environments, PETRA 2012. ACM, New York (2012)

Low-Cost Whole-Body Touch Interaction for Manual Motion Control of a Mobile Service Robot

Steffen Müller, Christof Schröter, and Horst-Michael Gross*

Ilmenau University of Technology, Thuringia, Germany,
steffen.mueller@tu-ilmenau.de
http://tu-ilmenau.de/neurob

Abstract. Mobile service robots for interaction with people need to be easily maneuverable by their users, even if physical restrictions make a manual pushing and pulling impossible. In this paper, we present a low cost approach that allows for intuitive tactile control of a mobile service robot while preserving constraints of a differential drive and obstacle avoidance. The robot's enclosure has been equipped with capacitive touch sensors able to recognize proximity of the user's hands. By simulating forces applied by the touching hands, a desired motion command for the robot is derived and combined with other motion objectives in a local motion planner (based on Dynamic Window Approach in our case). User tests showed that this haptic control is intuitively understandable and outperforms a solution using direction buttons on the robot's touch screen.

1 Introduction

This paper deals with the problem of intuitive human control of a mobile service robot. In our current SERROGA-project (SERvice RObotics for health (Gesundheits) Assistance) [14], we are developing a service robot for elderly that is supposed to maneuver in the narrow home environment of elderly users. The robot's main communication channel is a tiltable touch display, which can be used from a standing or sitting position. For service delivery in a home environment, the robot in many cases has to find and approach a user autonomously to engage in interaction. However, even with the best person detection and navigation algorithms, the user might want the robot to take a different position, making a manual control interface necessary, especially if the robot is too heavy to be pushed and pulled directly. In our case, the robot with a weight of about 40kg (see Fig. 1(a)) can only be pushed manually with a high degree of internal friction in the geared motors, requiring high force and even a suitable bent over handling position to prevent the robot from tilting. The idea presented in

---

* This work has received funding from the Federal State of Thuringia and the European Social Fund (OP 2007-2013) under grant agreement N501/2009 to the project SERROGA (project number 2011FGR0107).

G. Herrmann et al. (Eds.): ICSR 2013, LNAI 8239, pp. 229–238, 2013.

this paper is a low-cost tactile interface allowing for intuitive local control of the robot's position by means of touching the enclosure, which is usable from a standing or a sitting position. This physical interaction should give the user the impression he/she is still simply pushing the robot, while the friction and perceived mass is reduced to an easily manageable amount. In order to prevent from accidental collisions with furniture and other obstacles, the amplified motion command is filtered by the local obstacle avoidance, which is also used for autonomous navigation. Therefore, the robot will evade the obstacles, while the user coarsely pushes in the desired direction. We know, that highly sophisticated force measuring sensors exist, that simulate an artificial skin, but for a mass market product they are too expensive. We could show, that even with a simple sensor, a very intuitive motion control interface could be realized, that improved suitability for daily use a lot.

In the following, the paper will introduce our robot and discuss the state-of-the-art regarding tactile sensors on robots briefly. Subsequently, the touch sensor we use is explained, before the physical model for generating the manual motion command is described. After that, an overview of the software architecture, especially the navigation framework, is presented, showing how the manual command is integrated with obstacle avoidance. Finally, some experiments with users show the benefit against display interaction for navigation.

2   The Robot

The platform used (see Fig. 1(a)) is a Scitos-G3, which was developed in the EU-funded project CompanionAble [1]. It is a mobile service robot based on a differential drive with a castor that limits the possible motion to two degrees of freedom. It can move in combinations of rotation and forward/backward translation. Fortunately, the small ground shape of the robot nearly allows for rotation in place and provides excellent maneuverability even in narrow environments. The two driven wheels have a diameter of 20 cm and a distance of 36 cm. They are driven by geared DC motors, which have a high internal inertia making manual moving difficult. Besides the drive, the robot has a tiltable touch display and a head with two OLED displays as eyes.

The robot's covering is formed by a free form enclosure that is made from 4mm polyamide plastic in a rapid prototyping process. It consists of three removable covers for the base and the rear head as well as two side panels and the forehead that are load-bearing parts.

For navigation and user perception, the robot is equipped with a SICK laser range finder at a height of 21cm and an array of 14 ultra sonic sensors at the base. Furthermore, a Kinect sensor and a 180°fish-eye camera in the frontal head assist obstacle and person detection. For user interaction, the robot additionally is equipped with a microphone and a sound system.

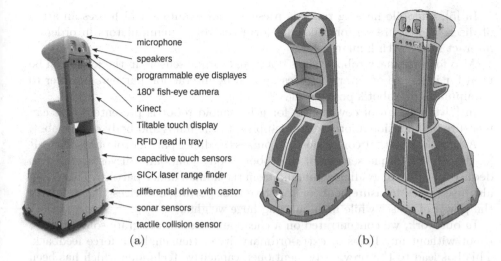

- microphone
- speakers
- programmable eye displays
- 180° fish-eye camera
- Kinect
- Tiltable touch display
- RFID read in tray
- capacitive touch sensors
- SICK laser range finder
- differential drive with castor
- sonar sensors
- tactile collision sensor

(a)   (b)

Fig. 1. a) Scitos-G3 (by Metralabs GmbH Ilmenau) robot with the main interaction devices and sensors, b)Placement of capacitive touch sensor electrodes at the inside of the enclosure (red areas)

3   Touch Sensors and Related Work

Before the implementation of our touch sensitive robot enclosure is described, the state-of-the-art regarding tactile sensors will be discussed briefly here.

Tactile sensors comprise various devices that can either measure the position and quality of a contact of objects (mainly the human fingers or hands) with a technical system like a robot, or additionally give quantitative information on the force applied on the mechanical parts of a robot. Sometimes even directional information can be perceived.

[2] identified optical, capacitive and resistive effects as possible sources for information. These effects have been applied for realization of touch displays for many years and can also be brought to other surfaces, e.g. the enclosure of robots. Also acoustic approaches using surface acoustic waves are known from touch displays. An interesting approach is the usage of a flexible textile material, able to cover non-planar surfaces [3].

All these sensors are comparatively cost efficient, but come along with some disadvantages. Capacitive sensors are limited to conductive contact objects, and only qualitative contact events can be observed omitting the direction of applied forces and power of contact.

Alternatively, more expensive techniques are emerging, able to measure force and pressure at a tactile surface. In 1993, a force sensitive transducer technology was applied in [4] for measuring contact forces with a robot. Nowadays, various projects work on artificial skin for robots, that should combine perception capabilities for various stimuli qualities.

In [5], a mobile nursing robot is presented for example, which uses an array of discrete pressure sensors allowing for a control of manipulators in order to interact safely with human users.

Also for stationary robotic arms, force and torque sensors in the joints can be used for inference of external forces applied by interaction partners in order to manipulate the robot's position [6].

In [7] the design of cover parts for a humanoid robot is presented, that can measure 3-dimensional force vectors applied to the limbs and body of the robot.

An application that comes closer to ours is the dance partner robot Ms DanceR [8]. Force and torque sensors in the robot's body joints and drives are used to deduce an external guiding command from the human dance partner. This group also used force measurement for assistive transportation devices that reinforce the power of a user while manipulating huge weights.

In our work, we concentrated on a cheap and easy to integrate solution for a robot without any limbs or extra-ordinary drives that enable a force feedback. This has lead to the previously mentioned capacitive technique which has been implemented finally.

When a touch sensor is designed, a compromise on the spacial resolution of sensitive area and the number of simultaneously distinguishable contact points needs to be found. If the quality of contact is known, like finger tips on a touch screen, it is possible to interpolate the position of a contact between specially arranged electrodes, which on the one hand is very exact in the position as known from touch screens, but on the other hand reduces the number of simultaneous contact points. In our case, the touch events are more versatile and of larger area than finger tips, making interpolation difficult. Touching the robot with a full hand should be recognized as well as just sliding fingers over it.

Furthermore, the complex shape of the electrodes makes it difficult to integrate it into a non-planar shape of a robot enclosure. Since we do not need a high spatial resolution, and because we want to superimpose touch events at multiple points on the robot's enclosure, a simple layout of the electrodes has been chosen for testing, that can improved in the future.

4   Capacitive Touch Sensors in Robot's Enclosure

For human-robot interaction, the parts of the robot's enclosure have been equipped with 12 laminar electrodes made from aluminum foil, that are attached to the inside of the plastic covers: four electrodes in the head, two in the side panels and six in the base. Fig. 1(b) illustrates the position of the electrodes as red areas. The electrodes are connected to an Atmega16 micro controller implementing the Atmel QTouch® technology [9] to measure the capacitance of the electrodes, that is affected by a finger or hand near the surface. QTouch needs a minimum of additional circuit elements and thus is very simple to implement. The principle of meassurement is a reference capacitor that is charged in steps, where the sensor electrode's capacity determines the number of steps needed. By means of the 12 independent channels, contact to multiple areas of the robot's cover can be evaluated simultanously due to a cyclic sheme of measuring. The sensor's sensitivity ranges over about two decades, beginning when the hand is hovering 1 to 2 cm over the surface, significantly indicating contact of fingers close to the electrodes, and saturating when the full hand is placed on the electrode's area. The sensors' measurements are sampled periodically at 100Hz. This rather high sampling frequency allows for a recursive band-pass filtering of the raw sensor readings reducing noise to a minimum (low-pass) and enabling a self-calibration (high-pass). This high-pass filter removes drift in the sensor readings and is useful to find the individual operating point of each sensor.

This preprocessing is done directly on the micro-controller, which is sending the resulting touch signals to the robot PC via USB, as floating point numbers in [0, 1] range and at a frequency of 10Hz.

The electrodes have been placed to cover regions of low curvature and span the whole surface beginning from the opening for the laser range finder up to the robot's head. The movable display has been omitted, because of the difficult cable routing and the ability to notice touch at the display anyway. Effectively, almost the complete orange area of the robot (see Fig. 1(a)) is able to perceive contact to human fingers.

5 Touch Based Motion Control

The idea behind the manual motion control is to simulate a physical model of the robot as an inertial mass that is subject to restrictions of the differential drive and external forces and moments that result from the touch interaction. This model has own damping and friction parameters that are magnitudes lower than the real robot's friction. Using that model to compute a velocity, we control the active drives of the robot to achieve a motion that corresponds to the tactile user inputs. By means of this approach, the mass and friction of the robot appear to be reduced and an intuitive control is facilitated. As we will show later, the simulated robot's velocity command is combined with local obstacle avoidance and other navigation tasks to enable safety and integration of internal and external tasks.

For the virtual robot model, we assume that touching the robot at a certain point is associated with a fixedly oriented force vector F that is pointing approximately perpendicularly to the robots outer shape at the offset r. These offsets and directions have been configured manually in order to find the most convenient motion behaviour. Fig. 2 is illustrating the principle for a touch sensor at the back of the robot. Touching it there will cause a slight rotation to the left due to the large angle between F and r, as well as a forward motion.

It should be mentioned, that the assumption of perpendicular force is not fully eligible in all situations, since people do not push the robot exclusively. In user studies, we observed some intents to rotate the robot by grasping the robot's head and turning it. This would result in forces tangential to the robot's enclosing. In fact, this causes irritations sometimes, if the resulting motion does not match the users intention. We will discuss a possible improvement of our

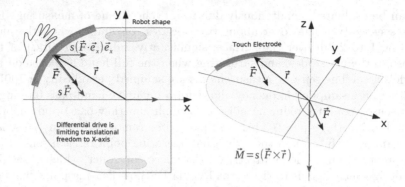

Fig. 2. Illustration of force model used for determining the robot model's acceleration and angular acceleration. Both are using the D shaped footprint of the robot. Left: projection of the unit force vector F scaled by sensor reading s on the x-axis e, Right: cross product of offset vector r and force vector F lead to a rotational moment M.

approach relating to this problem in the outlook chapter at the end of this paper.

The value of the virtual force applied to the robot model is assumed to be proportionally to the touch sensor reading s, which is representing the contact area and proximity of the user's hand.

From the modeled force vectors and the respective offset position on the robot, we can compute a summarized turning moment M^* and a translational force F^*. Considering the limitations of the differential drive (no sidestepping), the force vectors F_i are projected onto the internal x-axis before summarizing over all available sensors: y and z components of the force vectors are absorbed by the wheels in reality.

$$F^* = e_x \sum_i s_i \left(F_i \cdot e_x \right) \tag{1}$$

Here, e_x is the unit vector in x direction. Turning moments result from forces directed not exactly towards the center of mass. The overall turning moment is:

$$M^* = \sum_i s_i F_i \times r_i \tag{2}$$

From the total translational force and turning moment, the accelerations and subsequently the resulting translation and rotation velocity can be simulated, by assuming an artificial friction term f, as well as a hypothetical mass m and moment of inertia L. The model of internal friction is quadratic in the current velocity, which helps limiting the maximum velocity.

Since we model the motion in the robot's internal coordinates, the system state is reduced to a scalar velocity v (translational velocity along the x-axis) and rotational velocity ω. To keep the model in synchronization, we use the actual robot velocity $\tilde{v}_{t-\Delta t}$ and $\tilde{\omega}_{t-\Delta t}$ from the last time step for the prediction of the new target state v_t, ω_t

$$v_t = \tilde{v}_{t-\Delta t}(1 - f|\tilde{v}_{t-\Delta t}|) + \frac{|\boldsymbol{F}^*|}{m}\Delta t \qquad (3)$$

$$\omega_t = \tilde{\omega}_{t-\Delta t}(1 - f|\tilde{\omega}_{t-\Delta t}|) + \frac{|\boldsymbol{M}^*|}{L}\Delta t \qquad (4)$$

In the most simple case, the resulting model state variables v_t and ω_t can be sent to the robot hardware directly in order to generate a reinforced motion. In our case, the translational and rotational velocity need to be transformed into two wheel speeds v_r and v_l, which are the targets for the hardware controllers. Using the wheel distance d this is straight forward:

$$v_r = v_t + \omega_t \frac{d}{2} \qquad (5)$$

$$v_l = v_t - \omega_t \frac{d}{2} \qquad (6)$$

However, for the intended application it is not sufficient to directly send the motion command to the hardware, since we want to combine the manual motion command from our robot model with the obstacle avoidance and other internal navigation tasks running in background. The next section, therefore, introduces our navigation architecture in an overview.

6 Navigation Architecture

Besides the hardware design, considerable effort has been spent on a versatile reusable modular software. The robot's navigation software is based on the MIRA middleware [10]. Furthermore, the generic navigation concept of [11] is used in order to combine aims of different navigation tasks.

The navigation concept uses a Dynamic Window Approach [12] for motion planning, that is operating on a two-dimensional search space, spanned by the velocities of the robot's left and right wheel. This way, the differential drive can be modeled very efficiently. The adaptive dynamic window consists of a cell discretization of variable size, for which a set of so-called objectives yields a cost estimation for the respective wheel speed combination. If an objective decides that a specific speed is not allowed, the respective cell is marked as forbidden. Otherwise, the costs of all active objectives are summed up, before the cell with minimum cost is used to select the desired wheel velocities, which is then sent to the hardware controller as a target value.

The flexible and extensible set of objectives is a benefit of the used navigation concept. For autonomous operation, there are a path following objective, a distance objective for avoiding dynamic obstacles, a speed and no-go objective, as well as a heading direction objective. The tactile control is enabled by a further objective. Purpose and mechanisms of these objectives are briefly described in the following:

The *path following objective* is based on the E* planning algorithm [13] that operates on a hierarchical occupancy map, which is updated online using the laser range sensor and the Kinect depth data. The planner yields a map containing the distance to a target position, which is sampled with the predicted position for each wheel speed combination. Reducing distance to the target corresponds to low costs and vice versa.

The *distance objective* is based on the occupancy grid map of the local vicinity, including obstacles perceived by the robot's sensors. Applying a distance transformation, the distance to the next obstacle is stored in the map for fast query. Considering the robot's maximum physical acceleration, the cells in the dynamic window can be forbidden if a collision would be unavoidable using the respective speed. That way, we can keep large distance to obstacles when driving fast, but are also able to pass through narrow doors with a low speed.

The *no-go and speed objectives* also contain a global map of allowed speed vectors. This enables realization of one way areas and thus e.g. socially acceptable navigation behaviors. By means of setting the allowed speed to zero, it is also possible to completely prevent the robot from entering forbidden regions.

The *heading objective* is used to orient the robot when arriving at a target by scoring the difference of the predicted orientation and the target orientation. A very low weight for the heading objective makes it only relevant when the target is reached and the other objectives do not produce significant differences in their vote anymore.

Tactile motion commands are incorporated into the motion planner by a *tactile control objective*, which calculates costs $C(\hat{v}_r, \hat{v}_l)$ for a wheel speed combination that increases with the distance to the desired speed v_r and v_l of the robot model.

$$C(\hat{v}_r, \hat{v}_l) = 1 - exp\left(\frac{-(v_r - \hat{v}_r)^2 - (v_l - \hat{v}_l)^2}{\sigma^2}\right) \tag{7}$$

The parameter σ determines the steepness of the radial cost function.

7 Resulting Motion Behavior

In order to give an impression on the usability of the touch-based navigation control, we conducted a short contest with 10 members of our staff in the age of 25 to 35. The subjects were tech-savvy but did not navigate the robot in before. The aim was steer the robot to three distinct positions in our living lab that have been marked on the floor. The test users first were offered a display-based navigation GUI, consisting of four direction buttons and a stop button on the touch screen. These buttons set the desired velocity similarly to the tactile control objective. Afterwards, the users were asked to play the same game using the body touch control. It turned out, that this was much faster than using the buttons on the screen, the average time for the tour was only one minute compared to 1:43 minutes for Button based control.

That confirms what users reported in the interview after the test: Most of them liked the touch control and would prefer it over the screen button based interaction. Nevertheless, as described above, some strange situations occurred, when the users did not respect the assumption of perpendicular forces (pushing only). Once informed on the rules, also these users managed to navigate the robot safely with the touch interface.

Table 1. Durations of navigation task for the test users in the experiment using GUI buttons compared to touch control

	duration of tests for individual users in minutes
GUI	1:45 2:00 1:48 2:34 1:50 1:22 1:19 1:29 1:15 1:50
touch	0:51 0:52 0:54 1:01 1:10 1:00 0:52 1:06 1:00 1:18

The benefits of the touch-based control result from the analoguous speed control and that it is reachable from nearly any position. In many situations with the GUI based control, the user had to walk around the moving robot when turning it around, which is avoided by the touch based navigation.

Recently, additional usability studies have been conducted with elderly users, that besides the haptic control also comprise a remote control mode using a tablet pc and the autonomous navigation behaviours of the robot. Analysis of this study is still ongoing work.

8   Conclusion and Outlook

It could be shown that an intuitive input modality for local manual robot control can be implemented with very simple capacitive touch sensors placed within the enclosure of a mobile robot. The existing obstacle avoidance capabilities of the robot can easily be combined with the manual control due to a modular navigation concept, which is based on a dynamic window approach.

User studies showed that people favor the touch motion control over a GUI-button-based local motion control.

Some drawbacks could be observed due to the coarse spatial resolution of the only 12 sensor areas. There are parts of the robot's cover that are oriented in different directions but are in the sensing range of the same electrode. In these cases, the direction of the virtual force does not necessarily correspond to the surface normal of the enclosure, which causes a wrong rotation direction in some cases. Also the assumption that the force applied by the user is always perpendicular to the surface is inappropriate. Depending on the relative position of the user to the desired direction of movement, the robot may be pulled sometimes instead of being pushed only. In these cases, the resulting force is tangential to the surface, which is not modeled yet.

Thus, one aspect for optimization is the number of sensor electrodes, which needs to be increased in order to reflect the different parts of the robot's surface better. The second option to overcome the drawbacks of the undirected observations of contact is to spend more effort in the mapping from touch sensor readings to the desired motion command. We plan to apply machine-learning approaches in order to learn the mapping from example data. People's interaction behaviour is to be observed while being instructed to move the robot in certain directions. A function approximation can be trained with the values of the touch sensors, the relative position of the user, and the current velocity of the robot as input and the motion vector to the desired position as a target.

References

1. Gross, H.-M., et al.: Progress in Developing a Socially Assistive Mobile Home Robot Companion for the Elderly with Mild Cognitive Impairment. In: Proc. IEEE/RSJ Int. Conf. on Intelligent Robots and Systems (IROS 2011), pp. 2430–2437 (2011)
2. Weiss, K., Worn, H.: Resistive tactile sensor matrices using inter-electrode sampling. In: Proc. 31st Conf. IEEE Industrial Electronics Society (IECON 2005), Raleigh, North Carolina, USA, pp. 1949–1954 (2005)
3. Pan, Z., Cui, H., Zhu, Z.: A flexible full-body tactile sensor of low cost and minimal connections. In: Proc. IEEE Intl. Conf. on Systems, Man and Cybernetics (IEEE-SMC 2003), Washington, D.C., USA, vol. 3, pp. 2368–2373 (2003)
4. McMath, W.S., Colven, M.D., Yeung, S.K., Petriu, E.M.: Tactile pattern recognition using neural networks. In: Proc. Intl. Conf. on Industrial Electronics, Control, and Instrumentation (IECON 1993), Lahaina, Hawaii, vol. 3, pp. 1391–1394 (1993)
5. Mukai, T., et al.: Development of the Tactile Sensor System of a Human-Interactive Robot 'RI-MAN'. IEEE Transactions on Robotics 24(2), 505–512 (2008)
6. Cannata, G., Denei, S., Mastrogiovanni, F.: Contact based robot control through tactile maps. In: Proc. 49th IEEE Conference on Decision and Control (CDC 2010), Atlanta, Georgia, USA, pp. 3578–3583 (2010)
7. Iwata, H., Sugano, S.: Whole-body covering tactile interface for human robot coordination. In: Proc. IEEE Intl. Conf. on Robotics and Automation (ICRA 2002), Washington, D.C., USA, vol. 4, pp. 3818–3824 (2002)
8. Kosuge, K., Hayashi, T., Hirata, Y., Tobiyama, R.: Dance partner robot - Ms DanceR. In: Proc. IEEE/RSJ Intl. Conf. on Intelligent Robots and Systems (IROS 2003), Las Vegas, Nevada, USA, vol. 4, pp. 3459–3464 (2003)
9. QTouch technology website, http://www.atmel.com/products/touchsolutions/bsw/qtouch.aspx
10. Einhorn, E., Stricker, R., Gross, H.-M., Langner, T., Martin, C.: MIRA - Middleware for Robotic Applications. To appear: Proc. IEEE/RSJ Int. Conf. on Intelligent Robots and Systems (IROS 2012), Vilamoura, Portugal (2012)
11. Einhorn, E., Langner, T.: Pilot - Modular Robot Navigation for Real-World Applications. In: Proc. 55th Int. Scientific Colloquium, Ilmenau, Germany, pp. 382–387 (2010)
12. Fox, D., et al.: The Dynamic Window Approach to Collision Avoidance. IEEE Robotics & Automation Magazine 4(1), 23–33 (1997)
13. Philippsen, R., Siegwart, R.: An Interpolated Dynamic Navigation Function. In: Proc. IEEE Intl. Conf. on Robotics and Automation (ICRA 2005), Barcelona, Spain, pp. 3782–3789 (2005)
14. http://www.serroga.de

Human-Robot Interaction between Virtual and Real Worlds: Motivation from RoboCup @Home

Jeffrey Too Chuan Tan[1], Tetsunari Inamura[2], Komei Sugiura[3],
Takayuki Nagai[4], and Hiroyuki Okada[5]

[1] Institute of Industrial Science, University of Tokyo, Japan
jeffrey@iis.u-tokyo.ac.jp
[2] Principles of Informatics Research Division, National Institute of Informatics, Japan
inamura@nii.ac.jp
[3] National Institute of Information and Communications Technology, Japan
komei.sugiura@nict.go.jp
[4] The University of Electro-Communications, Japan
tnagai@ee.uec.ac.jp
[5] Tamagawa University, Japan
h.okada@eng.tamagawa.ac.jp

Abstract. The main purpose of this work is to investigate on the HRI issue based on recent RoboCup @Home competitions, in order to formulate an alternative research framework for HRI studies. Based on the survey of the current competition mechanism and benchmarking approach, and the analysis of the actual team performances, the shortcoming of current approach in motivating high level cognitive robot intelligence development is identified. A new proposal of an alternative research framework with HRI between virtual and real world is illustrated based on the SIGVerse system. The implementations of the research framework in various developments are discussed to show the feasibility of this framework in supporting HRI development. The level of abstraction in embodiment and multimodal interaction can be adjusted in the simulation based on the HRI requirements, to reduce the complexity from the integration problem. The framework is motivating subject experts to connect high level intelligence (e.g. dialogue engine, task planning) to low level robot control for "deeper" HRI. The HRI simulation is reproducible and scalable by the distributed system architecture that enables participation of a large group of human users (over the network/internet) and robots in a complex multi-agent multi-user social interaction.

Keywords: Human-Robot Interaction, Virtual Reality, RoboCup @Home.

1   Introduction

RoboCup, founded in 1997, is a well-established international effort to promote artificial intelligence and robotics development via a set common platforms motivated by a competition setting. In 2006, a new initiative was formed to extend this effort into service robotics and human-robot interaction with the creation of the RoboCup

G. Herrmann et al. (Eds.): ICSR 2013, LNAI 8239, pp. 239–248, 2013.

@Home league [1]. Distinguished from the original soccer game concept, RoboCup @Home focuses on domestic real world applications with human-robot interaction, without a fixed set of robot task and operation environment. The goal of RoboCup @Home is to promote the development of a versatile domestic service robot that can assist human in everyday life [2].

Starting from 2006, in the first four years of the RoboCup @Home competitions, the main focus on the competitions was to benchmark the integration of the physical capabilities of the robots in navigation, object manipulation, human recognition and basic human-robot interaction. However, based on the task requirements in the first four years of competitions, the robots did not seem to require intelligence with high level of cognitive capabilities in order to perform the tests in the competitions [1]. Only after 2009, a new test was introduced into RoboCup @Home 2010 competition to challenge the cognitive capabilities of the robots [2].

Although the aim of RoboCup @Home is to foster the development of all-rounded service robots, is the current competition mechanism motivating high level robot intelligence development for future service robots with strong human-robot interaction capabilities? The main purpose of this work is to investigate this issue based on the last two years of competitions, RoboCup @Home 2011 and 2012, in order to formulate an alternative research framework for HRI studies. Section 2 is the survey of the current competition mechanism and benchmarking approach, and analysis of the actual team performances to study the actual impact of the competitions towards robot intelligence development in HRI. Based on the analysis findings, the shortcoming of current approach in motivating high level cognitive robot intelligence development is discussed and summarized in Section 3. It is followed by a new proposal of an alternative research framework. Based on several previous and current implementations [3]-[5], the significant features of this proposal for HRI development are illustrated in Section 4. Section 5 concludes the overall writing in this work.

2   Motivation from RoboCup @Home

2.1   The Current Competition Mechanism

In general, the RoboCup @Home competition consists of an initial *Robot Inspection and Poster Session*, a series of *Tests* on basic service robot functionalities, an Open Challenge for the team to demonstrate its own recent development and its robot's best abilities, a *General Purpose Service Robot* test, an advanced test in real environment (e.g. shopping mall, restaurant), and a final *Demo Challenge*, which is similar to *Open Challenge* but with a scoped theme.

In 2012, the *Stage I* competition consisted the tests of *Robot Inspection and Poster Session* (RIPS), *Follow Me* (FM), *Who Is Who* (WW) and *Clean Up* (CU) and *Open Challenge* (OC). In *Stage II*, the tests were *General Purpose Service Robot* (GPSR), *Restaurant* (Res), and *Demo Challenge* (DC). The five teams with the highest total scores from *Stage I* and *Stage II* competed in *Finals* with the final open demonstration.

2.2   The Distribution of the Functionality-Score

Based on the latest RoboCup @Home Rules & Regulations[1,2], the desired robot technical abilities, which the competition tests will focus on, were quoted as follows:

- *Navigation* in dynamic environments
- Fast and easy *calibration and setup* – The ultimate goal is to have a robot up and running out of the box.
- *Object recognition*
- *Object manipulation* – Manipulation is essential for almost any future home applications.
- *Recognition of humans*
- *Human-robot interaction* – An aim of the competition is to foster natural interaction with the robot using speech and gesture commands.
- *Speech recognition* – For intuitive interaction, it is essential to come up with solutions that do not require headsets in the future.
- *Gesture recognition*
- *Robot applications* – RoboCup @Home is aiming for applications of robots in daily life.
- *Ambient intelligence* – Communicate with surrounding devices, getting information from the Internet, e.g. asking the robot about the weather, reading/writing emails.

In the benchmarking approach, based on these desired robot technical abilities, a set of *functional abilities* [1] is determined. In each year of competition, the weightage of the *functional abilities* is readjusted to determine a desired score system for the year. Table 1 shows the functionality-score table of 2010 competition [1]. The *Cognition* ability was firstly introduced in 2010 with a weightage of 13% of the overall score. This shows that the robot cognition ability had started to get the attention of the organizer but the weightage was still relatively low as compared to other physical capabilities, such as navigation and object manipulation.

Table 1. Functionality-score table of 2010 [1]

Ability	Weightage (%)
Navigation	22
Mapping	9
Person Recognition	12.5
Person Tracking	3
Object Recognition	7.5
Object Manipulation	14
Speech Recognition	15
Gesture Recognition	3.5
Cognition	**13**

[1]   Robocup @Home Rules & Regulations. Version 2012, Revision 286:288 (2012)
      http://purl.org/holz/2012_rulebook.pdf
[2]   Robocup @Home Rules & Regulations. Version 2011, Revision 156:158M (2011)
      http://www.ai.rug.nl/robocupathome/documents/rulebook2011.pdf

2.3   The Actual Competitions and Team Performances

The team performances of the top five teams with the highest scores in 2011 and 2012 are analyzed to investigate the actual impact of the competitions towards robot intelligence development.

In 2011, the top five teams were *NimbRo*, *WrightEagle*, *b-it-bots*, *RobotAssist* and *ToBI*. Table 2 summarizes the team performance analysis in all tests. GPSR I and GPSR II were two tests that were required more robot cognition capabilities [1]. The tests have 6.67% and 13.33% maximum score weightages, which are pretty much agreed to the functionality-score (Table 1). However, it is also observed that the average achievement percentages of these tests by the top teams are only 6.00% and 34.38%.

Table 2. Analysis of the team performance 2011[3]

Test	Max. Score	Score Weightage (%)	Achieved Score [Average]	Achievement Percentage (%)
RIPS	1000	6.67	890.4	89.04
FM	1000	6.67	680	68.00
WW	1000	6.67	255	25.50
OC	2000	13.33	1262.6	63.13
GPSR I	**1000**	**6.67**	**60**	**6.00**
GGI	1000	6.67	240	24.00
SM	2000	13.33	362.5	18.13
GPSR II	**2000**	**13.33**	**687.5**	**34.38**
EWW	2000	13.33	595	29.75
DC	2000	13.33	679.4	33.97
Total	15000	100	5502.4	36.68

On the other hand, the top five teams in 2012 were *NimbRo@Home*, *eR@sers*, *ToBI*, *WrightEagle@Home*, and *homer@UniKoblenz*. The team performance analysis in all tests is summarized in Table 3. GPSR was the only test that was focused on robot cognition capabilities [1]. The test has 19.23% maximum score weightage, which was similar to 2011 (in total). However, it is also observed that the average achievement percentage of this test by the top teams is only 8.20%, much lower compared to 2011.

From the survey and analysis above, it is clear that the RoboCup @Home competition is "slowly" introducing challenges that require high level robot cognition capabilities but it was not resulting promising developments by the participated teams.

---

[3]   RIPS (Robot Inspection and Poster Session); FM (Follow Me); WW (Who Is Who); OC (Open Challenge); GPSR (General Purpose Service Robot); GGI (Go Get It); SM (Shopping Mall); EWW (Enhanced Who Is Who); DC (Demo Challenge).

Table 3. Analysis of the team performance 2012[4]

Test	Max. Score	Score Weightage (%)	Achieved Score [Average]	Achievement Percentage (%)
RIPS	1000	7.69	721.8	72.18
FM	1000	7.69	195	19.50
WW	2000	15.38	655	32.75
CU	1000	7.69	366.6	36.66
OC	2000	15.38	1249.8	62.49
GPSR	**2500**	**19.23**	**205**	**8.20**
DC	1500	11.54	644	42.93
Res	2000	15.38	260	13.00
Total	13000	100	4297.2	33.06

3   Human-Robot Interaction between Virtual and Real Worlds

The main concern of HRI studies is to analyze the details of interaction, e.g. the dynamics properties in physical interaction, and the information flow in verbal (text, speech) and non-verbal communication (body gesture, eye gaze). The most direct approach to these studies is to conduct experiments (empirical studies) with real human subjects and measure the desired elements. However, this approach required great effort to prepare the robot, bring in the human subjects and deal with real world noise in measuring the interaction.

3.1   Lesson from RoboCup @Home

From the above studies, it is evident that the current research approach in RoboCup @Home (empirical studies with real robot) is limiting higher robot intelligence development with the following two main reasons:

(A) The Integration Problem

The tests in RoboCup @Home are designed to require robot with integrated several functional abilities (navigation, object recognition, manipulation, etc.) in order to perform the tasks. It is obvious that a robot that only specializes in a particular ability but lacks of some other fundamental abilities like navigation might not able to perform many tasks. As a result, the balance among all functional abilities developments is far more critical than specialize in a particular ability. This is also meant that subject experts especially in the field of robot cognitive development are difficult to implement their work without a workable robot platform and HRI environment. Hence, low level robot control and mechanism have becoming the limiting factor for high level intelligence development.

---

[4] RIPS (Robot Inspection and Poster Session); FM (Follow Me); WW (Who Is Who); CU (Clean Up); OC (Open Challenge); GPSR (General Purpose Service Robot); DC (Demo Challenge); Res (Restaurant).

(B) The Limited HRI

In order to keep the mechanical feasibility of the robot, most of the tests in RoboCup @Home require challenging physical performance rather than high level of cognitive reasoning for the robot to perform the tasks. The only test that focuses on robot cognition is the integrated test, *General Purpose Service Robot* (GPSR). Based the descriptions from [1], the robot needs to be capable to:

1. *Understanding orders* – The order of the instructions corresponded to the required actions to perform the task is not predefined to reduce state machine-like behavior programming.
2. *Understand command* – The command given is not having all the information necessary to comprehend the task. The robot needs to ask questions to get the missing information.
3. *Understand environment* – The robot has to answer three questions about (the events happened) the environment. The first two questions are about the understanding of the world. The third question is where the robot is being "tricked".

The above approach seems to be a promising direction to motivate high level cognition but the depth of the interaction still limits to only 2-3 steps of question and answer between the human and robot.

3.2   HRI between Virtual and Real Worlds

The above findings motivate this work to present a new proposal of an alternative research framework with HRI between virtual and real worlds with a simulation platform. The fundamental concept of the proposal is to replace the physical robot with a virtual embodied agent in a three dimensional virtual environment while still enabling natural interaction with real world humans. The human subject in real world will be immersed into the virtual world as an avatar via various virtual reality (VR) interfaces, in order to perform HRI with the virtual robot. Although simulating a robot[5] or even a human avatar[6] is nothing new in the current robotics development, but what this work concerns are the HRI requirements and how simulation can assist or even bring forward HRI research, especially in response to the current research resistances mentioned above.

In this paper, based on the RoboCup @Home concept and the two shortcomings discussed above, the significant features of this proposal for HRI development are illustrated in the following section with the practical scenarios implemented in the developed research framework [3]:

1. Level of abstraction in embodiment and multimodal interaction
2. Connecting high level intelligence to low level control for "deeper" HRI
3. Reproducibility and scalability for more HRI

---

[5] http://gazebosim.org/
[6] http://www.openrobots.org/wiki/morse/

4   HRI Research Implementations

A simulation platform for human-robot interaction, SIGVerse [3] is developed for the implementations in this work. More details on the technical features of the SIGVerse system are available in [3]-[5].

4.1   Level of Abstraction in Embodiment and Multimodal Interaction

The main advantage of simulating a virtual robot rather than developing a real robot is the flexibility to adjust the level of abstraction in action-perception simulation of the whole system. The *integration problem* discussed above limits the subject expert development by having to deal with all other component systems. In this work, low level robot controls can be simplified (abstracted) to focus on the interaction and intelligence developments. However, having too much abstraction in the model will also make the simulation meaningless to HRI studies. Therefore in the following applications, the fidelity of the simulation can be adjusted based on the HRI requirements (task requirements of RoboCup @Home).

In the *Follow Me*, *Clean Up* and *Restaurant* tasks in RoboCup @Home, the robot is required to receive instruction from the human operator. The instruction given may not totally verbal but recognition of the human features (face, body gesture, etc.) is required. The HRI requirements in this condition involve embodiment and multimodal interaction.

Taking a simple "take it" HRI scenario as shown in Fig. 1, a human subject is interacting with a robot agent by pointing to an object (a plate with green apples) in the virtual world and giving a verbal message "take it" to instruct the robot to take the pointed object. The human subject is controlling his own avatar by having his body movement tracked by a motion sensing device (Microsoft Kinect) and the avatar's visual view displayed in a head-mounted display (HMD) on the human subject.

From the above example, embodied interaction can be achieved by having both the human and the robot as embodied agents interacting in a shared three dimensional virtual environment. The spatial information in the pointing gesture is simulated. The robot receives partial instruction (take object) in verbal, and the information of the object's location is obtained from the body gesture of the human avatar by visual perception. Hence, multimodal interaction is achieved in this scenario with verbal and body gesture.

Similarly, there are many other interface devices that can be added (as the SIGVerse extensible Service Provider [3]) for various research requirements, e.g. eye gaze for joint attention, haptic device for object manipulation [3], etc.

Fig. 2 illustrates a joint attention simulation [3] via visual perception to recognize human agent's eye gaze and finger pointing when danger arises (a book is dropping from the cabinet).

Fig. 1. Simple "Take it" HRI scenario in SIGVerse. (Top: Simulation displayed in SIGViewer; Bottom left: The avatar's visual view displayed in HMD on the human subject; Bottom right: The human subject in real world).

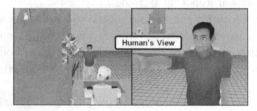

Fig. 2. Joint attention simulation via visual perception and gesture behaviors recognition [3]

4.2 Connecting High Level Intelligence to Low Level Control for "Deeper" HRI

The only test in RoboCup @Home that focuses on robot cognition is the GPSR, but the current depth of the interaction is still limits to only 2-3 steps of question and answer between the human and robot. In order to have a "deeper" HRI, the following application (Fig. 3) illustrated a robot that is conversing verbally with the human to obtain the information of the objects and the surroundings in a *Clean Up* task. Conversational intelligence and dialogue management system [4] are developed to understand the context (including the temporal and spatial information of the objects and environment) in order to deduce the full meaning from the partial verbal instruction giving by the human.

Taking another example, in a collaborative cooking task described in a previous development [5], the embodied interaction is integrated into lengthy work operation. The high level work planning (Fig. 4) is grounded to the local low level embodied and multimodal interaction for high level robot intelligence development in collaborative manipulation.

Fig. 3. Conversation between human and robot in a Clean Up task

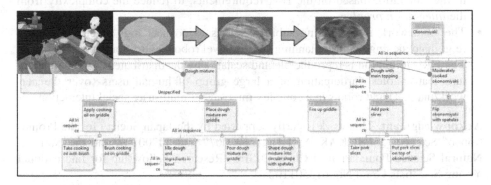

Fig. 4. Hierarchical task model of the *okonomiyaki* (Japanese pancake) cooking task [5]

4.3   Reproducibility and Scalability for More HRI

An undeniable advantage of the simulation system is the ability to reproduce the simulation result after the experiment. This feature is particularly important for HRI studies to reexamine certain interaction parameters that are not able to be recorded during experiment.

A future vision of this proposal is to extend HRI challenge in a larger (and longer) scale. The distributed system architecture of the simulation platform and the potential to be hosted on the cloud as presented in [3], the current system is designed to cater a large group of human users (over the network/internet) and robots in a complex multi-agent multi-user social interaction. Continuous robot learning can also be achieved with this framework by using the crowdsourcing HRI strategy, where many human users from different locations and different time, can log into the simulation and interact with robots. This feature is built in this framework, for example in the *chatterbot* system with multimodal interaction [4] and the applications of the *okonomiyaki* collaborative cooking task [5], that are capable for large scale and long hour experiments. This feature is promising to extend HRI experimental work to address challenges that involve much larger scale in order to study HRI in a more complex social domain.

5   Conclusions

The aim of this work is to investigate on the HRI issue based on recent RoboCup @Home competitions, in order to formulate an alternative research framework for HRI studies. Based on the survey of the current competition mechanism and benchmarking approach, and the analysis of the actual team performances, the short-coming of current approach in motivating high level cognitive robot intelligence development is identified. A new proposal of an alternative research framework with HRI between virtual and real world is illustrated based on the SIGVerse system. The implementations of the research framework in various scenarios are discussed to show the feasibility of this framework to assist HRI development:

- The level of abstraction in embodiment and multimodal interaction can be adjusted in the simulation based on the HRI requirements, to reduce the complexity from the *integration problem*.
- The framework is motivating subject experts to connect high level intelligence (e.g. dialogue engine, task planning) to low level robot control for "deeper" HRI.
- The HRI simulation is reproducible and scalable by the distributed system architecture that enables participation of a large group of human users (over the network/internet) and robots in a complex multi-agent multi-user social interaction.

Acknowledgment. This work is partly supported by the Japan Society for the Promotion of Science (JSPS) KAKENHI No. 23300077 and 24700199, and the National Natural Science Foundation of China (NSFC) Research Fellowship for International Young Scientists Grant No. 61350110240.

References

1. van der Zant, T., Iocchi, L.: Robocup@Home: Adaptive Benchmarking of Robot Bodies and Minds. In: Mutlu, B., Bartneck, C., Ham, J., Evers, V., Kanda, T. (eds.) ICSR 2011. LNCS, vol. 7072, pp. 214–225. Springer, Heidelberg (2011)
2. Wisspeintner, T., van der Zant, T., Iocchi, L., Schiffer, S.: RoboCup@Home: Scientific Competition and Benchmarking for Domestic Service Robots. Interaction Studies 10(3), 392–426 (2009)
3. Tan, J.T.C., Inamura, T.: SIGVerse - A Cloud Computing Architecture Simulation Platform for Social Human-Robot Interaction. In: IEEE Int. Conf. on Robotics and Automation, pp. 1310–1315 (2012)
4. Tan, J.T.C., Duan, F., Inamura, T.: Multimodal Human-Robot Interaction with Chatterbot System: Extending AIML Towards Supporting Embodied Interactions. In: IEEE Int. Conf. on Robotics and Biomimetics, pp. 1727–1732 (2012)
5. Tan, J.T.C., Inamura, T.: Integration of Work Sequence and Embodied Interaction for Collaborative Work Based Human-Robot Interaction. In: 8th Annual ACM/IEEE Int. Conf. on Human-Robot Interaction, pp. 239–240 (2013)

Habituation and Sensitisation Learning in ASMO Cognitive Architecture

Rony Novianto, Benjamin Johnston, and Mary-Anne Williams

Center of Quantum Computation and Intelligent Systems
University of Technology Sydney, Australia
rony@ronynovianto.com, benjamin.johnston@uts.edu.au,
mary-anne.williams@uts.edu.au

Abstract. As social robots are designed to interact with humans in unstructured environments, they need to be aware of their surroundings, focus on significant events and ignore insignificant events in their environments. Humans have demonstrated a good example of adaptation to habituate and sensitise to significant and insignificant events respectively. Based on the inspiration of human habituation and sensitisation, we develop novel habituation and sensitisation mechanisms and include these mechanisms in ASMO cognitive architecture. The capability of these mechanisms is demonstrated in the 'Smokey robot companion' experiment. Results show that Smokey can be aware of their surroundings, focus on significant events and ignore insignificant events. ASMO's habituation and sensitisation mechanisms can be used in robots to adapt to the environment. It can also be used to modify the interaction of components in a cognitive architecture in order to improve agents' or robots' performances.

Keywords: Habituation, Sensitisation, ASMO Cognitive Architecture.

1 Introduction

Social robots are designed to interact with humans in unstructured environments, such as the home. Unstructured environments are less well-defined and engineered than structured environments. As a consequence, social robots cannot have complete or prior knowledge about their surroundings. For example, people and objects can be occluded and how people interact with the objects cannot be predicted or anticipated in advance. In unstructured environments, social robots have to continuously monitor and be aware of their surroundings in order to acquire information necessary to support their decision making. However, they have limited physical resources that can be used to monitor their surroundings. They have limited head, arms and CPU resources to look at surroundings, interact with objects and process information from the environment. Therefore, social robots need to adapt to the environments by focusing on significant events and ignoring insignificant events.

A human is a good example of an agent that adapts to unstructured environments. People have the ability to habituate and sensitise to insignificant

G. Herrmann et al. (Eds.): ICSR 2013, LNAI 8239, pp. 249–259, 2013.

and significant events respectively [1, p.99]. Inspired by this ability in humans, we develop computational habituation and sensitisation mechanisms that can learn to focus and ignore events based on the significance of the events. These mechanisms are implemented in Attentive and Self-Modifying (ASMO) cognitive architecture [2, 3] in order to improve agents' or robots' performances.

In this paper, Section 2 first explores the definition of habituation and sensitisation. Section 3 discusses the existing computational models of habituation and sensitisation proposed in the literature and how they are different to this work. Section 4 describes the design and implementation of the habituation and sensitisation mechanisms in ASMO cognitive architecture. Section 5 evaluates ASMO's habituation and sensitisation mechanisms in the 'Smokey robot companion' experiment and shows that the robot can adapt to the environment. Finally, Section 6 summarises the benefit and future work of ASMO's habituation and sensitisation mechanisms.

2   Definition of Habituation and Sensitisation

Habituation is the decrease in response to a repeated stimulus as a result of a lack of significance or reinforcement [1, p.99] [4] [5, p. 1411]. It involves the elimination of unnecessary behaviours, which are not needed by the individual. The stimulus will not be attended to or responded to (i.e. it is given less attention) if it does not hold any significance for the individual even though it may be repetitive. For example, when tracking an object, an individual who initially pays attention to (and distracted by) repeated movements of other people can become habituated to (thus ignore) these movements if these movements do not hold any significance to the object tracking task.

Habituation is not a form of sensory adaptation (e.g. fatigue). While both habituation and sensory adaptation involve the decrease in response to a repeated stimulus, habituation occurs in the brain as attention phenomenon and it can be consciously controlled whereas sensory adaptation occurs directly in the sense organ and it cannot be consciously controlled [6, p. 138]. For example, in habituation, we can still pay attention and force ourselves to hear a loud noise although we get habituated to the noise. In sensory adaptation, we cannot force ourselves to smell an odor once our smell organ has adapted to the smell regardless how hard we try to pay attention to the odor. Thompson and Spencer [7] and Rankin et al. [8] have described characteristics of habituation that can distinguish habituation with sensory adaptation.

In contrast to habituation, sensitisation is the increase in response to a repeated stimulus because the stimulus holds a significance for the individual [1, p.99] [4] [5, p. 1413]. The stimulus will be attended to or responded to (i.e. it is given more attention) as long as it holds any significance for the individual although it may not be intense. For example, when tracking an object, an individual who initially ignores and does not pay attention to movements of other people can become sensitised to (thus focus on) these movements if these movements hold any significance to the object tracking task (e.g. the object is often occluded by the movements).

Habitation and sensitisation are an automatic process that is developed as a result of an adaptation to the environment. They are a form of *non-associative learning*, which is a learning that does not involve an association with other stimuli or responses. They occur without reinforcement or feedback from the environment. Individuals do not receive a positive or negative feedback for their responses from the environment. Instead, they measure the effectiveness of their responses based on their internal judgement of the significance of the repeated stimulus.

3   Existing Computational Models

In the literature, there have been more computational models of habituation than sensitisation. The majority of the computational models of habituation are based on neural networks [9–14], which range from feed-forward neural networks and recurrent neural networks to artificial spiking neural networks. Some of these models have been implemented in robots [11–14]. While these networks have various details to simulate habituation effects, all these networks share a common fundamental approach. They have to be trained based on some prior input stimuli. Thus, they cannot be learned online and they can only habituate to stimuli that have been trained.

Sirois and Mareschal [15] have reviewed a range of computational models of habituation that include neural-based and non neural-based models. They classify the models into six categories, namely function estimators, symbolic, simple recurrent networks, auto-encoders, novelty filter and auto-associators. The function estimators and symbolic categories are not based on neural network whereas the other four are based on neural network. Sirois and Mareschal assess these models against seven features, namely temporal unfolding, exponential decrease, familiarity-to-novelty shift, habituation to repeated testing, discriminability of habituation items, selective inhibition and cortical-subcortical interactions. Their analysis discovers that none of these models accommodates the familiarity-to-novelty shift feature. Thus, Sirois later proposes a hebbian-based neural network model of habituation that can accommodate this familiarity-to-novelty shift feature [12].

While as to our understanding, there is no existing computational model designed specifically for sensitisation, there are some computational models that can accommodate both habituation and sensitisation. The most commonly found model of both habituation and sensitisation is the model that estimates the exponential strength of a response over time. This model defines the change in response as a single differential equation, as follow:

$$\tau \frac{dy(t)}{dt} = \alpha(y_0 - y(t)) - S(t) \tag{1}$$

where $y(t)$ is the strength of the response at current time t, τ is the time constant that adjusts the exponential decay or growth, α is the rate of recovery, y_0 is the initial strength before habituation or sensitisation and $S(t)$ is the intensity of the stimulus. This model is often credited to Stanley [16], although in his paper, he used few different equations to describe his model. He refers to his model as a simplification of the dual-process theory of Groves and Thompson [17].

This model accounts for the exponential decay or growth in responses as found in biological habituation and sensitisation respectively. It also accounts for recovery of responses (i.e. dishabituation or desensitisation). However, it lacks a mechanism to accommodate the significance and specificity of the stimulus. The response is a function of stimulus intensity. A stimulus is represented by its strength. The response decays when a stimulus is on and recover when a stimulus is off regardless of the significance of the stimulus.

Damper, French and Scutt have proposed alternative computational model of habituation and sensitisation based on a spiking neural network [18]. They have implemented this model on a Khepara robot called ARBIB. This model is connected such that if neuron A synapses onto neuron B then the repetition of fire of neuron A causes the synaptic strength to decrease, which causes the response of neuron B to decrease too. In this model, habituation is achieved by decreasing the *weight* of a synapse every time it fires such that the the weight will return toward the *base weight* with a time constant determined by *recovery*. In addition, sensitisation is achieved by using a facilitatory interneuron I with synapse-on-synapse connection to an $A \to B$ synapse.

The work in this paper differs from previous works in following: (i) it accounts for both habituation and sensitisation, (ii) it is not based on neural network, (iii) it accommodates the significance of the situation and (iv) it is implemented in a robot.

4   Design and Implementation in ASMO Cognitive Architecture

In this section, we describe the design and implementation of ASMO's habituation and sensitisation mechanisms based on the inspiration of human habituation and sensitisation. We first review the overview of ASMO cognitive architecture. We follow by describing the mechanisms and how they fit in the architecture.

4.1   Overview of ASMO Cognitive Architecture

ASMO [2, 3] is a flexible biologically-inspired cognitive architecture that integrates a diversity of artificial intelligence components based on attention weighting in order to control intelligent agents and robots. It does not aim to mimic or imitate the models of human intelligence and cognition, instead it aims to produce practical and effective solutions based on the inspiration of human intelligence and cognition.

ASMO cognitive architecture contains a set of independent and *self-governing* processes (also called modules) that can run concurrently on separate threads (see Fig.1). Each module requests 'actions' to be performed. An action can be a low-level command to actuators, such as move head to a ball or walk to a

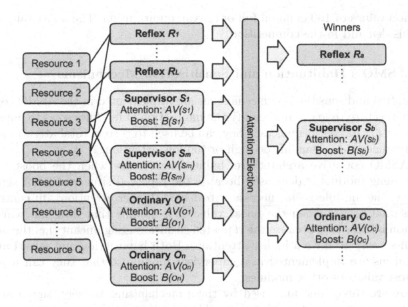

Fig. 1. Attention election in ASMO cognitive architecture

specific location, or it can be a high-level function, such as store data to a memory, recognise objects (i.e. percept) or find the shortest path (i.e. plan).

Actions can only be performed if the resources required by the actions are available (e.g. hand, leg, CPU, memory, virtual resource, etc). They can be performed simultaneously when there is no conflict in using the resources. Otherwise, modules with actions that require the same resources have to compete for 'attention' to use the resources. The winner of the competition is chosen based on the modules' types and attention levels.

Modules are divided into three types, namely ordinary, supervisor and reflex modules. The ordinary and supervisor modules are modules that have a 'attention value' attribute and 'boost value' attribute to determine their total attention levels (used to compete for attention). The 'attention value' attribute captures the degree of attention the module seeks based on the demand of the tasks whereas the 'boost value' attribute represents the bias associated with the module as a result of learning or subjective influence [3]. Supervisor and ordinary modules are similar, except that supervisor modules can influence the total attention levels of ordinary modules but not vice versa. Reflex modules are modules that have urgent needs and do not have the 'attention value' and 'boost value' to compete for attention, instead they are given higher priority than ordinary and supervisor modules to use the resources immediately.

Currently, modules that have the highest total attention levels will win the competition. The total attention level is given by the sum of boost and attention values. Under ordinary operation, attention values are (by convention) bounded between 0.0 and 100.0, equivalent to scaled values between 0.0 and 1.0. Modules with attention values of 0 demand the least attention whereas modules with attention values of 100 demand full or maximum attention. The boost value will bias this demand in the competition.

4.2 ASMO's Habituation and Sensitisation Mechanisms

Habituation and sensitisation mechanisms are implemented in the ASMO cognitive architecture to learn and modify the interaction of modules so as to improve agents' or robots' performances. They can be used to correct total attention levels of modules that demand too high or too low attention.

In ASMO cognitive architecture, habituation occurs when the boost value of a winning module is decreased because the repeated situation is not significant (i.e. the module is *not* necessary to have a higher attention). In contrast, sensitisation occurs when the boost value of a module that does not win the attention is increased because the repeated situation is significant (i.e. the module is necessary to have a higher attention). Both habituation and sensitisation mechanisms are implemented in a supervisor module so that they can modify the boost values of other modules.

There are three functions used by these mechanisms, namely 'significance', 'delta (δ)' and 'termination' functions. The 'significance' function measures the significance of the repeated stimulus. It provides a value between -1.0 to 1.0. A value of less than zero is insignificant whereas a value of greater than zero is significant. The 'delta' function determines the amount of the boost value that will be decreased or increased (i.e. for habituation and sensitisation respectively). The boost value of a module is modified in proportion to the delta function (see 2). The 'termination' function specifies the condition for the learning to stop.

These three functions can be defined either locally in each of the competing modules or globally in the habituation and sensitisation supervisor module. Locally defined functions allow each competing module to define its own significance measurement, its own amount of boost value to be modified and its own stopping condition. In contrast, globally defined functions allow competing modules that do not specify their 'significance', 'delta' and 'termination' functions to habituation and sensitise based on these common global functions.

Among the three functions, only the 'significance' function is required by the mechanisms whereas the 'delta' and 'termination' functions can be provided optionally. The mechanisms will not modify the boost value of a module (i.e. skip the module) if the 'significance' function is not provided. In addition, they will use the default 'delta' function (see 3) and/or the default 'termination' function (see 4) when the 'delta' and/or 'termination' functions are not provided respectively. Using the default 'delta' function, the mechanisms will modify the boost value of a module in proportion to the significance function. Using the default 'termination' function, the mechanisms will terminate when the root mean square deviation of the previous 20 consecutive boost values is less than or equal to a defined threshold (see 5).

$$B_t(p) = B_{t-1}(p) + \delta(p) \tag{2}$$

$$\delta(p) = Significance(p) \times C \quad -1.0 \geq Significance(p) \leq 1.0 \tag{3}$$

$$Termination(p) = \begin{cases} True & \text{if } RMSD(p) \leq T_{\text{RMSD}} \\ False & \text{otherwise} \end{cases} \tag{4}$$

$$RMSD(p) = \sqrt{\frac{1}{n} \sum_{i=t-n}^{t} (B_i(p) - B_{i-1}(p))^2} \tag{5}$$

Where:

$B_t(p)$ is the boost value of process p at time t

$\delta(p)$ is the delta function of process p

$Significance(p)$ is the significance function of process p

C is the maximum value of change

$Termination(p)$ is the termination function of process p

$RMSD(p)$ is the root-mean-square deviation of process p

T_{RMSD} is the threshold to stop learning

n is the number of previous boosts

5   Evaluation

This section demonstrates the proof of concept of ASMO's habituation and sensitisation mechanisms in an experiment as part of the 'Smokey robot companion' project (see Sect. 3 for comparison of ASMO's habituation and sensitisation mechanisms with other mechanisms). The 'Smokey robot companion' project aims to bring a bear-like robot (called Smokey) to 'life' and explores the meaning of life by interacting socially with people. It has potential applications in nursing, healthcare and entertainment industries to accompany people with disability, people with autism, elderly and children.

The experiment involves a simplified scenario where Smokey has to accompany or entertain a person (i.e. target user) while simultaneously regulating the person's rest (see Fig.2). It can either play a ball game to accompany the person or go to sleep to encourage the person to rest (since it will not interact with the person when it is sleeping). When playing the ball game, Smokey will track and search for the ball while pay attention to any motion in the environment.

There are three ordinary modules and one supervisor module created in this experiment to govern Smokey's behaviours, namely 'attend_motion', 'play_ball' and 'go_sleep' ordinary modules and 'habituate-sensitise' supervisor module.

The 'attend_motion' module proposes an action when Smokey is not sleeping to look at the fastest motion in the environment. Its attention value is set to the average speed of the motion scaled between 0.0 and 100.0. The faster the motion, the more attention demanded by the module to look at the motion. The significance of this module is measured based on the distance of the motion to the ball, as follow:

$$significance(\text{attend\_motion}) = \begin{cases} \frac{b-d}{b} & \text{if } d \leq b \\ \frac{-d}{l} & \text{if } d > b \end{cases} \tag{6}$$

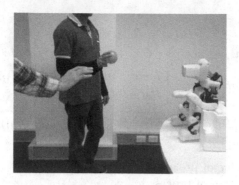

Fig. 2. Smokey robot companion

where $-1.0 \leq significance(\text{attend\_motion}) \geq 1.0$, d is the distance between the motion and the ball, b is the length of the significant boundary and l is the length of the vision. This significance function returns a positive value for a distance less than or equal to b and a negative value for a distance greater than b. Motion closer to the ball (i.e. object of interest) has higher significance than motion far from the ball since Smokey's goal is to track the ball when it is not sleeping.

The 'play_ball' module proposes an action when Smokey is not sleeping either to track or to search for the ball depending on whether the location of the ball is known or not respectively. Its attention value is set to a constant value of 50.0. This module is not provided with a 'significance' function, hence its boost value will not be modified by the habituation and sensitisation mechanisms.

The 'go_sleep' module proposes an action to go to sleep and wake up in every defined period. Its attention value is set to 100.0 (i.e. maximum value of attention) in order to ensure that the sleeping and wake up behaviour can be performed every time the action is proposed. This module is also not provided with a 'significance' function.

The 'habituate-sensitise' module proposes two actions to habituate and sensitise respectively. The first action is to evaluate the significance of the 'attend_motion' module when it wins the attention competition and decrease its boost value if the significance is negative (i.e. not significant). The second action is to evaluate the significance of the 'attend_motion' module when it does not win the attention competition and increase its boost value if the significance is positive (i.e. significant). The attention value of the 'habituate-sensitise' module is set to 10.0.

Figure 3 shows the results of the experiment with and without the habituation and sensitisation learning mechanisms for two situations. The first situation is when people other than the target user are moving fast but far from the ball whereas the second situation is when they are moving slow but near to the ball. The attention value of the 'attend_motion' (thus the respond to the motion) is changing based on the speed of the motion.

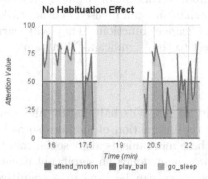

(a) Faster & Far Motion, No Learning

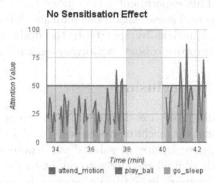

(b) Slower & Near Motion, No Learning

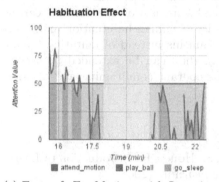

(c) Faster & Far Motion, with Learning

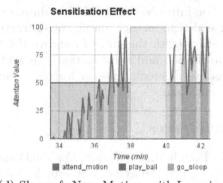

(d) Slower & Near Motion, with Learning

Fig. 3. Smokey's habituation and sensitisation to different motions

Without habituation and sensitisation learning (i.e. without accommodating the boost value of the 'attend_motion' module), Smokey tends to look at the fast motion although it is far from the ball (see Fig.3a). In addition, Smokey also tends to ignore the slow motion although it is near to the ball (see Fig.3b). Any fast motion can easily distract Smokey from tracking the ball and any slow motion can easily be ignored by Smokey.

With habituation and sensitisation learning (i.e. with accommodating the boost value of the 'attend_motion' module), Smokey tends to ignore the motion far from the ball although they are fast (see Fig.3c). In addition, Smokey also tends to focus on the motion near to the ball although they are slow (see Fig.3d). Any fast motion does not simply distract Smokey from tracking the ball and any slow motion is not easily ignored by Smokey. Thus, Smokey has habituated and sensitised to the motion that is far from and near to the ball respectively.

This experiment has demonstrated the proof of concept of ASMO's habituation and sensitisation mechanisms. These mechanisms can be used in other tasks or situations by specifying the desired 'significance' functions. They will learn to improve robots' performances with respect to the 'significance' functions.

6   Conclusion

This paper has demonstrated the capability of ASMO's habituation and sensitisation mechanisms to learn and modify the interaction of modules so as to improve agents' or robots' performances. These mechanisms allow social robots with limited resources to adapt to the environment, focus on significant events and ignore insignificant events. In addition, they can be used in a cognitive architecture or robot architecture to modify the structure of a diversity of components. Structuring components can be challenging in complex and open environments where the 'correct' structure is unknown or uncertain.

For future work, ASMO's habituation and sensitisation mechanisms can be extended to further match the characteristics of human habituation and sensitisation (with the aim to improve the mechanisms instead of imitating human habituation and sensitisation). In addition, they can be extended to accommodate different types of learning, such as operant conditioning and classical conditioning.

References

1. Powell, R., Symbaluk, D., MacDonald, S., Honey, P.: Introduction to Learning and Behavior. Wadsworth Cengage Learning (2008)
2. Novianto, R., Johnston, B., Williams, M.A.: Attention in the asmo cognitive architecture. In: Proceedings of the First Annual Meeting of the BICA Society (2010)
3. Novianto, R., Williams, M.A.: Innate and learned emotion network. In: Samsonovich, A.V., Jóhannsdóttir, K.R. (eds.) Proceedings of the Second Annual Meeting of the BICA Society. Frontiers in Artificial Intelligence and Applications, vol. 233, pp. 263–268. IOS Press (2011)
4. Eisenstein, E., Eisenstein, D.: A behavioral homeostasis theory of habituation and sensitization: Ii. Further developments and predictions. Reviews in the Neurosciences 17(5), 533–558 (2006)
5. Seel, N.: Encyclopedia of the Sciences of Learning. Springer (2011)
6. Sternberg, R.J., Mio, J., Mio, J.: Cognitive Psychology. Cengage Learning (2009)
7. Thompson, R., Spencer, W.: Habituation: a model phenomenon for the study of neuronal substrates of behavior. Psychological Review 73(1), 16 (1966)
8. Rankin, C.H., Abrams, T., Barry, R.J., Bhatnagar, S., Clayton, D.F., Colombo, J., Coppola, G., Geyer, M.A., Glanzman, D.L., Marsland, S., et al.: Habituation revisited: an updated and revised description of the behavioral characteristics of habituation. Neurobiology of Learning and Memory 92(2), 135–138 (2009)

9. Kohonen, T.: Self-organization and associative memory, 3rd edn. Springer-Verlag New York, Inc., New York (1989)
10. Mareschal, D., French, R.M., Quinn, P.C., et al.: A connectionist account of asymmetric category learning in early infancy. Developmental Psychology 36(5), 635–645 (2000)
11. Chang, C.: Improving hallway navigation in mobile robots with sensor habituation. In: Proceedings of the IEEE-INNS-ENNS International Joint Conference on Neural Networks, IJCNN 2000, vol. 5, pp. 143–147. IEEE (2000)
12. Sirois, S.: Hebbian motor control in a robot-embedded model of habituation. In: Proceedings of the 2005 IEEE International Joint Conference on Neural Networks, IJCNN 2005, vol. 5, pp. 2772–2777. IEEE (2005)
13. Marsland, S., Nehmzow, U., Shapiro, J.: On-line novelty detection for autonomous mobile robots. Robotics and Autonomous Systems 51(23), 191–206 (2005)
14. Cyr, A., Boukadoum, M.: Habituation: a non-associative learning rule design for spiking neurons and an autonomous mobile robots implementation. Bioinspiration & Biomimetics 8, 016007 (2013)
15. Sirois, S., Mareschal, D.: Models of habituation in infancy. Trends in Cognitive Sciences 6(7), 293–298 (2002)
16. Stanley, J.C.: Computer simulation of a model of habituation. Nature 261, 146–148 (1976)
17. Groves, P.M., Thompson, R.F.: Habituation: a dual-process theory. Psychological Review 77(5), 419 (1970)
18. Damper, R.I., French, R.L., Scutt, T.W.: Arbib: an autonomous robot based on inspirations from biology. Robotics and Autonomous Systems 31(4), 247–274 (2000)

Effects of Different Kinds of Robot Feedback

Kerstin Fischer[1], Katrin S. Lohan[2], Chrystopher Nehaniv[3],
and Hagen Lehmann[3]

[1] University of Southern Denmark
Department of Design and Communication Sonderborg, Denmark
kerstin@sdu.dk
[2] Instituto Italiano di Tecnologia
iCub Facility Italy, Genova
katrin.lohan@iit.it
[3] University of Hertfordshire Adaptive Systems, United Kingdom

Abstract. In this paper, we investigate to what extent tutors' behavior is influenced by different kinds of robot feedback. In particular, we study the effects of online robot feedback in which the robot responds either contingently to the tutor's social behavior or by tracking the objects presented. Also, we investigate the impact of the robot's learning success on tutors' tutoring strategies. Our results show that only in the condition in which the robot's behavior is socially contingent, the human tutors adjust their behavior to the robot. In the developmentally equally plausible object-driven condition, in which the robot tracked the objects presented, tutors do not change their behavior significantly, even though in both conditions the robot develops from a prelinguistic stage to producing keywords. Socially contingent robot feedback has thus the potential to influence tutors' behavior over time. Display of learning outcomes, in contrast, only serves as feedback on robot capabilities when it is coupled with online social feedback.

1 Introduction

In this paper, we address different types of robot feedback and their impact on human tutors' interaction strategies. Understanding under what conditions people adjust to the robot and produce tutoring strategies is important since attempts at adapting to the robot's capabilities may lead to strategies that can be exploited to improve human-robot interaction. For instance, with respect to language, tutoring strategies may be used to improve understanding and to bootstrap language [1]. Especially in the case when human tutors provide simplifications similar to child-directed speech, these cues can be exploited to facilitate language learning [2]. In order to understand under what conditions human tutors adjust to robots' behavior, we investigate the impact of different robot feedback strategies which we expect to influence tutors' linguistic and gazing behaviors, namely increasing robot capabilities on the one hand and different types of contingent feedback on the other hand. Here, we mostly consider contingency with reference to a temporal dependency between behavior and reaction

G. Herrmann et al. (Eds.): ICSR 2013, LNAI 8239, pp. 260–269, 2013.

[3]. The term can however also be applied to spatial patterns; for instance, when somebody claps their hands, the sound will be perceivable from the same spatial source as the moving of the hands. Contingent feedback is an important mechanism in interactions between humans, and Watson [4], for instance, assumes that children are born equipped with a contingency detector module that allows them not only to detect contingency, but also to expect contingent behavior and to try to elicit it. Csibra [5] argues that contingency is a prerequisite to discerning sequential organization of a co-activity and thus for learning to interact. While contingent robot feedback is suspected to play a considerable role in human-robot interaction, it is an open question what aspects of the human tutor's behavior a robot's feedback should be contingent with. Here, we compare socially contingent with object-contingent behavior.

Besides contingent online feedback concerning the tutors' current actions, robots may also produce indirect, delayed feedback on the success of the tutoring sessions. These displays may potentially provide the tutor with very helpful information on the robot's learning progress and cognitive capabilities. If the robot demonstrates attention to the tutor's behavior by displaying newly acquired knowledge, this may have a considerable impact on how participants view the interactions and thus may lead them to attend to the robot and to tailor their instruction strategies to what the robot has learned. Therefore, we investigate the effect of such displays on tutors' teaching strategies, as well as the interaction of this kind of feedback with contingent robot response.

2   Previous Work

Evidence that people respond positively to contingent (i.e. timely) robot response to their own actions has been indicated by several studies. For instance, Nourbakhsh et al. write that "the single most successful way for a robot to attract human interest is for the robot to demonstrate awareness of human presence" [6]. Furthermore, Sidner et al. [7] investigate the effects of contingent robot gestures in response to human action. The results of the study show that human communication partners do not only attend to the responsive robot more and interact longer with it, but they also judge a robot that is producing contingent nonverbal feedback to be more reliable and its movements to be more appropriate. Contingent robot response has also been found to influence people's perception of mutual eye gaze [8], their judgment of robot acceptability and appropriateness [9] and their willingness to interact with the robot [10].

In our own previous work, we found contingent robot behaviors to lead people to increased tutoring behavior because the contingent displays are taken as signals of understanding and thus as unmediated clues to the robot's competence [11] [12]. Concerning robot feedback based on learning, in a previous study in which the robot learned words from the participants' tutoring, we found that people used fewer attention getting signals, shorter utterances, talked more and produced more linguistic feedback over time, indicating that the robot's increasing capabilities made them choose different strategies over the course of several sessions [13].

To sum up, previous work suggests that the robot's behavior in response to people's behavior may have an impact on their perception of the quality of the interaction. What is open, though, is what types of robot feedback have what social effects on human-robot interactions. The question raised here concerns the effects of responsive behaviors that respond either to the human tutor's social behavior or to the object presented. In addition, both types of contingent robot feedback are combined with time delayed feedback (displays of learning) in an instructional scenario. Thus, we address the timing and kinds of responses to feedback behaviors and their role of eliciting tutoring strategies in HRI.

2.1 The Robot

The robot used is the humanoid robot iCub [14], which is designed to resemble a young child (see figure 1).

Fig. 1. The iCub Robot

Data were elicited in two conditions (between subjects), representing two different robot feedback strategies. In one condition, the robot was reacting based on the tutoring spotter ([11]; see section 3). In this condition, the robot was responsive to the gaze direction and the demonstrating behavior of the human tutor, that is, to social cues. In the other condition, the robot's movement was controlled by an implementation based on object tracking. In both conditions, the robot's behavior changed over the course of the three sessions, being initially nonverbal and consisting of words learned in previous sessions in the second and third sessions. The robot's behavior consists of three components: First, in both conditions the robot was capable of looking at the face of the interaction partner, at the objects presented to it, or elsewhere in the room. Second, the robot was able to point at an object. Third, the robot learned the names of relevant objects over the three sessions and uttered these in order to give feedback on the tutors' presentations. The language feedback was controlled by the same system in both conditions.

2.2 Setup

For the technical realization of the tutoring spotter, we equipped the robot with a Kinect sensor. Before the experiment, the participants took part in initializing the kinect and face tracking systems. Then, participants were seated across the robot in front of a table. They were asked to wear a headset. A separate webcam was used for the face- and object-tracking module.

3 Implementation

For the robot behavior, we used two different but well studied controll approaches. For the language acquisition, we used the same system in both conditions.

3.1   Tutoring Spotter

The tutoring spotter system was created in the ITALK project and is capable of giving socially contingent feedback by means of eye gaze and pointing [11], [15].

Gazing Feedback. The robot detects three gazing classes (gaze at the robot, gaze away and gaze at an object) in the human tutor's behavior and responds with the same behavior. Thus, the robot would look at you when you are looking at it, the robot would look around when you are looking elsewhere and when you are looking at the object, the robot would follow your gaze to the object.

Pointing Feedback. Child-directed action, and in particular object presentation, has been shown to facilitate learning in a tutoring situation. In particular, Matatyaho and Gogate [16] find that the demonstrating action, in which a tutor moves an object towards a learner's face, is likely to highlight a novel word-object relation [17] and thus serves as a reliable cue for word learning. The movement is a motion away from the body of the presenter. Using the tutoring spotter, the iCub robot responds to a demonstrating gesture by pointing towards the object that the tutor is presenting.

3.2   Object-driven Contingency

For the setup in the second condition, we used an implementation in which the robot's behavior is based on the tracking of the objects or the face of the participant, in order to control the iCub's behavior. As in the tutoring spotter condition, the robot was able to look at the participant's face, at the object or somewhere else, and it was using pointing gestures. The object-driven implementation, which focuses on tracking objects and, if not available, switches the robot's gaze to the tutor's face, seems to correspond to infants' behavior [18].

Gazing Feedback. In the second condition, the robot's gaze is controlled by a 'boredom' filter. If the same face or object is being seen for too long, the robot switches to random gaze. This means that the robot is changing the gazing behavior based on timing.

Pointing Feedback. In the second condition, the robot tracked the object and occasionally (on a random basis) pointed at the object. The robot made no use of the tutor's social behavior in this condition.

3.3   Language Acquisition System

The system uses prosodic salience and shared belief to bootstrap word learning via interactions with the iCub robot. We automatically analyzed intonational contours from the transcribed sessions in order to extract words which are prosodically salient. The association of these words with the sensorimotor experiences of the robot allows the robot to derive rudimentary meaning and to bootstrap the word learning process. The robot produced feedback on the basis of each interactional session. A learning algorithm related the object-detection input with the spoken utterances and resulted in the robot expressing words that were salient and information-wise relevant, based on its sensorimotor perceptions (for a detailed description of this work see [19], [20]). These words were uttered when that object reappeared in the following session. Thus, there was no verbal feedback in the first session, but only in the second and third. A previous study of the effects of this method shows that participants adjusted their linguistic behaviors based on the robot's verbal behavior [13].

4   Experimental Study

The participants' task in this study was to teach the child-like robot iCub the names of colors and shapes on cubes. The participants were 19 naive subjects none of whom was familiar with robots or issues in robotics. All participants were asked to come to three different sessions. Between these sessions, the robot was trained on the speech it had received from the respective participant the session before. Unfortunately, not all participants showed up for all three sessions. Of the 19 participants, 9 interacted with the robot over all three sessions and 9 participants interacted with the robot in two sessions. Data elicitation took place in two conditions between subjects so that every participant only got to see only one version of the robot (socially contingent versus object-driven). Thus, in Condition 1, eight people interacted with the socially contingent robot, controlled by the tutor spotter. Altogether, there were 18 interactions in Condition 1. In Condition 2, the robot's behavior was object-driven and controlled by the 'boredom' filter. This condition comprises of eleven speakers in 22 interactions.

5   Language and Gaze Analysis

As our language acquisition system needs phonemically transcribed input, the speech produced by the participants recorded in the videos was manually transcribed by two scientific assistants at the University of Hertfordshire between the sessions. The transcription was based on the phoneme inventory of the computer-readable Speech Assessment Methods Phonetic Alphabet (SAMPA). These transcriptions formed the basis for the linguistic analysis. The transcripts of the participants' utterances were analyzed for a set of linguistic features that have been found to be informative concerning participants' understanding of the robot [21]. In particular, we wanted to know whether participants took the robot's task, to learn the linguistic labels of the shapes presented to it, into account. Thus, we analyzed how appropriate tutors' linguistic utterances were for the robot to bootstrap language from interaction. For this reason, we were interested in the number and proportion of keywords participants produced for the robot. The list of keywords comprises the colors *red, blue* and *green*, the shapes *cube, arrow, heart* and *moon* and the sizes *small, large, little* and *medium* relevant for describing the target object. Another linguistic feature we investigated are personal pronouns, which function as indicators for interpersonal relationships [22]. In particular, we investigated the occurrences of *you, I* and *we*. In addition, we looked at attention getting. Attention getting devices serve on the one hand as an indicator for the smoothness of the interaction and on the other hand, they are means to involve the robot in the interaction. Here, we looked at the robot's name *Deechee* and at the imperative *look*. Finally, we studied the feedback that participants gave to the robot, in particular the feedback signals *yes, yeah, okay* and *good*. For the analysis of the participants' gaze behavior, we used the automatic detection of the gazing classes to determine the tutors' gaze behavior. For all linguistic features, the numbers reported are occurrences per 100 words; in the case of gaze direction, we report the number of words uttered while looking in a particular direction compared to the total number of words.

6 Results

Below, we present the results for people's gaze behavior in general, for the two conditions and over time separately.

6.1 Differences between the Two Conditions

When we compare the two conditions in order to estimate the effect of the robot's behavior, we find significantly more instances of *look* in Condition 2, as well as a tendency towards more feedback in Condition 1 feedback and tendencies towards different distributions of pronouns in the two conditions. The two conditions do not differ with respect to tutors' gaze behavior. Table 1 presents the results of a multivariate MANOVA.

Table 1. ANOVA of the two Conditions; *= p<.05; +=p<.10

	Mean (sd) Condition 1	Mean (sd) Condition 2	F (1,18)	p
gaze at face	2.91 (1.95)	4.11 (5.09)	0.57	.45
gaze at object	90.66 (3.69)	82.71 (17.23)	0.61	.44
you	0.86 (0.79)	1.57 (1.77)	3.38	.07+
we	2.52 (2.31)	1.62 (2.90)	4.01	.05+
I	0.08 (0.16)	0.23 (0.27)	3.75	.06+
look	0.00 (0.00)	0.30 (0.48)	5.21	.03*
feedback	0.92 (0.85)	0.26 (0.39)	3.05	.09+

Investigating correlations within the two conditions shows that very different kinds of relations can be observed. In Condition 1, there are several statistically significant correlations between the pronouns, such that tutors who use *you* also use *I* often (r= .79) but they don't use *we* (r= -.71). Furthermore, tutors who use *you* also call the robot by its name (r= .78), whereas tutors who use *we* often do not (r= -.74). In contrast, in Condition 2 we find significant correlations between pronoun use and gaze such that tutors who use *we* gaze at the robot (r= .80) and not at the object (r= -.87). Thus, the two conditions differ, not with respect to speakers' gaze behaviors, but with respect to the amount of feedback given and the number of attempts at getting the robot's attention. Moreover, gaze behavior and pronoun use show different patterns in the two conditions.

6.2   Differences over Time

The analysis of changes over time reveals only one significant change between the sessions, and that concerns the amount of feedback given. While tutors made use of 0.16 feedback signals per 100 words on the average in session 1 (sd = 0.27), they used 1.05 (sd = 1.34) and 1.59 (sd = 1.52) on the average in sessions 2 and 3 (F(2,41)=6.185, p < .005). All other linguistic and non-linguistic behaviors remain the same over the course of the sessions.

6.3   Differences over Time within the Two Conditions

While there is only one feature that changes over time in both conditions together, the separate analysis of tutors' behavior in the two conditions reveals that there is a strong impact of the robot's contingent feedback on tutors' behavior over the course of the three sessions.

Table 2. ANOVA results concerning differences between sessions, Condition 1; *= p<.05; +=p<.10

	Mean (sd) Session 1	Mean (sd) Session 2	Mean (sd) Session 3	F (2,17)	p
gaze at face	0.03 (0.03)	0.02 (0.02)	0.04 (0.4)	0.76	.49
gaze at object	0.93 (0.06)	0.85 (0.04)	0.89 (0.07)	3.06	.07+
gaze at elsewhere	0.04 (0.06)	0.13 (0.05)	0.07 (0.08)	3.11	.07+
you	0.63 (0.85)	0.53 (0.62)	1.23 (0.87)	1.67	.22
we	2.87 (2.04)	4.76 (4.42)	0.68 (1.35)	3.56	.05+
I	0.16 (0.32)	0.00 (0.00)	0.00 (0.00)	1.47	.26
look	0.09 (0.28)	0.00 (0.00)	0.00 (0.00)	0.73	.50
feedback	0.24 (0.35)	1.68 (1.67)	1.90 (1.53)	3.87	.04*
keywords	18.70 (11.75)	13.60 (3.99)	28.51 (13.65)	0.17	.09+

Table 2 presents the changes over time in Condition 1, the condition in which the robot reacted contingently to the tutor's social behaviors, driven by the tutor spotter. There is a significant difference concerning the use of feedback signals, as well as statistical tendencies towards less gaze at objects, an increase in gaze elsewhere between sessions 1 and 2 and a decrease between sessions 2 and 3, an increase concerning the use of the personal pronoun *we* between sessions 1 and 2 and a decrease between sessions 2 and 3, and an increase concerning the use of keywords. In contrast, in Condition 2 there is not a single significant difference (nor a statistical tendency (p<.10)) between the sessions. Thus, participants in the object-driven feedback condition did not change their behavior towards the robot even though the robot was learning to produce new words between sessions. The differences in amount of feedback given over time described above is thus entirely due to the behavior by tutors in Condition 1, the socially contingent feedback condition. Moreover, the two conditions tend to differ with respect to the proportion of keywords to words. While in Condition 1 the number of keywords increased slightly while the total number of words decreases, there was no higher density of keywords observable in Condition 2, as illustrated in Figure 2.

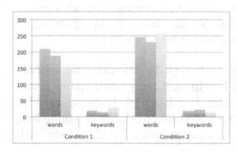

Fig. 2. Number of words and keywords used in the three sessions

Thus, based on the robot's verbal feedback, tutors adjusted their linguistic productions to the robot's output in Condition 1, yet, not in Condition 2. To sum up, the fact that the robot has learned words from the respective tutor's instructions in previous sessions did not by itself have an effect on the participants' gaze and language behavior. Only in combination with socially contingent online robot feedback, the robot's displays of developing linguistic capabilities had an impact on tutors' behaviors.

7 Discussion

The analysis has shown significant differences between the two contingency conditions, which carry over into people's attention to linguistic knowledge displayed by the robot. Even though more words were uttered to the object-driven robot, the proportion of keywords used was higher over time in the socially contingent version. This relationship can be exploited, for instance, for keyword spotting or language learning. We furthermore identified significant differences in the use of feedback over time, yet only in the condition in which tutors interacted with a socially contingent robot. That there were relatively few differences over time was unexpected since in a previous study [13], we had found that people adjust the amount of attention getting signals, mean length of utterance and general verbosity, in addition to giving more linguistic feedback. However, in that study, participants interacted with the robot over a course of five sessions and thus had more opportunity to get acquainted with the robot.

The most important finding however concerns the fact that the robot's increasing linguistic capabilities were only taken into consideration by participants if it was complemented with socially contingent robot response. Thus, direct displays of learning and thus indicators of the robot's cognitive processes do not by themselves serve as clues that tutors exploit for partner adaptation. This is quite unexpected given the focus on partners' knowledge in the literature on audience design [23]. However, the findings are in fact consistent with findings on child-directed speech that show that children's online receptive ability is the most important determinant factor for caregivers to make relevant adjustments [24], more important than the age of the child of his/her cognitive development.

8　Conclusion

From the perspective of social robotics we can conclude first that what the robot is modelled to respond to is of crucial importance for human interactants. In particular, only if the robot takes people's social behaviors into consideration do people adjust their behaviors to the robot. Second, display of robots' cognitive capabilities has no effect on tutors' online choice of tutoring strategies unless it is paired with online social feedback. Thus, people choose tutoring strategies almost exclusively on the basis of contingent social robot feedback.

Acknowledgment. The authors gratefully acknowledge the financial support from the FP7 European Project ITALK. Katrin S. Lohan furthermore gratefully acknowledges the financial support from the EU FP7 Marie Curie ITN RobotDoc, Contract No. 235065. Finally, we would like to thank Joe Saunders, Christian Dondrup, Britta Wrede and Katharina Rohlfing for their help with data elicitation and for fruitful discussion.

References

1. Wrede, B., Schillingmann, L., Rohlfing, K.J.: Making Use of Multi-Modal Synchrony: A Model of Acoustic Packaging to Tie Words to Actions, ch. 10, pp. 224–240. IGI Global, Hershey (2013)
2. Fernald, A., Weisleder, A.: Early language experience is vital to developing fluency in understanding. In: Neuman, S., Dickinson, D. (eds.) Handbook of Early Literacy Research, vol. 3. Guiltford Publications, New York (2011)
3. Gergely, G., Watson, J.: Early socio-emotional development: Contingency perception and the social-biofeedback model. In: Early Social Cognition: Understanding Others in the First Months of Life, pp. 101–136 (1999)
4. Watson, J.: Contingency perception in early social development. Social Perception in Infants, 157–176 (1985)
5. Csibra, G.: Recognizing communicative intentions in infancy. Mind & Language 25(2), 141–168 (2010)
6. Nourbakhsh, I., Kunz, C., Willeke, T.: The mobot museum robot installations: A five year experiment. In: Proceedings of the 2003 IEEE/RSJ International Conference on Intelligent Robots and Systems, Las Vegas, Nevada, pp. 3636–3641 (October 2003)
7. Sidner, C., Lee, C., Kidd, C., Lesh, N., Rich, C.: Explorations in engagement for humans and robots. Artificial Intelligence 166(1-2), 140–164 (2005)
8. Yoshikawa, Y., Shinozawa, K., Ishiguro, H.: Social reflex hypothesis on blinking interaction. In: Proceedings of the 29th Annual Conference of the Cognitive Science Society, pp. 725–730 (2007)
9. Breazeal, C.: Toward sociable robots. Robotics and Autonomous Systems 42(3), 167–175 (2003)
10. Kose-Bagci, H., Broz, F., Shen, Q., Dautenhahn, K., Nehaniv, C.L.: As time goes by: Representing and reasoning about timing in human-robot interaction studies. In: AAAI Spring Symposium: It's All in the Timing (2010)
11. Lohan, K.S., Rohlfing, K.J.R., Pitsch, K., Saunders, J., Lehmann, H., Nehaniv, C.L., Fischer, K., Wrede, B.: Tutor spotter: Proposing a feature set and evaluating it in a robotic system. International Journal of Social Robotics 4(2), 131–146 (2012)

12. Fischer, K., Lohan, K.S., Saunders, J., Nehaniv, C., Wrede, B., Rohlfing, K.: The impact of the contingency of robot feedback on HRI. In: International Conference on Cooperative Technological Systems (CTS 2013), San Diego, May 20-24 (2013)
13. Fischer, K., Saunders, J.: Getting acquainted with a developing robot. In: Salah, A.A., Ruiz-del-Solar, J., Meriçli, Ç., Oudeyer, P.-Y. (eds.) HBU 2012. LNCS, vol. 7559, pp. 125–133. Springer, Heidelberg (2012)
14. Metta, G., Natale, L., Nori, F., Sandini, G., Vernon, D., Fadiga, L., Von Hofsten, C., Rosander, K., Lopes, M., Santos-Victor, J., et al.: The icub humanoid robot: An open-systems platform for research in cognitive development. Neural Networks 23(8), 1125–1134 (2010)
15. Lohan, K., Pitsch, K., Rohlfing, K., Fischer, K., Saunders, J., Lehmann, H., Nehaniv, C., Wrede, B.: Contingency allows the robot to spot the tutor and to learn from interaction. In: ICDL-EpiRob 2011 (2011)
16. Matatyaho, D., Gogate, L.: Type of maternal object motion during synchronous naming predicts preverbal infants' learning of word–object relations. Infancy 13(2), 172–184 (2008)
17. Gogate, L., Bolzani, L., Betancourt, E.: Attention to maternal multimodal naming by 6-to 8-month-old infants and learning of word–object relations. Infancy 9(3), 259–288 (2006)
18. de Barbaro, K., Johnson, C.M., Forster, D., Deak, G.O.: Temporal dynamics of multimodal multiparty interactions: A micrognesis of early social interaction. In: Spink, A., et al. (eds.) Proceedings of Measuring Behavior 2010, Eindhoven, The Netherlands, pp. 247–249 (2010)
19. Saunders, J., Nehaniv, C.L., Lyon, C.: Robot learning of lexical semantics from sensorimotor interaction and the unrestricted speech of human tutors. In: Proc. Second International Symposium on New Frontiers in Human-Robot Interaction, AISB Convention, Leicester, UK (2010)
20. Saunders, J., Lehmann, H., Sato, Y., Nehaniv, C.L.: Towards using prosody to scaffold lexical meaning in robots. In: Proceedings of ICDL-EpiRob 2011. IEEE (2011)
21. Fischer, K., Lohan, K., Foth, K.: Levels of embodiment: Linguistic analyses of factors influencing HRI. In: Proceedings of HRI 2012, Boston, Mass., MA (March 2012)
22. Chung, C.K., Pennebaker, J.W.: The psychological function of function words. In: Fiedler, K. (ed.) Social Communication: Frontiers of Social Psychology, pp. 343–359. Psychology Press, New York (2007)
23. Horton, W.S., Gerrig, R.J.: The impact of memory demands on audience design during language production. Cognition 96, 127–142 (2005)
24. Cross, T.G., Nienhuys, T.G., Kirkman, M.: Parent–child interaction with receptively disabled children: Some determinants of maternal speech style. In: Nelson, K. (ed.) Children's Language, vol. 5, pp. 247–290. Erlbaum, Hillsdale (1985)

The Frankenstein Syndrome Questionnaire – Results from a Quantitative Cross-Cultural Survey

Dag Sverre Syrdal[1], Tatsuya Nomura[2], and Kerstin Dautenhahn[1]

[1] Adaptive Systems Research Group, School of Computer Science, University of Hertfordshire, Hatfield, UK
{d.s.syrdal,k.dautenhahn}@herts.ac.uk
[2] Department of Media Information, Faculty of Science and Technology Ryukoku University, Japan
nomura@rins.ryukoku.ac.jp

Abstract. This paper describes the results from a cross-cultural survey of attitudes towards humanoid robots conducted in Japan and with a Western sampe. The survey used the tentatively titled "Frankenstein Syndrome Questionnaire" and combined responses both from a Japanese and Western sample in order to explore common, cross-cultural factor structures in these responses. In addition, the differences between samples in terms of relationships between factors as well as other intra-sample relationships were examined. Findings suggest that the Western sample's interfactor relationships were more structured than the Japanese sample, and that intra-sample characteristics such as age and gender were more prevalent in the Western sample than the Japanese sample. The results are discussed in relation to the notion of the Frankenstein Syndrome advanced by Kaplan [1].

1   Introduction

This paper reports recent findings from our continued work in developing a tool for examining attitudes towards humanoid robots that is valid across Western and Japanese Cultures. As described in Syrdal et al.[2] and Nomura et al.[3], these findings inform our investigation into how members of society may respond to the possibility of humanoid robots being used and encountered in their everyday lives.

Previous cross-cultural studies have found conflicting results and [4,5], particularly when considering comparisons between absolute scores on scales intended to measure specific constructs related to participants' attitudes towards robots. In order to further investigate cross cultural differences and similarities along such attitudes, we conducted an open-ended survey of attitudes towards humanoid robots both in Japan and in the UK [2] from which statements representative of different categories from each sample were selected and made into the Frankenstein Syndrome Questionnaire (FSQ) [3].We have based our theoretical approach in terms of cultural differences on that of Kaplan's [1] description of the "Frankenstein Syndrome". This approach posits that the act of creation, particularly innovative creation is seen as a taboo in Western cultures. In these cultures, the use of novel technologies is often seen as

G. Herrmann et al. (Eds.): ICSR 2013, LNAI 8239, pp. 270–279, 2013.

Table 1. Factor Loadings on the FSQ (*item removed from Subscale)

F1	F2	F3	F4	F5	Variable
.692	-.042	-.105	.080	.121	I would feel uneasy if humanoid robots really had emotions or independent thoughts.
.491	.005	-.007	.337	-.126	If humanoid robots cause accidents or trouble, I believe that the people and organizations developing of them will provide sufficient compensation to the victims.
.417	.017	.108	-.004	-.162	Widespread use of humanoid robots would lead to high maintenance-costs for them.
.380	-.182	.210	.111	.250	I am concerned that humanoid robots would be a bad influence on children.
.447	-.070	.098	.059	.158	I would hate the idea of robots or artificial intelligences making judgements about things.
.832	.093	-.166	-.094	-.008	I feel that if we depend on humanoid robots too much, something bad might happen.
.570	-.069	.310	-.120	-.080	I don't know why, but humanoid robots scare me.
.574	-.078	.209	.093	-.022	Many humanoid robots in society will make it less warm.
.545	.083	.054	-.011	.113	Something bad might happen if humanoid robots developed into human beings.
.371	.216	.011	-.088	.238	Widespread use of humanoid robots would take away jobs from people.
-.141	**.539**	-.021	.199	.320	Humanoid robots can create new forms of interactions both between humans and between humans and machines.
.277	**.414**	.131	-.194	.285	Humanoid robots may make us even lazier.*
.154	**.466**	-.125	-.024	.059	Humanoid robots can be very useful for caring the elderly and disabled.
-.110	**.493**	.112	.050	.125	Humanoid robots should perform repetitive and boring routine tasks instead of people.
-.055	**.573**	.011	.149	-.175	I don't know why, but I like the idea of humanoid robots.
-.219	**.537**	.363	-.010	-.327	Humanoid robots can be very useful for teaching young kids.
-.129	**.389**	.060	.025	.138	Humanoid robots are a natural product of our civilization.
.119	**.723**	-.113	.051	-.021	Humanoid robots can make our life easier.
.307	**.499**	-.214	.039	-.013	Humanoid robots should perform dangerous tasks, for example in disaster areas, deep sea, and space.
.002	.000	**.511**	-.090	.318	I am afraid that humanoid robots make us forget what it is to be human.
-.071	-.196	**.759**	.111	.165	The development of humanoid robots is a blasphemy against nature.
.023	.139	**.640**	-.107	.027	I feel that in the future, society will be dominated by humanoid robots.
.351	-.073	**.524**	.076	-.047	The technologies needed for developing humanoid robots are amongst those fields that humans should not advance too far in.
.095	-.131	**.731**	.063	.007	The development of humanoid robots is blasphemous.
-.040	.081	-.172	**.656**	.122	The people and organizations that develop humanoid robots can be trusted.
-.079	.280	.033	**.603**	.065	The people and organizations that develop humanoid robots seem sincere.
-.001	.077	.241	**.569**	-.256	I trust the people and organizations that develop humanoid robots to disclose sufficient information to the public, including negative information.
.196	.214	-.126	**.411**	-.053	Persons and organizations related to development of humanoid robots will consider the needs, thoughts and feelings of their users.
.053	.107	.203	-.009	**.606**	Interacting with humanoid robots could sometimes lead to problems in relationships between people.
.180	.113	.298	-.052	**.457**	I am afraid that humanoid robots will encourage less interaction between humans.

potentially problematic in itself, while other cultures, such as that of Japan, may have a more pragmatic view, judging innovations on their own merits. This phenomenon could manifest as an underlying factor in attitudes towards humanoid robots in a much greater extent in Western cultures than one would see in a Japanese population.

The presence of such a factor, in addition to differences in how demographic factors interact with culture on the impact scores from the different factors formed from the questionnaire that we are using, will form a foundation for our further effort in examining the role of the Frankenstein syndrome in cross-cultural studies in social robotics.

2   Methodology

2.1   Sampling

The Japanese sample consisted of 1000 persons recruited through a professional survey company. The Western sample consisted of 146 participants(61 male and 85 females; age range 20-64, Mean age 28, Median age 25) , recruited through adverts in social media and through the University of Hertfordshire intranet. Exclusion criteria for the Western sample was (a) not having a European or Middle Eastern native language, and not living in Europe, the Middle East, The Americas or Australia/New Zealand. Due to the disparity in size between the Western sample and Japanese sample, a subsample was extracted from the Japanese sample using a stratified random sampling technique, where the strata were based on gender and age-category. This random sample was combined with the Western sample in order to create a joint sample for analysis. A second random sample was also taken from the Japanese sample to assess the generalizability of the findings from the Japanese subsample.

2.2   Survey

The survey was presented as a series of webpages, with a cover page displaying images of a wide range of humanoid robots. The survey itself consisted of the statements presented in Table 1, inviting the participants to indicate their agreement with each on a 7-point likert scale. See [3] for a more in-depth description of the survey, including the pictures used.

3   Results

3.1   Factor Analysis

The joint sample data was assessed using a maximum likelihood, exploratory factor analysis, which found 5 factors using the Cattell extraction criteria[6], explaining 54.32% of the variance in the sample. The promax rotation Factor Loading Matrix can be found in Table 1. The items loading into the different factors were combined into scales.

The items in Factor 1 had a Cronbach's α of .84 for the sample as a whole, .84 for the Western sample and .85 for the Japanese sample. This factor was tentatively named **General Negative Attitudes towards Robots**.

The items in Factor 2 had a Cronbach's α of .75 for the sample as a whole, .65 for the Western sample and .77 for the Japanese sample. Due to the low reliability the western sample, subsequent investigation of Item-Scale correlation found that this was caused by the item: "*Humanoid robots may make us even lazier*", loading negatively on this subscale for the western sample while it was positively correlated with the subscale for the Japanese sample. After this item was removed, the sample as a whole had a Cronbach's α of .75, with .73 for the Western sample and .78 for the Japanese sample It was tentatively named **General Positive Attitudes towards robots**.

The items in Factor 3 had a Cronbach's α of .83 , with .81 for the Western sample and .85 for the Japanese sample. It was tentatively named **Principal Objections to Humanoid Robots**.

Table 2. Differences in Subscale Correlations according to Sample

Factor	General Negative	General Positive	Principal Objections	Trust in Creator
General Negative	1			
General Positive	r= .21 r_w=-.39, r_j=.01 z=3.63,p<.01**	1		
Principal Objections	r=.59 r_w=.65, r_j=.53 z−1.55,p=.06	r=-.39 r_w=-.41, r_j=-.22 z=1.71,p<.05*	1	
Trust in Creators	r=-.13 r_w=-.20, r_j=.01 z=1.67,p<.05*	r=.54 r_w=.45, r_j=.51 z=0.67,p=.25	r=-.25 r_w=-.34, r_j=.05 z=3.39,p<.01**	1
Interpersonal Fears	r=.56 r_w=.58, r=.52 z=0.75,p=.23	r=-.18 r_w=-.36, r_j=.05 z=3.59,p<.01	r=.53` r_w=.57, r_j=.52 z=0.6,p=.27	r=-.22 r_w=-.34, r_j=.05 z=2.63,p<.01

The items in Factor 4 had a Cronbach's α of .73 for the sample as a whole, 68 for the Western sample and .75 for the Japanese sample. Subsequent Item-Scale correlations suggested that this was caused by overall lower reliability along this scale for the Western sample. It was tentatively titled **Trust in Robot Creators**.

The items in factor 5 had a Cronbach's α .68, with an α of .64 for the Western sample and .75 for the Japanese sample. This subscale was tentatively named **Interpersonal Fears** .

3.2   Intra-sample Correlation between the Subscales

In order to further investigate the relationship between culture and subscale scores, intra-sample relationships between the subscales were investigated using correlations.

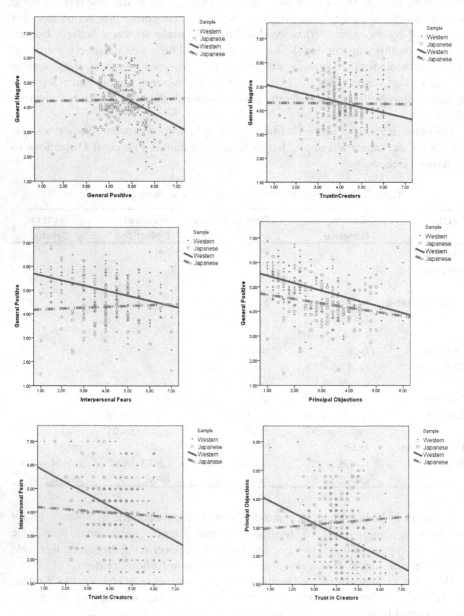

Fig. 1. Differences in Subcale Correlations according to sample

The correlations between the subscales for the sample as a whole can be found in Table 2, which shows which suggests a high degree of inter-correlation between all the subscales in the sample as a whole (r), but as Fig. 1, illustrates, this relationship is more complicated, however, as there are significant differences between the samples in terms of subscale correlations. These differences all manifest as stronger correlations between the subscales in the Western sample (r_w), suggesting that scores on one subscale predict scores on the other subscales well in this sample, while this is not the case for the Japanese sample(r_j). This would in turn suggest that the Western sample's attitudes is to a much larger extent dependent on one underlying factor that impacts overall attitudes towards robots

3.3 Subscale Score ANOVAs

A series of 2x2x3 ANOVAs were run for each subscale, in order to investigate the relationship between Sample (Western or Japanese), Gender (Male or Female), Age Category (Participants in their twenties, thirties or above 40) and subscale scores. The overall mean for each subscale score by each variable is presented visually in XXX and when responsible for a significant main effect, is described under the description of each subscale.

General Negative Attitudes towards Humanoid Robots

There were significant main effects for Sample $(F(1,278)=7.48,\ p<.01,\eta^2=.03)$, Gender $(F(1,278)=19.60,\ p<.01,\eta^2=.07)$ and Age Category $(F(2,278)=5.86,$

Table 3. Subscale Means by Sample

Subscale	Western Mean(SD)	Japanese Mean (SD)
General Negative	4.23(1.14)	4.31(0.88)
General Positive	4.97(.081)	4.29(0.78)
Principal Objections	2.55(1.09)	3.06(0.88)
Trust in Creators	4.56(1.07)	3.88(0.90)
Interpersonal Fears	4.01(1.54)	4.01(1.16)

(Bold scores represent significant main effects)

$p<.01,\eta^2=.04)$. There was also an interaction effect for Sample and Age Category $(F(2,278)=8.00,\ p<.01,\eta^2=.05)$

The Descriptive Statistics in Table 3 suggest that overall, the Japanese sample scored higher in this subscale. Table 4 suggest females scored higher than males and according to Table 5, participants in their 20s scored higher than the other two age categories $(t>2.29,p<.05)$. The descriptive statistics for the Age Category and Sample interaction effect can be found in Table 6A which suggest that the main effect observed was caused by the differences between the participants in their 20s in the Western sample scoring higher in this subscale than the other two categories $(t>3.16,p<.05)$ while this effect is not observed in the Japanese sample, which was more uniform across the different age categories $(t<.49,p>.6)$.

General Positive Attitudes towards Humanoid Robots

There was a main effect for Sample (F(1,278)=35.12, p<.01,η^2=.11) and Gender (F(1,278)=9.01, p<.01,η^2=.03). There was also a significant interaction effect for Sample and Gender (F(1,278)=6.72, p<.01,η^2=.02).

The Descriptive Statistics for the Main Effects can be found in Table 3-5 and Fig. 2 and suggest that overall, Western participants scored higher on this subscale than the Japanese, and that male participants scored higher than female. Table 6B describes the interaction effect for General Positive Attitudes, and suggest that in the Western Sample, male participants score higher along this subscale(t=3.96,p<.01), while in the Japanese sample, this effect is not evident(t=.29,p=77).

Table 4. Subscale Means by Gender

Subscale	Male Mean(SD)	Female Mean(SD)
General Negative	3.91(0.98)	**4.53(0.96)**
General Positive	**4.88(0.94)**	4.48(0.90)
Principal Objections	2.58(1.17)	3.07(1.16)
Trust in Creators	4.25(0.93)	4.19(1.08)
Interpersonal Fears	3.85(1.20)	4.13(1.46)

(Bold scores represent significant main effects)

Table 5. Subscale Means by Age Category

Subscale	20s Mean (SD)	30s Mean (SD)	40+ Mean (SD)
General Negative	**4.43(0.98)**	3.94(1.10)	3.96(0.85)
General Positive	4.58(0.99)	4.72(0.85)	4.92(0.74)
Principal Objections	**3.06(1.19)**	2.45(1.08)	2.45(1.11)
Trust in Creators	4.22(1.05)	4.16(0.97)	4.29(0.97)
Interpersonal Fears	**4.22(1.38)**	3.53(1.17)	3.66(1.34)

(Bold scores represent significant main effects)

Principal Objections

There were significant main effects for Sample (F(1,278)=24.66, p<.01,η^2=.08), Gender (F(1,278)=5.17, p<.05,η^2=.02), and Age Category (F(1,278)=7.89, p<.01,η^2=.05). These effects are described in Table 3-5 and suggest that the Japanese sample scored higher on this subscale than the Western sample, the Female sample higher than the Male sample, and participants in their 20s scored higher than the other two age categories.

Trust in Creators

There was a significant main effect for Sample (F(1,278)=13.59, p<.01,η^2=.05) and an interaction effect for Sample and Age Category (F(2,278)=4.06, p<.05,η^2=.02). The main effect is described in Table 3-5 and suggests that participants in the Western sample scored higher than participants in the Japanese sample on this subscale while the interaction effect described in Table 6C suggests that this was caused by participants in the the 40+ category scored significantly higher than the other categories on this subscale in the Japanese sample (t>2.17, p<.05), but that this was not the case in the Western sample(t<1.15,p>.25). In fact the trend in the Western sample was in the opposite direction.

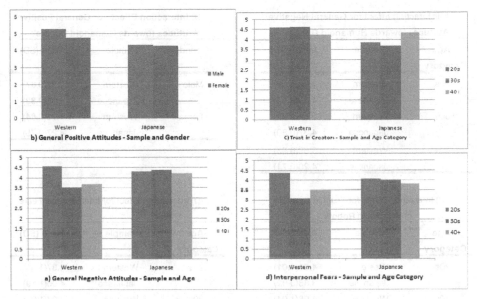

Fig. 2. Interaction Effects for Subscale Scores

Interpersonal Fears

There was a significant main effect for Age Category (F(2,278)=6.02, p<.01,η^2=.04) and an interaction effect for Sample and Age Category (F(2,278)=4.72, p<.01,η^2=.03).

The interaction effect was is described in Table 6D and suggest that in the Western sample, participants in their 20s scored higher along this subscale than other age categories, while this was not the case for the Japanese sample.

4   Discussion

4.1   Summary of Findings

The findings from the Factor Analysis using both samples found 5 factors that had a reasonably high degree of reliability, which were consistently higher for the Japanese sample than the Western sample. As discussed in Syrdal et al. [7], differences in absolute scores between the two populations are not as meaningful as exploring differences in the relationships between measures across the cultures, both in terms of the scales relate to each other as well as how they relate to demographic characteristics.

Subscale Correlations

Correlation between subscales suggested that overall, the relationship between subscales were more structured in the Western Sample than in the Japanese, with a

Table 6. Interaction Effect for Subscale Scores

A)Interaction effect for General Negative Attitudes towards Humanoid Robots

Age Category	Sample	Mean (SD)
20s	Western	4.55(1.02)
	Japanese	4.31(0.94)
30s	Western	3.51(1.21)
	Japanese	4.37(0.80)
40+	Western	3.69(0.93)
	Japanese	4.20(0.71)

B) Interaction Effect For General Positive Attitudes towards Humanoid Robots

Gender	Sample	Mean (SD)
Male	Western	5.28(0.67)
	Japanese	4.31(0.82)
Female	Western	4.76(0.84)
	Japanese	4.27(0.75)

C) Interaction Effect for Trust in Creators of Humanoid Robots

Age Category	Sample	Mean (SD)
20s	Western	4.60(1.01)
	Japanese	3.85(0.92)
30s	Western	4.62(1.04)
	Japanese	3.70(0.61)
40+	Western	4.25(1.00)
	Japanese	4.33 (0.84)

D) Interaction Effect for Interpersonal Fears

Age Category	Sample	Mean (SD)
20s	Western	4.37(1.58)
	Japanese	4.06(1.19)
30s	Western	3.07(1.24)
	Japanese	4.00(0.89)
40+	Western	3.50(1.26)
	Japanese	3.81(1.43)

higher degree of interfactor correlation. This can be taken as supporting the idea and construct of a culturally dependent Frankenstein Syndrome as advanced by Kaplan [1]. The Western sample tend to respond to the different subscales in a manner consistent with their responses being towards humanoid robots in and of themselves, rather than for specific issues related to their creation and adoption. This is in line with Kaplan's thesis of the Frankenstein Syndrome being an expression of a Western Taboo regarding the act of creation itself.

Intersample Differences

The results from this analysis replicates the emphasis of age from Nomura et al.[3] in that age seems important in Western Cultures as well, with the youngest group of participants being the most skeptical of humanoid robots both in terms of General Attitudes as well as Interpersonal Fears when compared to the older groups, however this effect was most pronounced in the Western sample with this particular factor structure. A similar effect for Trust in Creators was observed, but here age differences were most pronounced in the Japanese sample, where the 40+ group scored higher than the other age categories along this dimension. This suggests that these age differences are more closely related to changing views of technology in the Western sample, but while in the Japanese sample may be related to changes in how scientific and industrial authorities are viewed.

Finally, there were gender differences between the two samples terms of Positive Attitudes, in the Western sample, male respondents scored higher along this subscale than females, but this was not the case for the Japanese sample.

4.2   Conclusions and Future Work

As a first, cross-cultural use of the FSQ, the results are encouraging. The current structure of the FSQ has a high degree of reliability across both Japanese and Western samples and reveal interesting differences between the two groups in terms of intra-sample characteristics as well as in terms of subscale correlations. However, as previously pointed out, these now need to be supplemented by examining the role of FSQ Subscale scores and how they interact with related scales and behaviour within human-robot interactions. This will allow for a deeper validation of the FSQ and greater understanding of attitudes towards humanoid robots .

Acknowledgements. The research leading to these results has received funding from the European Union's Seventh Framework Programme (FP7/2007-2013) under grant agreement n°[287624], the ACCOMPANY(project, and   grant agreement n°[215554], the LIREC (LIving with Robots and intEractive Companions) project and from the Japan Society for the Promotion of Science, Grants-in-Aid for Scientific Research No. 21118006.

References

1. Kaplan, F.: Who is afraid of the humanoid? Investigating cultural differences in the acceptance of robots. International Journal of Humanoid Robotics 1(03), 465–480 (2004)
2. Syrdal, D.S., Nomura, T., Hirai, H., Dautenhahn, K.: Examining the frankenstein syndrome: an open-ended cross-cultural survey. In: Mutlu, B., Bartneck, C., Ham, J., Evers, V., Kanda, T. (eds.) ICSR 2011. LNCS, vol. 7072, pp. 125–134. Springer, Heidelberg (2011)
3. Nomura, T., Sugimoto, K., Syrdal, D.S., Dautenhahn, K.: Social Acceptance of Humanoid Robots in Japan: A Survey for Development of the Frankenstein Syndrome Questionnaire. In: 2012 IEEE-RAS International Conference on Humanoid Robots, Humanoids and Humans: Towards A New Frontier, November 29-December 1, Business Innovation Center Osaka, Japan (2012)
4. Bartneck, C., Nomura, T., Kanda, T., Suzuki, T., Kennsuke, K.: A cross-cultural study on attitudes towards robots. In: Proceedings of the HCI International 2005, Las Vegas (2005)
5. MacDorman, K.F., Vasudevan, S.K., Ho, C.-C.: Does Japan really have robot mania? Comparing attitudes by implicit and explicit measures. AI & Society 23(4), 485–510 (2009)
6. Cattell, R.B.: The Scree Test for the Number of Factors. Multivariate Behavioral Research 1(2), 245–276 (1966)
7. Syrdal, D.S., Dautenhahn, K., Koay, K.L., Walters, M.L.: The Negative Attitudes towards Robots Scale and Reactions to Robot Behaviour in a Live Human-Robot Interaction Study. In: Proceedings New Frontiers in Human-Robot Interaction, a Symposium at the AISB 2009 Convention. Heriot Watt University, Edinburgh, Scotland, April 8-9 (2009)

Region of Eye Contact of Humanoid Nao Robot Is Similar to That of a Human

Raymond H. Cuijpers and David van der Pol

Eindhoven University of Technology, Den Dolech 2,
5600 MB Eindhoven, The Netherlands
r.h.cuijpers@tue.nl, davidvdp@gmail.com

Abstract. Eye contact is an important social cue in human-human interaction, but it is unclear how easily it carries over to humanoid robots. In this study we investigated whether the tolerance of making eye contact is similar for the Nao robot as compared to human lookers. We measured the region of eye contact (REC) in three conditions (sitting, standing and eye height). We found that the REC of the Nao robot is similar to that of human lookers. We also compared the centre of REC with the robot's gaze direction when looking straight at the observer's nose bridge. We found that the nose bridge lies slightly above the computed centre of the REC. This can be interpreted as an asymmetry in the downward direction of the REC. Taken together these results enable us to model eye contact and the tolerance for breaking eye contact with the Nao robot.

Keywords: gaze perception, eye contact, cognitive robotics, social robotics.

1 Introduction

Robots are applied more and more in domestic environments within which they need to interact with people in increasingly sophisticated ways [14]. For example, in the KSERA project a humanoid robot is integrated with a smart home environment in order to support independent living of seniors in their own homes [15, 7]. Such applications place an increasing demand on robots with respect to their social and cognitive skills [18]. It is generally believed that robots with decent social and cognitive skill are more easily accepted [5]. Especially, for humanoid robots people expect a certain basic level of verbal and non-verbal behavioural intelligence to be present [14]. One such non-verbal social cue is gaze and eye contact: it has been shown that mutual eye contact with robots can add fluency to the engagement [1] and increase the level of engagement [9, 13]. Thus, eye contact is relevant to many interactive robotic applications such as care for elderly [16, 15].

Eye contact is especially relevant when using robots in therapy interventions for autistic spectrum disorder [2, 12], because eye contact facilitates shared attention and turn taking, which is known to be impaired in people with autism.

G. Herrmann et al. (Eds.): ICSR 2013, LNAI 8239, pp. 280–289, 2013.

However, much is still unclear about eye contact with robotic agents. As the facial features of humanoid robots vary considerably between platforms and deviate substantially from human facial features, it is necessary to investigate the perception of eye contact for robotic agents. A popular platform for studying non-verbal communication is Aldebaran Robotics' Nao robot [2, 16]. The Nao robot is a 57cm tall humanoid robot with a moveable head and facial features that resemble eyes and a mouth. Although Nao's physical appearance resembles that of a human, its detailed features still differ considerably. For one, its visual sensors differ from the human visual system: its main camera has a limited resolution and a limited field of view, and the 'eyes' of the Nao robot cannot move within the head nor do they correspond to the actual camera locations. Even so, it is intuitively clear that robot can 'look at you' or not by orienting its head, and, thus, it can establish eye contact.

In human-human interaction studies eye contact influences all sorts of social behaviours and many relevant parameters associated to gazing have been identified like gaze direction, gaze duration, gaze frequency, glancing, interactions with facial expressions and more [8]. For eye contact a particularly relevant feature is the region of eye contact (REC). It is a measure for the tolerance people have for gaze directions as being perceived as eye contact. This range of gaze directions maps onto a region within the observer's face (see Figure 1), and it is much larger than one would expect based on the accuracy with which people can distinguish gaze directions [6]. The REC is about 8.1 degrees across for a viewing distance of 1m, and about 3.9 degrees for a distance of 5m [6]. We found a REC that is 7 degrees across at a distance of 1m, both horizontally and vertically, for pictorial displays of human faces [10], which is very similar. We also found that it is slightly larger in the downward direction than in the other directions (see Figure 2). When uncertain, perceivers are more prone to judge gaze as eye contact, resulting in an increase of size of the REC with distance [6]. According to Chen [3] the vertical asymmetry of the REC occurs because people reduce their eyelid separation when looking down and not when looking up. Closing the eyelids obscures the pupil position inducing additional uncertainty resulting in a larger REC in the downward direction.

In this study we therefore measure the region of perceived eye-contact for the Nao robot. We expected that the shape of the REC would be similar to human-human interaction, but that the size may be larger because the facial features of the Nao robot only resemble that of a human leading to more uncertainty. In addition, we varied the relative height of the Nao robot with respect to the observer. As the Nao robot is small it will frequently happen that Nao will be looking up whereas the observer looks down. We suspected that this could negatively affect the tolerance for eye contact, and, thus , increase the size and/or asymmetry of the REC. With this information we can model 'eye contact' with the Nao robot and use this as a social cue to make human-robot interaction more natural and acceptable.

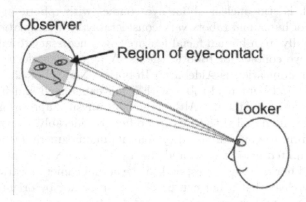

Fig. 1. Region of Eye-Contact (REC) is the range of gaze locations of looker on the face of the observer for which eye contact is perceived

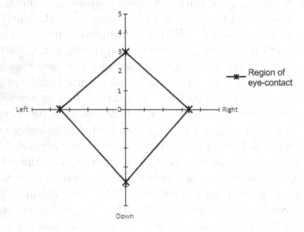

Fig. 2. The size of the REC in degrees for life size pictures of human heads at 1.25m distance [10]

2   Method

2.1   Participants

Six participants took part in the study. Two of them were female and four of them male. Al participants were employees of the university and knew about the KSERA project, but they were naive with respect to the specific details of the experiment. The right eye was the dominant eye of all participants. Participants had normal or corrected-to-normal vision and were between 1.65 m and 1.89 m tall.

2.2   Design and Task

The participants had to perform two tasks. The centring task and the nose-bridge task. In the centring task, the participant had to adjust the head orientation of Nao until it appeared to make eye contact. Thus, the measured head orientation corresponds to the edge of the Region of Eye-Contact (REC). There were four directions in which the participant had to adjust the head orientation (up, down, left, right) and for each of these centring tasks the robot's initial gaze direction was directed away in the opposite direction (see Figure 3). It is important to note that all gaze directions obtained in this way are perceived as making eye contact. In the nose-bridge task participants had to adjust the head orientation until the robot appeared to look straight at the observer's nose bridge.

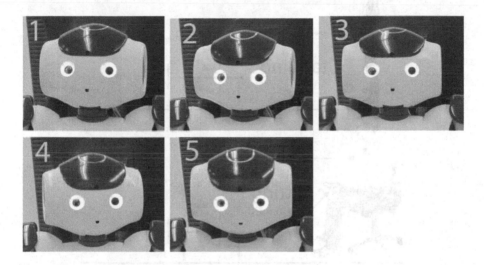

Fig. 3. Different initial gaze directions for the different tasks. (1) Nose-Bridge task, (2) Right task, (3) Down task, (4) Left task, (5) Up task.

A within-subjects design was used for the experiment. There were three conditions: (1) a standing condition, (2) a sitting condition, and (3) a eye height condition (see Figure 4). For each condition, the REC was measured from four directions (see Figure 3) and each direction was repeated four times. The location of the nose-bridge was also measured from four directions (upper left, upper right, lower left and lower right), but each direction was repeated only once. This results in a total of 3 (conditions) x 4 (repetitions) x 4 (directions) + 3 (conditions) x 1 (repetition) x 4 (directions) = 60 trials.

2.3   Experimental Setup

We used a 19" TFT screen to present the task instructions to the participants. The humanoid robot Nao (Aldebaran Robotics, France) was used as the anthropomorphised looker. We recorded the relative yaw and pitch angle in degrees, the completion time for each task, and the size and position of the detected face. We used OpenCV's implementation of Viola & Jones' object detection algorithm [17] to detect the face from the image of Nao's main camera that was recorded at the moment that the "enter" key was pressed. Nao was either located on a small 40 cm high table (standing and sitting condition), or an 80 cm high table (eye-height condition). During the sitting and eye-height condition participants were seated in a chair that was adjusted to the same height for all participants. In every condition Nao was placed at 160 cm from the participant (see Figure 4).

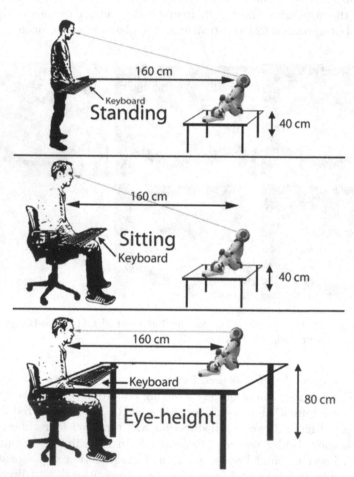

Fig. 4. Experimental setup. Upper pane shows the setup for the standing condition. Nao is located on a small table while the participant is standing. Middle pane shows the setup for the sitting condition. Nao sits on a small table while the participant is sitting on a chair. The lower pane shows the setup for the eye-height condition. The participant is sitting while Nao is on the tabletop. Nao is located at eye-height.

2.4 Procedure

Every participant's body height was noted, and a test was done to determine the visual acuity and the dominant eye. Participants were asked to sit down on the chair and perform a few practise trials. The experimenter was present during these practise trials to explain the different tasks. If the participant understood what to do for the different tasks, the actual experiment was started, and the participant was left alone. The experimenter could monitor the experiment's progress through Nao's camera.

In the centring task, Nao moved its head to one of the initial gaze directions and told the participant in which direction to move Nao's head (see Figure 3). The participant moved the robot's head by pressing the corresponding keys on the keyboard. As soon as the participant perceived eye-contact (s)he pressed the "enter" key and the robot's head pose and front camera image was recorded. In the nose-bridge task Nao looked away from the participant in one of four directions. The participant then moved Nao's gaze toward their nose-bridge adjusting both the yaw and pitch angles of Nao's head. Participants pressed the 'enter' key when satisfied. This was repeated until all trials were completed. The order of presentation was randomised for each participant.

3 Results

In Figure 5 the REC is shown for each of the four directions of approach of the robot's head (cornerpoints of the solid line) along with the 95% Confidence

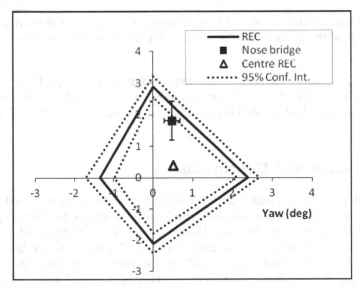

Fig. 5. Region of eye contact (REC, solid line) from the participants' point of view. The centre of the REC (triangle) does not coincide with the origin, but it is horizontally aligned with the gaze directed at the observer's nose bridge (square).

intervals (corner points of dotted lines). The width of the REC at its widest point was 3.7 ± 0.7 deg (SD = 1.3 deg) and its vertical span was 5.0 ± 0.8 deg (SD = 1.5 deg). We used the centre of the image coming from Nao's forward camera as its presumed gaze direction. Ideally, the nose bridge of the participant should be in the centre of the image in the nose bridge task. We found that this centre of the REC (triangle in Figure 5), lies approximately 0.52 degrees (2,5 pixels in a 320x240 window) to the left and 0.39 degrees (1.9 pixels in a 320x240 window) of the image centre. The robot's head orientation when looking straight at the participant's nose bridges is indicated by the square at ($0.5°, 1.8°$). The centre of the REC differed from the measured nose bridge location: the pitch angle was significantly less ($t(71) = -4.85$, $p < 0.5$), but the yaw angle was not significantly different ($t(71) = 0.30$, $p = 0.77$).

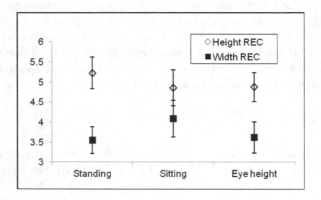

Fig. 6. The vertical (diamonds) and horizontal (squares) dimension of the REC plotted for every pose of the participant. The whiskers represent the 95% confidence intervals.

The effect of the conditions (standing, sitting or eye height) are shown in Figure 6. The height of the REC was larger than the width for all conditions, but the conditions had no significant effect on the size of the REC ($F(2, 71) > 0.2, p > 0.82$).

4   Discussion and Conclusions

The most important conclusion from the results of the experiment is that the perception of the gaze-direction and, specifically, eye contact with Nao as a looker, is not that different compared to a human looker. For a distance of 1.60m, we found a width of about 4° and a height of about 5°, which lies in-between the values of 8.1° for 1m distance and 3.9° for 5m distance as observed by Gamer and Hecht [6]. The similarity between the REC of Nao and a human looker is encouraging, because the Nao robot only has very basic facial features. This suggests that the realism of the facial features is not essential for the accuracy with which people judge spatial orientations of a (robotic) head. As a consequence, we expect that the REC is similar for other robots as long as their heads have a well-defined spatial orientation.

Unlike more sophisticated humanoid robots, the Nao robot does not have orientable eyes within the head. This may seem a serious drawback, but our results show that head rotations are sufficient to mimick eye contact with comparable tolerances. Using only head rotations for mimicking eye contact is not as unrealistic as one might initially believe. From a distance real human eyes are hardly visible let alone that their spatial orientations are discernable. Still, when the eyes are visible, people can very accurately discriminate pupil positions within the sclera of the eye [4]. It is possible that realism of facial features plays a much more important role for judgements of eye turn, also because the spatial orientations of real human eyes are typically not the same when focused on a target.

Chen [3] reported an asymmetry in the REC in the downward direction for human lookers. To check whether such an asymmetry is present we compared the centre of the REC with the gaze direction obtained when the robot is looking straight at the participants' nose bridge. We found that the computed REC (assuming symmetry) lies below the nose bridge gaze direction. Thus, if the nose bridge is considered the true centre of the REC, this difference implies that the REC is larger in the downward direction than in the other directions. Chen [3] proposed that this asymmetry is due to the eyelids making eye-movements more salient in the upward direction. However, this cannot explain the asymmetry of the REC of Nao, because it does not have eyelids. More likely, the asymmetry reflects the fact that the nose bridge is not the centre of the REC.

We used a face detection algorithm to centre the robot's gaze on a person with an adjustable offset. It may be that the centre of the head in the image of Nao's camera is misaligned with its perceived head orientation. To test this we determined the location of the central pixel in Nao's video frame when it was perceived as looking at the participant's nose bridge. We found a significant difference betweem the image centre and the position of the nose bridge in the image. This suggests the camera was misaligned about 0.5° in both pitch and yaw.

The obtained REC constrains the parameters for Nao's eye contact behaviour. In terms of eye contact, it is better to use the centre of the REC Nao's gaze direction because it divides the downward and upward direction in equal parts, making the chance of accidentally breaking eye contact as small as possible. The REC enables the use of different head orientations for signalling eye contact. We can use this property to signal emotions using head orientation. For example, looking down is perceived as having a sad emotional state, where looking up is perceived as having an "up" or alert emotional state.

The results from this study provide a good estimate of the tolerances for eye contact with the Nao. They show that these tolerances are similar whether the person is standing, sitting or at eye height with Nao. Our results also show that care should be taken to use the centre of Nao's camera images as looking straight ahead. Preferably, one should calibrate the straight ahead direction for each robot in order to obtain the best results. The results from this study also tell us how to break eye contact. The obtained tolerances give us the precision requirements for Nao's gaze behaviours and they allow us to signal emotion using head tilt without losing eye-contact with the observer. Since the Nao's facial features are very basic, we expect that our results extend to other robots as long as they have heads with clear spatial orientations.

Acknowledgements. The research leading to these results is part of the KSERA project (http://www.ksera-project.eu) and has received funding from the European Commission under the 7th Framework Programme (FP7) for Research and Technological Development under grant agreement n° 2010-248085.

References

[1] Breazeal, C.: Toward sociable robots. Robotics and Autonomous Systems 42(3-4), 167–175 (2003), doi:10.1016/S0921-8890(02)00373-1

[2] Gillesen, J.C.C., Barakova, E.I., Huskens, B.E.B.M., Feijs, L.M.G.: From training to robot behavior: Towards custom scenarios for robotics in training programs for ASD. In: 2011 IEEE International Conference on Rehabilitation Robotics (ICORR), pp. 1–7 (2011), doi:10.1109/ICORR.2011.5975381

[3] Chen, M.: Leveraging the asymmetric sensitivity of eye contact for videoconference. In: Proceedings of the SIGCHI Conference on Human Factors in Computing Systems Changing Our World, Changing Ourselves - CHI 2002, p. 49. ACM Press, New York (2002), doi:10.1145/503376.503386

[4] Cline, M.G.: The Perception of Where a Person Is Looking. The American Journal of Psychology 80(1), 41 (1967), doi:10.2307/1420539

[5] Dautenhahn, K.: Socially intelligent robots: dimensions of human-robot interaction. Philosophical Transactions of the Royal Society of London. Series B, Biological Sciences 362(1480), 679–704 (2007)

[6] Gamer, M., Hecht, H.: Are you looking at me? Measuring the cone of gaze. Journal of Experimental Psychology. Human Perception and Performance 33(3), 705–715 (2007)

[7] Johnson, D.O., Cuijpers, R.H., Juola, J.F., Torta, E., Simonov, M., Frisiello, A., Bazzani, M., Yan, W., Weber, C., Wermter, S., Meins, N., Oberzaucher, J., Panek, P., Edelmayer, G., Mayer, P., Beck, C.: Socially Assistive Robots: A comprehensive approach to extending independent living. International Journal of Social Robotics (submitted)

[8] Kleinke, C.: Gaze and eye contact: A research review. Psychological Bulletin 10, 78–100 (1986)

[9] Mutlu, B., Hodgins, J., Forlizzi, J.: A storytelling robot: Modeling and evaluation of human-like gaze behavior (2006)

[10] van der Pol, D.: The effect of slant on the perception of eye-contact. Master Thesis (2009)

[11] van der Pol, D., Cuijpers, R.H., Juola, J.F.: Head pose estimation for a domestic robot. In: Proceedings of the 6th International Conference on Human-robot Interaction (HRI 2011), pp. 277–278. ACM, New York (2011), doi:10.1145/1957656.1957769

[12] Robins, B., Dautenhahn, K., Dickerson, P.: From Isolation to Communication: A Case Study Evaluation of Robot Assisted Play for Children with Autism with a Minimally Expressive Humanoid Robot. In: Second International Conferences on Advances in Computer-Human Interactions, ACHI 2009, pp. 205–211 (2009), doi:10.1109/ACHI.2009.32

[13] Sidner, C.L., Kidd, C., Lee, C., Lesh, N.: Where to look: a study of human-robot engagement. In: Proceedings of the 9th International Conference on Intelligent User Interfaces, pp. 78–84. ACM, New York (2004)

[14] Tapus, A., Mataric, M., Scassellati, B.: The grand challenges in socially assistive robotics. Robotics and Automation Magazine, 1–7 (2007)

[15] Torta, E., Oberzaucher, J., Werner, F., Cuijpers, R.H., Juola, J.F. (2012). The attitude toward socially assistive robots in intelligent homes: results from laboratory studies and field trials. Journal of Human-Robot Interaction 1(2), 76–99. DOI: 10.5898/JHRI.1.2.Torta.

[16] Torta, E., van Heumen, J., Cuijpers, R.H., Juola, J.F.: How Can a Robot Attract the Attention of Its Human Partner? A Comparative Study over Different Modalities for Attracting Attention. In: Ge, S.S., Khatib, O., Cabibihan, J.-J., Simmons, R., Williams, M.-A. (eds.) ICSR 2012. LNCS, vol. 7621, pp. 288–297. Springer, Heidelberg (2012)

[17] Viola, P., Jones, M.: Rapid Object Detection using a Boosted Cascade of Simple Features. In: Proc. IEEE CVPR 2001, vol. 01 (2001)

[18] Wada, J., Shibata, T., Saito, T., Tanie, K.: Analysis of factors that bring mental effects to elderly people in robot assisted activity. In: Proceedings of the 2002 IEEE/RSJ Conference on Intelligent Robots and Systems, Lausanne, Switzerland, pp. 1152–1157 (2002)

Exploring Robot Etiquette: Refining a HRI Home Companion Scenario Based on Feedback from Two Artists Who Lived with Robots in the UH Robot House

Kheng Lee Koay, Michael L. Walters, Alex May, Anna Dumitriu,
Bruce Christianson, Nathan Burke, and Kerstin Dautenhahn

University of Hertfordshire, Hatfield AL10 9AB, UK
{K.L.Koay,M.L.Walters,B.Christianson,K.Dautenhahn}@herts.ac.uk,
annadumitriu@hotmail.com, {bigfug,natbur}@gmail.com

Abstract. This paper presents an exploratory Human-Robot Interaction study which investigated robot etiquette, in particular focusing on understanding the types and forms of robot behaviours that people might expect from a robot that lives and shares space with them in their home. The experiment was intended to tease out the participants' reasoning behind their choices preferences and suggestions for passive robot behaviours that could usefully complement the robot's active behaviours in order to allow the robot to exhibit considerate and socially intelligent interactions with people.

Keywords: Robot Etiquette, Human-Robot Interaction, Social Robot, Robotic Home Companion, Assistive Robotic Technology.

1   Introduction

Recent advances have led to the possible realisation of assistive robotic technology that is capable of providing support to elderly people to maintain independent living. Two European Commission funded FP7 projects, SRS [1] and ACCOMPANY [2] focus heavily on robotic assistive technology in the form of a Care-O-bot® 3 [3] robot to provide support for elderly persons to maintain independent living in their own home. Part of the work for the EU funded ACCOMPANY project focuses on investigating the use of a robot as a cognitive prosthesis to provide reminders for the users as well as offering physical assistance, including fetching and carrying objects for the user. The use of a robot in a domestic environment raises many issues related broadly to users' acceptance of robotic technology (cf. [4,5,6,7]). These include ethics and privacy concerns with regards to the robotic technology collecting user data [8,9], usability issues regarding Human-Robot Interaction [10,11,12,13] as well as annoyance and conflicts that may arise from a robot failing to meet user's expectation in teams of social behaviour. This last area of concern is broad and complex in relation to Human-Robot Interactions (HRI), but in this paper the term "robot etiquette" is used as a convenient term for all these aspects (cf. [14]).

This paper aims to explore some aspects of robot etiquette which are relevant to a robot companion or service robot operating in a domestic environment.

G. Herrmann et al. (Eds.): ICSR 2013, LNAI 8239, pp. 290–300, 2013.
© Springer International Publishing Switzerland 2013

An opportunity was presented when two artists volunteered to live in the University of Hertfordshire (UH) Robot House for a week and also consented to take part in various experiments with different robots [15]. This provided an opportunity to conduct robot etiquette experiments with the artists, who by the time of the experiment had habituated to the robot house and were familiar with its domestic setting, which were important requirements for this experiment. The UH Robot House is a typical three bedroom house, located in a residential area in the UK. It is used for conducting Human-Robot Interaction Studies in an ecologically realistic domestic environment domestic setting as compared to laboratory conditions.

Hüttenrauch et al. [16] involved an actual user in a long-term immersive HRI study to aid the design of their CERO service robot and associated robot behaviours. It is difficult to provide long-term continuous interaction in an HRI study between participants and robots, for both practical and organisational reasons, so long-term, immersive HRI studies are rarely carried out. As the artists would be living for a whole week in the Robot House, therefore, they would be able to provide insightful feedback related to human and robots sharing spaces in domestic environment. Additionally, the experienced gain from this experiment will guide the design for constructive robot etiquette studies which are suitable for long-term participants from various backgrounds and age groups.

1.1   The Resident Artists

The artists who were the participants in the study are Artists in Residence for the University of Hertfordshire, and their brief is to explore the boundaries between Art and Science. They also are tasked with producing artistic installations which raise public awareness, and promote discussion of issues raised by the application of new technologies. The artists are:

Anna Dumitriu: Her artistic research is based around the investigation and communication of emerging technologies in robotics, as a means of engaging non-scientific audiences in societal debates, as well as intervening artistically in the research process. She works with performance techniques to develop robot behaviour in order to create deeply felt, almost visceral (positive or negative) reactions in diverse audiences (see www.annadumitriu.co.uk).

Alex May: He uses low level programming, interactive video projection, and performance to explore the gaps between public understanding of the workings of digital systems and the reality of the processes involved, and reveals how such technologies can be used to manipulate our perception of reality in our digitally augmented world. He is currently developing non-accurate visual and physical memory systems for robot companions that will "get bored of the humans around them", and a robot servo control system that introduces micro-movements for increased non-verbal communication, and more emotional inverse kinematics (see www.bigfug.com).

Note, the artists that participated in this experiment have both 1) experience of working with robots, and 2) expertise from their work domain where they explored and manipulated technology for the purpose of engaging audiences (to elicit both positive and negative reactions from general public) through visual performance techniques. This makes them ideal candidates to have taken part in this exploratory experiment to design robot behaviours that are meaningful (intuitive), engaging, and pleasant.

2   Expressions Experiment

During their stay at the University of Hertfordshire (UH) Robot House, the resident artists were involved in a series of exploratory experiments with different robotic platforms concerning companion robots in domestic environment. However, unlike previous HRI studies in the UH Robot House [17,18,19], the artist participants were fully informed as to the procedure and purpose of all the experiments. This paper focuses on the set of experiments which investigated some aspects of expressive and social robot behaviour with a view to explore and define some initial social robot behaviours in order to understand these aspects of Robot Etiquette. This was con- ducted using the ACCOMPANY project robot platform (see Fig 1), the Care-O-bot® 3 (COB), a robotic companion and care robot.

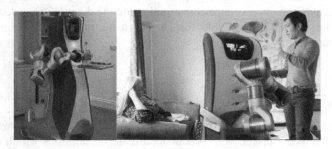

Fig. 1. Left: The Care-O-bot® LED colour display flashes red, while the robot is using its arm to put a bottle on its tray (which doubled as one of its GUIs). Right: The experimenter is guid- ing the artist through the scenario.

2.1   Motivation

The COB's social and expressive robot behaviour experiment aimed to improve the robot's social behaviour when in the vicinity of the user, especially including periods when the robot was not directly interacting with the user. This was to facilitate the robot being perceived by the user as friendly, considerate and socially aware when carrying out tasks in domestic environment. The overall objective for the experiment was to attempt to gain some understanding of people's expectations for the robot's social behaviour, with a long-term aim of developing "Robot Etiquette" – a code of robot behaviour that encompasses people's expectation for robot social behaviour in a domestic environment.

To explore robot behaviours within a domestic environment, one of the scenarios studied in the ACCOMPANY [2] project was used. This working scenario (see Ta- ble 1in Section 2.3) involved the COB robot providing cognitive assistance, such as approaching to remind a user to take medication or to have a drink, and physical as- sistance, such as accompanying the user to the kitchen and helping by carrying ob- jects to the living room. The experiment involved the experimenter collecting information from the artists with regard to how they preferred their COB to behave and interact with them within the scenario. The experimental procedure was designed to actively involve the artists in the process of interactively guiding the creation of the robot's social behaviour for the scenario in the University of Hertfordshire Robot House domestic environment. This included manipulating the robot's simple auditory and visual expressive channels (i.e. sound and LED colour display), as well as the robot's posture and position configurations (i.e. taking into account of the environment and the user), cf. [7,20,21,22,23]. Therefore it was essential that the participants were familiar with the experimental environment.

The combination of the robot's sound and LED colour display was an attempt to simplify how the robot might communicate or express its internal working states and status. The states and status selected for study were ones that were particularly relevant to the users as they related to user safety and convenience. Instead of expressing these states solely through the COB's Graphics User Interface (GUI), which also functions as a tray (see Fig 1), the robot displayed this status on its chest LED light panels (see Fig 1). This is much more visible compared to the GUI (i.e. without needing to get close to the robot or search for the robot's mobile tablet GUI to check the robot's status). In addition, simple beeps or a memorable melody could be used to attract or direct the user's attention to the robot's LED colour display if the user's attention is needed urgently.

The LED colour display used on the COB is based on an RGB colour system to show a wide range of colours, and has various display effects such as blinking and varying in intensity or brightness. The COB onboard sound system can also be used to play either simple beeps or more complicated midi melodies.

In this experiment the robot was programmed to utilise its expressive channels to express basic non-verbal behaviours that communicated the robot's task-based alert level which was focused on safety when sharing space with a user. This involved the robot using its two of the main expressive channels, sound (beep or tune) and the LED colour display to provide cues to signal a) the potential level of hazard of the current robot task/action, and b) the state of the current robot task. The robot behaviours were programmed without including any considerate companion behaviour towards the user. We expected the artist-participants to modify for themselves these default non-verbal behaviours and to design their own considerate companion behaviour for the robot. It was expected that this would highlight behaviour parameters and variables that the artists might deem important for improving the robot behaviour. Therefore, these can then be considered as important parameters and variables which might be customised by future users of the system for implementing their own robot etiquette preferences in future studies.

2.2   Procedure

The experimental procedure involved one artist at a time. The experimenter showed the artist the scenario in a step-by-step walk-through. At different stages within the scenario that involved the robot, the participant was asked specific questions about the robot's behaviour and their feedback was recorded. Methodologically, this is very similar to a methodology used in previous study [24] which included a structured interview technique to gain an open-ended insight into participants reasoning and feelings. The robot was present throughout the study and was remotely controlled by the experimenter in order to demonstrate its functionality and to act as a reference for the participants' creation of robot behaviours at the different stages. Fig 1 (right) shows a picture taken during the experiment where the experimenter and the artist were discussing about scenario step 3 (see section 2.3). The questions used are in the form of:

i)   How the robot should perform the particular task:
    a.   Should the robot make any sound and if so, what kind of sound?
    b.   Should the robot display something on its LED colour display and if so, what colour or display behaviour should it display?
ii)   Where and how should the robot position itself for this part:
    a.   Position, orientation and posture.
    b.   Reasoning behind this decision.

Note, that the day before this experiment, the artists were shown a live demo of the robot performing the scenario, but were not told any details of this particular experiment. This demo was provided to settle any initial curiosity the artists may have had regarding the robot's functionalities, capabilities and how it might utilise its embodiment within the domestic environment to achieve its goal of assisting the user. We believe this exposure is important for the experiment as it directs the artists' awareness to the various research issues related to a domestic robotic companion, such as robot embodiment, human-robot joint task and sharing space with robot in domestic environment, and to take that into account when providing feedback during the experiment. Figure 2 shows the environment layout of the UH Robot House. For this experiment, only the living room, dining area and kitchen area were used.

Fig. 2. Left: Picture of the UH Robot House ground floor environmental layout. Right: Photograph taken from the sofa area (i.e. top left corner of the ground floor layout).

2.3   Scenario and Feedbacks

The scenario used is described here in detail from the robot's perspective. This is to help researchers, engineers and designers to identify the functionalities required by

Table 1. Scenario and feedbacks from the artists

1.	**The user sits on sofa, while the robot is located in the default location charging.**
Alex	Amber colour when charging and green colour when fully charged.
Anna	Amber colour when charging (i.e. fading in and out, with adaptable intensity depending on environment's lighting condition), and green colour when it is fully charged.
2.	**The robot leaves the charging station and approaches the user.**
Alex	Make noise to indicate it is about to move, blinking green when it is moving.
Anna	Blinking green when it is moving and green colour when it stopped. White colour when it is ready for interaction.
3.	**The robot stops at a socially appropriate distance/orientation from user.**
Alex	Green color when the robot stopped.
Anna	Green color when the robot stopped.
4.	**The robot reminds the user that he/she has not had a drink for 3 hours, the user can select one of the following options:**
Alex	n/a
Anna	Blue colour before it speaks.
	Tilt head forward, speaks then tilt head back.
	a. Remind me later (robot will remind user after 5 minutes)
Alex	Stay where it is rather than going back.
Anna	Robot moves back to charging station.
	b. Let's go to the kitchen.
Alex	Noise to register it is about to move, blinking green when it is moving, and the robot should move first as not to block me.
Anna	Robot should move back, turn to make way for the user in a similar to an after you/inviting posture.
5.	**When arriving in the kitchen the user fetches a drink from the fridge and places it on robot.**
Alex	Tray always up.
Anna	Tray up.
6.	**Both robot and user move back to the sofa, while robot is carrying the drink.**
Alex	Robot should allow the user to go first.
Anna	Robot should moves back, turn to make way for the user in a similar to an after you/inviting posture.
7.	**Robot places the object on sofa table.**
Alex	Robot any orientation near table and user.
	Robot should start blinking green when moving its arm, tray and body.
Anna	Robot should flash Red, use speech "Putting the drink on table", and uses arm to grab and put cub on table, then tilt head forward and use speech "Here is your drink", tilt head back.
8.	**The robot observes the user to check if the user is drinking.**
	a. If he/she does the robot will go back to the default position (i.e. robot charging station).
Alex	Robot should stay where it is rather than going back.
Anna	n/a
	b. If the user is not drinking, the robot will wait and remind the user to drink within 10 minutes by displaying the related action possibility on the GUI and using expressive behaviour to attract the user's attention.
Alex	Robot should stay where it is rather than go back.
Anna	Robot should move to a side (not charging station).
	Robot should turn its head to user and say "Don't forget your drink" and turn back.

the robot in order to be able to assist its user by providing both cognitive and physical assistance. The robot's tasks were to remind the user to take their medication, keep the user hydrated and assist the user with fetch and carry tasks. These are the typical activities that an elderly or infirm person might require assistance with. Table 1 shows the main steps of the robot's working scenario, together with the artists' feedback, laid out in in a step-by-step sequential manner. The robot and the user starting location are shown in Fig 2.

The data collected from the experiment indicates that there are both similarities and also some marked differences in terms of how the artists preferred the robot to behave in the scenario. In general, both artists preferred their robot to behave in a consistent manner. However, Anna preferred her robot to be more animated while Alex preferred his robot to be less animated.

The data indicated both the artists preferred the robot to have the "user go first rule" when it comes to moving together to any location. For example, both artists indicated that they preferred the robot to make way for them to go first with the robot following them from behind, unless it is impossible for the robot to make way due to limited space (a typical situation in domestic environment!). In this situation, the robot should move first, if possible trying not to block the user's path.

The robot's posture played an important part of the robot's behaviour for Anna. She preferred the robot to take an inviting or welcoming posture when making way for her to go first. Similarly, she preferred the robot to tilt its head forward when using speech. Both the artists also preferred the robot to remain in the living room with them, rather than moving away and coming back later if the robot has an immediate upcoming task such as to remind them to drink in the next five minutes.

The differences noted from the artists' feedback are mainly related to their personal preferences and include things such as different colours or display patterns used for the LED colour display status indicator.

3   Summary of the Resulting Robot Behaviours

Figure 3 demonstrates how the new robot behaviours, which were created by integrating the artists' feedback, will perform in the scenario. Starting from Fig 3-a) where the robot is in charging mode (its LED colour display shows amber colour). The robot makes a short beep, changes its LED colour display from amber to flashing green and starts moving towards the user (See Fig 3-b). It takes human-robot proxemics [18,23,25] into account when approaching the user. Its LED colour display changes from flashing green (navigating) into a solid green as the robot stops moving, and starts flashing blue as it tilts its head forward and reminds the user to have a drink using speech (see Fig 3-c). As the user agrees to go to the kitchen with the robot to fetch a bottle of water, the robot's LED colour display starts flashing green as it slowly makes room for the user. The robot takes an "after you" posture and changes its LED colour display to solid green (see Fig 3-d and 3-e). The user gets up and starts walking towards the kitchen while the robot LED colour display flashes green and follows the user from behind (see Fig. 3-f). As the user opens the fridge to fetch a

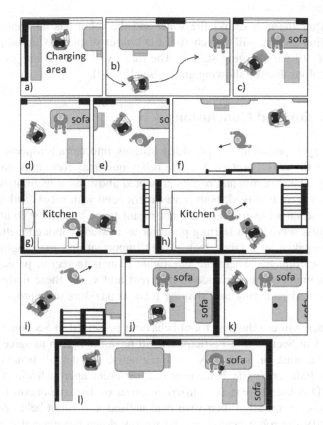

Fig. 3. Demonstrates how the new robot behaviour, created by integrating the artists' feedbacks, will act in the scenario

bottle of water, the robot slowly approaches the user taking up a position close to the user, but not blocking the kitchen entrance. It lifts its tray, as a gesture to offer assistance to carry the bottle, and then switches its LED colour display from flashing to solid green (see Fig 3-g). The user then places the bottle on the robot's tray and the robot's LED colour display starts flashing green. The robot slowly moves back to make way for the user. It takes an "after you" posture and stops flashing its LED colour display (see Fig 3-h). The user starts walking towards the sofa area, while the robot LED colour display starts flashing green and follows the user from behind (see Fig 3-i). The user sits on the sofa while the robot slowly approaches and stops next to the table then keeps its distance from the user while switching its LED colour display from flashing green to flashing red. The robot moves its arm slowly to grab the bottle from its tray and places it on the table. It then parks its arm at its back, as shown in Fig 3-j then switches its LED colour display from flashing red to flashing green. The robot then moves to take up a position next to the user, taking into consideration of not blocking the user's view of the television located in the living room, or from other areas of the experimental area (see Fig 3-k). The robot switches its LED colour display to solid green and stays there, waiting to remind the user to have their drink if they forget. After the user drinks the water and the robot has no immediate task to interact with the user, it will switch its LED colour display to flashing green then slowly navigate to the charging station. The robot will then go into charging mode with its LED colour display showing amber (see Fig 3-1).

4   Discussion and Conclusions

This exploratory experiment has provided insights into certain aspects of the robot behaviours that the artists preferred. These behaviours may form important aspects of robot etiquette. The experimental results presented above have its limitation (i.e. only two participants were involved, both have experienced with robots and are not in the age group of the target users), and are not meant to be generalize to all robots and users. The results served as a starting point for us to study robot etiquette and at best should only be taken as a guide to highlight important and meaningful factors for designing robot etiquette. Therefore our future plan is to recruit participants' from various backgrounds and experience to take part and verify these findings, with the goal of further understanding and shaping robot behaviours to improve etiquette for robots.

Future studies will use the new robot behaviour presented in Section 3 and the scenario presented in Section 2.3. Participants will have the option to agree with the default robot behaviour, or to modify and personalise the default behaviour settings. These may include proxemic behaviour (i.e. different approach direction and distance) and LED colour display (i.e. different colour or display pattern) to match their preferences, or even redesign their own personalised new robot behaviour. This will help us to verify the robot behaviour and narrow down the general consensus with regards to the most preferred way for the robot to behave in the vicinity of their user, and also for the user to personalise the robot's social behaviours. This will help us to assess the most effective and meaningful context based LED colour display signals for participants.

We will also study participants' preferences with regards to the level of animation and arousal exhibited by the robot's behaviours and how these levels of animation have an effect on the robot's perceived social behaviour by users (i.e. characteristics that influence peoples' perceptions of the robot). This experiment has also enabled us to improve the design of our next study, where the robot will have specific behaviours, based on high consistency feedback and will also incorporate personalisation options which will be selected based on the participants discretionary data.

Acknowledgements. The work described in this paper was conducted within the ACCOMPANY (Acceptable robotiCs COMPanions for AgeiNg Years) project and is partially funded by the European Commission under contract numbers FP7-287624.

References

1. SRS Project, http://srs-project.eu/ (accessed July 10, 2013)
2. ACCOMPANY Project, http://accompanyproject.eu/ (accessed July 10, 2013)
3. Care-O-bot®, http://www.care-o-bot.de/english/ (accessed July 10, 2013)
4. Khan, Z.: Attitude towards intelligent service robots. TRITA-NA-P9821, NADA, KTH (1998)
5. Bartneck, C., Forlizzi, J.: A Design Centred Framework for Social Human-Robot Interaction. In: Proceedings of the RO-MAN 2004, Kurashiki, Okayama, Japan, pp. 591–594 (2004)
6. Broekens, J., Heerink, M., Rosendal, H.: Assistive Social Robots in Elderly Care: A Review. Gerontechnology 8(2), 94–103 (2009)
7. Leite, I., Martinho, C., Paiva, A.: Social Robots for Long-Term Interaction: A Survey. International Journal of Social Robots 5(2), 291–308 (2013)
8. Scopelliti, M., Giuliani, M.V., D'Amico, A.M., Fornara, F.: If I Had a Robot at Home: Peoples' Representation of Domestic Robots. In: Keates, S., Clarkson, J., Langdon, P., Robinson, P. (eds.) Designing a More Inclusive World, pp. 257–266. Springer (2004)
9. Syrdal, D.S., Walters, M., Otero, N.R., Koay, K.L., Datenhahn, K.: He knows when you are sleeping - Privacy and the Personal Robot. Technical Report from the AAAI 2007 Workshop: W06 on Human Implications of Human-Robot Interaction, pp. 28–33. AAAI Press (2007)
10. Goetz, J., Keisler, S., Powers, A.: Matching Robot Appearance and Behavior to Tasks to Improve Human-Robot Cooperation. In: Proceedings of the RO-MAN 2003, Berkeley, CA, USA, pp. 55–60 (2003)
11. Yanco, H.A., Drury, J.L., Scholtz, J.: Beyond Usability Evaluation: Analysis of Human-Robot Interaction at a Major Robotics Competition. Human-Computer Interaction 19(1 & 2), 117–149
12. Bethal, C.L., Murphy, R.R.: Affective Expression in Appearance-Constrained Robots. In: Proceedings of HRI 2006, Salt Lake City, Utah, US, pp. 327–328 (2006)
13. Heerink, M., Kröse, B., Evers, V., Wielinga, B.: The Influence of a Robot's Social Abilities on Acceptance by Elderly Users. In: Proceedings of RO-MAN 2006, pp. 521–526 (2006)
14. Walters, M.L., Dautenhahn, K., Woods, S.N., Koay, K.L.: Robotic Etiquette: Results from User Stud-ies Involving a Fetch and Carry Task. In: Proceedings of the HRI 2007, USA, pp. 317–324 (2007)
15. Lehmann, H., Walters, M.L., Dumitriu, A., May, A., Koay, K.L., Saez-Pons, J., Syrdal, D.S., Wood, L., Saunders, J., Burke, N., Duque, I., Christianson, B., Dautenhahn, K.: Artists as HRI Pioneers: A Creative Approach to Developing Novel Interactions for Living with Robots. In: Herrmann, G., Pearson, M., Lenz, A. (eds.) ICSR 2013. LNCS (LNAI), vol. 8239, pp. 391–400. Springer, Heidelberg (2013)
16. Hüttenrauch, H., Green, A., Norman, M., Oestreicher, L., Severinson-Eklundh, K.: Involving Users in the Design of a Mobile Office Robot. IEEE Transactions on Systems, Man and Cybernetics, Part C: Applications and Reviews 34(2), 113–124 (2004)
17. Walters, M.L., Dautenhahn, K., te Boekhorst, R., Koay, K.L.: An Empirical Framework for Human Robot Proximity. In: Proc.of Artificial Intelligence and Simulation of Behaviour Convention (AISB 2009), New Frontiers in HRI Symposium, Edinburgh, Scotland, pp. 144–149 (2009)

18. Koay, K.L., Syrdal, D.S., Walters, M.L., Dautenhahn, K.: Living with Robots: Investigating the Habituation Effect in Participants' Preferences During a Longitudinal Human-Robot Interaction Study. In: RO-MAN 2007, pp. 564–569 (2007)
19. Syrdal, D.S., Dautenhahn, K., Walters, M.L., Koay, K.L.: Sharing Spaces with Robots in a Home Scenario – Anthropomorphic Attributions and their Effect on Proxemic Expectations and Evaluations in a Live HRI Trial. 2008 AAAI Fall Symposium Technical Report FS-08-02, pp. 116–123 (2008)
20. Koay, K.L., Lakatos, G., Syrdal, D.S., Gácsi, M., Bereczky, B., Dautenhahn, K., Miklósi, A., Walters, M.L.: Hey! There is someone at your door. A Hearing Robot using Visual Communication Signals of Hearing Dogs to Communicate Intent. In: Proc. IEEE Int. Symp. on Artificial Life, Singapore, pp. 90–97
21. Bethel, C., Murphy, R.: Survey of Non-facial/Non-verbal Affective Expressions for Appearance-Constrained Robots. IEEE Transactions on Systems, Man, & Cybernetics, C 38(1), 83–92 (2008)
22. Dautenhahn, K.: Socially Intelligent Robots: Dimensions of Human - Robot Interaction. Philosophical Transactions of the Royal Society B: Biological Sciences 362(1480), 679–804 (2007)
23. Walters, M.L., Dautenhahn, K., te Boekhorst, R., Koay, K.L., Syrdal, D.S., Nehaniv, C.L.: An Empirical Framework for Human Robot Proximity. In: Proceedings of the New Frontiers in Human-Robot Interaction, AISB 2009 Convention (2009)
24. Koay, K.L., Sisbot, E.A., Syrdal, D.S., Walters, M.L., Dautenhahn, K., Alami, R.: Exploratory study of a robot approaching a person in the context of handing over an object. In: AAAI 2007 Spring Symposia, pp. 18–24. AAAI Press (2007)
25. Dautenhahn, K., Walters, M.L., Woods, S.N., Koay, K.L., Nehaniv, C.L., Sisbot, E.A., Alami, R., Simeon, T.: How may I serve you? A robot companion approaching a seated person in a helping context (2006)

The Good, The Bad, The Weird:
Audience Evaluation of a "Real" Robot in Relation to Science Fiction and Mass Media

Ulrike Bruckenberger, Astrid Weiss, Nicole Mirnig,
Ewald Strasser, Susanne Stadler, and Manfred Tscheligi

ICT&S Center, University of Salzburg, Austria
firstname.lastname@sbg.ac.at

Abstract. When researchers develop robots based on a user-centered design approach, two important questions might emerge: How does the representation of robots in science fiction and the mass media impact the general attitude naïve users have towards robots and how will it impact the attitude towards the specifically developed robot? Previous research has shown that many expectations of naïve users towards real robots are influenced by media representations. Using three empirical studies (focus group, situated interviews, online survey) as a case in point, this paper offers a reflection on the interrelation of media representations and the robot IURO[1] (Interactive Urban Robot). We argue that when it comes to the evaluation of a robot, "good" and "bad" media representations impact the attitude of the participants in a different way. Our results indicate that the previous experience of fictional robots through the media leads to "weird", double-minded feelings towards real robots. To compensate this, we suggest using the impact of the mass media to actively shape people's attitude towards real robots.

Keywords: fictional robots, perception of robots.

1   Introduction

Meeting service robots in everyday life, still only happens in rare occasions in the Western world, e.g. at science and trade fairs. More often, people come in touch with robots as intelligent, pre-programmed machines in an industrial context or with robotic vacuum cleaners and lawn mowers in a domestic context. People hardly get in touch with anthropomorphized service robots, besides in the mass media and science fiction. Nevertheless, empirical studies showed that people without real life experience with real robots have expectations towards them [1, 7]. Where do these expectations come from? Expectations towards and opinions about robots, especially anthropomorphic ones, are presumably generated through something else than practical experiences with such systems - most likely through the media [1]. For example through science fiction movies, which promote stereotypical "good" and "bad" concepts of human-robot interaction, such as robots as super-heroes that save

---

[1] www.iuro-project.eu

G. Herrmann et al. (Eds.): ICSR 2013, LNAI 8239, pp. 301–310, 2013.

the planet or as evil intelligence that enslaves mankind. These stories leave strong pictures in people's mind. Therefore, it is likely to presume that they affect the participants of user studies with service robots too [2], in a "good", "bad", or "weird" way. Subsequently, the following important question emerges, when we are developing a new service robot based on a user-centered design approach: How does the representation in science fiction and the mass media impact the general attitude naïve users have towards robots and how will it impact the attitude towards the specifically developed robot? Using a focus group, situated interviews, and an online survey as a case in point, this paper offers a reflection on the interrelation of media representations and the IURO robot. Thus, in this paper, we firstly outline related work on the topic "media representations and robots". After laying out each of the three studies in detail, we conclude with an overall discussion focused on the question: "How can we take into account the audience evaluation of "good" and "bad" media representations of robots for the user-centered design of novel service robots?" Our purpose is to provide some insights on how researchers can take advantage of positive media biases and limit negative ones.

2   Related Work

Several researchers described in previous studies that the reception of movies about fictional robots influences the attitude of people towards real robots [1, 2, 3, 4]. The attitude of people towards robots fundamentally depends on their assumptions about robots [7]. As most people never interacted with real robots, their expectations about robot capabilities must come from other sources, e.g. from science fiction movies [1]. The interaction between humans and robots - as presented in many of these movies - utilizes people's emotions about the unknown; therefore it is likely that people keep those stories in mind, which might have an influence on the outcome of empirical studies [2]. For example the fear of participants to loose control over highly developed robots is most likely based on the influence of science fiction movies, which promoted this scenario in various ways [6]. Science fiction movies have often been inspired by real science and in turn inspired scientists to keep up with the presented fictional inventions [3]. Furthermore, those movies shape the idea of human-robot relationships that might occur in the future and therefore support the conception of how future societies including robots might look like [4]. Personal contact with real robots has a positive influence on the attitude of people, e.g. in the context of field studies with service robots [5] or in a private context with a robot pet (e.g. AIBO) [8].

On the basis of the outcomes of these previous studies, we tried to gain deeper insights into the way the representation of robots in the media influences people's attitude towards real robots. Knowing the pictures that are in the mind of our participants might enable Human-Robot Interaction (HRI) to actively shape people's opinion about real robots in a positive way.

3   The Three Studies

The goal of our research was to gain a deeper understanding of how "good" and "bad" representations of robots in science fiction and the mass media impact the evaluation of a real robot by naïve users. The platform we used for our studies was the IURO robot. The goal of the IURO project is to develop a robot that is capable of navigating in densely populated human environments using only information obtained from encountered pedestrians. The robot can serve for fetch and carry tasks in unknown environments (for details on the scenarios see [9]). We started our investigations on the impact of media representations with a qualitative approach by means of a video-based focus group, followed by situated interviews with people who had interacted (or at least observed someone else interacting) with the IURO robot in a field study. Finally, we conducted a video-based online survey to support our qualitative findings with quantitative data and gain additional insights how the IURO robot is affected. The main research question in all three studies was: "How can we take into account the audience evaluation of "good" and "bad" media representations of robots for the user-centered design of novel service robots?"

3.1   Focus Group – Understanding the General Impact

Study Setup. The focus group was conducted with five participants (2 females, 3 males) in August 2012. The average age of the group was 28.4 years (SD = 8.85) ranging from 23 to 44 years. In the beginning, the participants were asked to fill in a questionnaire about their general attitude towards robots and their previous knowledge about fictional robots. Thereafter, the participants were shown four short sequences out of popular robot movies and a short video about the IURO robot interacting with a human. We have chosen movie scenes in which robots are depicted either "good" or "bad", in terms of behavior. The good ones included T-800 (Terminator 2 – Judgment Day) and Wall-E (Wall-E) whereas David (A.I. – Artificial Intelligence) and Ratchet (Robots) counted as bad robots. The sequence about the IURO robot shows it asking a human for the way and interpreting the answer. It uses arm gestures and a pointing device on its head to accompany its utterances. Furthermore, the participants were asked to freely discuss their opinion about real and fictional robots following 13 guiding questions. In the end, the participants were asked to fill in a second questionnaire about their attitude towards the IURO robot.

Results. Except for one, all participants stated that they know robots from science fiction movies. One participant mentioned real life experience with robots in the context of an automobile factory but media coverage about real robots was not mentioned at all. Only two participants thought that the previous experience with fictional robots did influence their attitude towards real robots in any way. The results of the group discussion about robots in general showed that the participants have expectations towards real robots. These expectations cannot be based on reality, although the participants felt that they do so. In their ratings of the robots all of the participants followed the pre-assumed categories of "good" and "bad". The IURO robot was rated rather negative. In comparison to the fictional robots, it could not meet the high expectations of the participants in terms of speech and interaction capabilities. Nevertheless, all of the participants stated that they would wish to see the IURO robot "in person" to interact with it. Altogether, the participants of the focus group were very double-minded in their attitude towards robots in general. On the one hand, robots were called "iron idiots" and, on the other hand, they were expected to be capable of skills and abilities like humans and even more. During the discussion four major topics emerged: design, personality, skills & behavior, and prospects & risks.

In terms of *design,* the participants clearly preferred robots that look machine-like. The functionality of robots was more important than their appearance. *"If it has a function then it should fulfill it the best way possible and I do not think that human-like appearance is feasible for all aims."* It was especially mentioned that robots, which are indistinguishable from humans, might cause problems due to a transfer of emotions towards them. *"I think the more a robot looks like a human, the higher is the risk to fall in love with it".*

In terms of *personality,* all of the participants agreed that robots could not possess an independent personality. They stated that emotions in general are a dangerous thing to have for a machine. *"There has to be a difference between humans and robots. Emotions are that difference in my opinion".* Still, some of the participants were sure that they could develop feelings towards a robot.

Regarding *skills & behavior,* the scope of tasks robots might execute for humans in the future ranges from helping in the household to driving a car, to being a sports partner. *"We could use robots in all classical sports disciplines".* A skill an autonomous robot should never be capable of was *"to program itself".* One essential expectation was: *"There should always be a possibility to turn a robot off".* Furthermore, it was very important for the participants that a robot is able to act politely. *"It should be more polite and say 'please' and 'thank you'".*

Regarding *prospects & risks*: In terms of prospects, all of the participants could imagine that robots assist them in everyday life in the future. Robots as a part of our society in the future were seen as given by all of our participants. The possibility of robots that help old and handicapped people was especially discussed. On the one hand, the participants expected robots to be companions with sophisticated skills and abilities. *"It should be a companion that listens and understands".* On the other hand, one major concern was a possible emotional dependency between humans and robots. *"The imagination that my grandmother develops some kind of feeling towards an iron thing because I do not have time for her is terrible."* In terms of risks, nearly all of the participants were sure that it is impossible that highly developed robots might turn against mankind someday. In addition, the participants stated that they thought about autonomous robots rather positively. Furthermore, it was elaborated that the deployment of robots might lead to a reduction of jobs for humans.

Reflection. The participants are straightforward in their opinion that robots will be part of our everyday life in the future. As neither experience with real robots in an everyday life context nor media coverage about real robots were mentioned, we assume that science fiction movies paved the way for this perception and the human-robot relationships that might occur, as Weiss et al. [4] also stated. Although the participants believed that their previous knowledge about fictional robots did not affect their attitude towards real robots, their fears are clearly based on the "good" and "bad" robot representations shown in science fiction movies. The participants desire the skills and abilities fictional robots represent and they expect them from real robots too. Concerning social skills like behavior and emotions, they do not want real robots to be similar to fictional robots. Basically, the possibility of robots being capable of emotions like humans causes anxiety. Robots that look like humans arouse a strong feeling of fear of being replaced in terms of personal relationships. The IURO robot could not meet the expectations of the participants in terms of skills and abilities in comparison with the fictional robots. Nevertheless, all of the participants were interested in it and had the wish to interact with it. This shows the "weird", double-minded component in the attitude of the participants towards real robots. On the one hand, real robots do not meet the participants' expectations but, on the other hand, people are very interested and want to get in touch with this technology.

3.2 Situated Interviews

Study Setup. We executed situated interviews with 15 passers-by (5 females, 10 males) that interacted or watched an interaction with the IURO robot during a field trial in October 2012. The participants were aged between 21 and 73 years (M=45.5, SD=19.94). We compiled guidelines for the situated interviews on the basis of the four main topics identified in the focus group. Furthermore, we added the topic "safety aspects" because of the mentioned need that there has to be a possibility to turn a robot off anytime.

Fig. 1. The IURO robot in interaction with a pedestrian during a field study

Results. In terms of *design,* the participants rated the IURO robot basically very well except its size (too large) and its missing hands. *"All in all it appears really felicitous".* In general, the participants expect robots to look anthropomorphic, but want to be able to distinguish them from humans at first sight. Nearly all of our participants stated that they do not like to watch science fiction movies. However, nearly all of them mentioned the robots "R2D2" and "C3PO" from the science fiction movie "Star Wars".

In terms of *personality,* the participants were undecided if the IURO robot is capable of having one. But due to its design and the ability to display facial expressions it appeared *"rootedly friendly".* The participants presumed that robots in general would become more and more anthropomorphic. This assumption is based on the belief that robots need to be human-like in part to execute the given tasks. Concerning fictional robots, the most favored character trait was "intelligence".

Regarding *skills & behavior,* all of the participants could imagine that the IURO robot is able to support them in the future. Robots in general should be able to help in the household and can be imagined to assist old and handicapped people. *"My grandfather is blind, I think that a robot could support him".* When talking about the nearer future, the participants were not quite sure if it is possible that real robots can be built as perfect as the fictional robots we know from movies nowadays.

Regarding *prospects & risks,* the participants were sure that the IURO robot could assist them, if it had arms with grippers. They liked the idea of being supported by a robot. *"I think it is cute and robots are great".* Robots should be applied in areas where humans could be injured, would be too cost-intensive, and could be challenged too little. All of the participants agreed that the fictional storyline about highly developed robots that turn against mankind is impossible because humans will be superior to robots for all time. *"Humans cannot be replaced".*

In terms of *safety aspects,* none of our participants had any concerns when meeting the IURO robot on the street. *"Safety? No, I do not have any concerns about that".* Talking about robots in general, the participants had either no safety concerns or they saw possibility to increase the personal safety in terms of being accompanied. None of the participants expected robots to be that highly developed to accompany people in the near future. When it comes to fictional robots the participants did not mention any safety concerns because they are fictional and therefore not able to injure a human in any way.

Reflection. It could be revealed that all participants knew fictional robots, also the participants who do not like to watch science fiction movies. In contrast, no one mentioned to have seen real robots in the media or in an everyday life context before. This shows how ubiquitous the representations of fictional robots are in our society. Again, the participants had "weird", double-minded feelings towards real robots in terms of skills and abilities. The IURO robot could not meet their expectations but all of them were very interested in it and wanted to interact with it. Furthermore, the personal contact with the real robot revealed that the expectations of the participants are not fixed yet. It seems like the participants are willing to accept that robotic technology nowadays is not as perfect as it is shown in science fiction movies. We assume that people understand that it takes time to develop this technology. The missing safety concerns towards real robots and the refusal of the possibility that highly developed robots could turn against mankind someday, might be based on the fact that the participants of this study exclusively referred to "good" robot characters.

3.3   Online Survey

Study Setup. We conducted the online survey to support our qualitative data from the two previous studies with quantitative data. The online survey was conducted in January 2013 with 58 participants (36 females, 22 males), aged between 18 and 47 years (M = 27.7/SD = 6.82). We have sent out the request to participate over various mailing lists, mainly to students from the University of Salzburg. The participants were shown a video recorded at the field trial (see Figure 1) about the IURO robot in interaction with a pedestrian. Afterwards, they were asked about their previous knowledge about fictional robots, their previous knowledge about real robots from the media, their general attitude towards robots, and their attitude towards the IURO robot.

Results. The overall majority of the participants (91.4 %) knew robots from science fiction movies and clearly more than half (60.3 %) of the participants declared that the previous knowledge of fictional robots shaped their opinion about robots in general (science fiction movies: 50 %, cartoon series: 8.6 %, books: 1,7 %). More than half (60.3 %) of the participants knew real robots from media coverage, but only one fourth (24.1 %) of them declared that previous knowledge of real robots from the media shaped their opinion about robots in general (documentations: 22.4 %, newspaper articles: 1.7 %). One fifth (20.7 %) of the participants knew real robots from work, but not all of them (15.5 %) stated that previous experience with real robots shaped their opinion about robots in general (personal experience with real robots: 8.6 %, toy robots: 6.9 %).

We divided the answers to the question "What shaped your opinion about robots the most?" into three different groups: group 1 - fictional robots from the media; group 2 - real robots from the media; group 3 - experience with real robots. Participants from group 1 had only experience with fictional robots through science fiction movies, cartoons, and books. Participants from group 2 had in addition experience with real robots through watching documentations on TV and reading newspaper articles about the actual state of the art in robotics. Participants from group 3 had, in contrast to the other groups, interacted with real robots before, either with robot toys or with real robots in a factory context. These three groups were compared (by one-way ANOVA) regarding their answers to the question "Which kind of feelings do you link with fictional robots?" (positive, neutral, negative). The groups differed only by tendency ($F(2,55)=2.87$, $p=.065$) but the Post-Hoc test (LSD) revealed a significant difference ($p=.039$) between group 1 (M=2.40, SD=0.60) and group 3 (M=2.00, SD=0.00). A look at the means of these groups showed that the participants from group 1 link more positive feelings to fictional robots than the participants from group 3. All of the participants from group 3 have chosen the same option: "neutral feelings". Group 2 (M=2.14, SD=0.36) did not significantly differ from group 1 and group 3. This result points towards the fact that participants who knew real robots have a more distant view towards fictional robots because they can reach back in mind towards a real model of a robot.

From the answers on the question "Which characteristics should a robot possess in your opinion?", we calculated a "positive attitude score" (kindness, reliability) and subtracted a "negative attitude score" (aggressiveness, carelessness). A one-way ANOVA showed a significant ($F(2,55)=3.50$, $p=.037$) difference between the three user groups. The Post-Hoc LSD test further showed a significant difference ($p=.013$) between group 1 (M=1.49, SD=.78) and group 2 (M=.86, SD=.66). These results show that group 1 applies more positive characteristics to real robots than group 2. Group 3 (M=1.11, SD=0.93) did not significantly differ from groups 1 and 2. This outcome shows that participants who consider fictional robots as their model for robots in general, apply more positive characteristics to real robots than participants who consider real robots as their model for robots in general.

We divided the answers from the question "Wherefrom do you know robots?" into two groups: group 1 ("media experience") knew fictional and real robots from the media only; group 2 ("real life experience") knew in addition to fictional and real robots seen in the media real robots from professional life. From the question "Which robot personalities do you prefer in movies?" we calculated the "positive personality score" (helpful & friendly) and subtracted the "negative personality score" (aggressive & frightening). This resulted in a score for "attitude towards fictional robots". On this data we conducted a t-test between the group "media experience" (M=1.46, SD=0.78) and the group "real life experience" (M=0.94, SD=0.97), which showed a significant difference (t (56)=2.17, p=.035). This result reveals that participants who had only experience with robots through the media prefer more positive fictional robot personalities than participants who had experience with real robots.

About half (55.2 %) of the participants wanted robots to look human-like while the other (44.8 %) participants wanted robots to look machine-like. More than half (56.9 %) of the participants thought that it is possible that we will have robots in the future that will be as perfect as those in movies (22.4 % disagreed, 20.7 % were undecided). Nearly half (46.6 %) of the participants believed that the IURO robot could help them in the future. More than half (58.6 %) of the participants did not think that it is possible that highly developed robots turn against mankind someday, whereas 41.4 % considered that possible. When thinking about meeting the IURO robot in public space, the overall majority of the participants (81 %) would not have any safety concerns.

Most of the participants connected neutral emotions to robots (65.5 % fictional robots, 74.1 % IURO robot, 46.6 % real robots in general). This is followed by positive emotions towards robots (31 % fictional robots, 19 % IURO robot, 41.4 % real robots in general). Only a few participants connected negative emotions to robots (3.4 % fictional robots, 6.9 % IURO robot, 6.9 % real robots in general).

Reflection. Science fiction movies are clearly the influence that shapes the attitude of the participants towards real robots the most. In contrast to the two previous studies, most of the participants of this study were aware of that, but the outcomes are the same. The influence of "good" and "bad" representations of fictional robots leads to "weird", double-minded feelings towards real robots. On the one hand, the movie scenario about highly developed robots that turn against mankind someday is considered possible but, on the other hand, nearly no one is afraid of real robots, instead a clear majority connects neutral emotions to them. The IURO robot could again not meet the expectations of the participants in terms of skills and abilities, although half of the participants can imagine that it will help them in the future. It seems like the expectations of the participants towards real robots are not fixed yet, as they do not expect robotics to provide as perfect robots as we know from movies nowadays in the near future. Previous experience with real robots through the media leads to a more distant view towards real robots and seems to decrease the belief that robots are more than highly elaborated technology. Furthermore, it appears to reduce the fear of robots in general because it decreases the influence of previous experience with fictional robots through the media. It is noticeable that only a little amount of the participants who knew real robots from media coverage considered them as their model for robots in general. This might be due to the fact that media coverage about real robots is hard to find and therefore consumed only in a little amount whereas fictional representations of robots are ubiquitous.

4   Conclusion and Recommendations

We have presented findings from three different studies, which aimed at exploring how the representation of robots in science fiction and the mass media may affect the audience evaluation of real robots in general and the IURO robot in specific. At this point, the reader might ask: "How can we take into account the audience evaluation of "good" and "bad" media representations of robots for the user-centered design of novel service robots?" Thus, we recommend fellow researchers who develop service robots and follow a user-centered design approach to consider the following.

The findings of our three studies support the results from other studies [1, 2] in terms of the influence of previous experience with fictional robots through the media on the attitude of participants. Previous experience with fictional robots through the media not only increases expectations towards real robots, but it also seems to lead to the fact that all of the participants see a future society including robots as given. Science fiction established a picture of robots that will be a great advantage for our everyday life in the minds of the participants. So, on the one hand, science fiction made it harder for real robots to be accepted and, on the other hand, science fiction paved the way for a society including robots in the mind of the participants [4].

It seems not to be of importance if fictional robots are represented as "good" or "bad", because both representations lead to the same outcome: Robots behave like humans and are therefore considered as being doubtful, in any case. Either they might replace humans in personal relationships or they might enslave mankind. Both fears can be seen in the attitude of the participants towards real robots, although not everyone is aware of the fact that they are based in science fiction and not in reality. People's opinions are indifferent when it comes to physical appearance, human-like and machine-like appearances are nearly preferred in the same way. The development of human-like appearance is only favored until some point, if robots look alike humans then again pre-existing fears based on fictional representations of robots are aroused.

When the previous experience with fictional robots through the media clashes with the experience of interacting or watching an interaction with a real robot then the "weird", double-minded feelings of the participants become obvious. Nearly no one is afraid of real robots, in general the participants connect neutral emotions to them and believe that this technology will support them in the future. Based on this fact, we assume that the negative attitude and the fear some participants have towards real robots can be decreased through information about real robots through the media and the possibility to interact with real robots. Interesting media coverage about real robots is hard to find but fictional representations of robots are present in the media for every age cohort, e.g. cartoon series for children and action movies for adults. Robotics has to cope with science fiction and to reach the same perfection in real technology will be a long way (and is utopian in some cases). But despite the high expectations people have in mind because of their previous experience with fictional robots through the media they are very interested in real robots and their expectations seem to be not fixed yet: they are willing to give robotics time to develop.

We strongly suggest using the impact of the mass media to present the robotic technology we own today. The media coverage concerning this topic should focus on the benefits that service robots could provide to our society in the near future. Media coverage about real robots might lead to a discussion about technical limitations robotics has to face today. This may eventually arouse new fears among people with no personal experience with robots which, however, can be seen as a chance to generate realistic expectations towards real robots in the public and raise the interest of researchers from other disciplines to be part of the solution. HRI research should take the chance to shape the expectations of people towards service robots in a way that the actual state of the art in robotics can cope with. Robotics must not fear the challenge with science fiction, but it has to leave the labs and must go out and present itself to the broad public [5].

Acknowledgments. This work is supported within the European Commission as part of the IURO project.

References

1. Kriz, S., Ferro, T.D., Damera, P., Porter, J.R.: Fictional robots as a data source in HRI research: Exploring the link between science fiction and interactional expectations. In: 2010 IEEE RO-MAN, pp. 458–463 (2010)
2. Bartneck, C.: From Fiction to Science – A cultural reflection of social robots. In: Proceedings of the CHI 2004 Workshop on Shaping Human-Robot Interaction, Vienna (2004)
3. Lorencik, D., Tarhanicova, M., Sincak, P.: Influence of Sci-Fi films on artificial intelligence and vice-versa. In: 2013 IEEE Applied Machine Intelligence and Informatics (SAMI), pp. 27–31 (2013)
4. Weiss, A., Igelsböck, J., Wurhofer, D., Tscheligi, M.: Looking forward to a "Robotic Society"? Notions of Future Human-Robot Relationships. International Journal of Social Robotics 3, 111–123 (2011)
5. Mirnig, N., Strasser, E., Weiss, A., Tscheligi, M.: Studies in Public Places as a Means to Positively Influence People's Attitude towards Robots. In: Ge, S.S., Khatib, O., Cabibihan, J.-J., Simmons, R., Williams, M.-A. (eds.) ICSR 2012. LNCS (LNAI), vol. 7621, pp. 209–218. Springer, Heidelberg (2012)
6. Ray, C., Mondada, F., Siegwart, R.: What do people expect from robots? In: IROS 2008, pp. 3816–3821 (2008)
7. Nomura, T., Kanda, T., Suzuki, T., Kato, K.: Prediction of Human Behavior in Human-Robot Interaction Using Psychological Scales for Anxiety and Negative Attitudes Toward Robots. IEEE Transactions on Robotics 24(2), 442–451 (2008)
8. Bartneck, C., Suzuki, T., Kanda, T., Nomura, T.: The influence of people's culture and prior experiences with Aibo on their attitude towards robots, pp. 217–230. Springer Verlag London Limited (2006)
9. Złotowski, J., Weiss, A., Tscheligi, M.: Interaction Scenarios for HRI in Public Space. In: Mutlu, B., Bartneck, C., Ham, J., Evers, V., Kanda, T. (eds.) ICSR 2011. LNCS, vol. 7072, pp. 1–10. Springer, Heidelberg (2011)

Systems Overview of Ono

A DIY Reproducible Open Source Social Robot

Cesar Vandevelde[1], Jelle Saldien[1], Maria-Cristina Ciocci[1], and Bram Vanderborght[2]

[1] Ghent University, Dept. of Industrial Systems and Product Design, Campus Kortrijk, Belgium
{cesar.vandevelde,jelle.saldien,maria-cristina.ciocci}@ugent.be
[2] Vrije Universiteit Brussel, Dept. of Mechanical Engineering, Brussels, Belgium
bram.vanderborght@vub.ac.be

Abstract. One of the major obstacles in the study of HRI (human-robot interaction) with social robots is the lack of multiple identical robots that allow testing with large user groups. Often, the price of these robots prohibits using more than a handful. A lot of the commercial robots do not possess all the necessary features to perform specific HRI experiments and due to the closed nature of the platform, large modifications are nearly impossible. While open source social robots do exist, they often use high-end components and expensive manufacturing techniques, making them unsuitable for easy reproduction. To address this problem, a new social robotics platform, named Ono, was developed. The design is based on the DIY mindset of the maker movement, using off-the-shelf components and more accessible rapid prototyping and manufacturing techniques. The modular structure of the robot makes it easy to adapt to the needs of the experiment and by embracing the open source mentality, the robot can be easily reproduced or further developed by a community of users. The low cost, open nature and DIY friendliness of the robot make it an ideal candidate for HRI studies that require a large user group.

Keywords: Do-It-Yourself, emotions, facial expressions, human-robot interaction, maker movement, open hardware, open source, rapid prototyping, robotic user interface, social robot.

1 Introduction

To study human-robot interaction (HRI), one needs appropriate systems with social capabilities. Multiple studies and applications in HRI make use of facial expressions since people rely on face-to-face communication in daily life. The face plays a very important role in the expression of character, emotion and/or identity [1]. Mehrabian [2] showed that only 7% of affective information is transferred by spoken language, that 38% is transferred by paralanguage and 55% of transfer is due to facial expressions. Facial expressions are therefore a major modality in human face-to-face communication, especially in fields such as robot-assisted therapy (RAT), where emotions play a crucial role in the communication process.

G. Herrmann et al. (Eds.): ICSR 2013, LNAI 8239, pp. 311–320, 2013.
© Springer International Publishing Switzerland 2013

A major problem in HRI studies is that many HRI platforms are very expensive: Kobian [3], HRP-4C [4], Waseda Emotion Expression Humanoid Robot WE-4RII [5], iCub [6], Kismet [7], Probo [8], … are very high performing social robots, but with a high degree of complexity, hence the high cost. For many studies such performance is not always required and the ability to have many robots for a broader scale of experiments can have benefits. Cheap social robots that are potentially usable for HRI studies are platforms like My Keepon (a toy based on the more expensive Keepon platform) [9], KASPAR [10], or Furby (Tiger Electronics), but their hardware and software is not open-source, giving little possibilities to adapt the platform to the specific needs of the research. Therefore a novel design of a social robot named Ono is presented (figure 1), with the aim to obtain a DIY reproducible open source social robot. Ono is built with the following requirements in mind:

1. Open source hardware and software
2. Do-It-Yourself
3. Modular
4. Reproducible
5. Social expressiveness

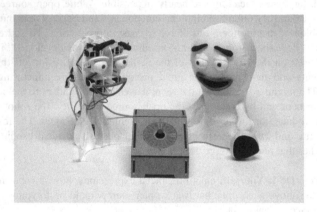

Fig. 1. The Ono prototype with control unit

1.1   Open Source Hardware and Software

One well-known open source platform is the European humanoid robot iCub [6]. Other projects are the Dora Opensource Robot Assistant [11] and the successful e-puck educational robot [12]. The open source community has made new technologies accessible not only to professionals, but also to many hobbyists. Examples include 3D printers such as Fab@Home [13] and RepRap [14] and microcontroller platforms such as Arduino [15]. A lot of initiatives focus on the development and distribution of open source software for robot systems, popular systems are ROS [16], the Player/Stage project [17], the Orocus project [18], and the Urbi project [19]. The aim for Ono is to distribute both the open hardware and the open software. The source files of the robot can be found in a public Github repository [20], however because the design is still being worked on, no final assembly instructions have been made available yet .

By promoting the free redistribution and access to hardware and software, other researchers have the opportunity to easily extend the capabilities of the robots and make the platform suitable for their particular applications.

1.2   Do-It-Yourself

In contrast with iCub [6], which is complex and expensive to build, our goal is that Ono can be built without the aid of paid experts or professionals. The goal is that the robot can be built using easy to understand instructions, in an Ikea-wise manner. The interest and wider adoption of DIY is facilitated by (1) easy access to and affordability of tools and (2) the emergence of new sharing mechanisms, which plays a major role in motivating and sustaining communities of builders, crafters and makers [21].

1.3   Modular

By dividing the robot into small, independent modules, each containing a set of related sensors and/or actuators, newer versions can be developed and distributed easily. If a module is damaged or broken, it can be replaced without needing to disassemble the entire robot. Also, the development of independent modules makes it easy to reuse these modules within other social robots. This means that different types and forms of social robots can be developed more quickly and that the improvements made to modules by other research groups can be reincorporated into the Ono project. Another advantage is that a degree of customization will be possible. Not all research requires the same degree of complexity of the social robot used; some applications may only require a smaller subset of modules while others may need a camera and additional sensors. By allowing this customization, the Ono social robot can become accessible to a larger group of users.

1.4   Reproducible

Another important factor of this new platform is the ease of making or reproduction. Developing this new prototype with low volume manufacturing techniques in mind has several consequences.

The advantage over traditional methods is that more and more rapid prototyping machines, such as laser cutters and 3D plastic printers, are available in different labs or at low cost over the internet. Low-cost 3-D printers could very well be the next big trend in home robots [22]. These machines can be operated by non-skilled users, which in contrast with e.g. CNC milling machines, where specialized technicians are required.

1.5   Social Expressiveness

Bartneck and Forlizzi propose the following definition of a social robot [23]: *"A social robot is an autonomous or semi-autonomous robot that interacts and communicates with humans by following the behavioral norms expected by the people with whom the robot is intended to interact."* Communication and interaction with humans is a critical point in this definition. This definition also implies that a social robot requires a physical embodiment and a social interface. For Ono we concentrated on the facial expressions and gaze as social interface. For the display of the emotions most of the DOFs in the face are based on the Action Units (AU) defined by the Facial Action Coding System (FACS) developed by Ekman and Friesen [24]. AU express a motion of mimic muscles as 44 kinds of basic operation, with 14 AU to express the emotions of anger, disgust, fear, joy, sadness, and surprise, which are often supported as being the 6 basic emotions from evolutionary, developmental, and cross-cultural studies [25]. Several other robots rely on this principle, such as Kismet [7], Probo [26] and EDDIE [27].

2   Structural Makeup

The Ono robot consists of 5 main groups:

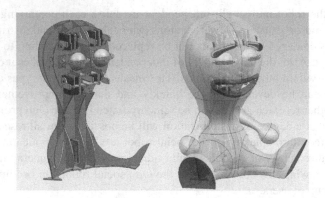

Fig. 2. Skeletal structure and modules

2.1   Skeletal Frame

A 3D model of the intended appearance of the robot was used as a starting point. After accounting for the thickness of the foam covering, this model was sliced into multiple cross-sections. These sections were then used as the basis for the skeletal structure of the robot. The sections of the frame are connected to each other using interlocking slots and tabs. The frame also has 5 openings in order to accommodate they eye-, eyebrow- and mouth-modules (figure 2). Finally, mounting holes and slots were made so that the electronics and cables could be connected to the frame. We chose a sitting pose for Ono to improve the robot's stability. The robot head is made disproportionally large in order to make facial expressions easier to see and to make it easier to integrate the modules into the head.

2.2   Modules

Sets of related sensors and/or actuators are grouped into modules. The current prototype has 3 distinct types of modules: an eye module, an eyebrow module and a mouth module. Each module uses a set of cantilever snaps to connect the module to the frame. In addition, cross-shaped mounting holes allow the modules to be connected to Lego Technic bricks, so that new robots can be built and tested rapidly. For the current prototype each eye module was given 3 DOFs, each eyebrow was given 2 and the mouth was given 3. These DOFs allow Ono to express the 6 basic emotions [25] and additionally move the eyes both horizontally and vertically, which allows the robot to gaze.

2.3   Foam and Skin Covering

The skeletal frame is wrapped in a protective cover made from polyurethane foam. This gives the robot a soft exterior for interaction with children and protects the inner components from potential damage. The foam covering is made from flat pieces of laser-cut foam. This technique requires that the shape of the robot is smooth and continuous, so that the body can easily be split in two-dimensional patterns. The soft foam covering is in turn covered by a sewn lycra suit. This covers up the inner components and provides a visually pleasant appearance. The color yellow was chosen because of the association between yellow and positive emotions, as described by N. Kaya et al. [28]. We chose to give Ono an exaggerated, cartoon appearance, so as to avoid the effects of the uncanny valley [29].

2.4   Electronics and Interface

The main power supply, logic processing and interface are contained within a separate unit. This unit provides power and movement instructions through a cable to a servo controller inside the robot, which in turn powers and controls the servos individually. A joystick interface allows the operator to select the correct emotion and intensity from Russell's circumplex model of affect [30] or alternatively, the robot can be controlled through a USB interface.

3   Production Techniques

The construction of the robot relies heavily upon the use of laser cutting. The main advantages of laser cutting are that (1) it is fast, (2) the files can be edited easily, (3) it is well suited for larger components and (4) the machine is easy to operate.

Both the hard, mechanical parts and the soft, protective foam covering are cut using a laser cutter. The structural parts are made from 3 mm thick ABS plastic (figure 3). This material was chosen because it is a readily available, low cost material and because it is flexible, making the robot more resistant to damage and allowing design features such as snap connectors. To give the robot a soft, huggable appearance, a protective foam cover is placed over the skeletal structure. In early prototypes, this foam cover was made by casting a flexible PU resin in a mold. The main disadvantage of this approach is that a mold is required, which makes producing the parts

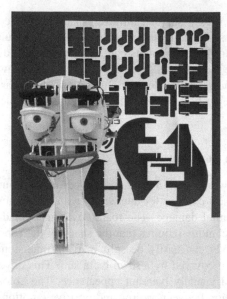

Fig. 3. Prototype with one of 2 laser-cut sheets in the background

from digital files much more labor intensive. We solved this problem by recreating the 3D foam shell using 2D laser cut pieces of foam. The cost of materials to build one robot is around €310. This cost may vary depending on location and on what components the user already owns. Table 1 provides a rough breakdown of the costs.

Table 1. Cost Overview

3mm polystyrene sheets	€20
20mm polyurethane foam	€5
Arduino Uno microcontroller	€25
SSC-32 servo controller	€40
PC power supply	€25
RC servos	€80
Nuts, bolts, cable ties	€10
Connectors and electric components	€30
Textile supplies	€30
Laser cutting cost	€45
Total	**€310**

4   User Tests

An important aspect during the development of Ono was that building the robot has to be as easy as possible so that even users with little technical skill can complete the

Fig. 4. Child interacting with Ono during user tests

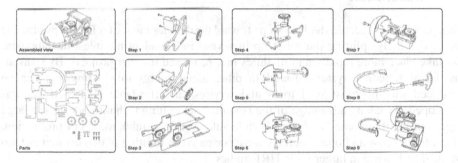

Fig. 5. Assembly instructions for an eye module

project successfully. A small-scale user test with 5 users (aged 18-55) was done to identify possible roadblocks during the assembly. The tests showed that the visual instructions work well to guide inexperienced users through the assembly process. Users noted that while the drawing of the finished module looks very complex, the actual assembly process is greatly simplified by the step-by-step instructions. One user compared the module kit to a set of Lego bricks and that completing the assembly of the eye module gave him a sense of satisfaction. These tests also revealed a number of pitfalls for both the instructions and the parts themselves. The instructions should include more color to identify new or dissimilar components. Each component should also be labeled with an identifying number and required tools – if applicable – should be shown in each step. Problems with the laser-cut components include snap cantilevers being too rigid or soft, distinct parts being too similar to one another and parts being used in mirror position. Another problem was that nut & bolt connections were hard to make due to the inability to properly grip the nut while fastening the bolt. Changes to solve these problems were then reincorporated into the newer versions of those modules: parts were reduced and simplified, bolts are screwed directly into the ABS sheet and the snap cantilevers were readjusted.

Additionally, a pilot study was performed with 5 autistic children (aged 3-10), without the presence of the engineers and without explicit training of the operator, to test the overall interaction of the children with the robot and to test the recognition of emotions. Children were asked to identify the emotion expressed by the robot, they were then asked to mimic the robot's facial expressions and were finally allowed some time to freely interact with Ono. These tests show that Ono has an overall inviting appearance that elicits interaction, but that there are still several issues that need to be solved. During testing happiness and sadness were recognized correctly 15 times out of 16, while anger was only recognized only 3 times out of 16 and surprise only 6 times out of 16. The facial expressions for these emotions need to be adjusted. Additionally, the control box interface proved to be suboptimal: the joystick distracts the children and the (short) cable means that the whole setup can be unwieldy at times. Another required feature is the addition of idle animations, to make the robot seem more lifelike when the operator is not actively controlling it.

5   Conclusion

This paper has presented the first steps toward the development of a DIY reproducible social robot. By splitting the robot up in several independent modules, development can take place more rapidly and modules may be reused in other projects. By translating facial actuation systems found in other social robots to systems that can be produced with hobbyist-level tools and services, an affordable HRI platform was developed. By taking advantage of the benefits laser cutting technology offers, a large degree of flexibility can be obtained while still offering a quick means of production. The ultralow cost and open source nature compared to existing platforms makes Ono an ideal tool to use in larger scale HRI studies.

The next steps include the further development of the robot's electronics. Tests have shown that the current control box setup needs improvement. One option is to integrate all the electronics within the body of Ono and to make the control of the robot wireless. Newer, more advanced modules will also be developed to accommodate a wider range of possible applications. The eye module in particular needs to be improved further; the current one is difficult to assemble and is not robust enough. Current modules can be connected to Lego bricks using cross-shaped axle holes, however this interfacing system should be extended to electronic Lego bricks. This should allow fast prototyping of social robots using Lego Mindstorms. For example, the eye modules of Ono can be connected to the Tribot robot of Mindstorms NXT in order to create a rudimental social robot. With further development these modules may complement Lego Mindstorms, providing an inexpensive and easy to use platform for HRI studies. This would solve the lack of sufficient HRI capabilities of the Mindstorms platform, as noted by Murphy et al. [31].

Acknowledgment. We would like to thank Carl Grimmelprez for his guidance and support during the project. We would also like to thank Protolab and Bart Grimonprez in particular for his help with the rapid production techniques. Part of this work has been funded by the Industrial Design Center and the Dehousse scholarship of Ghent University.

References

1. Cole, J.: About Face. MIT Pr. (1998)
2. Mehrabian, A.: Communication without words. Psychology Today 2, 53–56 (1968)
3. Zecca, M., Endo, N., Momoki, S., Itoh, K., Takanishi, A.: Design of the humanoid robot KOBIAN-preliminary analysis of facial and whole body emotion expression capabilities. In: IEEE-RAS International Conference on Humanoid Robots (Humanoids 2008), pp. 487–492 (2008)
4. Kaneko, K., Kanehiro, F., Morisawa, M., Miura, K., Nakaoka, S., Kajita, S.: Cybernetic Human Hrp-4c. In: IEEE-RAS International Conference on Humanoid Robots (Humanoids 2009), pp. 7–14 (2009)
5. Miwa, H., Itoh, K., Matsumoto, M., Zecca, M., Takanobu, H., Roccella, S., Carrozza, M.C., Dario, P., Takanishi, A.: Effective emotional expressions with emotion expression humanoid robot WE-4RII. In: IEEE/RSJ International Conference on Intelligent RObots and Systems (IROS 2004), pp. 2203–2208 (2004)
6. Tsagarakis, N.G., Metta, G., Sandini, G., Vernon, D., Beira, R., Becchi, F., Righetti, L., Santos-Victor, J., Ijspeert, A.J., Carrozza, M.C., et al.: iCub: the design and realization of an open humanoid platform for cognitive and neuroscience research. Advanced Robotics 21, 1151–1175 (2007)
7. Breazeal, C.: Toward sociable robots. Robotics and Autonomous Systems 42, 167–175 (2003)
8. Goris, K., Saldien, J., Vanderborght, B., Lefeber, D.: Mechanical design of the huggable robot Probo. International Journal of Humanoid Robotics 8, 481 (2011)
9. Kozima, H., Michalowski, M.P., Nakagawa, C.: Keepon. International Journal of Social Robotics 1, 3–18 (2009)
10. Dautenhahn, K., Nehaniv, C.L., Walters, M.L., Robins, B., Kose-Bagci, H., Mirza, N.A., Blow, M.: KASPAR–a minimally expressive humanoid robot for human–robot interaction research. Applied Bionics and Biomechanics 6, 369–397 (2009)
11. http://www.dorabot.com/
12. Mondada, F., Bonani, M., Raemy, X., Pugh, J., Cianci, C., Klaptocz, A., Magnenat, S., Zufferey, J.C., Floreano, D., Martinoli, A.: The e-puck, a robot designed for education in engineering, pp. 59–65 (Year)
13. Malone, E., Lipson, H.: Fab@ Home: the personal desktop fabricator kit. Rapid Prototyping Journal 13, 245–255 (2007)
14. Jones, R., Haufe, P., Sells, E., Iravani, P., Olliver, V., Palmer, C., Bowyer, A.: RepRap–the replicating rapid prototyper. Robotica 29, 177–191 (2011)
15. Mellis, D., Banzi, M., Cuartielles, D., Igoe, T.: Arduino: An open electronic prototyping platform. In: Proc. CHI, vol. 2007 (2007)
16. Quigley, M., Conley, K., Gerkey, B., Faust, J., Foote, T., Leibs, J., Wheeler, R., Ng, A.Y.: ROS: an open-source Robot Operating System (Year)
17. Gerkey, B., Vaughan, R.T., Howard, A.: The player/stage project: Tools for multi-robot and distributed sensor systems, Portugal, pp. 317–323 (Year)

18. Bruyninckx, H.: Open robot control software: the OROCOS project. In: IEEE International Conference on Robotics and Automation, ICRA 2001 (2001)
19. Baillie, J.C.: URBI: towards a universal robotic low-level programming language. In: IEEE/RSJ International Conference on Intelligent Robots and Systems (IROS 2005), pp. 820–825 (2005)
20. https://github.com/cesarvandevelde/Ono
21. Kuznetsov, S., Paulos, E.: Rise of the expert amateur: DIY projects, communities, and cultures, pp. 295–304. ACM (Year)
22. Guizzo, E., Deyle, T.: Robotics Trends for 2012. IEEE Robotics & Automation Magazine (2012)
23. Bartneck, C., Forlizzi, J.: A design-centred framework for social human-robot interaction, pp. 591–594. IEEE (Year)
24. Ekman, P., Friesen, W.V., Hager, J.C.: Facial action coding system 160 (1978)
25. Ekman, P.: Are there basic emotions? Psychological Review 99, 550–553 (1992)
26. Saldien, J., Goris, K., Vanderborght, B., Vanderfaeilli, J., Lefeber, D.: Expressing Emotions with the Huggable Robot Probo. International Journal of Social Robotics, Special Issue on Social Acceptance in HRI 2, 377–389 (2010)
27. Kuhnlenz, K., Sosnowski, S., Buss, M.: Impact of Animal-Like Features on Emotion Expression of Robot Head EDDIE. Advanced Robotics 24(8), 1239–1255 (2010)
28. Kaya, N., Epps, H.H.: Relationship between color and emotion: a study of college students. College Student Journal 38, 396–405 (2004)
29. Mori, M.: The uncanny valley. Energy 7, 33–35 (1970)
30. Russell, J.A.: A circumplex model of affect. Journal of Personality and Social Psychology 39, 1161 (1980)
31. Murphy, R., Nomura, T., Billard, A., Burke, J.: Human–Robot Interaction. IEEE Robotics & Automation Magazine 17, 85–89 (2010)

Sharing Spaces, Sharing Lives – The Impact of Robot Mobility on User Perception of a Home Companion Robot

Dag Sverre Syrdal, Kerstin Dautenhahn, Kheng Lee Koay,
Michael L. Walters, and Wan Ching Ho

Adaptive Systems Research Group, School of Computer Science,
University of Hertfordshire, Hatfield, UK
{d.s.syrdal,k.dautenhahn,k.l.koay;m.l.walters}@herts.ac.uk,
wanchingho@gmail.com

Abstract. This paper examines the role of spatial behaviours in building human-robot relationships. A group of 8 participants, involved in a long-term HRI study, interacted with an artificial agent using different embodiments over a period of one and a half months. The robot embodiments had similar interactional and expressive capabilities, but only one embodiment was capable of moving. Participants reported feeling closer to the robot embodiment capable of physical movement and rated it as more likable. Results suggest that while expressive and communicative abilities may be important in terms of building affinity and rapport with human interactants, the importance of physical interactions when negotiating shared physical space in real time should not be underestimated.

1   Motivation

The study presented in this paper is part of the on-going work being done in the University of Hertfordshire (UH) Robot House to perform early prototyping of long-term Human-Robot Interaction (HRI) in domestic environments. There is a growing interest in the use of robots as *assistive companions* in domestic environments [1], however, there are many challenges that need to be overcome in order for robots to be not only useful in these settings, but also acceptable to their users. While all adoption of novel technologies in domestic settings may be disruptive, autonomous robots may be especially so, in particular if they are mobile [2]. The ability to move autonomously makes them qualitatively different from other household appliances. This ability allows for a wider range in functionalities, such as the ability to assist their use when moving [3], and transporting objects [4,5]. It may also confer advantages when obtaining information from its environment when compared to static technologies, which may make them more suitable for safe-guarding the health of their users [6,2].

However, having technological artefacts that move in a shared space with human residents is not without its problems. Hüttenrauch and Severinson Eklundh's [4] study highlighted instances where negotiation of space caused annoyance and discomfort in

G. Herrmann et al. (Eds.): ICSR 2013, LNAI 8239, pp. 321–330, 2013.
© Springer International Publishing Switzerland 2013

a professional environment, and movement in a human-centred environment have inherent safety concerns [7]. In addition, having to share spaces with robots in domestic environments may cause discomfort if the robot does not conform to social norms for proxemics behaviour [8]. This may be of particular concern if the robot is perceived to be a social actor [9].

1.1   Socially Assistive Still Robots

There are several robots with functionalities that do not rely on the ability to navigate around the environment of its user. In terms of market-ready products, the zoomorphic PARO has been shown have an impact on the emotional well-being of its users. Studies using the similarly zoomorphic Philips iCat [10] found that it could be used to engage elderly people in a care home for in conversation. The Autom weight loss coach produced by Intuitive Automata offers advice and attempts to persuade its users in making decisions that are conducive to a healthier life style [11].

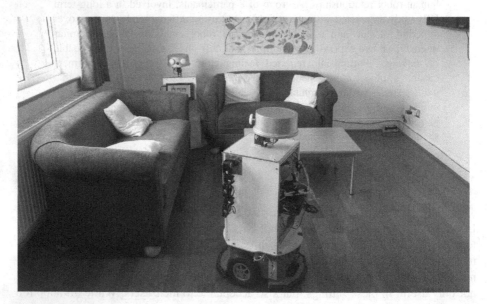

Fig. 1. Robot House Living Room Area. It shows one mobile Sunflower robot in the front, and a stationary Sunflower robot in the back.

1.2   The Importance of Being Mobile

There are however, important benefits for a social robot in being able to move in a shared space with human interactants. If one was to consider human-robot relationships in terms of their anthropomorphic counterparts, one could most certainly make this case. In human-human interactions, proxemic behavior and interpersonal spacing is a highly communicative act [12]. Kendon [13] gives several examples of how humans manage and signal the quality and nature of their interactions through continuous maintenance of appropriate spatial behaviour. Even when we engage in it in an unconscious manner, we continuously validate and define the nature of our relationship with those we interact with. Hall [14] and Mehrabian [15] both offer evidence of proxemic behaviour as indicative of the interactants' relationship, mutual attitude and relative status to each other. In fact, Burgoon and Walther [16] suggest that proxemics behavior can dramatically alter the nature of our relationships, and that changes in how we feel or reason about the people we interact with depend on responses to such changes in proxemic behaviour to take effect.

With such richness in human-human interaction being dependent on this spatial interactional dimension, it seems that even for robots that may not strictly be required to navigate autonomously in order to perform their functions, this ability may be of benefit from a purely interactional point of view.

Stienstra and Marti [17], when considering the possibility of emergent human-machine empathy, highlight the importance of going beyond the explicitly symbolic interactions that are most commonly associated with human-machine interactions and instead focus on interaction modalities that allow for a rich and continuous loop of mutual action and perception. Instead of sharing one's intentionality through constrained voice commands or navigating menus, moving within the same space as the robot is a continuous interaction that allows for synchronization between human and robot interactant in an authentic manner, based on actual behavioral affordances for both parties, through a series of small-scale epistemic actions [18] that occur in addition to the large-scale tasks. In this perspective, the ability to negotiate shared spaces allows for experiences that are seen as shared, both in terms of perception as well as behaviour, which in turn allows for greater feelings of mutual understanding and empathy in the human interactant.

The above argument is supported by HRI research regarding the role of physical embodiment. Wainer et al. [19] found that both task performance and social perceptions of a robot benefited from interacting in shared physical space. Kose-Bagci et al. [20] found a similar effect using a humanoid robot in a synchronization task. In socially assistive robotics, Tapus et al. [21], provided some evidence that changes in proxemic behaviour was an effective way to allow for personalization in stroke rehabilitation therapy.

This suggests that an examination into the role of the ability to move and share the physical space with the user in the formation of human-robot relationship is a valid avenue of investigation, in particular when comparing to robotic embodiments that have the same physical expressive capabilities, with the exception of gross physical movement, which is only possessed by one robot embodiment.

2   Method

2.1   Participants

Eight participants were recruited via advertisements on the University Intranet, 3 males and 5 females. The mean age was 25, the oldest participant was 32 and the youngest was 21.

2.2   Apparatus

UH Robot House. The UH Robot house was built as a residential house in a neighbourhood near the University of Hertfordshire campus, which has been adapted for Human-Robot Interaction studies, including low-cost, resource-efficient sensor systems to inform robots about user-activities and other events in the environment [22]. In the course of the current study, participants would spend time in the Kitchen, Dining, and Living Room areas of the house.

Sunflower Robots. This study used two UH Sunflower robots. The UH Sunflower is built on top of a commercially available Pioneer P3-DX mobile base. Its custom-built superstructure includes as an expressive head with a static face, a speaker capable of playing midi tunes and a diffuse color LED display panel. In addition it has a slide-out carrying tray, which can be used for transporting objects, and an integrated touch screen user interface for menu-driven interaction displaying messages. Apart from gross body movements such moving back and forwards and moving its base from side to side, the UH Sunflower has several expressive channels for expressing its internal states utilizing head motions, sound tunes, and color LED display panel. These expressive cues have been designed based on inspirations from dog-owner interactions [23]. In this study, both Sunflower robots used all the expressive modalities, however, the stationary Sunflower could not use gross body movements. **Fig. 1** shows the Sunflowers within the robot house setting.

2.3   Long-Term Study

The results from this study were obtained as part of an ongoing long-term HRI study in the UH Robot house involving complex human-robot interaction scenarios. Throughout this study, a large amount of data was gathered and a full description and analysis of the general results from this study is currently being prepared, but for the purposes of this paper, a brief introduction follows:

The long-term study in the robot house aimed to convey the experience of long-term human robot interaction, by exposing participants to the robots in a series of episodic interactions. The user played the part of someone living in the robot house, with an Artificial Agent who inhabited different robot embodiments, one at a time. The agent's 'mind' could migrate between these different embodiments, see [24,25] for details. Each episode was framed a part of a specific day, with the user looking forward to performing some specific activities, within which the agent would assist by reminding the user of activities and previously inputted preferences; alerting the user to events in the environments like the doorbell going off, kettles and toasters being finished; as well as function as a platform for communicating via Skype. In addition, the mobile robot was available for assisting in transporting objects.

Participants interacted with the agent in its robot embodiments in 9 sessions, two sessions a week, and filled in the questionnaire at the beginning of the 10th, debriefing session. Participants interacted with the stationary embodiment in all sessions. For the mobile embodiment, however, there were 2 sessions in which they did not interact with the mobile embodiment at all.

2.4   Design and Experimental Control

This study used a within-groups design, all participants interacted with both robot embodiments throughout the long-term study, using them for a variety of tasks. Both robots were capable of performing most socially assistive tasks, such as reminders and providing information but only the stationary robot could be used for communicating with another person via Skype, while the mobile robot could follow and guide the participant when walking around the robot house. Participants would have interactions through touch-screens with both robots for approximately the same amount of time

2.5   Measures

There were three means of data-capture. The first was the pictorial Inclusion of Other in the Self scale (IOS) [26]. This validated scale has been shown to correlate with feelings of closeness in human-human relationships and has also previously been used in HRI studies [27,28]. The second was a 5-point Differential Scale asking how close a participant felt two contrasting robot embodiments. These items were intended to explicitly make participant contrast the embodiments while considering their impressions of them.

Also used was the GODSPEED questionnaire[29], which was chosen as it is a robot specific scale, which addresses issues directly related to both evaluations of robotic embodiments as well as subjective impressions of robots. The final means of data capture was the use of open-ended questions in order to examine the specifics of participants' impressions of the robot. The relatively small sample-size made this a highly relevant approach.

2.6   Hypotheses

H1: Participants will rate their relationship with the mobile robot as closer on the IOS scale compared to the stationary robot.
$H1_0$: Participants will not distinguish between the two robot embodiments on the IOS Scale.
H2: Participants will report feeling close to the mobile robot on the Differential Scale item compared to the stationary Robot
$H2_0$: Participants will not distinguish between the two robots on the Differential Scale item.

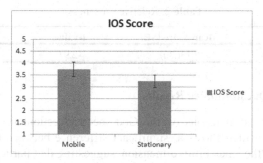

Fig. 2. IOS Scores

3   Results

3.1   Inclusion of Self in the Other Scale

In the IOS scale, a score of 1 is indicative of no closeness, while a score of 6 is indicative of a high degree of closeness. While not significant for this number of participants, there was a trend in which participants rated their relationship with the mobile robot as closer than that with the stationary robot. This trend is shown in Fig. 2 and Table 1, and had an effect size of d= .49 observed power of .33, with a critical N of 49. One should also note that only 1 of the 8 participants rated the stationary robot closer than the mobile robot on the IOS Scale. While encouraging, however, this result did not allow us to fully reject null hypothesis 1.

Table 1. IOS Scores

Robot	Mean IOS (SE)	Mean difference	t-value (7 DF)	P
Mobile	3.75(.61)			
		.50(.33)	1.27	.17
Stationary	3.25(.56)			

3.2   Relative Closeness

Participant responses to the Semantic Differential Item contrasting the two robot embodiments in terms of relative closeness was assessed using a one-sample t-test, testing for significant deviation from the middle value (3) of the contrast. A low score on this item was associated with feeling closer to the mobile robot and a high score with feeling closer to the stationary robot. The result is presented in Table 2 and suggests that the scores not only deviated significantly from the neutral value, but that the direction of this deviation was caused by participants responding that they felt closer to the mobile robot, allowing us to reject null hypothesis 2.

Table 2. Relative Closeness

	Mean Score (SE)	t-value	P
Relative Closeness	2.13(.30)	-2.97	.02

3.3   Godspeed Questionnaire Results

The results from the dimensions on the God speed Questionnaire can be found in Table 3 and Fig. 3, which shows a significant difference between the stationary and mobile robot along the Likeability dimension, suggesting that participants viewed the mobile robot as 'nicer' and more 'sympathetic' than the stationary robot. In addition, there was a trend in which participants tended to rate the mobile robot higher in terms of animacy. There were no differences between the robots in terms of anthropomorphism, intelligence and perceived safety. This allows us to reject null hypothesis 3.

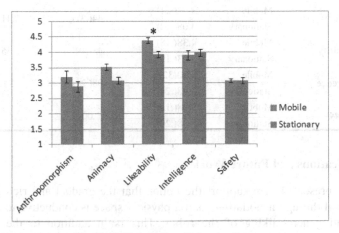

Fig. 3. Godspeed Questionnaire Scores

4   Discussion

4.1   Summary of Results

The results are, overall encouraging, in terms of the research hypotheses. While not significant, there was a salient trend suggesting that participants rated their relationship with the mobile robot as closer than that with the stationary robot on the Inclusion of Self in Other scale. This trend, combined with the significant result for the Relative Closeness semantic differential item, as well as the significant difference between robots along the Likability scale on the Godspeed Questionnaire, suggests that interactions that involve moving in, and negotiating shared physical space, and the investment that understanding and learning about the robot's spatial behaviour involves, do play a role in the experienced building of a relationship between a domestic robot and a human user, even when comparing the results from two robots that have very similar interactional capabilities. It also seems that participants scored the two robots very closely in terms of intelligence and safety, suggesting that it was only in terms of relational measures that they viewed the mobile robot more favourably.

This was echoed in terms of the open-ended comments that many participants offered, where they described the relationship between themselves and the stationary embodiment, as on in which they received and gave instructions to perform particular tasks, while the ability of the mobile embodiment to move and follow them as they were going about their business, conveyed a sense that they were sharing the experience or that they were collaborating to a larger extent, even if they conceded that there was no actual added practical benefit from this functionality.

Table 3. Scores on the Godspeed Questionnaire

Dimension	Robot	Mean(SE)	Mean Diff (SE)	t-value (7 DF)	p
Anthropomorphism	Mobile	3.20(.38)	.33(.53)	0.61	.56
	Stationary	2.88(.35)			
Animacy	Mobile	3.52(.21)	.44(.31)	1.41	.20
	Stationary	3.08(.22)			
Likeability	Mobile	4.38(.18)	.45(.19)	2.35	.05
	Stationary	3.93(.19)			
Intelligence	Mobile	3.90(32)	.08(.21)	.36	.73
	Stationary	3.98(.24)			
Safety	Mobile	3.08(.12)	.00	.29	1
	Stationary	3.08(.20)			

4.2 Implications and Future Work

The findings presented here support the notion that the gradual and rich interaction that can be had through negotiating shared physical space is conductive to feelings of closeness and general liking of the robot. This is in addition to the impact of embodiment as reported by Wainer et al. [19]. While there are risks and difficulties associated with the use of large-scale physical movement for robots, the advantages beyond that of the purely task-oriented should be considered.

The next step in this work is a thorough qualitative analysis of participants' interview responses to examine their conscious reasoning, and whether or not the ability of the robot to negotiate spaces were referenced in the debrief interviews that were held. While such an analysis is beyond the scope of this paper, early findings from this analysis are promising. In addition, further findings from the on-going analysis of this long-term study will also shed further light on the relationship between proxemic interactions and relationship building in Human-Robot Interaction.

Acknowledgements. The research leading to these results has received funding from the European Union's Seventh Framework Programme (FP7/2007-2013) under grant agreement n°[287624], the ACCOMPANY project, and grant agreement n°[215554], the LIREC (LIving with Robots and intEractive Companions) project.

References

1. Dautenhahn, K.: Robots We Like to Live With? - A Developmental Perspective on a Personalized, Life-Long Robot Companion. In: Proceedings of the 13th IEEE International Workshop on Robot and Human Interactive Communication (RO-MAN 2004), pp. 17–22 (2004)
2. Cesta, A., Cortellessa, G., Giuliani, M.V., Pecora, F., Scopelliti, M., Tiberio, L.: Psychological Implications of Domestic Assistive Technology for the Elderly. PsychNology Journal 5(3), 229–252 (2007)

3. Montemerlo, M., Pineau, J., Roy, N., Thrun, S., Verma, V.: Experience with a Mobile Robotic Guide for the Elderly. In: National Conference on Artificial Intelligence. AAAI (August 2002)
4. Huettenrauch, H., Severinson Eklundh, K.: Fetch-and-carry with CERO: Observations from a long-term user study with a service robot. In: Proceeding of the 11th IEEE International Workshop on Robot and Human Interactive Interactive Communication (RO-MAN 2001), Berlin, Germany, September 25-27, pp. 158–163 (2002)
5. Danish Technological Institute, James - Robot Buttler (2012), http://robot.dti.dk/en/projects/james-robot-butler.aspx (accessed December 19, 2012)
6. Roy, N., Baltus, G., Fox, D., Gemperle, F., Goetz, J., Hirsch, T., Margaritis, D., Montemerlo, M., Pineau, J., Schulte, J.: Towards personal service robots for the elderly. In: Workshop on Interactive Robots and Entertainment (WIRE 2000), p. 184 (2000)
7. Fraichard, T.: A short paper about motion safety. In: 2007 IEEE International Conference on Robotics and Automation, pp. 1140–1145. IEEE (2007)
8. Syrdal, D.S., Dautenhahn, K., Woods, S.N., Walters, M.L., Koay, K.L.: 'Doing the right thing wrong': Personality and tolerance to uncomfortable robot approaches. In: Procs. of 15th IEEE Int Symp. on Robot and Human Interactive Communication, RO-MAN 2006, pp. 183–188 (2006)
9. Syrdal, D.S., Dautenhahn, K., Walters, M.L., Koay, K.L.: Sharing spaces with robots in a home scenario - anthropomorphic attributions and their effect on proxemic expectations and evaluations in a live HRI trial. In: Proceedings of the 2008 AAAI Fall Symposium on AI in Eldercare: New Solutions to Old Problems (2008)
10. Heerink, M., Kroese, B., Evers, V., Wielinga, B.: The Influence of a Robot's Social Abilities on Acceptance by Elderly Users. In: Proceedings of the 15th IEEE International Symposium on Robot and Human Interactive Communication (RO-MAN 2006), Hatfield, UK, September 6-8, pp. 521–526 (2006)
11. Kidd, C.D., Breazeal, C.: Robots at home: Understanding long-term human-robot interaction. In: 2008 IEEE/RSJ International Conference on Intelligent Robots and Systems (IROS 2008), September 22-26, pp. 3230–3235 (2008), doi:10.1109/iros.2008.4651113
12. Huettenrauch, H., Eklundh, K.S., Green, A., Topp, E.A.: Investigating Spatial Relationships in Human-Robot Interaction. In: Proceedings of the IEEE/RSJ International Conference on Intelligent Robots and Systems (IROS 2006), Beijing, China, pp. 5052–5059 (2006)
13. Kendon, A.: Conducting Interaction: Patterns of behavior in focused encounters, vol. 7. Cambridge University Press, Cambridge (1990)
14. Hall, E.T.: The Hidden Dimension. Doubleday, New York (1966)
15. Mehrabian, A.: Significance of posture and position in the communication of attitude and status relationships. Psychological Bulletin 71(5), 359 (1969)
16. Burgoon, J.K., Walther, J.B.: Nonverbal Expectancies and the Evaluative Consequences of Violations. Human Communication Research 17(2), 232–265 (1990)
17. Stienstra, J., Marti, P.: Squeeze me: gently please. In: Proceedings of the 7th Nordic Conference on Human-Computer Interaction: Making Sense Through Design, pp. 746–750. ACM (2012)
18. Cowley, S.J., Macdorman, K.F.: What baboons, babies and Tetris players tell us about interaction: A biosocial view of norm-based social learning. Connection Science 18(4), 363–378 (2006)
19. Wainer, J., Feil-Seifer, D.J., Shell, D.A., Mataric, M.J.: Embodiment and human-robot interaction: A task-based perspective. In: The 16th IEEE International Symposium on Robot and Human Interactive Communication, RO-MAN 2007, pp. 872–877. IEEE (2007)

20. Kose-Bagci, H., Ferrari, E., Dautenhahn, K., Syrdal, D.S., Nehaniv, C.L.: Effects of embodiment and gestures on social interaction in drumming games with a humanoid robot. Advanced Robotics 23(14), 1951–1996 (2009)
21. Tapus, A., Țăpuş, C., Matarić, M.J.: User—robot personality matching and assistive robot behavior adaptation for post-stroke rehabilitation therapy. Intelligent Service Robotics 1(2), 169–183 (2008)
22. Duque, I., Dautenhahn, K., Koay, K.L., Willcock, I., Christianson, B.: Knowledge-driven user activity recognition for a Smart House. Development and validation of a generic and low-cost, resource-efficient system. In: The Sixth International Conference on Advances in Computer-Human Interactions (ACHI 2013), Nice, France, February 24-March 1 (in press, 2013)
23. Koay, K.L., Lakatos, G., Syrdal, D., Gácsi, M., Bereczky, B., Dautenhahn, K., Miklósi, A., Walters, M.L.: Hey! There is someone at your door. A Hearing Robot using Visual Communication Signals of Hearing Dogs to Communicate Intent. In: IEEE ALIFE 2013 (The 2013 IEEE Symposium on Artificial Life), part of the IEEE Symposium Series on Computational Intelligence, Singapore, April 16-19. IEEE (2013)
24. Koay, K.L., Syrdal, D.S., Dautenhahn, K., Arent, K., Małek, Ł., Kreczmer, B.: Companion Migration – Initial Participants' Feedback from a Video-Based Prototyping Study. In: Wang, X. (ed.) Mixed Reality and Human-Robot Interaction. Intelligent Systems, Control and Automation: Science and Engineering, vol. 47, pp. 133–151. Springer (2011)
25. Koay, K.L., Syrdal, D.S., Walters, M.L., Dautenhahn, K.: A User Study on Visualization of Agent Migration between Two Companion Robots. In: 13th International Conference on Human-Computer Interaction (HCII 2009), San Diego, CA, USA (2009)
26. Aron, A., Aron, E.N., Smollan, D.: Inclusion of Other in the Self Scale and the structure of interpersonal closeness. Journal of Personality and Social Psychology 63(4), 596 (1992)
27. Mutlu, B., Osman, S., Forlizzi, J., Hodgins, J., Kiesler, S.: Perceptions of ASIMO: an exploration on co-operation and competition with humans and humanoid robots. In: Proceedings of the 1st ACM SIGCHI/SIGART Conference on Human-Robot Interaction, pp. 351–352. ACM (2006)
28. Segura, E.M., Cramer, H., Gomes, P.F., Nylander, S., Paiva, A.: Revive!: reactions to migration between different embodiments when playing with robotic pets. In: Proceedings of the 11th International Conference on Interaction Design and Children, pp. 88–97. ACM (2012)
29. Bartneck, C., Kulic, D., Croft, E., Zoghbi, S.: Measurement Instruments for the Anthropomorphism, Animacy, Likeability, Perceived Intelligence, and Perceived Safety of Robots. International Journal of Social Robotics 1(1), 71–81 (2008)

Qualitative Design and Implementation
of Human-Robot Spatial Interactions

Nicola Bellotto[1], Marc Hanheide[1], and Nico Van de Weghe[2]

[1] School of Computer Science, University of Lincoln, LN6 7TS, United Kingdom
{nbellotto,mhanheide}@lincoln.ac.uk
[2] Department of Geography, Ghent University, 9000, Belgium
nico.vandeweghe@ugent.be

Abstract. Despite the large number of navigation algorithms available for mobile robots, in many social contexts they often exhibit inopportune motion behaviours in proximity of people, often with very "unnatural" movements due to the execution of segmented trajectories or the sudden activation of safety mechanisms (e.g., for obstacle avoidance). We argue that the reason of the problem is not only the difficulty of modelling human behaviours and generating opportune robot control policies, but also the way human-robot spatial interactions are represented and implemented. In this paper we propose a new methodology based on a *qualitative* representation of spatial interactions, which is both flexible and compact, adopting the well-defined and coherent formalization of Qualitative Trajectory Calculus (QTC). We show the potential of a QTC-based approach to abstract and design complex robot behaviours, where the desired robot's motion is represented together with its actual performance in one coherent approach, focusing on spatial interactions rather than pure navigation problems.

1 Introduction

In the context of this paper, human-robot spatial interaction (HRSI) is defined as a set of relative motion events between two or more (possibly coordinated, cooperative and/or communicative) agents, which are executed according to particular social rules, agents objectives and safety constraints. In this paper we focus particularly on the 2D free-motion case, i.e. the trajectories followed by humans and robots on a planar space without obstacles, generally associated with the actions of walking towards something or someone, standing still, moving away, etc. The interpretation of such motion behaviours, as well as the capability of performing them in a social context, are essential skills for a mobile robot aiming at interacting and providing services to humans.

Typically, the trajectories of HRSIs are treated as geometrical relations in a metric frame, which is often complicated or dependent on specific training datasets. Such an approach has proven only partially effective so far, preventing the implementation of more complex and meaningful spatial behaviours. In contrast to the majority of existing approaches, we propose a *qualitative* approach

G. Herrmann et al. (Eds.): ICSR 2013, LNAI 8239, pp. 331–340, 2013.

to represent and implement HRSIs, which offers the following advantages with respect to the typical quantitative solutions: it is possible to design and implement social rules of spatial behaviours without necessarily having to learn them from huge datasets; human-human and human-robot spatial interactions can be easily mapped into semantic descriptions close to natural language. The key novelty presented in this paper, founded on our previous work on qualitative robot control and human behaviour analysis [1,2], is the combination of coarse and fine QTC representations for a comprehensible ad-hoc definition of HRSIs, which is both flexible and compact. This approach allows for the rapid design of complex spatial behaviours with varying resolution of qualitative description.

Numerous solutions for HRSI have been developed in the past. For example, in [3] the authors developed a geometrical-based algorithm for a mobile robot to enter a queue of people according to their (static) position and orientation; in [4], a solution to learn typical motion patterns of people in an office environment was proposed, which was used to estimate the metrical location of a person tracked across different rooms. Recent works have considered the motion activity of people in relation to their spatial location, so that a social robot can predict the position of potential users and approach them more effectively [5]. None of these approaches, however, take into account the robot's influence in estimating and modifying current human trajectories. Among the solutions that have considered explicitly the effect of a robot's action on human motion behaviours, the "social force model" proposed by [6] is often used to provide a quantitative description of pedestrian behaviours [7]. This model describes human motion according to forces driven by internal objectives, such as the desire of reaching a target or avoiding an obstacle, although some recent work suggests the model is not suitable for dealing with individual pedestrians during evasive manoeuvres [8]. The authors proposed instead to learn a model of human motion, based on the principle of maximum entropy, from pedestrian observations. A probabilistic framework was also proposed in [9] to generate collision-free trajectories with a robot in dynamic human environments. Differently from others, the last two solutions take into account mutual human-robot interactions to estimate and plan joint trajectories. However, all these works are associated with a numerical representation of the agents position, which might be not the most accessible approach for programming social behaviours with robots.

A qualitative interpretation of motion activities seems to be a more tractable way of dealing with HRSI, as shown by relatively simple but effective QTC representations implemented in our previous work [1]. There, QTC was adopted as a formalism for representing and implementing HRSIs. Initial simulation results using QTC Basic (QTC_B) suggested it was possible to abstract simple human and robot trajectories to generate motion commands based on qualitative terms. The work in [2] extended the spatial behaviours representation to analyse more complex trajectories with QTC Double Cross (QTC_C). To this end, however, no solutions have been implemented that exploit both QTC_B and QTC_C representations and generate, from real observations, robot control policies within the same qualitative framework. This is explored by the current research.

2   The Qualitative Trajectory Calculus

When implementing HRSI behaviour in autonomous robots, the designer is usually not interested in the exact trajectory of the robot, but rather on how it *qualitatively* moves in relation to the human, obeying implemented rules and conventions. To accommodate that need, we propose a qualitative framework based on the analysis of *relative* position and movement direction between two interacting agents on a 2D environment. In order to reduce the space domain and focus only on those terms relevant to HRSI, we adopt the well-defined set of symbols and relations provided by QTC, which is an elegant formalism to deal with the relative motion of two points in space [10]. QTC belongs to the broad research area of qualitative spatial representation and reasoning, inheriting some of its properties and tools [11]. There are several versions of QTC, depending on the number of factors considered (e.g. distance, speed, direction, etc.) and on the dimensions, or constraints, of the space where the points move.

QTC_B represents the relative motion of two points k and l (Fig. 1), with respect to the reference line connecting them, with a 3-tuple of qualitative relations (a b c), where each element can assume any of the values $\{-, 0, +\}$ as follows:

a) movement of k with respect to l
  $-$: k is moving towards l
  0 : k is stable with respect to l
  $+$: k is moving away from l
b) movement of l with respect to k
    as above, swapping k and l
c) relative speed of k with respect to l
  $-$: k is slower than l
  0 : k has the same speed of l
  $+$: k is faster than l

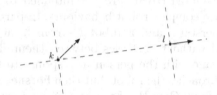

Fig. 1. Example of moving points k and l. The respective QTC_B and QTC_C relations are $(-+)$ and $(-+-0)$.

Depending on the application, a simplified version of QTC_B without the speed relation can be adopted, considering only the 2-tuple (a b). All the different combinations and relative motion description for two points are illustrated in Fig. 2. In this case there are 9 (3^2) possible states, the transitions of which can be represented by a Conceptual Neighbourhood Diagram (CND) [10]. In the QTC framework, a CND restricts the number of legally possible transitions. This helps to reduce the complexity in building temporal sequence of QTC states.

A variant of QTC_C extends the previous calculus to specify which side the two points are moving on, with respect to the reference line connecting them (see Fig. 1). In addition to the previous relations, the following ones are included:

d) movement of k with respect to $\overline{k\,l}$
  $-$: k is moving to the left side of $\overline{k\,l}$
  0 : k is moving along $\overline{k\,l}$
  $+$: k is moving to the right side of $\overline{k\,l}$
e) movement of l with respect to $\overline{l\,k}$
    as above, swapping k and l

QTC_C has 81 (3^4) states in total and 1088 possible transitions in the relative CND. It can be combined with QTC_B to represent and reason about HRSIs.

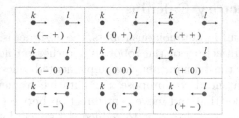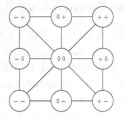

Fig. 2. Graphical representation of QTC_B (without speed) and relative CND. Due to the original formulation [10], there are no direct transitions in the CND between some of the states that, at a first glance, appear to be adjacent (e.g. (-0) and $(0-)$).

3   Application Case Studies

The proposed approach facilitates the design of a variety of HRSIs. Here we will study two cases that stem from real-world scenarios of robots moving and interacting with people in public spaces (i.e. to approach potential users or share narrow passageways) extending and improving our previous models in [1,2]. The considered HRSIs are not intended to be comprehensive, but just a means to more complex robot behaviours. In particular, we consider the scenario in which a person k and a robot l are in front of each other, just a few meters away and without obstacles between them. The robot is programmed to proactively engage with the person in response to two possible actions: Case I) the person approaches the robot, but then he/she stops and moves away from it before being reached; Case II) the person move towards the robot, but then he/she deviates from the initial trajectory to pass on its left-hand or right-hand side. Both the situations, illustrated in Fig. 3, terminate with the robot standing still.

3.1   Approach and Withdraw (Case I)

The first scenario refers to the following temporal sequence, which extends a previous QTC_B-based only example discussed in [1]:

$$(-000) \rightsquigarrow (--00) \rightsquigarrow (0-) \rightsquigarrow (+-) \rightsquigarrow (+0) \tag{1}$$

The person initially approaches the robot, triggering the same response on it. A QTC_C representation is necessary to specify that the person is moving straight towards the robot, and not aside of it. The initial state of the sequence is therefore (-000), which reads "(person) k moves towards (robot) l, while l stands still;

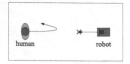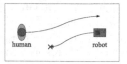

Fig. 3. HRSI in Case I (left) and II (right). The arrows indicate the trajectories of the agents. The crosses correspond to the locations where the robot stops.

neither k nor l moves laterally". The robot should then approach the person and stop only when this moves away (see Fig. 3). A command must be issued to the robot to facilitate the first transition $(-000) \rightsquigarrow (--00)$, as discussed in Sec. 4. In the remaining part of the sequence, the robot moves and stops regardless of the human (and robot) motion side. A simpler QTC_B representation is therefore sufficient to generate the opportune robot behaviour. In particular, in the following $(0-)$ state, the robot keeps approaching the person as long as he/she does not move away. Should that happen, the robot stops, as indicated by the transition $(+-) \rightsquigarrow (+0)$.

The shift in resolution from a detailed to a coarser QTC representation is particularly interesting. In reasoning language, this can be formally presented as follows [10]: there is a change from a fine relation in QCT_C to a coarse relation in QTC_B; coarse relations are specific unions of fine relations; in this case, the union of the QTC_C relations $(0-??)$ gives the QTC_B relation $(0-)$. The possibility to switch between resolutions is important in order to deal with the computational complexity arising from the many possible interactions between two or more moving agents. In the QTC_C space and relative CND, this is somehow equivalent to switch between neighbouring subsets. The transitions between QTC_C subsets correspond to the transitions between the associated QTC_B states, and the choice between one or the other representation depends on the level of accuracy required to model the particular spatial interaction.

3.2 Approach and Avoid (Case II)

In the second case, we consider a variant of the scenario discussed in [2], where the robot followed predefined trajectories to let a person pass in a narrow corridor, but without reacting to the human movements. Our qualitative approach can encode the same situation, more flexibly, accommodating the situative behaviour of the human. The interaction is described by the following temporal sequence, in which the agents keep the left-hand or right-hand side, following the respective top or bottom branch after the second state (see also Fig. 3):

$$(-000) \rightsquigarrow (--00) \begin{array}{c} \nearrow (---0) \rightsquigarrow (----) \searrow \\ \\ \searrow (--+0) \rightsquigarrow (--++) \nearrow \end{array} (00) \rightsquigarrow (++) \rightsquigarrow (+0) \quad (2)$$

The complexity of the manoeuvre, in this case, is reflected by the increased length of the sequence. It differs from the previous scenarios also by the fact that the robot, besides executing "towards" and "stop" actions, has to perform additional "left" and "right" movements, depending on the convention chosen by the person. These actions are represented in Eq. (2) by the transitions $(--+0) \rightsquigarrow (--++)$ and $(---0) \rightsquigarrow (----)$ on the top and bottom branch of the sequence respectively. In practice, the robot gives way to the person, who has the priority in deciding which side of the corridor to carry on. Two more robot commands are generated next, independently of the motion side: one to move away the robot from the person when they are side-by-side, i.e. $(00) \rightsquigarrow (++)$;

another one to stop the robot when the person walks away, i.e. $(++) \rightsquigarrow (+0)$. For this last part of the sequence, a QTC_B representation is sufficient.

It is important to note that, although in the above cases the human always does "the first move", it does not imply that the HRSI designer has to specifically program robot commands *in response* to particular human actions. With our qualitative approach, indeed, robot commands "emerge" naturally from the QTC description of the interaction scenario. One could design HRSIs in which the robot is the first agent taking the initiative, for example, to prompt some desired human motion behaviour. The command generation process is transparent from the designer's point of view.

4   Implementation

As in a previous work [1], the implemented system consists of three intercommunicating modules: a laser-based people tracker; a high-level reasoner for QTC-based inference and robot commands; and a control module that converts high-level commands to low-level instructions for the robot. The last module feeds a simple motion planning algorithm with obstacle avoidance provided by the robot's middleware. In a typical run of the system, the tracking module provides positions and velocities of the agents (i.e. person and robot) in form of string messages for the reasoner. The information is converted in QTC relations by the reasoner and included as new evidence, upon which an inference process is run. The output is a status message (e.g. a QTC string '(-000)') and, if available, a command for the robot control module (e.g. '`moveRightOf(robot,Agent)`'). The tracking and control modules are based on standard algorithms available in the literature [12,13]. The high-level reasoner, instead, is based on an inference engine, which is discussed next.

4.1   QTC Relations

The QTC relations described in Sec. 2 are implemented with the logic constructs of F-Limette[1], an inference engine based on rules that are written in a Prolog-like language for Fuzzy Metric-Temporal Horn Logic (FMTHL). In particular, we extended and improved the QTC_B implementation in [1] to include also QTC_C, following the specifications suggested in [14]. The "fuzzyness" is used to accommodate some of the ambiguities in the qualitative characterization of the motion properties. It is implemented assigning *degree of validities* (i.e. functions of the difference between current speed/orientation and some nominal value) to QTC relations, which are utilized by F-Limette during the inference process.

4.2   Situation Graph Trees

Knowledge about particular motion behaviours, expressed in terms of QTC sequences, is encoded in F-Limette using a schematic representation called Situation Graph Tree (SGT) and created with a dedicated editor [15]. During the

---

[1] http://cogvisys.iaks.uni-karlsruhe.de/Vid-Text/f_limette/

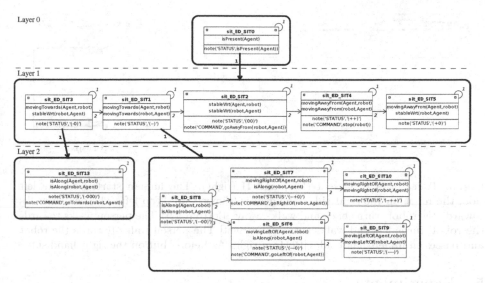

Fig. 4. SGT of Case II. Small boxes are situations with QTC relations and associated actions. Situations are temporally connected to others by thin arrows to form a graph. Situation graphs at the bottom are specializations of situations at the top.

inference process, the SGT is traversed in a depth-first fashion to find instantiable situations, each one corresponding to a particular QTC state (except the root, which is used simply to instantiate the presence of a human Agent). When available in a situation, the action COMMAND of a successful traversal is sent to the robot controller, enabling the potential instantiation of the next situation. Details about robot commands using F-Limette and SGTs are discussed in [1].

Since the SGTs are conceptually similar, we describe only the one relative to Case II, which is shown in Fig. 4. The nine QTC states of the sequence in Eq. (2) are encoded by the situations in the middle and bottom layer (Layer 1 and 2 respectively) of the SGT. Thin arrows between situations indicate prediction edges, while thick arrows from Layer 1 to Layer 2 point to specializations, respectively, from QTC_B to QTC_C. When a new inference process starts, the SGT traversal tries to instantiate the logic predicates in the first situation of Layer 1 (i.e. movingTowards(Agent,robot) and stableWrt(robot,Agent)), which correspond to the QTC_B state (-0). An attempt is also made to satisfy the respective specialization in Layer 2 (i.e. isAlong(Agent,robot) and isAlong(robot,Agent)), which, combined with the previous, gives the QTC_C state (-000). If successful, the command goTowards(robot,Agent) is issued, enabling the following transition $(-000) \rightsquigarrow (--00)$ at the next traversal.

The traversal proceeds with the next situation in Layer 1. Since to the new QTC_B state $(--)$ may have five QTC_C extensions, according to the sequence in Eq. (2), the respective specialization in Layer 2 is another temporal graph with five situations. Note that two of them, $(--+0)$ and $(--+0)$, include also the new robot commands goRightOf(robot,Agent) and goLeftOf(robot,Agent) respectively. The process continues in a similar way to the next situations.

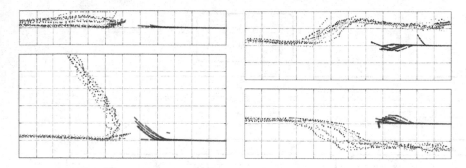

Fig. 5. Trajectories of Case I (left) and II (right). The human started on the left side, the robot on the right side. The grid is 0.5m x 0.5m. [Top left] The person walks towards the robot, turns back and walks away. [Bottom left] The person walks towards the robot, turns left and walks away. [Top right] The person walks towards the robot and passes on its left-hand side. [Bottom right] As before, but on the right-hand side.

5   Experiments

To validate the proposed approach, a set of human trajectories have been collected in an office environment using a laser-based tracking algorithm for mobile robots [12]. The trajectories, consisting of 2D coordinates and velocities, were performed according to the descriptions of Case I and II, and used to feed a real-time robot simulation. The robot was controlled by the system described in Sec. 4. Four trajectory types have been recorded, two for Case I and two for Case II (see Fig. 5), each one consisting of 10 runs (i.e. 20 trajectories per case).

The robot trajectories in Fig. 5 show that the system performed mostly correctly, stopping or driving the robot towards the desired direction according to the planned HRSI. Only in a couple of occasions, during the execution of the avoid behaviour (Case II), the interaction was unsuccessful, with the robot failing to stop after passing the human. The errors were caused by the rotation in place (no translation) of the robot, which unfortunately is not captured by the current versions of QTC. The problem is sensitive to the particular robot motion planner and it should be addressed in future implementations of the system.

To analyse the actual performance of HRSIs, the observed QTC sequences have been modelled using Markov chains (Fig. 6), the transition probabilities of which were derived from the total number of trials per case (i.e. 20). It is interesting to note that, although the behaviours generally followed the expected ones, in both cases a QTC_B state that was not in the original sequence (i.e. $(--)$ for Case I and (-0) for Case II), enabled the successful completion of the interaction replacing missing transitions between other states, mostly QTC_C. This evidence further supports the advantage of using hybrid QTC representations.

Real-world case studies have also been carried out with our MetraLabs SCITOS G5 platform using a Sick S300 laser scanner restricted to 180° field of view to detect the subject's pair of legs and obstacles. The robot employed DWA local path planning [13] to implement safe QTC movement patterns. A subject

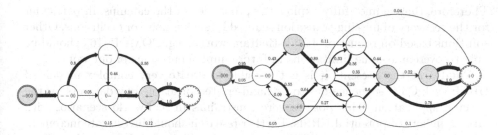

Fig. 6. Markov chains of Case I (left) and II (right). Thicker edges correspond to higher transition probabilities. Initial and final states are marked, respectively, by continuous and dashed inner circles. Blue states indicate the robot execution of motion commands.

Fig. 7. Case II with real robot. The robot gives space to the subject shifting on its right-hand side. This is illustrated also by the QTC states under each snapshot.

was instructed to either approach and turn around for the withdraw behaviour (Case I) or to pass the robot on its left side for the avoid behaviour (Case II). Five runs of each case were analysed. In all of them, the robot behaviour was according to our model, evidencing that both tracking and reasoning are applicable in a real-world setting. However, the actual sequence of QTC states varied in a similar fashion as it did in the simulation runs. One prototypical sequence of the Case II interaction taken from our study is illustrated in Fig. 7 with photos of the subject augmented by the corresponding QTC states as they have been recognised by the robot. Supplementary material is available on our website: http://robots.lincoln.ac.uk/research/icsr13

6   Conclusions

In this paper, we proposed a novel approach for the qualitative design of HRSIs, which is based on a hybrid QTC representation of human and robot trajectories. Our solution allows for compact and flexible descriptions of spatial interactions, which are enabled, within the same qualitative framework, by implicit robot control policies. Several experiments with real data validated the proposed solution, but highlighted also some limitation of the current implementation. Although fuzzy sets are useful to represent ambiguities of (top-down) HRSI models, they seem unsuitable, at least in our case, to deal with real-world uncertainties. Indeed, F-Limette strongly relies on correct tracking estimates and reliable robot motion planners. Also, the hierarchical structure of SGTs does not allow for a hybrid representation of QTC transitions always consistent with the CNDs.

Therefore, they cannot fully exploit the properties of the calculus, in particular for the recovery of missing transitions caused by sensor noise or occlusions. Other solutions based on robust probabilistic frameworks, e.g. POMDP [16], should be able to overcome some limitations in future implementations of our work.

Finally, at the present, it is not possible to create very complex models of HRSI: each QTC sequence has to be hand-crafted or learned by datasets for the specific application. This poses severe constraints in real-world scenarios with large numbers of potential HRSIs. Further research should investigate incremental approaches, where initial QTC models are provided as skeletons to be refined and extended autonomously by the robot, based on actual experience.

References

1. Bellotto, N.: Robot control based on qualitative representation of human trajectories. In: AAAI Spring Symposium – Designing Intelligent Robots (2012)
2. Hanheide, M., Peters, A., Bellotto, N.: Analysis of human-robot spatial behaviour applying a qualitative trajectory calculus. In: Proc. of IEEE RoMan, pp. 689–694 (2012)
3. Nakauchi, Y., Simmons, R.: A social robot that stands in line. Autonomous Robots 12(3), 313–324 (2002)
4. Bennewitz, M., Burgard, W., Cielniak, G., Thrun, S.: Learning motion patterns of people for compliant robot motion. IJRR 24(1), 31–48 (2005)
5. Kanda, T., Glas, D.F., Shiomi, M.F., Hagita, N.F.: Abstracting people's trajectories for social robots to proactively approach customers. IEEE Trans. on Robotics 25(6) (2009)
6. Helbing, D., Molnár, P.: Social force model for pedestrian dynamics. Physical Review E 51(5), 4282–4286 (1995)
7. Ikeda, T., Chigodo, Y., Rea, D., Zanlungo, F., Shiomi, M., Kanda, T.: Modeling and prediction of pedestrian behavior based on the sub-goal concept. In: Proc. of RSS (2012)
8. Kuderer, M., Kretzschmar, H., Sprunk, C., Burgard, W.: Feature-based prediction of trajectories for socially compliant navigation. In: Proc. of RSS (2012)
9. Althoff, D., Kuffner, J., Wollherr, D., Buss, M.: Safety assessment of robot trajectories for navigation in uncertain and dynamic environments. Autonomous Robots 32(3), 285–302 (2012)
10. Van de Weghe, N.: Representing and Reasoning about Moving Objects: A Qualitative Approach. PhD thesis, Ghent University (2004)
11. Cohn, A.G., Renz, J.: Qualitative Spatial Representation and Reasoning. In: van Harmelen, F., Lifschitz, V., Porter, B. (eds.) Handbook of Knowledge Representation, vol. 3, ch. 13, pp. 551–596. Elsevier (2008)
12. Bellotto, N., Hu, H.: Multisensor-based human detection and tracking for mobile service robots. IEEE Trans. on SMC – Part B 39(1), 167–181 (2009)
13. Fox, D., Burgard, W., Thrun, S.: The dynamic window approach to collision avoidance. IEEE Robotics and Automation Magazine 4(1), 23–33 (1997)
14. Delafontaine, M., Cohn, A., de Weghe, N.V.: Implementing a qualitative calculus to analyse moving point objects. Expert Systems with Applications 38(5) (2011)
15. Arens, M., Nagel, H.-H.: Behavioral knowledge representation for the understanding and creation of video sequences. In: Günter, A., Kruse, R., Neumann, B. (eds.) KI 2003. LNCS (LNAI), vol. 2821, pp. 149–163. Springer, Heidelberg (2003)
16. Ong, S., Png, S., Hsu, D., Lee, W.S.: POMDPs for robotic tasks with mixed observability. In: Proc. of RSS (2009)

Unsupervised Learning Spatio-temporal Features for Human Activity Recognition from RGB-D Video Data

Guang Chen[1], Feihu Zhang[1], Manuel Giuliani[2],
Christian Buckl[2], and Alois Knoll[1]

[1] Institut für Informatik VI, Technische Universität München, Boltzmannstr. 3,
85748 Garching, Germany
`guang,zhang,knoll@in.tum.de`
[2] fortiss GmbH, Guerickestr. 25, 80805 Munich, Germany
`giuliani,buckl@fortiss.org`

Abstract. Being able to recognize human activities is essential for several applications, including social robotics. The recently developed commodity depth sensors open up new possibilities of dealing with this problem. Existing techniques extract hand-tuned features, such as HOG3D or STIP, from video data. They are not adapting easily to new modalities. In addition, as the depth video data is low quality due to the noise, we face a problem: does the depth video data provide extra information for activity recognition? To address this issue, we propose to use an unsupervised learning approach generally adapted to RGB and depth video data. we further employ the multi kernel learning (MKL) classifier to take into account the combinations of different modalities. We show that the low-quality depth video is discriminative for activity recognition. We also demonstrate that our approach achieves superior performance to the state-of-the-art approaches on two challenging RGB-D activity recognition datasets.

Keywords: activity recognition, unsupervised learning, depth video.

1 Introduction

Human action recognition has been widely studied in computer vision. Its applications include video surveillance, content-based video search, robotics and a variety of systems that involve interactions between persons and computers. Traditional research mainly concentrates on learning and recognizing human activities from video data captured by a single visible light camera. The video data is a sequence of 2D frames with RGB or gray channels. There is extensive literature on action recognition for such video. Most methods for activity recognition with RGB video use hand-designed features like STIP [1], or use the local features like HOF [2] or HOG [3] to represent the spatio-temporal pattern. In these methods, human activities can be interpreted by a set of interesting points. However, there is no universally best hand-designed features for different RGB video [4]. In addition, it is difficult and time-consuming to extend these features

G. Herrmann et al. (Eds.): ICSR 2013, LNAI 8239, pp. 341–350, 2013.

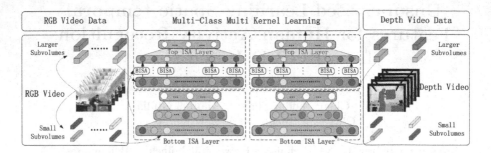

Fig. 1. An overview of our model: We randomly sample the subvolumes from RGB and depth video data. The small subvolumes are given as input to the Bottom ISA network. The learned bottom ISA model are copied to the Top ISA network. The stacked ISA learn the final features for each modality. The multi-class multi kernel learning is employed to learn a combination of different modalities and classify the activities (best viewed in color).

to other sensor modalities, such as laser scans or depth cameras. The depth camera can record RGB and depth video data has now become affordable and could be combined with standard vision system in social robot. The depth modality provides useful extra information to the complex problem of activity recognition since depth information is invariant to lighting and color variations.However, because there is no texture in the depth data,the extending hand-designed features from RGB data to depth modality are not discriminative enough for classifications. In addition, the depth video data is full of noises. There are large shadows or holes in the depth data. Hence the discrimination of the low quality depth video is considered doubtful.

Recently, there is a growing interest in unsupervised feature learning methods such as Deep Belief Nets [5], Sparse Coding [6, 7], Stacked Autoencoders [8], Independent Component Analysis (ICA) and Independent Subspace Analysis (ISA) [9]. These biologically-inspired learning algorithms show promise in the domain of the computer vision, such as object recognition with RGB-D images and action recognition with RGB video data [10–12]. Although many deep learning methods exist for learning features from RGB image or video data, none has yet been investigated for depth video data. In this paper, we provide an unsupervised learning method inspired by [9, 11]. We learn spatio-temporal features directly from RGB and depth video data independently. Fig. 1 outlines our approach. Our model starts with raw RGB and depth video data and extracts unlabeled space-time subvolumes from each modality. The subvolumes are then given to the stacked ISA network to learn hierarchical representations for each modality. We employ the multi kernel learning to combine the learned representations of different modalities. Our experiments show that although the low quality depth video data is full of noises, we could learn discriminative spatio-temporal features from depth data for activity recognition. We also achieve superior performance on the task of human activity recognition using RGB-D video data. Compared to other recent activity recognition methods for RGB-D video data [13–17], our approach is generalizable, does not need additional input channels such as skeleton joints and surface normals.

In this paper, we first briefly describe the ISA algorithm. Next we give details of how deep learning techniques such as convolution and stacking can be used to obtain hierarchical representations of the different modalities. Then, we learn the combinations of different modalities by multi kernel learning. The proposed features and models are evaluated on two RGB-D benchmark datasets: Pioneer-Activity dataset [13] and UTKinectAction3D dataset [14]. In our experiments, we show quantitative comparisons of different methods and different modalities.

2   Independent Subspace Analysis

In this section we describe the background of ISA algorithm [9]. ISA is an unsupervised learning algorithm that learns features from unlabeled subvolumes. First, random subvolumes are extracted into two sets, one for each modality(RGB and depth video data). Each set of subvolumes is then normalized and whitened. The pre-processed subvolumes are feed to ISA networks as input units. An ISA network [9] is described as a two-layer neural network, with square and square-root nonlinearities in the first and second layers respectively (see Fig. 1).

We start with any input unit $x^t \in \mathbb{R}^n$ for each random sampled subvolumes. We split each subvolume into a sequence of image patches and flatten them into a vector x^t with the dimension n. The activation of each second layer unit is

$$p_i(x^t; W, V) = \sqrt{\sum_{k=1}^{m} V_{ik} \left(\sum_{j=1}^{n} W_{kj} x_j^t \right)^2} \tag{1}$$

ISA learns parameters W through finding sparse feature representations in the second layer by solving

$$\min_{W} \sum_{t-1}^{T} \sum_{i=1}^{m} p_i(x^t; W, V)$$
$$s.t. WW^T = \mathbf{I} \tag{2}$$

Here, $W \in \mathbb{R}^{k \times n}$ is the weights connecting the input units to the first layer units. $V \in \mathbb{R}^{m \times k}$ is the weights connecting the first layer units to the second layer units; n, k, m are the input dimension, number of the first layer units and second layer units respectively. The orthonormal constraint is to ensure the features are diverse.

The model so far has been unsupervised. The learned ISA filters for each modality could be used for activity recognitions. The first layer of our ISA model learns spatio-temporal features that detect a moving edge in time as shown in Fig. 2. It shows that the learned feature (each row) is able to group similar features in a group thereby achieving spatial invariance. When the method is applied to depth video data, the resulting filters have shaper edges which arise due to the strong discontinuities at object boundaries. The experiments in section 5.3 show that this property may contribute to a better recognition within depth video data compared to RGB video data. We also study the sensitivity of the

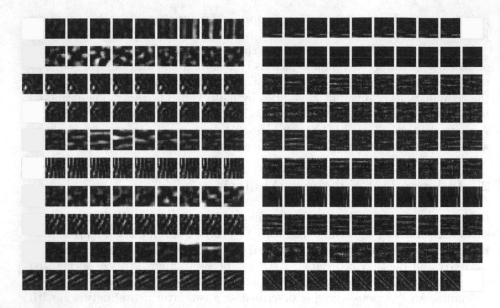

Fig. 2. Visualization of 20 ISA filters learned from PioneerActivity dataset. 10 filters (**left**) are from RGB video data and 10 filters (**right**) are from depth video data. These filters capture a moving edge in time. The filters from the depth video have sharper edges compared to filters trained on the RGB video data.

learned features to motion and orientation. In a control case, we limit this ability by using a temporal size of 4 frames instead of 10 frames and the recognition rate drops by 7.33% for the PioneerActivity dataset. If the temporal size is set to 2, the recognition rate drops by 2.56% again.

3   Stacked Convolutional ISA

3.1   Convolution and Stacking

In order to scale up to ISA algorithm to large input, we use a convolutional neural network architecture similar to [11, 18] for each modality. The network progressively makes use of PCA and ISA as sub-units for unsupervised learning as shown in Fig. 1.

We train the first layer of the networks with standard ISA algorithm on small input subvolumes for each modality. We randomly extract larger subvolumes from each modality. We then copy the learned bottom ISA filters and convolve with the larger subvolumes of the input video data (see Fig. 3). The responses of the convolution step are given as the input unit to the next ISA layer. As is common in neural network, we stack another ISA layer with PCA on top of the bottom ISA. We use PCA to whiten the data and reduce the dimensions of the input unit. The model is trained greedily layerwise in the same manner as other algorithms described in [5, 11, 19].

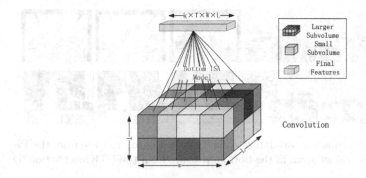

Fig. 3. Convolutional step of stacked ISA network. For clarity, the convolutional step is shown here non-overlapping, but in the experiments, convolution is done with overlapping.

3.2 Vector Quantization

As each activity is represented by a RGB video and an depth video, we perform the vector quantization by clustering the spatio-temporal features for each modality. We follow the state-of-the-art bag-of-words(Bow) paradigm. We construct the BoW features based on the dense spatio-temporal features for each modality.

4 Learning Multi-modality Combination

For each modality represented by the features of stacked convolutional ISA model, an SVM model on it defines a joint feature map $\Phi(x, y)$ on data $\mathcal{X}$ and labels $\mathcal{Y}$ as a linear output function $f_k(x, y) = \langle \omega_k, \Phi(x, y) \rangle + b_k$, parameterized with the hyperplane normal ω_k and bias b_k. The predicted class y for x is chosen to maximize the output $f_k(x, y)$.

Multi kernel learning considers a convex combination of n kernels, $K(x_i, x_j) = \sum_{k=1}^{n} \alpha_k K_k(x_i, x_j)$ where each kernel corresponds to a modality. We consider the following output function

$$f_{com}(x, y) = \sum_{k=1}^{n} [\alpha_k \langle \omega_k, \Phi(x, y) \rangle + b_k] \qquad (3)$$

MKL learns the coefficient α, the weight ω and the bias b. For a multi class problem, different α, ω and b are learned for each class. In our case, we choose one-against-rest to decompose a multi-class problem. As MKL can not give a posterior class probability $P(y = 1|x)$, we propose approximating the posterior by a sigmoid function

$$P_m(y = 1|x) \approx P_{A_m, B_m}(f_{com}) \equiv \frac{1}{1 + \exp(A_m f_{com} + B_m)} \qquad (4)$$

Fig. 4. Some example frames of two datasets. Samples in the top row are from the PioneerActivity dataset and samples in the bottom row are from the UTKinectAction3D dataset.

We follow Platt' method to learn A_m and B_m [20]. For each MKL-SVM model m, we learn a sigmoid function $P_{A_m,B_m}(f_{com})$. The maximum probability $l = \max_m P_m(y = 1|x)$ corresponds to the predicted label of x.

5   Experimental Results

We choose PioneerActivity dataset [13] and UTKinectAction3D dataset [14] to evaluate the proposed human activity recognition approach. Both datasets include the RGB video data and depth video data. The empirical results show the low-quality depth data provides more useful information for the task of human activity recognition. The results also show the proposed framework outperforms the state of art methods. We first give a brief overview of the datasets, followed by the detail of our stacked ISA model, and the experimental results.

5.1   Datasets and Experimental Setup

The PioneerActivity dataset is an human activity dataset of RGB and depth video data captured by a depth camera [13]. The dataset presents several challenges due to illumination change, dynamic background and variations in human motions. It contains six types of human activities: lifting (LF), removing (RM), pushing (PS), waving (WV), walking (WK), signaling (SG). Each activity has 33 samples for each modality. We follow the experimental setup of [13]. We divide the database into three groups. One group as the training set, and the remaining groups are used as the testing sets. The experiment results are reported as the average over 20 runs.

The UTKinectAction3D dataset contains 10 types of human activities in indoor settings. The 10 actions include: sit down, stand up, walk, pick up, carry, throw, push, pull, wave and clap hands. Each action was collected from 10 different persons for 2 times. We evaluate our approach on this dataset using leaving one out cross validation. We run the experiment 20 times. Some samples of activities from the two datasets are shown in Fig. 4.

5.2   Model Details

We focus on the problem of human activity recognition from RGB and depth video data. We train stacked ISA model for each modality. For the PioneerActivity dataset, the input units to the bottom layer are of size $16 \times 16 \times 10$, 16, 16, 10 means spatial and temporal size of the subvolumes. The larger subvolumns to the top layer are of size $20 \times 20 \times 14$. The model parameters for the RGB and depth video data are the same. For the UTKinectAction3D dataset, the subvolumes to the bottom layer of stacked ISA network are of size $16 \times 16 \times 6$. The larger subvolumes to the top layer are of the size $20 \times 20 \times 8$. Finally, we performs vector quantization by K-means on the learned spatio-temporal features and classifies by multi-class MKL classifiers by χ^2 kernel.

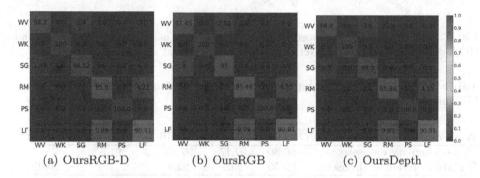

(a) OursRGB-D   (b) OursRGB   (c) OursDepth

Fig. 5. The confusion matrices for the proposed method on PioneerActivity dataset with different modalities. Rows represent the actual classes, and columns represent predicted classes. OursRGB-D, OursRGB, OursD are our proposed method using both of RGB and depth data, using RGB data only, and using depth data only respectively. (best viewed in color).

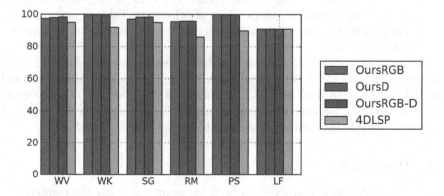

Fig. 6. The comparison between the average accuracy of the proposed method and 4DLSP [13] on the PioneerActivity dataset with different modalities

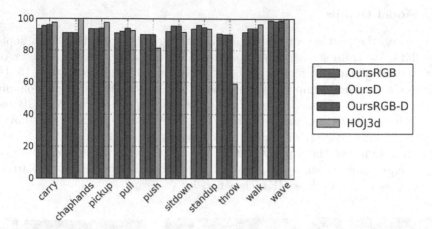

Fig. 7. The comparision between the average accuracy of the proposed method and HOJ3d [14] on the UTKinectAction dataset with different modalities

Table 1. Summary of the experimental results for PioneerActivity dataset and UTKinectAction3D dataset

Ave. acc on two Datasets	OursRGB	OursD	OursRGB-D	4DLSP [13]	HOJ3d [14]
PioneerActivity dataset	96.80	97.17	97.23	91.5	-
UTKinectAction3D dataset	92.55	93.6	93.8	-	90.95

5.3   Experimental Results

We shows the accuracies of our method on the PioneerActivity dataset in Fig.5 and Fig 6. The confusion matrices of our approach using different modalities are given in Fig. 5. The confusion matrix shows that the largest confusion lies between "removing" and "lifting". This is consistent with the bag-of-word paradigm as it assumes each word is independent of others. In Table 1, we compare our test results with the state of the art method [13]. Our method significantly outperforms 4DLSP [13].

We further compare our approach with the best published result on UTKinectAction3D dataset [14] in Fig. 7 and Table 1. The average accuracy of our method using RGB-D video data is 93.8%. Notice that the work of [14] only uses the depth video data for activity recognition. Our approach with the depth video data achieves 93.6% accuracy which is still better than HOJ3D [14].

5.4   Discussion

In the above experiments, we compared our approach using the combination of the RGB video data and depth video data. This raises some questions: "How much does the combination help?" and "Does the depth video data provide extra information for activity recognition? ". In Table 1, the accuracies of the methods using both modalities are just little better than the approach using only one modality. This result shows that the learned features from different modalities exhibit some similar patterns. One possible explanation would be the features learned from RGB video data is trained on gray scale versions of the RGB video data.

In general, the accuracies of the approach using depth video data are better than the accuracies of the method using RGB video data. Although the depth data is full of noise, such as shadows and holes, the result indicates that the depth video data provides more useful information for the task of human activity recognition than the RGB video data on our datasets. A possible explanation is that RBG video data is sensitive to the illumination variations and the dynamics background while the depth video data is not. As illustrated in Fig. 4, the computer monitors lead to a dynamic background for the RGB images, which distract the learned features from capturing useful human motions. But the depth sensor is not sensitive to the dynamic background. Another explanation would be the learned features by the depth video data capture more edge informations. Because the edge detectors like Gabor filters show great performance in a lot of computer vision domains [21]. This is consistent with Fig.2 that the resulting filters learned by depth data have shaper edges.

6   Conclusion

We introduced a method that learns spatio-temporal features from RGB-D video data. The stacked ISA network learns the hierarchical representations in an unsupervised way. The multi-class MKL learns the combinations of different modalities. This architecture could leverage the plethora of the unlabeled data and adapt easily to new modalities. The experiment results were carried out with PioneerActivity and UTKinectAction datasets. We observed that the low-quality depth video data provides more useful information for the task of human activity recognition on our datasets. The learned features from RGB and depth video data exhibit some similar patterns. Our method also outperforms the state-of-the-art methods for activity recognition.

References

1. Laptev, I.: On space-time interest points. Int. J. Comput. Vision 64(2-3), 107–123 (2005)
2. Laptev, I., Marszałek, M., Schmid, C., Rozenfeld, B.: Learning realistic human actions from movies. In: Conference on Computer Vision & Pattern Recognition (June 2008)
3. Dalal, N., Triggs, B.: Histograms of oriented gradients for human detection. In: CVPR, pp. 886–893 (2005)
4. Wang, H., Ullah, M.M., Kläser, A., Laptev, I., Schmid, C.: Evaluation of local spatio-temporal features for action recognition. In: British Machine Vision Conference, p. 127 (September 2009)

5. Hinton, G.E., Osindero, S., Teh, Y.W.: A fast learning algorithm for deep belief nets. Neural Computation 18(7), 1527–1554 (2006)
6. Olshausen, B.A., Field, D.J.: Emergence of simple-cell receptive field properties by learning a sparse code for natural images. Nature 381, 607–609 (1996)
7. Lee, H., Battle, A., Raina, R., Ng, A.Y.: Efficient sparse coding algorithms. In: NIPS, pp. 801–808 (2007)
8. Bengio, Y., Lamblin, P., Popovici, D., Larochelle, H.: Greedy layer-wise training of deep networks. In: Schölkopf, B., Platt, J., Hoffman, T. (eds.) Advances in Neural Information Processing Systems 19, pp. 153–160. MIT Press, Cambridge (2007)
9. Hyvrinen, A., Hurri, J., Hoyer, P.O.: Natural Image Statistics: A Probabilistic Approach to Early Computational Vision, 1st edn. Springer Publishing Company, Incorporated (2009)
10. Socher, R., Huval, B., Bath, B.P., Manning, C.D., Ng, A.Y.: Convolutional-recursive deep learning for 3d object classification. In: Bartlett, P.L., Pereira, F.C.N., Burges, C.J.C., Bottou, L., Weinberger, K.Q. (eds.) NIPS, pp. 665–673 (2012)
11. Le, Q., Zou, W., Yeung, S., Ng, A.: Learning hierarchical invariant spatio-temporal features for action recognition with independent subspace analysis. In: 2011 IEEE Conference on Computer Vision and Pattern Recognition (CVPR), pp. 3361–3368 (2011)
12. Ji, S., Xu, W., Yang, M., Yu, K.: 3d convolutional neural networks for human action recognition. IEEE Transactions on Pattern Analysis and Machine Intelligence 35(1), 221–231 (2013)
13. Zhang, H., Parker, L.: 4-dimensional local spatio-temporal features for human activity recognition. In: 2011 IEEE/RSJ International Conference on Intelligent Robots and Systems (IROS), pp. 2044–2049 (2011)
14. Xia, L., Chen, C.C., Aggarwal, J.: View invariant human action recognition using histograms of 3d joints. In: 2012 IEEE Computer Society Conference on Computer Vision and Pattern Recognition Workshops (CVPRW), pp. 20–27 (2012)
15. Yang, X., Tian, Y.: Eigenjoints-based action recognition using nave-bayes-nearest-neighbor. In: CVPR Workshops, pp. 14–19. IEEE (2012)
16. Wang, J., Liu, Z., Wu, Y., Yuan, J.: Mining actionlet ensemble for action recognition with depth cameras. In: 2012 IEEE Conference on Computer Vision and Pattern Recognition (CVPR), pp. 1290–1297 (2012)
17. Oreifej, O., Liu, Z.: Hon4d: Histogram of oriented 4d normals for activity recognition from depth sequences (June 2013)
18. LeCun, Y., Boser, B., Denker, J.S., Henderson, D., Howard, R.E., Hubbard, W., Jackel, L.D.: Backpropagation applied to handwritten zip code recognition. Neural Computation 1, 541–551 (1989)
19. Bengio, Y., Lamblin, P., Popovici, D., Larochelle, H.: Greedy layer-wise training of deep networks, pp. 153–160 (2007)
20. Platt, J.C.: Probabilistic outputs for support vector machines and comparisons to regularized likelihood methods. In: Advances in Large Margin Classifier, pp. 61–74. MIT Press (1999)
21. Kamarainen, J.K., Kyrki, V., Kälviäinen, H.: Invariance properties of Gabor filter based features - overview and applications. IEEE Transactions on Image Processing 15(5), 1088–1099 (2006)

Head Pose Patterns in Multiparty Human-Robot Team-Building Interactions

Martin Johansson, Gabriel Skantze, and Joakim Gustafson

Department of Speech, Music and Hearing, KTH, Stockholm, Sweden
vhmj@kth.se, {gabriel,jocke}@speech.kth.se

Abstract. We present a data collection setup for exploring turn-taking in three-party human-robot interaction involving objects competing for attention. The collected corpus comprises 78 minutes in four interactions. Using automated techniques to record head pose and speech patterns, we analyze head pose patterns in turn-transitions. We find that introduction of objects makes addressee identification based on head pose more challenging. The symmetrical setup also allows us to compare human-human to human-robot behavior within the same interaction. We argue that this symmetry can be used to assess to what extent the system exhibits a human-like behavior.

Keywords: focus of attention, human-robot interaction, turn-taking.

1 Introduction

Robots of the future are envisioned to help people perform tasks, not only as mere tools, but as autonomous agents interacting and solving problems together with people. These future robots could interact with humans in a way similar to the way humans interact with each other, or they could make use of more simple, machine-like interaction patterns. Some situations could call for the robot to be a human-like artificial conversational partner capable of participating in a multiparty dialogue. One way of measuring human-likeness is by measuring the behavior of the human interlocutor[1]. The closer the behavior of the human interlocutor in the interaction with the robot to that of interaction with other humans, the more human-like this interaction is.

Aside from being human-like or not, there are some fundamental problems a robot will have to deal with to engage in dialogue with multiple humans. Two of these are to manage turn-taking in order to know when it is acceptable to say something, and to figure out to whom an utterance is directed. Previous studies have found head pose[2] to be a useful indicator to identify the addressee in three-party interaction, especially in combination with acoustic cues[3]. The problems future robots could help solve might however include visible objects competing for attention, possibly affecting head pose behavior.

Edlund et al.[1] proposed to evaluate human-likeness in human-machine interaction using a *two-way mimicry target*. In order for the machine's behavior to

G. Herrmann et al. (Eds.): ICSR 2013, LNAI 8239, pp. 351–360, 2013.

be deemed human-like, it should behave like a human interlocutor in a human-human conversation, and the human speaking to it should behave like when speaking to another human. Any difference in the human's behavior is seen as a result of the machine's behavior. In a multiparty setting with two humans and a robot in similar roles, a two-way mimicry target can be evaluated using, for example, symmetry in behavior between the human-human and human-robot interactions.

This paper presents a data collection setup for three-party conversations with two humans and one robot, designed to allow comparisons between the robot and human participants through symmetry. The purpose of this study is to test the setup by exploring head pose patterns surrounding turn changes when we involve more targets for visual attention than just the participants. We also evaluate the human-likeness of the robot by comparing head pose patterns between human-human and human-robot interactions.

2   Background

The function of gaze in interaction has been found to serve multiple functions, one of them being turn-taking control. Kendon[4] found that speakers look away at beginning of turns, and look back at their interlocutors towards end of turns. Gaze behavior has also been found to provide information about the target of attention. Vertegaal et al.[5] found eye-gaze to be a good predictor of conversational attention in multiparty conversations, while Katzenmaier et al.[3] found head pose to be a cue for identifying addressee in human-human-robot interaction. Stiefelhagen and Zhu[6] showed that head pose is a reliable cue to estimate focus of attention in a small meeting scenario. Ba and Odobez[7] expanded the small meeting scenario to include more targets for attention, and concluded that good separation of targets is essential for accuracy. Automated means of recording gaze in conversational settings are available, using, for example, eye trackers[8], or head pose tracking[7] for estimated gaze. Both methods could conceivably be used by a robot to monitor interlocutors. Estimating gaze through head pose instead of tracking eye-gaze, however, has the advantage of being more robust in regard to head movement and blinking.

Argyle and Graham[9] studied dyadic interactions involving additional targets for visual attention. Objects relevant to the task at hand were found to attract visual attention at the expense of the other subject.

3   Method

We conducted a laboratory experiment in a human-human-robot setup for this exploratory study. The intention was to gain an understanding of symmetry in collaborative human-robot problem solving when objects relating to the task that is discussed are present. In order to elicit engagement in interaction, the task needs to be fun and interesting. One way of achieving this is to use games. The Speech group at KTH has initiated collections of a series of multi-party games for interactional corpus collections, the KTH games corpora[10], of which the current study is the first human-human-robot corpus.

We selected an adaption of the "Desert Survivor" team-building game[11] as the task in this study. The team is to collaboratively prioritize a set of items based on their usefulness for survival in a desert after a hypothetical plane crash. We chose to use the Desert Survival game since it is an engaging social task that elicits intra-group communication, thus allowing us to study group dynamics and multi-party interaction phenomena.

A *Wizard-of-Oz* setting was selected for initial data collection to give the robot some sense of intelligence in the dialogue via the wizard, while maintaining the physical appearance of the robot and limiting its range of actions compared to a human.

3.1   Experiment Setup

The experiment setup, overviewed in Fig. 1, was designed around a round table at which the two human subjects and the robot were placed. Subjects were seated on static chairs at fixed locations to have the triad placed in an equilateral triangle pattern, with equal distances and angles between all participants. The objective of this placement was to allow for conclusions regarding attention behavior between human-human and human-robot interactions. This setup also has the advantage that the predicted target areas for visual attention are as separated as possible from a head pose rotation frame of reference, as suggested by Ba and Odobez[7]. The complete setup in action is shown in Fig. 2 with a snapshot from one of the sessions.

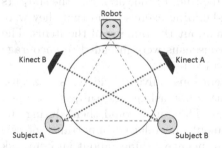

Fig. 1. Spatial configuration   **Fig. 2.** The robot and two subjects

Each subject was monitored by a Microsoft Kinect sensor, responsible for tracking the head pose of its subject as well as the angle to the most prominent audio source. The use of one sensor per subject limits the expected head pose yaw to around ±30° from the Kinect's point of view, avoiding problematic extreme angles yielding low accuracy. Another benefit is the symmetry of the recorded data for both subjects.

High quality audio was recorded using headsets worn by the subjects, keeping the table available for use by the subjects as they saw fit when discussing the items in the scenario. In this adapted exercise, the items to discuss were given physical presence in the form of stylized pictures, each on an individual sheet of paper. Video was recorded using two high-definition video cameras. The first camera oversaw the interaction from behind the human participants, whereas the second one recorded the interaction from the robot's point of view.

Furhat[12] was chosen as the robot participant in the experiment. Furhat is a back-projected human-like robot head using speech synthesis[1] and state-of-the-art facial animation, mounted on a robotic neck. This makes it capable of combining head pose and eye gaze to direct attention[13]. The facial animation architecture allows for speech with accurate synchronized lip movements, as well as for control and generation of non-verbal gestures, eye movement and facial expressions.

3.2   Methodology and Experimental Design

The robot in this experiment was partly automated and partly controlled by a wizard. The gaze of the robot was automated, whereas the speech was controlled by a wizard. The wizard decided when the robot should say something, and what it should say, by selecting one of several predefined utterances from an interface on a networked computer. When speaking, the gaze was directed either towards a subject or, while speaking about an item, towards the table. When not speaking, the robot's gaze was directed towards the participant coinciding with the most prominent audio source detected by the Kinect sensors.

Each session started with the robot giving instructions about the task, placing emphasis on its collaborative nature, before proceeding to the first iteration of items to rank. No instructions concerning the roles of the robot and the subjects were given, other than that it was a team-building exercise, and that they were to discuss and reach a unanimous decision about the ranking of the items. The instructions regarding collaboration and consensus were intended to encourage the subjects to affiliate with the robot as a team[14].

The adapted exercise comprised three iterations of five unique items to discuss and rank, each iteration starting with the robot asking the subjects to open a numbered envelope containing the items. The robot could then, during the discussion that followed, have opinions about the relative importance of two items, say one of two predefined positive or negative things about an item, ask the subjects for their opinions, answer questions with a yes or a no or confess that it did not know.

During the course of the three-party conversation, the goal was to have all interlocutors actively involved in order to collect comparable data for both human-human and human-robot interactions. The wizard's strategy to keep the robot involved in the discussion and, if necessary, both human subjects involved, was to try to answer directed or open questions and to either make a statement or pose a question when given opportunity.

---

[1] CereProc ltd: http://www.cereproc.com/

3.3   Participants

Eight subjects aged 23-41, divided into four pairs, were used in the data collection. The mean age was 30.5, with a standard deviation of 5.42. Three of the participants were female and five male. All subjects were employees at the Department of Speech, Music and Hearing, KTH.

3.4   Measurement

We captured the subjects' behavior using an audio recorder with headsets, Kinect sensors and high-definition cameras. Two Kinect sensors were used to record head positions and rotations in 3D space, as well as a set of face expression parameters and the angle to the most prominent sound source. The video recordings could be used when annotating the dialogues in the future, and the audio recordings can be processed with, for example, automatic speech recognition or prosody extraction.

For this initial analysis, we employed an automated approach where Kinect sensor data was used to estimate target of visual attention based on head position and rotation. The regions of interest as potential target of attention for a subject were the robot, the other subject and the table. No attempts were made to distinguish between different individual items on the table.

We employed automated segmentation using a voice activity detector to extract utterances from the recorded audio. The extracted utterances were then used to locate instances of turn changes. A change of turn was defined as having two consecutive non-overlapping utterances, not shorter than one second each, belonging to two different speakers. Utterances shorter than one second were not included in this analysis of turn changes for robustness reasons. In our case, this means that no change of turn took place; the current speaker continued to speak as the interlocutor with the very short utterance was not claiming the floor.

We recorded twelve iterations, three per pair of subjects. The recorded interactions, excluding instructions, lasted a total of 78 minutes with an average iteration length of 6.5 minutes (standard deviation 1.37). Next, we segmented the recorded audio through automated means, resulting in a total of 1312 segments of human subject speech encompassing 38 minutes.

4   Results

4.1   Observations

All subjects interacted with the items on the table during the dialogues, even though the only instruction given related to the physical items was to open the envelope containing them. The interaction ranged from active spatial organization of the items to signaling which items were considered, for example by picking them up or pointing at them.

4.2   Target of Attention Around Turn Changes

For each change of turn, the participants were labeled with one of the following roles: *current speaker*, the speaker who finished the utterance in the ending turn; *next speaker*, the one who was taking the turn; *other*, the one not speaking. The targets of attention for each subject in a time frame of three seconds before and after the end of the last utterance in a turn were estimated. Head pose data was split into intervals of fifty milliseconds, each with one single target defined by majority classification. The possible targets were either one of the other participants in their current role or the *table*, harboring the items.

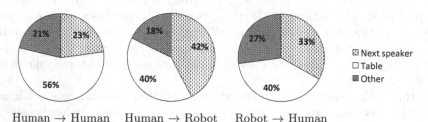

Fig. 3. Visual attention of the current speaker near the end of turn, all instances

We evaluated the overall distribution of the speakers' visual attention in the specified window at end of turns for the different combinations of robot and human speakers (Fig. 3). The robot as the current speaker exhibited an overall distribution of attention resembling the humans', suggesting that the robot did not behave completely different as a speaker. There was no dominant visual attention towards the next speaker in any of the three combinations of interlocutors. The largest share of looking at the next speaker was when the next speaker was the robot.

4.3   Speakers Looking at the Next Speaker

Many turn changes occurred without the current speaker looking at the next speaker. To compare symmetry in the situation where the current speaker looked at the next speaker, we analyzed the instances where the current speaker did look at the next speaker at least once in a time frame spanning one second before to one second after the end of the turn.

First we investigated the estimated target of visual attention for a human ending a turn while looking at the next speaker (Fig. 4). Results indicate that humans address the robot clearly. However, since the robot's decision on when to speak was made by a wizard, the human-to-robot patterns are likely affected by the turn-taking strategy employed by the wizard. Due to the criterion used to select instances, the peaks in attention towards the next speaker around the end of turn were expected.

Next we investigated the estimated target of visual attention for the human taking the turn, when looked at by the speaker who previously had the

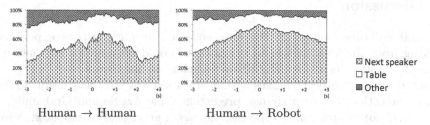

Human → Human   Human → Robot

Fig. 4. Visual attention of the current speaker near the end of turn. Instances where the current speaker was looking at the next speaker.

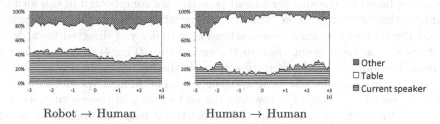

Robot → Human   Human → Human

Fig. 5. Visual attention of human taking the turn near the end of preceding turn. Instances where the current speaker was looking at the next speaker.

turn (Fig. 5). The majority of the turn taker's visual attention was directed towards either the current speaker or the table. More visual attention was directed towards the speaking robot than when the speaker was human.

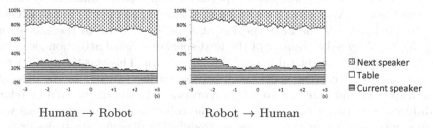

Human → Robot   Robot → Human

Fig. 6. Visual attention of human neither ending nor taking the turn, near the end of turn. Instances where the current speaker was looking at the next speaker.

Finally we investigated the estimated target of visual attention for the human neither ending nor taking the turn (Fig. 6), when the next speaker was looked at by the current speaker. The distributions appeared more similar in shape, with more attention to the first speaker towards the end of turn and more attention to the next speaker afterwards.

5   Discussion

We analyzed turn changes in the data collected during the pilot experiment, working from the idea that head orientations could be used for estimation of focus of attention as was concluded by Stiefelhagen and Zhu[6].

Introduction of objects makes visual attention tracking less useful for addressee detection and turn change prediction. Like Argyle and Graham[9], we found that objects relevant to the task attract a great deal of attention. Visual attention close to turn changes (Fig. 3) was placed on the table in a large portion of the collected instances. The head pose being directed towards the table in many cases, at the expense of the other participants, indicates that detection of addressee based on the speaker's head pose is more complicated in this situated dialogue than in, for example, the one evaluated by Skantze and Gustafson[2]. When the human speaker's visual attention actually was directed towards the next speaker at some point close to the end of turn (Fig. 4), our data show an increased amount of head orientation towards the next speaker around the end of turn, no matter who the next speaker was. The human speaker looking at the next speaker increasingly towards the end of turn in these situations is in line with the findings of Kendon[4], but the trend may be predisposed by the employed selection criterion.

Human-robot and robot-human turn changes were clearer than human-human turn changes. Comparing the head pose distribution patterns between human-human and human-robot turn changes (Fig. 4), the human-robot distribution was more consistent with fewer peaks, and had the main peak located at the end of turn. The difference could be related to the wizard, as the robot's decisions on when to speak were made by the wizard. It could also be related to the robot, due to, for example, the subjects' expectations on a robot dialogue system, or the robot not making use of visual cues. When the speaker's visual attention was directed towards the next speaker at some point close to the end of turn (Fig. 5), we observed a transfer of the next speaker's visual attention away from the current speaker to the table near the end of turn. This was the case for both human and robot current speakers, albeit an earlier transfer in the case of a human speaker and differing overall proportions. The disparity could be related to differences between the humans and the robot in signaling intentions with, for example, prosody or visual cues, or to differing dialogue strategies or types of utterances.

The behavior of the human speakers was not symmetrical between robot and human interlocutors (Fig. 4 and 5). The human not involved as one of the speakers, on the other hand, exhibited a similar distribution of attention (Fig. 6) close to turn changes regardless of the robot being the first or the second speaker. Visual attention towards the table remained fairly constant, while a transfer of attention from the current speaker to the next speaker took place. Using the mimicry target[1] to define human-likeness, we employ the distributions of visual attention as a measure of human-likeness. This gives us a measurable target to work towards in order to make the robot more human-like in the aspect of visual attention by a human interlocutor. As mentioned, we found differences in the human speakers' head pose patterns surrounding turn changes when the turn taker was another human, compared to when it was the robot. Thus, we need to modify the robot's behavior in order to bring the distribution of visual attention of its human interlocutor closer to the one exhibited for two human interlocutors. Matching distributions is however only a first step towards human-likeness. The distributions provide an overall view, they do not reveal if any individual actions were human-like or not.

Our observation that all subjects interacted with the objects on the table, combined with a large part of the interlocutors' attention directed towards the table, suggests that exploration of more detailed targets of attention could be worthwhile. A dialogue system could conceivably make use of estimations about which objects the interlocutors are paying attention to. Additionally, comparing, for example, the dialogue acts leading to the change of turn for different targets of attention might also be useful when designing a dialogue system for this setting.

6   Future Work

With the continued goal of exploring the symmetry of turn-taking, the next step is to adjust the robot's behavior to see if we can get it to trigger more human-like turn-taking behavior from humans talking to it, compared to when the humans talk to each other. The long-term goal is to build an automated system capable of improving the symmetry of an ongoing dialogue by adjusting the robot's behavior.

Another interesting goal is to replace the wizard with an autonomous system that appears to be intelligent. One first step in that direction is to improve the robot's ability to deduce where the visual attentions of the interlocutors are. Having a fine-grained sense of attention to task-related object could help provide the robot with valuable insights on the intentions of its interlocutors.

7   Conclusions

In this paper we presented a setup for collecting multimodal data from three-party human-robot interaction, and used the setup to create an initial corpus collected in a Wizard-of-Oz setting. We explored the symmetry of head pose patterns in turn-taking between human and robot participants, finding both similarities and differences. The robot attracted different head pose patterns from humans during turn changes than what humans used between each other. The differences can be useful when implementing a dialogue system, but also indicate a disparity between robot and human interlocutors. In other words, the robot's behavior needs to be changed to make it more human-like in this aspect. We can use symmetry in behavior as a target to evaluate progress. We also investigated the question about implications of introducing task-related objects into the dialogue. Objects attract visual attention at the expense of interlocutors, possibly affecting the usefulness of head pose as an indicator for addressee identification.

Acknowledgments. This work is supported by the Swedish research council (VR) project "Incremental processing in multimodal conversational systems" (2011-6237), as well as the SAVIR project (Situated Audio Visual Interaction with Robots) within the Strategic Research Area ICT - The Next Generation, funded by the Swedish Government.

References

1. Edlund, J., Gustafson, J., Heldner, M., Hjalmarsson, A.: Towards human-like spoken dialogue systems. Speech Communication 50(8-9), 630–645 (2008)
2. Skantze, G., Gustafson, J.: Attention and interaction control in a human-human-computer dialogue setting. In: Proceedings of SigDial, London, UK, pp. 310–313 (2009)
3. Katzenmaier, M., Stiefelhagen, R., Schultz, T.: Identifying the addressee in human-human-robot interactions based on head pose and speech. In: Proceedings of International Conference on Multimodal Interfaces, pp. 144–151 (2004)
4. Kendon, A.: Some functions of gaze-direction in social interaction. Acta Psychologica 26, 22–63 (1967)
5. Vertegaal, R., Slagter, R., van der Veer, G., Nijholt, A.: Eye gaze patterns in conversations: there is more to conversational agents than meets the eyes. In: Proceedings of the SIGCHI Conference on Human Factors in Computing Systems, pp. 301–308 (2001)
6. Stiefelhagen, R., Zhu, J.: Head orientation and gaze direction in meetings. In: Conference on Human Factors in Computing Systems, pp. 858–859 (2002)
7. Ba, S.O., Odobez, J.-M.: Recognizing visual focus of attention from head pose in natural meetings. IEEE Transactions on Systems, Man, and Cybernetics, Part B: Cybernetics 39(1), 16–33 (2009)
8. Jokinen, K., Nishida, M., Yamamoto, S.: Collecting and annotating conversational eye-gaze data. In: Proceedings of Workshop Multimodal Corpora: Advances in Capturing, Coding and Analyzing Multimodality, Language Resources and Evaluation Conference, Malta (2010)
9. Argyle, M., Graham, J.A.: The central Europe experiment: Looking at persons and looking at objects. Environmental Psychology and Nonverbal Behavior 1(1), 6–16 (1976)
10. Al Moubayed, S., Edlund, J., Gustafson, J.: Analysis of gaze and speech patterns in three-party quiz game interaction. In: Interspeech, Lyon, France (to appear, 2013)
11. Burgoon, J.K., Bonito, J.A., Bengtsson, B., Cederberg, C., Lundeberg, M., Allspach, L.: Interactivity in human-computer interaction: a study of credibility, understanding, and influence. Computers in Human Behavior 16(6), 553–574 (2000)
12. Al Moubayed, S., Skantze, G., Beskow, J.: The Furhat back-projected humanoid head - Lip reading, gaze and multiparty interaction. International Journal of Humanoid Robotics 10(1) (2013)
13. Al Moubayed, S., Skantze, G.: Perception of gaze direction for situated interaction. In: Proceedings of the 4th Workshop on Eye Gaze in Intelligent Human Machine Interaction. The 14th ACM International Conference on Multimodal Interaction, Santa Monica, CA, USA (2012)
14. Nass, C., Fogg, B.J., Moon, Y.: Can computers be teammates? International Journal of Human-Computer Studies 45(6), 669–678 (1996)

I Would Like Some Food: Anchoring Objects to Semantic Web Information in Human-Robot Dialogue Interactions

Andreas Persson, Silvia Coradeschi, Balasubramanian Rajasekaran,
Vamsi Krishna, Amy Loutfi, and Marjan Alirezaie

Center for Applied Autonomous Sensor Systems (AASS)
Dept. of Science and Technology, Örebro University
Örebro, Sweden
{andreas.persson,silvia.coradeschi,amy.loutfi}@oru.se

Abstract. Ubiquitous robotic systems present a number of interesting application areas for socially assistive robots that aim to improve quality of life. In particular the combination of smart home environments and relatively inexpensive robots can be a viable technological solutions for assisting elderly and persons with disability in their own home. Such services require an easy interface like spoken dialogue and the ability to refer to physical objects using semantic terms. This paper presents an implemented system combining a robot and a sensor network deployed in a test apartment in an elderly residence area. The paper focuses on the creation and maintenance (anchoring) of the connection between the semantic information present in the dialogue with perceived physical objects in the home. Semantic knowledge about concepts and their correlations are retrieved from on-line resources and ontologies, e.g. WordNet, and sensor information is provided by cameras distributed in the apartment.

Keywords: anchoring framework, semantic web information, dynamic system, human-robot dialogue, sensor network, smart home environment.

1 Introduction

Socially assistive robots in combination with smart home environments are increasingly considered as a possible solution for providing support to elderly and persons with disabilities [3]. Such ubiquitous robotic systems present a number of interesting application areas that aim at improving quality of life and at allowing people to live independently in their own home longer. These technological solutions can be used to monitor health condition via physiological sensors and activities recognition, to remind about medicines and appointments, to raise alarms, and to assist in everyday tasks like finding objects, detect if objects are misplaced, and guiding in the execution of tasks.

An important facet to many of these applications is the ability to communicate about and interact with objects that are present in the home. A key challenge is therefore to connect the information provided via the dialogue with the sensor

G. Herrmann et al. (Eds.): ICSR 2013, LNAI 8239, pp. 361–370, 2013.

information gathered by the robot and/or a sensor network. The sensor network is able to assist the robot in performing tasks which are otherwise difficult on its own, e.g. localization, while the robot also serves as a useful point of interaction. Studies in human-robot interactions such as [5] have even shown that smart homes with sensor networks are more readily acceptable when a mobile robot is present as an interaction, i.e. receives and relays information to the inhabitants. The challenge to connect semantic information (e.g. general concepts and object names) to the sensor data that refers to the object (e.g. feature descriptors) has been called the anchoring problem. This connection should be first established and then maintained over time. Anchoring has been initially defined in [4] and then extended to consider large knowledge base system like Cyc [6]. The specific challenge of anchoring objects in the context of social robotics in domestic environments, is both the large possibility/variety of objects that need to be anchored and the fact that each object can be referred to in a number of ways in a dialogue. To allow for the possibility to have a dynamic system without the constraints of the dialogue being defined a-priori, open sources of information such as the web can be used. Spoken dialogue is a natural and effortless medium of communication between humans. This together with the evolution of speech related on-line services, makes spoken dialogue a more and more prominent solution for human-robot interaction [7]. It has also been proven that a robot which is capable of interacting in natural language would be more convenient to use, even for users without technical experience [12].

This paper presents an implemented system in a real home environment that can establish a dialogue with a human user about common household objects. The system consists of a mobile robot together with a set of distributed cameras in the home. The robot can accept requests for finding objects via spoken dialogue with a human user and use stored information provided by the cameras about objects present in the environment to answer the requests. In order to create a robustness in the dialogue, ontologies are used to relate concepts employed by the human to the semantic information stored in the anchors that refer to the specific object present in the home. Using on-line ontologies, i.e. WordNet[1], this information is mined as it is needed rather than stored a-priori, as is the case in previously reported work on anchoring [6]. The novelty of this work is twofold. On one hand, the use of ontologies and in particular on-line ontologies is a novel and important contribution to allow for a flexible handling of objects in dialogue and task performance. On the other hand, the integration of the mobile robot in a smart home allows an efficient bottom up approach of anchoring, where all the significant objects in the environment are anchored and continuously and dynamically updated. A request for a specific object is then matched with the stored information. In previous approaches to anchoring, the focus was on a mobile robot that was processing an anchoring request and mainly finding a matching object to the current image and partly to the stored objects images. These approaches were mainly top down and where initiated by a specific request [9]. The new bottom up development presented in

---

[1] http://wordnet.rkbexplorer.com/

this paper has been made possible by the use of complementary static cameras and most importantly by the efficient use of database techniques allowing the storage and easy retrieval of thousands of objects. Section 2 presents a novel anchoring framework suitable for social interaction; Sect. 3 presents a scenario and experimental results while Sect. 4 concludes the paper.

2   Framework for Improved Anchoring via Social Interaction

The anchoring framework, seen in Fig. 1, is a distributed system consisting of several integrated nodes communicating through ROS (Robot Operating System)[2]. Each node has one or more designated task(s) triggered by system events, other nodes, or human users. The overall architecture is divided into three core modules (described in detail in the following sub-sections): 1) *perceptual anchoring module*, 2) *semantic knowledge module*, 3) *human-robot interface module*.

To facilitate persistent storage of information a MongoDB[3] database is used together with an upper generic query interface, which is integrated with both the *perceptual anchoring module* and the *semantic knowledge module*. Such a setup, with a NoSQL database, provides a scalable and dynamic solution suitable for storage of growing collections of both perceptual sensor data and semantic knowledge collected from on-line ontologies.

2.1   Perceptual Anchoring Module

The Perceptual Anchoring Module consists of both a *perceptual system* and an *anchoring system*, shown in Fig. 1 – No. 1. The perceptual system consists of distributed sensors, namely cameras, in the smart home environment, detecting objects in the environment. The output of the perceptual systems are percepts, which are then fed to the anchoring system. A percept is defined to be a structured collection of measurements assumed to originate from the same physical object. The measurements are described by a set of attributes. In our system, percepts are obtained only from visual features that are processed by the images obtained by the cameras. We denote a percept by π and the specific attributes by Φ, specifically, ϕ_1 refer to binary visual features (FREAK [1]), ϕ_2 refer to color for each percept. Each attribute contains a predicate grounding relation, e.g. (white, color, 255 255 255). However, for the visual features, a more complex method to extract the predicate grounding relation, based on existing image databases was used. This method is outside the scope of this paper but a description can be found in [11].

The anchoring system, follows a bottom-up approach described in [9]. The system receives a percept and invokes a matching algorithm in which the percept is compared against all previously stored anchors in the anchoring system. The

---

[2] http://www.ros.org/
[3] http://www.mongodb.org/

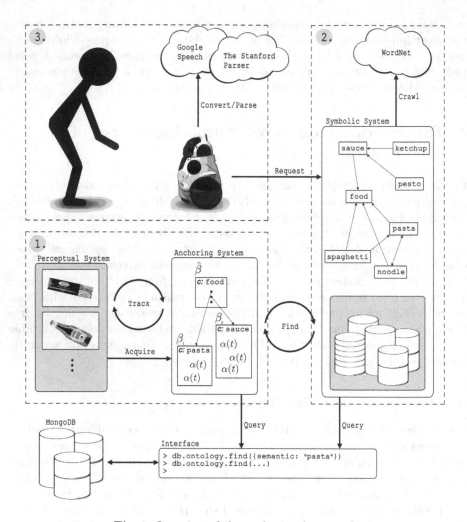

Fig. 1. Overview of the anchoring framework

details of the match algorithm can be found in [11]. Depending on the results of matching, anchors are either created or maintained through two functionalities:

– *Acquire* – creates a new anchor whenever a percept is received which currently does not match a percept of an existing anchor. The new anchor consists of the perceptual description, a symbolic description (which is the set of predicates from the grounding relation), current time t, and a unique identifier.
– *Track* – updates an existing anchor whenever a percept is received which matches the attributes of an existing anchor. Updates the `time` *field* from time $t - k$ to current time t, through the interface towards the database and with the use of a unique identifier of the existing matching anchor.

A third functionality called the *find* functionality can be invoked to find anchors top-down. It is this functionality which is used to find matching anchors requested by the user. This functionality could also provide new, refined or enriched semantic symbolic descriptions as result of a dialogue with a user or as the result of crawling on-line repositories. It is this particular facet which is the novel contribution of this work to the existing work on anchoring.

2.2   Semantic Knowledge Module

The core of this module is a *semantic system*, seen in Fig. 1 – No. 2, consisting of an *ontology* together with a *knowledge base*, both maintained as *collections* in the database. The knowledge base mirrors the result of crawling on-line repositories, while the ontology is stored in the form of hierarchies in a finite search space, where the relation between the various nodes is stored in the form of subsumption relation. A *field* in the *collection* contains a sequential list of all the nodes that have to be traversed in order to search the association between semantic concepts.

As an intermediate layer between the human-robot interface and anchors, this module must be able to receive, interpret and respond upon semantics descriptions about anchors as well as user requests. This is facilitated through three main functionalities:

- *Crawl* – initiates a web crawling in order to update the knowledge base whenever a user is requesting something outside the boundaries of existing knowledge. For this purpose, the WordNet on-line semantic repository is used. This functionality is created in such a way that all the hypernyms and hyponyms/troponyms of a noun/verb are explored recursively till there are no more nodes to explore. The result is then stored in *collections* **nouns**/**verbs** respectively in the database.
- *Request* – receives and parses user requests. Based on requested objects (nouns) and/or activities (verbs), a search for possible candidates in the ontology is conducted. If a match for the request is available, then a reply is sent immediately, otherwise an additional search is conducted for existing knowledge, and where a *crawl* is initiated in case there exists no knowledge about requested verb and/or noun. In the latter case, the crawling with its results is also synchronized with the regularly invoked *find* function so that the ontology is up to date before being re-searched.
- *Find* – the purpose of this is twofold: 1) update the ontology of the *semantic system* based on perceived anchors, 2) update semantics of the *anchor system* based on new knowledge as result of a dialogue with the user. Upon receiving semantic symbols of perceived anchors, the **ontology** *collection* of the database is updated so that the ontology only consists of 'is-a' hierarchies of concepts perceived at current time t. Furthermore, a comparison is made between the existing knowledge and semantic symbols of perceived anchor such that updated knowledge is sent as response to the *anchoring system*.

2.3   Human-Robot Interface Module

A Q.bo Robot[4] and a set of sensors connected via a network are present in the smart home. The robot is a commercial product using the Festival speech synthesis system [2] together with the Julius continuous speech recognition algorithm [8] for interacting with human users. However, the Julius algorithm requires a given grammar, which would limit the speech input, especially if the system is used for human natural conversations. Therefore, the robot's speech-to-text system is instead consisting of an on-line solution using Google Speech[5] together with the Stanford Parser[6] for parsing the text.

The human-robot interface module, seen in Fig. 1 – No. 3, is initiated by a voice command. The human voice is recorded and sent to the Google Speech service for conversion to text. The resulting text is then parsed using Stanford Parser. The result of the parser is further processed by the robot. Here, the robot tries to understand the meaning and context of the users request prior initiating the *semantic system*. Once the robot has confirmed that the user has made a request, the *request* functionality is called. Upon receiving response, results are interpreted, e.g. `there is 'milk' located in the 'kitchen'`, and synthesized as speech through the Festival algorithm.

3   Evaluation

The evaluation was carried out in a smart home used as test apartment (called Ängen), see Fig. 2 – No. 1, which is located in a unique building complex as part of an initiative to provide complete care facilitates for older people in Örebro, both elderly and independent seniors.

As a reference data for the anchoring system was a collection of 6,830 high resolution images (treated as percepts) of common products found in a typical (Swedish) grocery store used. Those images were used for prior training such that attributes were measured and stored in the database together with a semantic symbols for each set of attributes. Where stored attributes were later used for matching new percepts upon arrival at the anchoring system.

3.1   Scenario

The *perceptual system* – consisting of distributed cameras in the smart home environment – registers changes in the environment over a sequential series of frames, captured from one of the cameras located in the kitchen (Fig. 2 – No. 2–3). With the use of computer vision and morphological filters, an area of interest is selected and brought to attention (Fig. 2 – No. 4). Through zooming in on the selected area, five distinct regions of interest corresponding to five percepts are further selected (Fig. 2 – No. 5). Those percepts are passed on

---

[4] http://thecorpora.com/

[5] http://www.google.com/intl/en/chrome/demos/speech.html

[6] http://nlp.stanford.edu/software/lex-parser.shtml

to the *anchoring system* where specific attributes are measured, namely binary visual features (FREAK [1]) and color. Furthermore, locations of the percepts are given based on the locations of the cameras. The anchoring system performs matching between measured attributes and attributes of all previously stored anchors α_x (as result of prior training), where the system recognize the objects as previously known object and updates the time, t, of existing anchors, results seen in Table 1.

Table 1. Anchored objects perceived by the 'kitchen' camera

UID	Semantic	Color	Location
ketchup-1	ketchup	red	kitchen
milk-1	milk	white	kitchen
pasta-1	pasta	blue	kitchen
pasta-2	pasta	red	kitchen
sauce-1	pasta sauce	blue	kitchen

At arbitrary time, $t + 1$, the elderly resident of the smart home apartment initiates a conversation with the Q.bo robot in order to ensure him-/herself about that there is something to eat in the apartment (Fig. 2 – No. 6):

– *User:* Q.bo!
– *Robot:* Yes?
– *User:* I would like some 'food'.
– *Robot:* Let me see what I can find, just a second...

Once the robot has interpreted the parsed result of the conversation and concluded that the user has made a request (Fig. 2 – No. 7), the *semantic system* is triggered through the *request* function. Upon receiving the request, the *semantic system* searches the 'is-a' hierarchies of the ontology and/or the knowledge base of the concept food. In case there is no knowledge about the concept food, a search against WordNet is initiated through the *crawl* function. Results of crawling WordNet are further stored as knowledge before the system is synchronized with the *find* function in order to update both the ontology and to enrich the semantic description of perceived anchors based on the new knowledge. In this case, all five perceived objects are possible candidates related to food (Fig. 2 – No. 8). However, by looking at similarities between concepts (see following Section 3.2), there exists one semantic candidate corresponding to two objects (pasta-1 and pasta-2) which is more prominent (has higher similarity score) than the others. Information about those objects are sent as response (Fig. 2 – No. 9), where the result is spoken back to the user through the Festival speech synthesis:

– *Robot:* There are two 'pasta' located in the 'kitchen', one 'red' object and one 'blue' object.
– *User:* Thank you Q.bo.

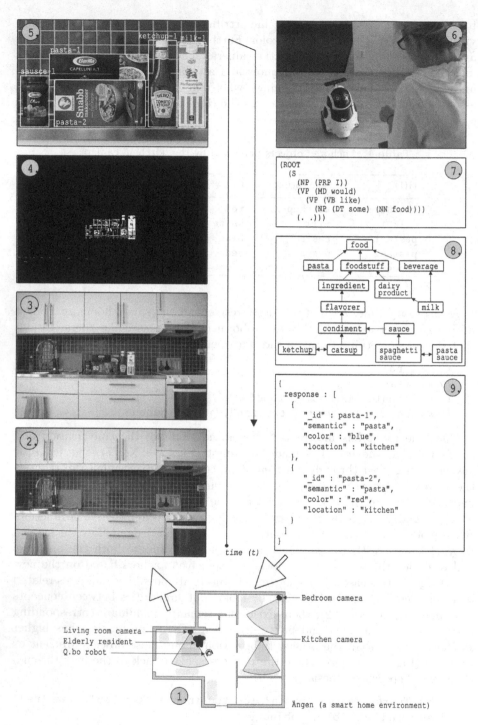

Fig. 2. Overview of test apartment and flow of the steps in evaluation scenario

3.2 WordNet Similarity Search

In a large scale setup, a user query could result in hundreds or thousands of matching candiates, and also ambiguous results, it is of importance to further process matching candiates to find the most prominent candiates. For this purpose was the WordNet::Similarity software package used [10] together with the shortest path algorithm: $y(s_1, s_2) = 1/length(s_1, s_2)$, where s_1 and s_2 are concepts and the *length* between the concpets is the shortest path between the concepts in the 'is-a' hierarchies of WordNet. The most prominent candidates are considered as the ones with highest similarity score y.

As stated previously, the use of on-line repositories, e.g. WordNet, promotes a more dynamic dialogue, and hence, the specific request can vary. The results of a few examples of possible requests like `I would like some 'X'` or `is there something 'X'?`, and the similarity between requested *noun* and known objects is shown in Table 2.

Table 2. Similarity score between a few concepts of example

	s_2 (nouns)				
s_1 (nouns)	`food`	`drink`	`liquid`	`eatable`	`fruit`
`ketchup`	0.1667	0.1429	0.1000	0.1429	0.0667
`milk`	0.3333	0.5000	0.3333	0.2500	0.0833
`pasta`	0.5000	0.2000	0.1667	0.2000	0.0909
`pasta sauce`	0.1429	0.1250	0.0909	0.1250	0.0625

4 Conclusions

The possibility to use technology to contribute to an improved quality of life of elderly and persons with disability and to allow them to live independently in their own homes has rised many expectations and hopes. However the fulfillement of such expectations will only be possible if robust and flexible systems are developed that are suitable for supporting everyday tasks. A key capability of such systems is the possibility to speak about physical objects present in the home using a natural vocabulary. In this paper we have presented a distributed perceptual anchoring framework in a smart home where an ontology can be extended based on requests given by human users and through crawling on-line repositories. This provides for not only an enriched dialogue, but also enhances the semantic symbols anchored to perceptual data in a anchoring system. An example of use of the system has been presented throughout a scenario, where an example of enrichment is shown.

Further developments of this work will consider how to initiate the framework. There exists a certain "chicken or the egg" -situation – percepts are needed in order to match and anchor percepts with semantic symbols, while semantic symbols are needed in the first place in order to initiate an on-line crawling. A solution to this problem is to a-priori train the framework (which was the case in the scenario presented in this paper), whereas a more complex solution, but also more challenging, would be to involve the user in a dialogue in order to learn objects. User experience and acceptable delays in a natural language dialogue is another area for further evaluation. Delays caused by crawling on-line repositories is highly dependent on the request, recursiveness in the search and bandwidth, and are therefore also difficult to predict.

Acknowledgement. This work has been supported by the Swedish Research Council (Vetenskapsrådet) under the grant number: 2011-6104.

References

1. Alahi, A., Ortiz, R., Vandergheynst, P.: Freak: Fast retina keypoint. In: CVPR, pp. 510–517. IEEE (2012)
2. Black, A.W., Taylor, P.A.: The Festival Speech Synthesis System: System documentation. Tech. Rep. HCRC/TR-83, Human Communciation Research Centre, University of Edinburgh, Scotland, UK (1997)
3. Coradeschi, S., Cesta, A., Cortellessa, G., Coraci, L., Gonzalez, J., Karlsson, L., Furfari, F., Loutfi, A., Orlandini, A., Palumbo, F., Pecora, F., von Rump, S., Štimec, A., Ullberg, J., Östlund, B.: Giraffplus: Combining social interaction and long term monitoring for promoting independent living. In: Proc. of the 6th Int. Conference on Human System Interaction (HSI 2013), Sopot, Poland (2013)
4. Coradeschi, S., Saffiotti, A.: Anchoring symbols to sensor data: preliminary report. In: Proc. of the 17th AAAI Conf., pp. 129–135. AAAI Press, Menlo Park (2000)
5. Cortellessa, G., Loutfi, A., Pecora, F.: An on-going evaluation of domestic robots. In: Robotic Helpers: User Interaction, Interfaces and Companions in Assistive and Therapy Robotics, pp. 87–91 (2008)
6. Daoutis, M., Coradeschi, S., Loutfi, A.: Cooperative knowledge based perceptual anchoring. International Journal on Artificial Intelligence Tools 21(3) (2012)
7. Kanda, T., Ishiguro, H., Ono, T., Imai, M., Mase, K.: Multi-robot cooperation for human-robot communication. In: IEEE Int. Workshop on Robot and Human Communication (ROMAN 2002), pp. 271–276 (2002)
8. Lee, A., Kawahara, T.: Recent development of open-source speech recognition engine julius. In: Proc. of APSIPA ASC, pp. 131–137 (2009)
9. Loutfi, A., Coradeschi, S., Saffiotti, A.: Maintaining coherent perceptual information using anchoring. In: Proc. of the 19th IJCAI Conf., Edinburgh, UK, pp. 1477–1482 (2005)
10. Pedersen, T., Patwardhan, S., Michelizzi, J.: Wordnet::similarity - measuring the relatedness of concepts. In: HLT-NAACL 2004: Demonstration Papers, May 2-7, pp. 38–41. Association for Computational Linguistics, Boston (2004)
11. Persson, A., Loutfi, A.: A hash table approach for large scale perceptual anchoring. In: Proc. of IEEE Int. Conference on Systems, Man and Cybernetics (SMC 2013), Manchester, UK (2013)
12. Spiliotopoulos, D., Androutsopoulos, I., Spyropoulos, C.D.: Human-robot interaction based on spoken natural language dialogue. In: Proc. of the European Workshop on Service and Humanoid Robots (ServiceRob 2001), Santorini, Greece, pp. 25–27 (2001)

Situated Analysis of Interactions between Cognitively Impaired Older Adults and the Therapeutic Robot PARO

Wan-Ling Chang[1], Selma Šabanović[1], and Lesa Huber[2]

[1] School of Informatics and Computing, Indiana University, Bloomington, IN, USA
[2] School of Public Health, Indiana University, Bloomington, IN, USA
{wanlchan,selmas,lehuber}@indiana.edu

Abstract. In order to explore the social and behavioral mechanisms behind the therapeutic effects of PARO, a robot resembling a baby seal, we conducted an eight-week-long study of the robot's use in a group activity with older adults in a local retirement facility. Our research confirms PARO's positive effects on participants by increasing physical and verbal interaction as evidenced by our behavioral analysis of video recorded interactions. We also analyzed the behavioral patterns in the group interaction, and found that the mediation of the therapist, the individual interpretations of PARO by different participants, and the context of use are significant factors that support the successful use of PARO in therapeutic implementations. In conclusion, we discuss the importance of taking the broader social context into account in robot evaluation.

Keywords: PARO, Older Adults, Dementia, Socially Assistive Robot.

1 Introduction

PARO, a socially assistive robot resembling a baby seal, has been commercially available since 2005 and is widely used as a social companion for older adults in institutional as well as domestic contexts. Studies in institutions all over the world (e.g., Japan [1], US [2], and Australia [3]) provide evidence of its therapeutic effect improving users' emotional status (e.g. [4]) and reducing stress levels (e.g., [1]). Evaluations of PARO have also shown that its use can increase social interaction among older adults. Most existing studies focus on examining PARO's effect in experimental settings or using quantitative measurements (e.g., [3]). While such studies substantiate PARO's therapeutic effects – i.e., show that PARO *works* – they give little information about the mechanisms producing the effects – i.e., *how* PARO works. The study we present corroborates PARO's positive effects on participants' activity levels, but focuses on describing observational findings regarding the specific behavioral, contextual, and personal factors that contribute to these effects.

In order to explore the behavioral, social, and contextual factors that support PARO's therapeutic effects, we conducted an eight-week-long observational study of ten older adults with dementia participating in Multi-Sensory Behavioral Therapy (MSBT) in a local eldercare institution. MSBT is widely used for people with

G. Herrmann et al. (Eds.): ICSR 2013, LNAI 8239, pp. 371–380, 2013.
© Springer International Publishing Switzerland 2013

dementia, and involves controlled sensory stimulation in a relaxing environment aimed at keeping the sensory systems of cognitively impaired people active [5]. We adapted the concept of MSBT by using PARO as a source of multimodal sensory stimuli – it makes seal-like vocalizations, has visible movement, and is covered with soft hair that encourages touch. We explored how older adults interacted with PARO and the therapists and other participants in the group, and used quantitative and qualitative approaches to study not only the incidence of particular behaviors but also their meaning. Our study shows increasing activity among participants in terms of incidence and duration, and suggests that mediation by therapists, environmental factors (e.g. change in venue, composition of the group), and participants' variable interpretations of PARO, are crucial mechanisms supporting PARO's therapeutic efficacy when used by older people with dementia.

2   Background

2.1   PARO's Use in Eldercare

PARO (see Fig.2) is a socially assistive robot developed by Japan's National Institute of Advanced Industrial Science and Technology (AIST). The built-in tactile, balance, and light sensors and microphones under its soft fur enable PARO to sense touch, sound, and changes in position (e.g. hugging), and the motor system and speakers allow PARO to react to human interaction with physical movement (e.g., wagging its tail) and vocalizations simulating a baby seal. PARO is categorized as a socially assistive or therapeutic robot and is often used in a manner similar to pet therapy and as a companion for children and older adults, especially seniors with dementia.

Numerous clinical studies of PARO discuss its positive psychological, physical, and social effects on older adults, including improving the mood of participants [4] and decreased stress levels after using PARO [1]. PARO's social effects have been substantiated by showing that seniors in a living center who had free access to PARO in a public space had increased contact with each other while the robot was present [6]. Moyle et al's [3] experimental study of PARO in a group setting showed using PARO could reduce agitation and anxiety caused by dementia.

Studies investigating the behavioral mechanisms of therapy using PARO have been more limited. Shibata [7] documents a variety of participant behaviors during sessions with PARO, including stroking, speaking, singing, and smiling to PARO, and communicating with caregivers. Existing studies have identified nurturing and sharing the robot among participants in group activities as significant emergent behaviors in interactions with PARO [2]. Taggart [2] and Shibata [7] have documented cases in which PARO incited users to reflect on their personal history and emotionally charged issues. Wada [8] developed guidelines for using PARO therapeutically by systematically filtering and categorizing observations of PARO's implementation in nursing institutions. While these guidelines suggest best practices, they focus on case studies and successful one-time observations, rather than providing a more comprehensive picture of situated interactions between older adults and PARO.

Our study contributes to existing evaluations of PARO with a systematic analysis of the robot's implementation in an MSBT group to understand the interaction mechanisms that lead to success and failure of PARO's use with older adults.

2.2   Socially Situating Human-Robot Interaction

Research in Human-Robot Interaction (HRI) suggests that interactions between humans and robots depend not only on the robot's capabilities, but also on the behaviors of human actors, the social and physical context, and daily routines. Forlizzi [9] suggests that robotic technologies designed for older adults are part of a "product ecology," which includes other people, technologies, and the environment, and need to fit this context. Sabanovic [10] argues that HRI studies need to take the broader social context into account to evaluate the meaning and consequences of robots. These conceptual points are supported by empirical studies. Mutlu's [11] study of an assistive robot used in a hospital showed that the physical environment, social/emotional tenor, and the organizational structure and workflow of the space significantly affected people's evaluations of robot coworkers.

Takanori Shibata, PARO's creator, describes PARO's design as underdetermined - - "it is not necessary for PARO to have all the functions, the interaction can enlarge the number of functions."[1] This suggests that "interpretive flexibility"—the ability of a technology to "sustain diverging opinions" from different user groups [12]—is at the foundation of PARO's design. Other HRI researchers have also noted interpretive flexibility plays an important role in the way users make sense of robots. Alač [13] suggested HRI is constituted through the interpretative actions of human participants. Shibata [7] described various cases in which older adults and caregivers adapted PARO in different ways to suit their needs. Taggart et al. [2] also described PARO as an "evocative object," provoking different reactions from different people.

Our study shows that the physical environment, and emotional status and behaviors had a significant impact on the interaction between participants and PARO, as well as the activity within the group. Our participants also used PARO's interpretative flexibility to develop various ways of interacting with PARO to meet their needs.

3   Study Design

3.1   Field Site and Participants

Our study was conducted in a local retirement community in Bloomington, IN, which provides both long-term and short-term care for older adults. Ten participants joined our research sessions. All of them were over 65 years old and had mild to severe dementia. Two were male and the rest were female. Most participants were in wheelchairs; only one could walk with limited assistance. The participants varied in terms of their social interactivity – some were eager to interact with others, while those with severe dementia had difficulty speaking and often fell asleep during sessions. Participants with severe dementia were sometimes confused about their whereabouts and displayed anxiety. Since participation in the sessions was voluntary,

---

[1] Talk at the Japan Society in New York, June 2007.

participants did not attend all sessions; we usually had six to seven participants in each session and only three participants showed up consistently throughout the study.

3.2   Methods

To get a comprehensive understanding of the interactions among participants, and between participants, PARO and the therapist during sessions, we performed both on-site coding and video recorded the sessions for more detailed coding later on. Results of on-site coding were published in a prior paper [14], so this paper focuses on the video data. We used one video camera to document the interaction with PARO in each session. Usually, the participants sat in a circle. To catch the best view of the interaction, the researcher moved the video and faced the participants interacting with PARO while the robot was passed around in the group. We ran two pilot sessions and eight study sessions from July 27th to September 14th, 2012. Due to technical issues, we lost video data for August 3rd so that date is excluded from our analysis. Each therapy session lasted around forty minutes.

Video Coding. Due to limitations of the camera's perspective, which focused on the participant who was directly interacting with PARO and did not include much of the interactions of other participants, coding focused on physical (e.g., petting, holding, kissing) and verbal interactions (e.g. talking, singing), without taking interpersonal gaze into account. We marked the interaction start and end times for each coded behavior, the interactor, the target interacted with, and interaction type using Anvil video coding software. The coders attended to three categories of behavior: "Interaction with PARO," "Interaction with other participants," and "Interaction with therapist." Physical interactions were mainly with PARO. Verbal interactions were limited and most of them occurred around PARO, though codes include verbal interaction with therapists, other participants, and other people (e.g., family members, researchers). We analyzed the interactions by aggregating the interaction durations across each behavioral category and interaction target.

Observational Field Notes and Post-interview. Besides coding the videos, we also took observational notes about the details of interaction to identify repeating behavioral patterns and interpret the meanings behind certain behaviors. We also conducted interviews with the therapists participating in the therapy sessions before and after the study. The therapists kept PARO in the retirement community for the duration of our study and could use it freely, so interviews focused on their general experiences with PARO. Furthermore, since the reported study is part of more long-term observation of the retirement institution, we use insights from continuing observation to interpret the context and meaning of the patterns found in this study.

4   Results

4.1   Overall Increase in Interactivity

In a previous publication reporting on the results of on-site behavioral coding [14], we showed that there was a consistent growth in the frequency of both the Primary (directly interact with PARO) and Non-Primary (not in direct contact with PARO) participants' interactions over the course of the study. Our analysis of the video-recorded interactions similarly shows an increase in the average duration of physical and verbal interactions with PARO over the course of the study (see Fig.1). Since verbal and physical interactions are at times exclusive of each other, we also added the duration of the two to get a more comprehensive view of the interaction. In Figure 1, the total duration of both physical and verbal interaction has a more obvious upward trend than physical or verbal interaction taken separately. There is also a slight decrease in interaction durations for the last two sessions. In interviews, the therapists mentioned that the mental and emotional status of participants had a significant impact on interaction during the research sessions. Both the general atmosphere and the activity levels of participants were low on 8/10, 9/7 and 9/14, with many participants sleeping, which may explain the decline in interaction. Finally, the average duration of physical interaction is always higher than verbal interaction, in line with the diminished verbal capabilities of some of the participants.

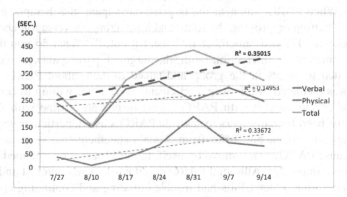

Fig. 1. Average duration of verbal and physical interactions

4.2   Mediation by Therapist

During our study, the therapist had full control over the conduct of the activity so that it would be most beneficial to participants. The two participating therapists were provided with Wada's Guidelines for Using PARO [8] prior to the study. The therapist ended up changing the way she ran the activity throughout the study, which allowed us to see the effects of a variety of social and physical arrangements of the group. In the beginning of our study, the therapist seated participants around a small table with PARO at the center, which made it difficult to reach for some residents (Fig. 2A). In this configuration, we found PARO did not engage the participants, even though some of them had interacted with it in the pilot study. Some of them looked at PARO without reaching for it, and others simply ignored it. After this session, the therapist started mediating the interaction by encouraging participants to interact with PARO. She passed PARO from one participant to the other, accompanying the robot and scaffolding its use. This led to participants interacting with PARO one-on-one, often with the help of the therapist, or sometimes sharing it with another person (Fig. 2B). During a session, each participant was offered to interact with PARO at least once and the more engaged elders had multiple chances to interact with the robot.

Fig. 2. (A) Initial interaction setting. (B) Therapist mediated interaction.

To begin the PARO interaction, the therapist put PARO in front of the resident or on their lap. When PARO was near the targeted participant, the therapist encouraged interaction by petting or stroking the robot and initializing conversation about PARO, saying things like "Do you remember it?" "Isn't it cute?" "What do you think about it?" Besides mediating interaction with the targeted participants, sometimes the therapist shifted to others in the group and talked to them about PARO. When the interaction was mediated in this way, participants were more willing to communicate with each other and interact with PARO. These results are in line with Wada's [8] findings on the need for caregivers to scaffold PARO's use in particular ways for it to have a therapeutic effect, but differ from Wada's statement that such facilitation is required because PARO cannot move by itself. In our observation, the mediator's role was much more important. Although PARO raised participants' interest and became a topic of conversation, older adults had trouble initiating and continuing conversation without mediation by the therapist due to their cognitive and physical impairments.

4.3   Effects of Social and Physical Environment

Aside from therapist mediation, we found that the environment significantly affected the residents' social interactions with PARO and with other people in the room. Normally, the MSBT activities were held in a closed room, but two sessions were held elsewhere. The session on August 17th was outside in the shadow of the building on a sunny day, while the session on August 24th was in the lobby. In both sessions, there was increased social interaction among participants. For example, participant P4 and P9 were both generally highly interactive with PARO, but they interacted with PARO individually most of the time. In the trial in the lobby, P9 moved next to P4 and shared PARO with her in the middle of the session. They would look at each other while both of them interacted with PARO, and conversed about PARO when PARO was passed to another participant. P3 usually had very short conversations with other people, if any. In the outdoor session, she spontaneously and continuously talked to P9 while she or P9 held PARO in their arms. P3 had sat next to P9 more than three times in previous sessions, but with they had not communicated.

The social composition of the group also affected the interactions. P10 attended our study on August 31st. She is talkative and humorous, and her presence had a positive effect on the group. Her conversations with PARO and the therapist caught other participants' attention, causing them to laugh and engage with each other. The frequency of participants' spontaneous conversation with their neighbors was higher at this session than any others and uncharacteristically involved multiple people.

4.4   Individual Interaction Styles

Our observations showed that participants made sense of PARO and defined its social roles in different ways that were related to their individual needs and personal histories. Following are three cases showing different interaction styles we observed.

P5 had memory problems caused by mild dementia, but was capable of fluent conversation with people. She enjoyed staying in the lobby and looked forward to talking to other residents' families when they visited them. In our study, she spent most of her time observing the researchers and events in the space. She did not initiate interaction with the robot, but would become extremely animated in interaction with it when the therapist asked her to participate. She talked to the therapist a lot about PARO, even showing appreciation for its cute appearance. She also talked to PARO sometimes, while attending to people's gaze (including therapist and researchers) at the same time, as if looking for social assurance. Her interactions with PARO seemed to mostly revolve round conversation with the therapists; soon after PARO was passed to another resident, she would lose interest in the robot and started observing the surroundings again.

P8 had severe dementia and had trouble conversing with people. She was often confused and anxious, asking "What should I do?" or " Did I do the right thing?" In our sessions, P8 usually slept, and watched the therapist when she was awake. She was friendly to PARO when the therapist gave it to her, even if she rejected PARO previously. For example, in one session when the therapist came to P8 and told her "You are going to visit him (PARO)," P8 said "No." Later, when the therapist put PARO in her arms, she treated it in a friendly way, petting, and exploring PARO's movement as she described it to the therapist. P8's interactions with PARO were usually encouraged by the therapist and she reacted very positively to the therapist's suggestions and social cues. On one occasion when she rejected holding PARO with a negative facial expression, the therapist laughed and said, "What is that look on your face for?" P9 then kissed PARO but did not hold it.

P9 was very excited to interact with PARO from the beginning, and she is the only person that had a consistently increasing trend in both physical and verbal interactions. She has very mild dementia and converses easily and fluently. She remembered PARO throughout the sessions. In the early stages of the study, P9 was nervous about physical contact with PARO. As she got more familiar with PARO, she was always happy when the therapist presented the robot and eager to interact with it. Previously a singer, she developed a habit of singing to the robot in addition to talking and petting. Because the therapist tried to let all participants have an equal chance to interact with PARO, she could not interact with PARO as long as she wanted. When

PARO wasn't with her, her gaze followed PARO and she laughed or smiled to the participant who had the robot. Because of her interest in PARO, she often shared PARO with other participants, or other participants talked about PARO with her.

These three cases show that PARO can be used by participants as a way to attract other people's attention, as a part of interacting with the therapist and following her suggestions, and as an engaging social companion on its own, depending on the participant's personality and interaction capabilities.

4.5   Post-interview with Therapists

We conducted interviews with two therapists in the nursing institution after the study was completed. One of them was the MSBT session coordinator in our study, while the other was her supervisor. They mentioned that it's difficult to identify steady progress in PARO's effect on the residents because their reactions to PARO depended on their daily emotional and physical status. Interviews with participating therapists helped us develop more informed interpretations of our observations, and also suggested that one-on-one interaction with PARO was more effective for older adults with cognitive impairments such as those in our participant pool.

The therapists used PARO in another MSBT group, and also sometimes let residents have individual PARO interaction time. For example, one resident did not want to get up in the morning so they had her interact with PARO to give her more motivation to leave her bed. Another case involved a resident with agitation and anxiety problems caused by her dementia. When the housekeeper changed her sheets, she usually became angry and screamed. The caregiver gave her PARO before changing her sheets to calm her down. In general, the therapists felt that one-on-one interaction was a more successful way to use PARO in this institution.

5   Discussion

The situational analysis of cognitively impaired older adults interacting with PARO reported in this paper supports prior findings of PARO's therapeutic effects, particularly an increase in both the physical and verbal interaction in the group. In addition, our findings suggest more attention should be paid to broader social structures, such as staff mediation, the social and physical context, and individual needs, that support the success of socially assistive robots in everyday use. We also showed that PARO's underdetermined design encouraged interpretive flexibility, with individual users and the therapist adapting the robot to their needs.

The observed patterns of therapist mediation, the importance of social and physical context, and differing individual interpretations of PARO's role emphasize the importance of studying human-robot interaction holistically across individual, group, and broader institutional frames. While most studies of PARO's use focused solely on the individual participants and PARO (e.g., [1][3][4]), our observations show specific ways in which a broader set of social, environmental, and personal factors contribute to the successful use of PARO in therapeutic settings. The guidelines developed by Wada [8] do not extend to specifying details about the social and physical contexts of use. We found the physical setting and placement of PARO and participants affected interactions, and key individuals (e.g. therapist, talkative elders) could change group dynamics. While substantiating PARO's effects in open-ended interaction can be challenging due to the numerous contextual factors researchers must take into account, further qualitative studies of PARO's use can provide a clearer picture of the interplay of different contextual factors in robot-mediated therapy.

Understanding the various ways in which therapists mediate interactions with PARO is also a crucial aspect of future studies. The therapist in our study adopted multiple methods of generating interaction in the group, including asking questions, showing the interaction with PARO, and directing participants' attention to others interacting with PARO, switching between different mediation methods based on the personal needs of each participant and the group dynamics. The described research suggests that PARO's use and therapeutic results must include the mediation and scaffolding given by therapist or other people to the interactions with the robot.

We observed that PARO's underdetermined design allowed for a diverse set of practices and meanings to evolve around its use, depending on the social context or personal needs of those interacting with it. Leaving space for interpretive flexibility to occur may therefore be a useful design guideline for assistive robots, since it can help counterbalance the limited technological abilities and situational awareness of robots, and accommodate the varied abilities of users. PARO's underdetermined design may also explain its use in more than 30 countries around the world [7].

6 Conclusion

Prior PARO studies showed the robot's positive therapeutic effects and increased social interactions in participants, but included very little information about how those effects were achieved. Our study of PARO focused on understanding the social, behavioral, and environmental factors that support PARO's function in the context of MSBT with cognitively impaired older adults. We suggest extending the focus of study to the physical and social contexts of PARO's use and taking advantage of interpretive flexibility to design therapeutic robots for different contexts. In the future, we plan to perform more extensive studies to understand PARO's use in different settings and focus on the broader social and cultural contexts.

Acknowledgements. This research was supported by NSF grant number IIS-1143712 and by Dr. Takanori Shibata's loan of a PARO robot. We are grateful to the participants and nursing home staff for their assistance, and to S. Hamner, P. Lopes, and H. Lee for their help in conceptualizing and implementing the study.

References

1. Wada, K., Shibata, T., Saito, T., Tanie, K.: Effects of robot assisted activity to elderly people who stay at a health service facility for the aged. In: Proceedings of the 2003 IEEE/RSJ International Conference on Intelligent Robots and Systems (IROS 2003), vol. 3, pp. 2847–2852 (2003)

2. Taggart, W., Turkle, S., Kidd, C.D.: An interactive robot in a nursing home: Preliminary remarks. In: Towards Social Mechanisms of Android Science: A COGSCI Workshop (2005)

3. Moyle, W., Cooke, M.L., Beattie, E., Jones, C., Klein, B., Cook, G., Gray, C.: Exploring the effect of companion robots on emotional expression in older people with dementia: A pilot RCT. Journal of Gerontological Nursing (2012)

4. Wada, K., Shibata, T., Saito, T., Sakamoto, K., Tanie, K.: Psychological and Social Effects of One Year Robot Assisted Activity on Elderly People at a Health Service Facility for the Aged. In: Proceedings of the 2005 IEEE International Conference on Robotics and Automation, ICRA 2005, pp. 2785–2790 (2005)

5. Lancioni, G.E., Cuvo, A.J., O'Reilly, M.F.: Snoezelen: an overview of research with people with developmental disabilities and dementia. Disability and Rehabilitation 24, 175–184 (2002)

6. Wada, K., Shibata, T.: Social Effects of Robot Therapy in a Care House - Change of Social Network of the Residents for Two Months. In: 2007 IEEE International Conference on Robotics and Automation, pp. 1250–1255 (2007)

7. Shibata, T.: Therapeutic Seal Robot as Biofeedback Medical Device: Qualitative and Quantitative Evaluations of Robot Therapy in Dementia Care. Proceedings of the IEEE 100, 2527–2538 (2012)

8. Wada, K., Ikeda, Y., Inoue, K., Uehara, R.: Development and preliminary evaluation of a caregiver's manual for robot therapy using the therapeutic seal robot Paro. In: 2010 IEEE RO-MAN, pp. 533–538 (2010)

9. Forlizzi, J., DiSalvo, C., Gemperle, F.: Assistive robotics and an ecology of elders living independently in their homes. Hum.-Comput. Interact. 19, 25–59 (2004)

10. Sabanovic, S.: It takes a village to construct a robot: A socially situated perspective on the ethics of robot design. Interaction Studies 11, 257–262 (2010)

11. Mutlu, B., Forlizzi, J.: Robots in organizations: The role of workflow, social, and environmental factors in human-robot interaction. In: 2008 3rd ACM/IEEE International Conference on Human-Robot Interaction (HRI), pp. 287–294 (2008)

12. Doherty, N.F., Coombs, C.R., Loan-Clarke, J.: A re-conceptualization of the interpretive flexibility of information technologies: redressing the balance between the social and the technical. Eur. J. Inf. Syst. 15, 569–582 (2006)

13. Alač, M., Movellan, J., Tanaka, F.: When a robot is social: Spatial arrangements and multimodal semiotic engagement in the practice of social robotics. Social Studies of Science 41, 893–926 (2011)

14. Sabanovic, S., Bennett, T., Chang, W., Huber, L.: PARO robot affects diverse interaction modalities in group sensory therapy for older adults with dementia. In: IEEE International Conference on Rehabilitation Robotics, ICORR 2013 (2013)

Closing the Loop: Towards Tightly Synchronized Robot Gesture and Speech

Maha Salem[1], Stefan Kopp[2], and Frank Joublin[3]

[1] Technical Faculty, Bielefeld University, Germany
msalem@cor-lab.uni-bielefeld.de
[2] Center of Excellence Cognitive Interaction Technology,
Bielefeld University, Germany
skopp@cit-ec.uni-bielefeld.de
[3] Honda Research Institute Europe, Offenbach, Germany
frank.joublin@honda-ri.de

Abstract. To engage in natural interactions with humans, social robots should produce speech-accompanying non-verbal behaviors such as hand and arm gestures. Given the special constraints imposed by the physical properties of a humanoid robot, successful multimodal synchronization is difficult to achieve. Introducing the first closed-loop approach to speech and gesture generation for humanoid robots, we propose a multimodal scheduler for improved synchronization based on two novel features, namely an experimentally fitted forward model and a feedback-based adaptation mechanism. Technical results obtained with the implemented scheduler demonstrate the feasibility of our approach; empirical results from an evaluation study highlight the implications of the present work.

Keywords: Multimodal Interaction and Conversational Skills, Robot Gesture, Speech-Gesture Synchrony, Social Human-Robot Interaction.

1 Introduction

Speech-accompanying hand and arm gestures are a fundamental component of human communicative behavior, often used to illustrate what is expressed in speech [12]. At the same time, human listeners are very attentive to information conveyed via such non-verbal behaviors [7]. Thus, gestures represent ideal candidates for extending the communicative capabilities of humanoid robots that are to interact socially with humans. In addition, providing multiple modalities helps to dissolve ambiguity and thus to increase robustness of communication.

In human communication, co-verbal gestures are tightly connected to speech as part of an integrated utterance, yielding semantic, pragmatic and temporal synchrony between both modalities [11]. A hand gesture typically consists of three phases ordered in a sequence over time to form a so-called gesture phrase: *preparation, stroke,* and *retraction* [7]. The *stroke* is the expressive phase of the gesture and bears meaning in relation to the corresponding part of speech, which is referred to as the *affiliate* [12]. Based on studies of human communication (e.g. [7,11]), gestures are found to be tightly synchronized with the accompanying linguistic affiliate to convey their intended communicative meaning: the gesture

G. Herrmann et al. (Eds.): ICSR 2013, LNAI 8239, pp. 381–391, 2013.

stroke typically precedes or ends at – but does not follow – the stressed syllable of the affiliate. To achieve this phonological synchrony, gesture may adjust to speech, e.g. by means of adequately timed preparation or additional hold phases, or speech may adapt to gesture, e.g. by means of pauses [7].

Ensuring such co-expressive synchrony between the two modalities poses a major challenge for artificial communicators such as virtual agents or social robots. Up to the present time, most technical systems dedicated to co-verbal gesture generation merely approximate crossmodal synchronization, typically by means of one modality adjusting to the other without any mutual adaptability. However, relevant literature (e.g. [2]) emphasizes the importance of accurate speech-gesture synchrony even for artificial agents, since asynchrony may result in conveying incorrect meaning.

2   Related Work

An increasing number of robotic systems incorporate both speech and gesture synthesis; however, in most cases the robots are equipped with a set of predefined gestures or even prerecorded behaviors that are not generated on-line but simply replayed during human-robot interaction (e.g. [18]). Crucially, to the best of our knowledge, no previous approach in robotics has realized exact (i.e. empirically sound), closed-loop synchronization of robot gesture and speech. For example, the system by Mead et al. [13] roughly attempts to synchronize the robot's gesture with speech at the phrase level rather than with the exact affiliate. Bremner et al. [1] define timings for speech-gesture synchronization based on estimations at the coding stage, i.e. gesture movements are preplanned in form and duration; during execution, movements of each gesture phases are then triggered by events in the speech output. Generally, in all existing robotic systems that aim to synchronize synthetic speech and gesture, running speech dictates the timing of the generated gestures, while gestures cannot affect the produced speech. While some approaches obtain exact speech timing information from the text-to-speech system, timing of gesture is typically based on estimations which lack accuracy. Another characteristic of related approaches is the open-loop control of the implementations, i.e. once speech and gesture have been planned and scheduled for the robot, both modalities are generated ballistically and cannot be further adjusted at run-time, e.g. based on sensory feedback. Such unidirectional synchronization as well as the open-loop production of multimodal behaviors are characteristic of all currently existing approaches to robot gesture generation (e.g. [14,10]). As a result, the present work stands out from other robotic systems for speech-gesture generation by incorporating reactive closed-loop feedback for a more flexible approach to multimodal robot expressiveness.

In contrast to the research area of robotics, the challenge of generating speech and co-verbal gesture has already been tackled rather extensively in various ways within the domain of embodied virtual agents (e.g. [2,3,15]). However, despite their elaborate architectures, these system are typically limited with regard to timing and synchronization of multimodal behaviors: to achieve crossmodal synchrony, gesture timing is heuristically adapted to the timing of generated speech, i.e. start and end times of phonemes are used to parametrize gesture time. In addition, speech and gesture are executed ballistically, i.e. no feedback-based adjustments can be made once behavior execution has been initiated.

Complementing these approaches, Kopp and Wachsmuth [8] introduced the Articulated Communicator Engine (*ACE*) as the first multimodal behavior realizer for virtual agents that provides for mutual adaptation mechanisms between speech and gesture. Inspired by theories from human gesture research, ACE assumes that the production of continuous speech and gesture is organized in successive segments. Each of these segments forms a multimodal *chunk* of speech-gesture production which, in turn, comprises a pair of an intonation phrase and a co-expressive gesture phrase [8]. That is, complex utterances with several gestures consist of multiple successive chunks. Given such a multi-chunk utterance, ACE coordinates the concurrent modalities at two levels. First, for *intra-chunk* scheduling, similar to the method used in the above discussed virtual agent systems, temporal synchrony is accomplished by adapting gesture to the timing of speech. Once scheduled and ready to be executed, speech and gesture run ballistically within a chunk, unaffected by the progress of the other respective modality. Second, for *inter-chunk* scheduling between two successive chunks, both speech and gesture can anticipate the forthcoming chunk and adapt to it if required. For gesture, this adaptation may range from the insertion of an intermediate rest pose to a direct transition movement skipping the retraction phase or parts thereof. For speech, a silent pause may be inserted between two intonation phrases for the duration of the preparation phase of the following gesture. Such adaptive inter-chunk synchrony is achieved by a global scheduler which plans the following chunk in advance while monitoring the chunk currently in execution. In summary, ACE overcomes some of the issues found in previous frameworks by providing a greater level of flexibility. However, despite mutual adaptation mechanisms at the inter-chunk level, its lacking adjustability within a chunk as well as the ballistic generation of complete gesture and intonation phrases conflict with findings from psychology (e.g. [7]).

The present work thus draws inspiration from the innovative quality of the ACE system regarding its incremental on-line scheduling of multimodal behavior at the *inter-chunk level*. Crucially, we expand the model of the underlying scheduler by countering its conceptual shortcomings at the *intra-chunk level* to realize more tightly synchronized robot gesture and speech. The concept of the proposed scheduler using the Honda humanoid robot[1] is presented in the following.

3   Closed-Loop Scheduler

An improved multimodal behavior scheduler needs to address two questions. First, in the behavior *planning* phase, how can a more reliable prediction of gesture execution time be obtained to allow for better multimodal scheduling? Second, in the *execution* phase, how can planning and scheduling errors be accounted for during execution if they would cause mistimed synchrony otherwise? Addressing the first issue, the proposed scheduler incorporates an empirically verified

---

[1] http://asimo.honda.com/downloads/pdf/honda-asimo-robot-fact-sheet.pdf

forward model that predicts an estimate of the gesture preparation time required by the robot prior to actual execution Addressing the second issue, an on-line adjustment mechanism was integrated into the synchronization process for cross-modal adaptation within a chunk based on afferent sensory feedback. Generation and continuous synchronization of gesture and speech are flexibly conducted at run-time. Fig. 1 illustrates the 'planning-execution' procedure of the scheduler.

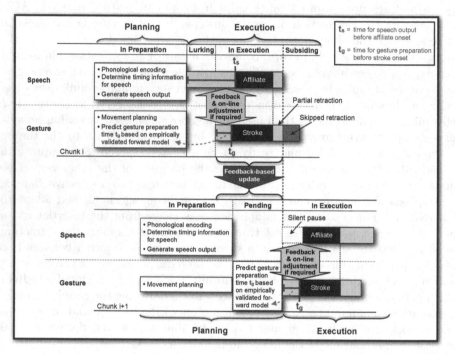

Fig. 1. Model of 'planning-execution' procedure of the proposed multimodal scheduler

3.1   Concept Overview

We specify multimodal utterances using the XML-based Multimodal Utterance Representation Markup Language (MURML [9]). It provides specific tags to label the speech affiliate based on time identifiers, while the posture of the gesture stroke can be specified in an overt form. An example MURML file is illustrated in Fig. 2. the proposed scheduler operates as follows.

Phase 1: Planning
Speech preparation. The planning phase begins with the phonological encoding of the designated speech output. During this process, the text-to-speech synthesis system MARY [17] establishes a complete list of phonemes and their respective durations, based on which the affiliate onset time is determined. Speech output is then generated in an audio file which is further processed depending on the position of the affiliate. If the affiliate is located at the beginning of the utterance, the audio file remains as it is. If the affiliate is in the middle or at the end of the utterance, the sound file is split into two parts: the first part contains speech

```
<definition>
  <utterance>
    <specification>
      <time id="t1"/> This one <time id="t2"/> is my favorite picture.
    </specification>
    <behaviorspec>
      <gesture id="gesture_1" scope="hand">
      <affiliate onset="t1" end="t2"/>
        <constraints>
          <parallel>
            <static slot="HandShape" value="BSffinger"/>
            <static slot="ExtFingerOrientation" value="DirA"/>
            <static slot="PalmOrientation" value="DirL"/>
            <static slot="HandLocation" value="LocShoulder LocRight LocNorm"/>
          </parallel>
        </constraints>
      </gesture>
    </behaviorspec>
  </utterance>
</definition>
```

Fig. 2. Example of a MURML specification for multimodal utterances

output to be uttered before the affiliate onset; the second part contains the speech affiliate and, if applicable, any subsequent remaining parts of speech.

Gesture preparation. Based on the specification of the designated location for the gesture stroke onset, a suitable trajectory is calculated, resulting in a detailed movement plan. Meanwhile, the predictive forward model further described in Section 3.2 computes the estimated execution time required by the robot for the gesture preparation phase before the stroke onset. Following a comparison of derived timing information for both modalities, start times for speech and gesture are determined so that temporal synchrony is expected to be achieved at the affiliate-stroke level.

Phase 2: Execution
Before the affiliate onset. Speech-gesture production begins as scheduled in the final step of the planning phase. If speech output is to precede the affiliate, the first sound file containing this part is replayed. While the robot performs the gesture preparation movement, variance between target and actual wrist position is constantly monitored utilizing afferent feedback from the robot controller.

Ensuring synchrony between affiliate and stroke onset. Once the robot's wrist has reached a position within a predefined range of the target location for the stroke onset, playback of the sound file containing the speech affiliate is triggered. If the predicted time was accurate, there should be no noticeable interruption in speech flow. If gesture preparation takes longer than scheduled, the feedback-based adaptation mechanism described in Section 3.3 becomes effective.

3.2 Predictive Forward Model

Neurophysiological findings suggest that human motor control relies more on sensory predictions than on sensory feedback, as feedback loops are considered too slow for efficient trajectory control [4]. Drawing inspiration from this neurobiological perspective, the concept of internal forward models is of particular interest to the realization of the proposed scheduler. Given the robot's physical properties and inability to move its arms with arbitrary speed, precise movement times are difficult to determine a priori. However, to successfully synchronize the robot's gesture with concurrent speech, the gesture preparation phase must be completed before the onset of the speech affiliate. Thus, the execution time of this gesture phase in particular needs to be estimated as accurately as possible so that the behavior onset times of the two modalities can be adequately scheduled.

Our approach to a robot-specific forward model was realized using the whole body motion (WBM) software controlling the Honda humanoid robot [6]. The robot's WBM controller allows for accelerated internal simulations of designated trajectories prior to – and temporally decoupled from – the actual movement process [5]. The WBM-based prediction model can account for the actual path of the simulated trajectory with regard to the robot's physical properties, e.g. joint limits. To simulate the robot's future state, the internal predictor iterates the robot model, but ten times faster than real-time control and at negligible computation time. As a result, it offers a viable option for the estimation of movement time required for the gesture preparation phase. The predicted trajectory time t_g is influenced by several parameters specific to the WBM software, such as speed response and maximal velocity. Thus, the controller-based forward model was fitted to determine appropriate values for these parameters. For this, a range of different values was empirically tested for each parameter on a set of experimental training data comprising 60 different target trajectories within a typical range of gesture space, and the combination yielding the smallest overall mean error was selected (see [16] for more details). For the given training data, the mean prediction error of this combination was 0.1458 seconds with a minimum deviation of 0.0024 seconds for the best predicted trajectory time, and a maximum deviation of 0.6180 seconds for the poorest trajectory time prediction.

3.3   Feedback-Based Adaptation Mechanism

Despite the improved accuracy of timing estimation for the gesture preparation phase based on the implemented forward model, actual timing of multimodal utterances might still deviate from the prediction during execution. Thus, the second feature realized in our proposed scheduler provides a feedback-based adaptation mechanism that allows for reactive crossmodal adjustment during the execution phase (see Fig. 1). Generally, two types of real-time deviation from scheduled timing plans are possible: the forward model may either *overestimate* or *underestimate* the time required for gesture preparation, resulting in either a premature or delayed gesture stroke onset. Findings from human gesture research state that the gesture stroke may precede but not follow the speech affiliate [11], thus the first type of deviation seems less problematic than the second. In fact, for the 60 tested training trajectories, the mean error values obtained for the WBM-based predictive model yielded a maximal overestimation error of only 0.276 seconds with a rare general occurrence of errors of more than 0.1 seconds (only 7 out of 60 cases); therefore, such cases may well be tolerated.[2] But given the model's inherent tendency to underestimate the time required by the robot, our approach focuses on this more problematic type of prediction error.

---

[2] If, however, the model exposed a pattern of frequent overestimation of gesture preparation time, the implementation of a pre-stroke hold phase would be reasonable.

Findings from human gesture research (e.g. [7]) demonstrate that speech may pause before the affiliate onset if gesture preparation has not yet been completed, ensuring temporal synchrony of the gesture stroke with speech. To account for the human ability to align speech to gesture also within a chunk, our proposed scheduler replaces the commonly used open-loop execution mechanism with a more flexible closed-loop approach. As outlined in Section 3.1, reactive on-line adjustment of the multimodal synchronization process within a chunk is achieved by utilizing afferent feedback from the robot's WBM controller as follows. Sensory feedback received from the controller provides information about the current wrist position of the robot's arm at each time step, which enables the scheduler to constantly check for variance between the target and actual position. This information is used to trigger the speech affiliate once the robot's hand has reached a position within the range of the stroke onset position.

If gesture preparation time is overestimated or correctly predicted, feedback about successful completion of this phase is transmitted to the speech processing module before or at the very latest at the scheduled affiliate onset time. As a result, speech output is uttered according to the previously established generation plan without any audible disruption. If, however, gesture preparation time is underestimated, the described crossmodal processing dependency results in a deviation of the actual multimodal production process compared to the scheduled behavior plan. That is, if the robot's hand has not yet reached the stroke onset location by the predicted time, the reactive feedback mechanism intervenes in the speech production process: if the affiliate is located at the beginning of the utterance, speech onset is delayed; if the affiliate is in the middle or at the end, a speech pause is inserted right before the affiliate. To re-establish speech-gesture synchrony in the latter case, speech is paused until sensory feedback confirming the completion of the gesture preparation phase triggers its continuation.[3] Such anticipation and reactive adaptation of the flow of speech to gesture timing complies with psycholinguistic models from gesture literature promoting an interactive view of the relationship between speech and gesture (e.g. [12]).

4 Technical Results

The following examples were generated using the proposed multimodal scheduler incorporating both the experimentally fitted WBM controller-based predictive model and the reactive adaptation mechanism. Ideally, the movement time required for gesture preparation is estimated precisely enough during the planning phase so that the actual execution is consistent with the scheduled multimodal plan and the speech production is not notably interrupted. However, to demonstrate the scheduler's crossmodal adjustment capability, we provide examples in which sensory feedback induced the insertion of pauses into the flow of speech.

---

[3] Alternatively, speech rate could be modulated (e.g. 'slowed down') until gesture preparation has completed, however, this is difficult to realize with a TTS engine that preproduces the audio files.

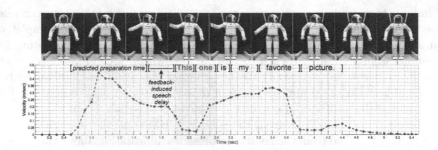

Fig. 3. Multimodal utterance generated with the proposed multimodal scheduler demonstrating a feedback-induced speech delay prior to speech onset

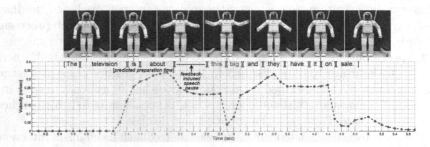

Fig. 4. Multimodal utterance generated with the proposed multimodal scheduler demonstrating a feedback-induced speech pause within a chunk

Such cases emerge when the gesture preparation time is considerably underestimated by the predictive model during planning.

Affiliate at the Beginning of a Chunk. Fig. 3 depicts an example of a multimodal utterance that illustrates the feedback-based adaptation mechanism effective at the beginning of the utterance. The plotted graph visualizes the velocity profile of the robot's right wrist during gesture execution; speech output is transcribed in temporal alignment to the generated gesture trajectory with words of the affiliate highlighted in red color. The figure illustrates the underestimation of gesture preparation time by ∼0.4 seconds resulting in a feedback-induced delay of the speech affiliate onset until the stroke onset position has been reached.

Affiliate in the Middle of a Chunk. Fig. 4 shows an example of a multimodal utterance which illustrates the feedback-based adaptation mechanism effective in the middle of the chunk. Initially, speech and gesture commence according to their assigned onset times as scheduled in the planning phase. In this case, playback of the first audio file containing speech to be uttered before the affiliate onset is initiated first and is accompanied by gesture movements after ∼0.72 seconds. Gesture preparation time is underestimated, resulting in a feedback-induced speech pause immediately preceding the affiliate onset. Once the preparation phase of the gesture has been completed, the feedback mechanism triggers the continuation of speech so that the affiliate and remaining speech output are replayed from the second audio file.

5   Evaluation

To evaluate the optimized generation of synchronized co-verbal robot gesture with our new scheduler, we conducted an empirical study with 20 participants (10m, 10f), ranging in age from 19 to 36 years ($M = 28.5$, $SD = 4.53$). Each participant was invited to view and rate a set of video clips showing the Honda humanoid robot while delivering various multimodal utterances (speech was generated in German language for the study, as all participants were German native speakers). For each utterance, two different versions were presented in randomized order: one version was recorded using the original ACE scheduler, while the other one employed the new extended scheduler. Participants were asked to identify the version in which they found the robot's gesture to be better synchronized with the speech output. Evaluation was based on twelve video sequences covering pairs of six different multimodal utterances (including the two examples presented in Section 4); in three of these utterances the speech affiliate was located at the beginning of the chunk, while the affiliate in the other three utterances was situated in the middle of the chunk. Participants were allowed to watch the video sequences as many times as needed to come to a decision.

In summary, participants showed a significant preference for the multimodal robot behavior generated by our proposed scheduler, $\chi^2(1) = 5.63$, $p = .018$, thus attesting a more positive perception of the speech-gesture synchronization resulting from the new system. These findings not only validate our technical approach but also contribute new insights into the role of speech-gesture synchrony by providing an HRI perspective for the systematic study of human perception of (a)synchrony in robotic agents. Despite the limited scope of the present evaluation, the results suggest that humans may indeed be aware of differences in the quality of synchronized multimodal behaviors generated by a humanoid robot.

6   Conclusion and Future Work

We presented a novel multimodal scheduler that overcomes the conceptual shortcomings of existing behavior schedulers based on two features. First, the scheduler incorporates a biologically inspired forward model that predicts a more accurate estimate of the robot's gesture preparation time prior to actual execution. Future work should extend the forward model to incorporate machine learning algorithms for a more accurate and, from a neurobiological perspective, more intuitive prediction model. The second feature of the extended scheduler incorporates an on-line adjustment mechanism into the synchronization process for crossmodal adaptation not only between two successive chunks, but also within a single chunk based on sensory feedback. In future, other strategies to modulate speech (e.g. using vocal timing instead of pauses) should be explored.

The functionality of the reactive adaptation mechanism provided by our scheduler at the intra-chunk level was illustrated by means of two examples. They show how synchrony between robot gesture and speech is ensured during execution in spite of prediction errors previously made during behavior planning and scheduling. In this way, the extended closed-loop scheduler enables the robot to plan, generate and continuously synchronize gesture and speech at run-time.

The realized scheduler was evaluated in a video-based user study. Our results show significantly higher ratings regarding better synchronization for the multimodal behaviors realized with the new scheduler compared to the utterances generated by the system that does not incorporate the extended features. The findings further suggest that humans are capable of perceiving differences in the quality of speech-gesture synchrony based on utterances generated by a humanoid robot. Since human perception of speech-gesture (a)synchrony has not yet been sufficiently investigated with regard to communicative robots, this observation should be systematically studied in more exhaustive studies in the future. Our generation system will provide a useful tool for this purpose.

Acknowledgement. The work described was supported by the Honda Research Institute Europe and the Center of Excellence Cognitive Interaction Technology.

References

1. Bremner, P., Pipe, A., Melhuish, C., Fraser, M., Subramanian, S.: Conversational Gestures in Human-Robot Interaction. In: Proceedings of the IEEE International Conference on Systems, Man and Cybernetics, pp. 1645–1649 (2009)
2. Cassell, J., Bickmore, T., Campbell, L., Vilhjálmsson, H., Yan, H.: Human Conversation as a System Framework: Desigining Embodied Conversational Agents. In: Embodied Conversational Agents, pp. 29–63. MIT Press, Cambridge (2000)
3. Cassell, J., Vilhjálmsson, H., Bickmore, T.: BEAT: the Behavior Expression Animation Toolkit. In: Proceedings of ACM SIGGRAPH 2001 (2001)
4. Desmurget, M., Grafton, S.: Forward modeling allows feedback control for fast reaching movements. Trends in Cognitive Sciences 4(11), 423–431 (2000)
5. Gienger, M., Bolder, B., Dunn, M., Sugiura, H., Janssen, H., Goerick, C.: Predictive Behavior Generation – A Sensor-Based Walking and Reaching Architecture for Humanoid Robots. In: Autonome Mobile Systeme 2007, pp. 275–281 (2007)
6. Gienger, M., Janßen, H., Goerick, S.: Task-Oriented Whole Body Motion for Humanoid Robots. In: Proceedings of the IEEE-RAS International Conference on Humanoid Robots, Tsukuba, Japan (2005)
7. Kendon, A.: Gesture: Visible Action as Utterance. Gesture 6(1), 119–144 (2004)
8. Kopp, S., Wachsmuth, I.: Synthesizing Multimodal Utterances for Conversational Agents. Computer Animation and Virtual Worlds 15(1), 39–52 (2004)
9. Kranstedt, A., Kopp, S., Wachsmuth, I.: MURML: A Multimodal Utterance Representation Markup Language for Conversational Agents. In: AAMAS 2002 Workshop on Embodied Conversational Agents - Let's Specify and Evaluate Them (2002)
10. Le, Q.A., Hanoune, S., Pelachaud, C.: Design and implementation of an expressive gesture model for a humanoid robot. In: Proceedings of the 11th IEEE-RAS International Conference on Humanoid Robots, Bled, Slovenia, pp. 134–140 (2011)

11. McNeill, D.: Hand and Mind: What Gestures Reveal about Thought. University of Chicago Press, Chicago (1992)
12. McNeill, D.: Gesture and Thought. University of Chicago Press, Chicago (2005)
13. Mead, R., Wade, E., Johnson, P., St. Clair, A., Chen, S., Matarić, M.: An Architecture for Rehabilitation Task Practice in Socially Assistive Human-Robot Interaction. In: Proceedings of IEEE RO-MAN 2010, pp. 404–409 (2010)
14. Ng-Thow-Hing, V., Luo, P., Okita, S.: Synchronized Gesture and Speech Production for Humanoid Robots. In: Proceedings of the IEEE/RSJ International Conference on Intelligent Robots and Systems, pp. 4617–4624 (2010)
15. Niewiadomski, R., Bevacqua, E., Mancini, M., Pelachaud, C.: Greta: An Interactive Expressive ECA System. In: Proceedings of 8th International Conference on Autonomous Agents and Multiagent Systems, pp. 1399–1400 (2009)
16. Salem, M.: Conceptual Motorics – Generation and Evaluation of Communicative Robot Gesture. Dissertation. Logos Verlag, Berlin (2012)
17. Schröder, M., Trouvain, J.: The German Text-to-Speech Synthesis System MARY: A Tool for Research, Development and Teaching. International Journal of Speech Technology, 365–377 (2003)
18. Sidner, C., Lee, C., Lesh, N.: The Role of Dialog in Human Robot Interaction. In: International Workshop on Language Understanding and Agents for Real World Interaction (2003)

Teleoperation of Domestic Service Robots: Effects of Global 3D Environment Maps in the User Interface on Operators' Cognitive and Performance Metrics

Marcus Mast[1,2], Michal Španěl[3], Georg Arbeiter[4], Vít Štancl[3], Zdeněk Materna[3], Florian Weisshardt[4], Michael Burmester[1], Pavel Smrž[3], and Birgit Graf[4]

[1] Stuttgart Media University, Germany
mast@hdm-stuttgart.de
[2] Linköping University, Sweden
[3] Brno University of Technology, Czech Republic
[4] Fraunhofer Institute for Manufacturing Engineering and Automation, Germany

Abstract. This paper investigates the suitability of visualizing global 3D environment maps generated from RGB-D sensor data in teleoperation user interfaces for service robots. We carried out a controlled experiment involving 27 participants, four teleoperation tasks, and two types of novel global 3D mapping techniques. Results show substantial advantages of global 3D mapping over a control condition for three of the four tasks. Global 3D mapping in the user interface lead to reduced search times for objects and to fewer collisions. In most situations it also resulted in less operator workload, higher situation awareness, and higher accuracy of operators' mental models of the remote environment.

Keywords: global environment map, service robot, teleoperation, user interface.

1 Introduction

Service robots may assist people in the home in the future. Frail elderly people and people with disabilities may particularly benefit from such robots. However, there are still numerous technical challenges associated with building reliable autonomous robots for unstructured environments. For example, robots can fail to find navigation paths through narrow passages or objects to be fetched can fail to be detected in cluttered scenes or under low illumination. Semi-autonomous robots [1] may be a viable shorter-term solution: when a robot cannot complete a task, a remote human operator takes control and tries to solve the local problem. The design of suitable teleoperation user interfaces therefore remains an important subject of research.

Studies have shown that user interfaces merely relying on the visualization of a video stream from a camera on the robot and a floor map lead to high collision rates [2], long task completion times [3, 2], high operator workload [4], and low situation awareness [5]. The camera's narrow field of view prevents operators from acquiring environmental cues from the surrounding. Also, people form mental models of spatial environments and rely on them for spatial reasoning [6]. In video-centric interfaces,

G. Herrmann et al. (Eds.): ICSR 2013, LNAI 8239, pp. 392–401, 2013.

operators need to infer a spatial mental model from two-dimensional data, which may result in less accurate models and negatively affect performance.

It has thus been established that teleoperators need more comprehensive information on the robot's environment to be able to control it effectively. Visualizing data from various sensors in separate areas of the screen has however shown to be problematic with regard to operators' task performance and cognitive load [2, 4]. To avoid divided attention and information overload, sensor fusion, augmentation, and ecological interfaces, where data from different sensors are integrated with direct spatial reference to the environment, have been proposed [7, 2]. To address the field of view problem, the use of multiple and panospheric cameras has been explored [8]. However, such approaches impose a high load on network bandwidth while not allowing, for example, assessment of distances between objects. Approaches where 2D map and video are enriched with a schematic 3D environment model have shown to lead to improvements in user performance [3, 4] but they require manual modeling of each apartment and can be misleading when environment features have changed.

With the advent of affordable depth sensors, techniques for generating complete 3D representations of the environment have emerged more recently. With infrared sensors, time-of-flight cameras, or 3D laser scanners, the robot acquires 3D data while moving around. Algorithms generate, and update in real-time, a global 3D environment map in form of a voxel-based point cloud [9, 10] or a set of geometric primitives [11, 12]. Research on global 3D environment mapping has so far focused on algorithms and applications in autonomous reasoning [9-12]. In this paper, we examine the suitability of displaying global 3D environment maps during teleoperation. The maps provide a teleoperator with recent and detailed information on objects all around the robot, on environment features further away, and on distances between objects.

2   User Interface, Robot, and Mapping Techniques

The user interface employed in this study (Figs. 1 and 2) was developed as part of a wider semi-autonomous human-robot interaction framework for assisting elderly people in the home in an iterative user-centered design process involving studies with a total of 241 participants [1]. The user interface's main intended user group is teleassistance staff but we do not regard its use limited to professional applications. The user interface is integrated with the Care-O-bot 3 service robot [13] (Fig. 2). Care-O-bot 3 has an omni-directional base, three laser range finders, and a Kinect RGB-D camera. During teleoperation, the robot autonomously avoids obstacles of up to 40cm above ground based on 2D laser range data but not currently higher objects.

The user interface is based on RViz as part of ROS (Robot Operating System) Electric [15]. Following the idea of ecological interfaces [2], it offers the following elements visualized in a unified way in a single 3D scene (Figs. 1 and 2): (1) laser-based 2D floor map (grey/black), (2) laser-based historic obstacle map (purple, active in Fig. 2 only), (3) live laser data (red), (4) live 3D colored point cloud in the robot's field of view, (5) global 3D environment map (voxel-based or geometric), (6) robot in accurate size and shape, (7) collision-relevant safety area around the robot ("footprint", yellow, Fig. 2), (8) a sectioned band of collision indicators around the robot lighting up when the robot slows down on approaching an obstacle or movement in a direction is not possible (not shown in Figs. 1 and 2). As a ninth element, a video image is overlaid. Users can freely rotate, pan, and zoom the 3D scene using the mouse. Navigation of the mobile platform is realized with the SpaceNavigator 3D mouse [14] in the coordinate system of the operator's current viewpoint on the scene.

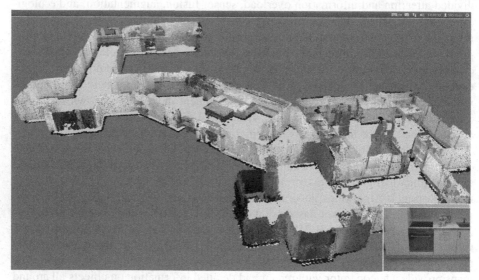

Fig. 1. User interface in zoomed-out perspective with apartment-sized environment mapped using voxel-based mapping technique

Two types of global 3D mapping techniques have been implemented. The voxel-based technique is based on OctoMap [9], a probabilistic approach that uses octrees to represent 3D occupancy grids. The geometric 3D mapping technique [11] segments consecutive point clouds into homogeneous regions, derives geometric primitives like planes or cylinders, and merges them into a map. From a technical point of view, its main advantages over the voxel-based technique are lower network bandwidth consumption and less required client-side computational resources for display.

3   Research Questions

We assumed that, due to the complete 3D representation of the environment, global 3D environment maps in the user interface should usually lead to improved user performance, decreased workload, increased situation awareness, and more accurate mental models of the remote environment. On the other hand, it is conceivable that they might not be needed in some situations or even be disadvantageous. For example, the visualization of a global 3D map might negatively impact user performance because close objects can obstruct the view on task-relevant objects behind them or lead to higher cognitive load due to additional interpretation effort. Concerning the suitability of voxel-based versus geometric mapping, likewise, arguments for differences in both directions are feasible. For example, the more simplistic, reduced visualization of the environment with a geometric approach might reduce required interpretation efforts. On the other hand, the algorithmic processing of the point cloud necessarily introduces artifacts that might be confusing.

Due to a wide range of factors that might affect results and a lack of directly comparable studies in the literature, we adopted an exploratory attitude towards this research. Our main goal was to identify usage scenarios where global 3D environment maps in the remote user interface would be advantageous as well as scenarios where this would rather not be the case. We were further interested in potential differences in the suitability of voxel-based versus geometric mapping approaches.

4   Method

Research Design. We adopted a between-subjects experimental design with three conditions: control condition without global 3D environment mapping (C), voxel-based global 3D mapping condition (V), geometric global 3D mapping condition (G). The user interface did not differ between conditions other than in the type (or absence) of global 3D environment mapping. All other visualization elements described in Section 2 were available in all conditions.

Participants. 27 people participated in the experiment (9 in each group). Their age ranged from 19 to 41 years (mean age group C: 27.4; V: 23.4; G: 26.7). All participants were male to avoid a confounding effect due to known gender differences in spatial problem solving [19]. Participants had no previous experience with robots, teleoperation user interfaces, or 3D mice. They used computers at least 12 hours per week but played 3D games or used professional 3D applications only up to 2 hours per week (mean 3D usage hours C: 0.5; V: 0.6; G: 0.5). All but two participants had university degrees or were currently pursuing such. Participants received €30 of compensation. Participants underwent a spatial ability test (abbr. Vandenberg Mental Rotations Test) [20] and were assigned to experimental conditions based on their score in a balanced way (mean score C: 4.4 out of 6; V: 4.3; G: 4.6). Participants also underwent a Snellen visual acuity test and all achieved a score of 5 or higher (mean score C: 6.7; V: 7.2; G: 6.1). According to a further color vision test, one participant in each group did not have full color vision.

Metrics. We employed objective and subjective metrics to investigate several constructs. Metrics mainly relating to performance were: time to complete the task, length of the path the robot traveled, and robot rotation. We assessed perceived workload with the NASA Task Load Index (TLX) in its unweighted (raw) form [16] post-task. We further measured perceived situation awareness post-task with the three-dimensional Situation Awareness Rating Technique (SART) [17]. We adjusted the SART scale to a range from 0 to 100. To assess perceived quality of the mental model of the remote environment, we employed (6) the subscale "spatial situation model" (SSM) from the Spatial Presence Questionnaire MEC-SPQ [18] post-experiment. For applicable tasks, we also asked participants to recall the number of obstacles around the robot as an indicator of the accuracy of their spatial mental model. For one task, we further recorded the number of collisions as a performance metric and an implicit indicator of situation awareness.

Procedure. The experiment took place in three rooms connected by two corridors. We installed 80 household furniture items and objects to simulate a home environment with a kitchen, a living room, and a bedroom. To control for illumination we covered all windows and used interior lighting only. Participants operated the robot from a separate fourth room and did not see the robot or its environment until after the trials.

After initial spatial ability and vision tests, participants underwent a 45-min training on robot hardware and sensors, user interface concept and visualization elements, SpaceNavigator device, teleoperation (in simulation). The type of global mapping visualized during training corresponded to the participants' experimental condition. Participants then carried out four teleoperation tasks (Fig. 2) with the real robot and filled in a questionnaire after each task. The tasks were designed to cover a range of different scenarios. At the beginning of each task, participants were confronted with a problem of the robot and asked to solve it quickly but without becoming stressed or sacrificing accuracy. The room's global map was always up to date, assuming the robot had moved around prior to contacting the teleoperator for assistance.

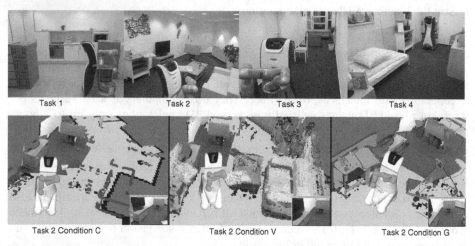

Fig. 2. Overview of the task setups with robot at starting positions (upper row); robot during task 2 at similar positions in the three experimental conditions (lower row)

Task 1: For this task in the kitchen only, we simulated darkness, so no video and only grayscale live depth data were available. The global maps were generated while the room was still illuminated. The robot's problem was that it could not find a medicine package to be fetched. The participant was told that the local elderly user stated to have left the package on the chair in the kitchen. The goal was to find the chair (which was hidden in a corner and there were three such corners) and approach it.

Task 2: This task took place in the living room. The robot's problem was that it could not find a navigation path to its destination due to a narrow passage.

Participants were asked to navigate the robot to the other side of the room. There were elevated protruding obstacles so it was possible for the robot to collide. Participants were asked to make the robot pass without collisions.

Task 3: This task took place in the corridor and the robot's problem was that it could not find a navigation path to the bedroom. All obstacles were on the floor. Towards the end of the task, participants had to push aside a carton with the robot.

Task 4: This task took place in the bedroom. As in task 1, participants had to find an object but it was small and without a characteristic shape (a rectangular pack of tea bags). The room was cluttered with objects and there were five possible hiding places.

After the last task, the session was recapitulated, participants were interviewed on their impressions, and then completed the post-experiment questions.

Data Analysis. We performed pairwise group comparisons using two-tailed independent samples *t*-tests, assuming normal distribution. For the collision metric only we used a two-tailed Wald test, assuming Poisson distribution. We used robust estimation of variance (Huber-White sandwich). Due to the exploratory nature of the experiment we did not adjust for multiple testing [21].

5   Results

All participants were able to complete all tasks, indicating effectiveness in all experimental conditions. Results on the investigated metrics are described subsequently.

Task 1: Finding a Large Object. As shown in Table 1, participants in both global map conditions were able to find the target object (chair) significantly faster with much shorter path lengths and less required robot rotation than in the control condition. They also experienced significantly lower workload and were subjectively more aware of the situation in conditions V and G. There was a tendency for condition V to be somewhat more efficient than condition G but statistical significance is only reached for the path length metric.

Task 2: Navigating Around Elevated Obstacles. Results in Table 2 show that task 2 was completed the fastest in the control condition. The difference to both, conditions V and G is statistically significant. Perceived workload was highest in condition G and the difference is significant over conditions C and V. In condition C, object collisions were most frequent and participants recalled significantly less obstacles than in the global map conditions. No significant differences were found for path lengths, robot rotation, perceived situation awareness.

Task 3: Navigating Around Obstacles on the Floor. Results for task 3 (Table 3) show no significant differences between conditions. There was a tendency to report more obstacles in condition V than in conditions G and C.

Task 4: Finding a Small Object. As shown in Table 4, participants in condition V found the pack of tea bags the fastest, with shortest path lengths, least rotation, and least cognitive load, statistically significant over conditions C and G. Perceived situation awareness was highest in condition V with a significant difference over C but just as a tendency over G. There was a tendency for better suitability of condition G than C for this task but the differences are not statistically significant.

Spatial Situation Model. Results of the post-experiment MEC-SPQ SSM subscale (Table 5) show that participants rated the perceived quality of their mental model of the situation significantly higher in condition V than in conditions C and G.

Table 1. Results of task 1 (conditions: C - control; V - voxel map; G - geometric map)

Metric	Means with 95% confidence intervals		Difference betw. means	
Task completion time	C: 235s		C vs. V:	p = 0.001 **
	V: 109s		C vs. G:	p = 0.004 **
	G: 126s		V vs. G:	p = 0.376
Robot path length	C: 8.4m		C vs. V:	p = 0.001 **
	V: 3.1m		C vs. G:	p = 0.004 **
	G: 4.2m		V vs. G:	p = 0.031 *
Robot rotation	C: 2.5rev		C vs. V:	p = 0.000 **
	V: 0.6rev		C vs. G:	p = 0.000 **
	G: 0.8rev		V vs. G:	p = 0.542
Perceived workload (NASA-TLX raw) Scale: 0-100	C: 40.4		C vs. V:	p = 0.002 **
	V: 16.8		C vs. G:	p = 0.029 *
	G: 23.1		V vs. G:	p = 0.115
Perceived situation awareness (SART) Scale: 0-100	C: 55.9		C vs. V:	p = 0.017 *
	V: 72.0		C vs. G:	p = 0.011 *
	G: 72.9		V vs. G:	p = 0.885

Table 2. Results of task 2 (conditions: C - control; V - voxel map; G - geometric map)

Metric	Means with 95% confidence intervals		Difference betw. means	
Task completion time	C: 121s		C vs. V:	p = 0.029 *
	V: 171s		C vs. G:	p = 0.001 **
	G: 206s		V vs. G:	p = 0.154
Robot path length	C: 5.6m		C vs. V:	p = 0.405
	V: 5.5m		C vs. G:	p = 0.406
	G: 5.4m		V vs. G:	p = 0.126
Robot rotation	C: 0.6rev		C vs. V:	p = 0.579
	V: 0.7rev		C vs. G:	p = 0.118
	G: 0.8rev		V vs. G:	p = 0.182
Perceived workload (NASA-TLX raw) Scale: 0-100	C: 36.8		C vs. V:	p = 0.754
	V: 34.9		C vs. G:	p = 0.010 *
	G: 52.0		V vs. G:	p = 0.003 **
Perceived situation awareness (SART) Scale: 0-100	C: 53.1		C vs. V:	p = 0.331
	V: 59.3		C vs. G:	p = 0.483
	G: 49.1		V vs. G:	p = 0.156
Object collisions	C: 1.7		C vs. V:	p = 0.044 *
	V: 0.6		C vs. G:	p = 0.013 *
	G: 0.9		V vs. G:	p = 0.352
Obstacles recalled	C: 2.1	correct	C vs. V:	p = 0.001 **
	V: 3.3		C vs. G:	p = 0.001 **
	G: 3.4		V vs. G:	p = 0.771

Table 3. Results of task 3 (conditions: C - control; V - voxel map; G - geometric map)

Metric	Means with 95% confidence intervals		Difference betw. means	
Task completion time	C: 139s		C vs. V:	p = 0.333
	V: 155s		C vs. G:	p = 0.195
	G: 160s		V vs. G:	p = 0.722
Robot path length	C: 6.1m		C vs. V:	p = 0.780
	V: 6.0m		C vs. G:	p = 0.902
	G: 6.1m		V vs. G:	p = 0.834
Robot rotation	C: 0.6rev		C vs. V:	p = 0.689
	V: 0.6rev		C vs. G:	p = 0.545
	G: 0.6rev		V vs. G:	p = 0.846
Perceived workload	C: 37.1		C vs. V:	p = 0.230
(NASA-TLX raw)	V: 30.9		C vs. G:	p = 0.515
Scale: 0-100	G: 33.2		V vs. G:	p = 0.737
Perceived situation	C: 60.7		C vs. V:	p = 0.493
awareness (SART)	V: 65.3		C vs. G:	p = 0.817
Scale: 0-100	G: 62.0		V vs. G:	p = 0.636
Obstacles recalled	C: 2.9	correct	C vs. V:	p = 0.264
	V: 3.3		C vs. G:	p = 0.748
	G: 2.8		V vs. G:	p = 0.057

Table 4. Results of task 4 (conditions: C - control; V - voxel map; G - geometric map)

Metric	Means with 95% confidence intervals		Difference betw. means	
Task completion time	C: 215s		C vs. V:	p = 0.001 **
	V: 97s		C vs. G:	p = 0.317
	G: 174s		V vs. G:	p = 0.014 *
Robot path length	C: 5.9m		C vs. V:	p = 0.020 *
	V: 2.8m		C vs. G:	p = 0.147
	G: 4.0m		V vs. G:	p = 0.020 *
Robot rotation	C: 2.2rev		C vs. V:	p = 0.013 *
	V: 1.2rev		C vs. G:	p = 0.518
	G: 2.0rev		V vs. G:	p = 0.018 *
Perceived workload	C: 42.5		C vs. V:	p = 0.017 *
(NASA-TLX raw)	V: 21.6		C vs. G:	p = 0.547
Scale: 0-100	G: 36.9		V vs. G:	p = 0.021 *
Perceived situation	C: 55.8		C vs. V:	p = 0.029 *
awareness (SART)	V: 72.8		C vs. G:	p = 0.514
Scale: 0-100	G: 60.1		V vs. G:	p = 0.133

Table 5. Results of MEC-SPQ SSM (conditions: C - control; V - voxel map; G - geometric map)

Metric	Means with 95% confidence intervals		Difference betw. means	
Perceived quality of	C: 3.6		C vs. V:	p = 0.006 **
spatial situation model	V: 4.3		C vs. G:	p = 0.190
5-point scale	G: 3.9		V vs. G:	p = 0.031 *

6   Discussion

When searching for a large object, there were strong benefits of using either type of global 3D map, voxel-based or geometric. Participants were around twice as fast in the mapping-enabled conditions than in the control condition when navigating the robot to the hidden chair in task 1. They also experienced less cognitive load and higher situation awareness. When searching for a smaller object (task 4), benefits were only clear in case of the voxel-based mapping approach. Likely, due to the more schematic and coarse environment representation in the geometric map, operators could not obtain a direct cue as to the target object's location. There still was some non-significant tendency for benefits of using the geometric map over not using a 3D map. Perhaps operators had better overall orientation than without a global 3D map.

When navigating around obstacles, benefits of using global 3D maps were only apparent for task 2, where obstacles were elevated. Interestingly, users in the control condition accomplished this task the fastest. However, they also had the highest number of collisions (C: 1.7, V: 0.6, G: 0.9 collisions on average). It seems that operators rushed through the obstacle course unaware of the existence or proximity to obstacles. This is corroborated by the fact that substantially too few obstacles were recalled in the control condition. A further interesting result for task 2 is that operators using the geometric map experienced higher workload than in the two other conditions. For other tasks too, there was a tendency for higher workload in the geometric condition than in the voxel condition. Perhaps this can be attributed to some ambiguity in visualization due to artifacts of algorithmic processing compared to the more unprocessed representation in a voxel map. The fact that results for all metrics were similar for task 3 suggests that this might be a scenario without benefits of using global 3D mapping. A likely reason is that all obstacles were on the floor and thus well visible through laser scanner visualization in the control condition too.

Participants using voxel-based maps rated the quality of their mental model of the remote environment highest on average. Results on the recall metric further suggest that whether or not 3D mapping leads to more accurate mental models may also depend on the environment or task. For task 2, participants in both mapping-enabled conditions were clearly much more accurate in recalling obstacles (C: 2.1, V: 3.3, G: 3.4 recalled; correct: 4). In task 3, where all obstacles were on the floor, no significant differences were found. Results thus suggest that global 3D maps tend to support the formation of accurate mental models but need not necessarily in every situation.

To summarize, visualizing global 3D environment maps in the teleoperation user interface in most situations showed to have positive effects on user performance, workload, situation awareness, and accuracy of users' spatial mental models. It substantially reduced search times for objects and reduced collisions when navigating around elevated obstacles. On a broad level, our results corroborate the findings of earlier studies on the benefits of 3D-enhanced teleoperation user interfaces [2, 3]. The automated, sensor-based, and fully three-dimensional mapping techniques evaluated in this paper can be regarded an evolution of those earlier approaches. Benefits of the voxel-based mapping technique were overall more pronounced than of the geometric mapping technique but the geometric mapping technique tended to show benefits over the control condition too. Due to its technical advantages, geometric mapping should be considered especially when identifying fine detail in the environment is not crucial.

Acknowledgements. This research was funded by the European Commission, FP7, Grant Agreement 247772. We would like to thank Thiago de Freitas Oliveira Araújo, Ali Shuja Siddiqui, Markus Noack, and Anne Reibke for supporting work.

References

1. Mast, M., Burmester, M., Krüger, K., Fatikow, S., Arbeiter, G., Graf, B., Kronreif, G., Pigini, L., Facal, D., Qiu, R.: User-Centered Design of a Dynamic-Autonomy Remote Interaction Concept for Manipulation-Capable Robots to Assist Elderly people in the Home. Journal of Human-Robot Interaction 1, 96–118 (2012)
2. Nielsen, C.W., Goodrich, M.A., Ricks, R.W.: Ecological Interfaces for Improving Mobile Robot Teleoperation. IEEE Transactions on Robotics 23, 927–941 (2007)
3. Labonté, D., Boissy, P., Michaud, F.: Comparative Analysis of 3-D Robot Teleoperation Interfaces With Novice Users. IEEE T. Syst. Man. Cyb. B 40, 1331–1342 (2010)
4. Bruemmer, D.J., Few, D.A., Boring, R.L., Marble, J.L., et al.: Shared Understanding for Collaborative Control. IEEE T. Syst. Man. Cyb. A 35, 494–504 (2005)
5. Drury, J.L., Scholtz, J., Yanco, H.A.: Awareness in Human-Robot Interactions. In: Proc. IEEE Int. Conf. Syst. Man. Cyb., pp. 912–918 (2003)
6. Kitchin, R.M.: Cognitive Maps: What Are They and Why Study Them? Journal of Environmental Psychology 14, 1–19 (1994)
7. Fong, T., Thorpe, C., Baur, C.: Advanced Interfaces for Vehicle Teleoperation: Collaborative Control, Sensor Fusion Displays, and Remote Driving Tools. Autonomous Robots 11, 77–85 (2001)
8. Fiala, M.: Pano-Presence for Teleoperation. In: Proc. IROS, pp. 3798–3802 (2005)
9. Hornung, A., Wurm, K.M., Bennewitz, M., Stachniss, C., Burgard, W.: OctoMap: an efficient probabilistic 3D mapping framework based on octrees. Autonomous Robots 34, 189–206 (2013)
10. Yguel, M., Aycard, O.: 3D mapping of outdoor environment using clustering techniques. In: Proc. IEEE International Conference on Tools with Artificial Intelligence, pp. 403–408 (2011)
11. Arbeiter, G., Bormann, R., Fischer, J., Hägele, M., Verl, A.: Towards Geometric Mapping for Semi-Autonomous Mobile Robots. In: Stachniss, C., Schill, K., Uttal, D. (eds.) Spatial Cognition 2012. LNCS (LNAI), vol. 7463, pp. 114–127. Springer, Heidelberg (2012)
12. Kakiuchi, Y., Ueda, R., Okada, K., Inaba, M.: Creating Household Environment Map for Environment Manipulation Using Color Range Sensors on Environment and Robot. In: Proc. ICRA, pp. 305–310 (2011)
13. Reiser, U., Connette, C., Fischer, J., Kubacki, J., Bubeck, A., Weisshardt, F., Jacobs, T., Parlitz, C., Hägele, M., Verl, A.: Care-O-bot 3 - Creating a product vision for service robot applications by integrating design and technology. In: Proc. IROS, pp. 1992–1998 (2009)
14. 3Dconnexion SpaceNavigator,
http://www.3dconnexion.com/products/spacenavigator.html
15. ROS documentation, http://www.ros.org/wiki/
16. Hart, S.G., Staveland, L.E.: Development of NASA-TLX (Task Load Index): Results of empirical and theoretical research. In: Hancock, P.A., Meshkati, N. (eds.) Human Mental Workload, pp. 139–183. North Holland, Amsterdam (1988)
17. Taylor, R.M.: Situational awareness rating Technique (SART): The Development of a Tool for Aircraft Systems Design. Proc. AGARD No. 478, pp. 3/1–3/17 (1989)
18. Wirth, W., Hartmann, T., Böcking, S., Vorderer, P., Klimmt, C., et al.: A Process Model of the Formation of Spatial Presence Experiences. Media Psychology 9, 493–525 (2007)
19. Jones, C.M., Healy, S.D.: Differences in cue use and spatial memory in men and women. Proceedings of the Royal Society, B 273, 2241–2247 (2006)
20. Peters, M., Laeng, B., et al.: A Redrawn Vandenberg & Kuse Mental Rotations Test: Different Versions and Factors That Affect Performance. Brain and Cogn. 28, 39–58 (1995)
21. Bender, R., Lange, S.: Adjusting for multiple testing – when and how? Journal of Clinical Epidemiology 54, 343–349 (2001)

Artists as HRI Pioneers: A Creative Approach to Developing Novel Interactions for Living with Robots

Hagen Lehmann, Michael L. Walters, Anna Dumitriu, Alex May, Kheng Lee Koay, Joan Saez-Pons, Dag Sverre Syrdal, Luke Wood, Joe Saunders, Nathan Burke, Ismael Duque-Garcia, Bruce Christianson, and Kerstin Dautenhahn

Adaptive Systems Research Group, University of Hertfordshire, Hatfield AL10 9AB, UK
{h.lehmann,m.l.walters,k.l.koay,j.saez-pons,
d.s.syrdal,L.Wood,j.l.saunders,n.m.burke,i.duque-
garcia,B.Christianson,k.dautenhahn}@herts.ac.uk,
annadumitriu@hotmail.com, alex@quadratura.info

Abstract. In this article we present a long-term, continuous human-robot co-habitation experiment, which involved two professional artists, whose artistic work explores the boundary between science and society. The artists lived in the University of Hertfordshire Robot House full-time with various robots with different characteristics in a smart home environment. The artists immersed themselves in the robot populated living environment in order to explore and develop novel ways to interact with robots. The main research aim was to explore in a qualitative way the impact of a continuous weeklong exposure to robot companions and sensor environments on humans. This work has developed an Integrative Holistic Feedback Approach (IHFA) involving knowledgeable users in the design process of appearances, functionality and interactive behaviour of robots.

Keywords: Human-Robot Interaction, Art, Human-Robot Co-Habitation, Robot Companions.

1 Introduction

With recent advances in robot technology, the integration of robotic home companions into the everyday lives of their users and the creation of mixed human-robot ecologies has become a concrete possibility for the near future. While these advances present opportunities for users, they also present research challenges. Many of these challenges are purely technical, but some are also concerned with Human-Robot Interaction (HRI) issues, ranging from robot ethics, privacy, companionship, social relationship and behaviour, and independent living issues [1-4].

Current HRI research already addresses many of these issues, with the overall aim of increasing peoples' acceptance of robotic companions, also including associated smart home environments [5]. The main findings in these areas tend to come from two types of HRI studies. The first is survey based research, where a (relatively) large sample of the general population is asked about their preferences and opinions about

G. Herrmann et al. (Eds.): ICSR 2013, LNAI 8239, pp. 402–411, 2013.

service robots and their associated (social) behaviour, appearance and task domains. The second type of HRI study is where people from a specific prospective user group (e.g. elderly, children, carers, office workers etc.) are exposed directly to actual (prototype) robots. This can be either as part of a real life situation, or in an experimentally controlled HRI scenario.

Relatively few live HRI studies have investigated long-term exposure of people to robots. Due to the practical and logistical difficulties involved, even in most long-term HRI studies, the duration of live robot exposure sessions has also been necessarily limited and either performed in a public arena or laboratory simulated home environment [6-8].

HRI studies are typically performed by researchers (robot experts) on robot-naive participants. Historically, the use of robot naive participants has been a logical choice for most HRI studies; most robots which have been designed for HRI were either research or demonstration type robots or prototypes. These were usually custom built, expensive and typically owned by a large company or institution and used specifically for HRI research or as demonstrations within a public setting, so unlikely to be encountered by the general population in their own homes. The focus for most HRI research is therefore on obtaining data from robot-naive individuals in order to help the development of natural ways of interacting with current robots. However, with the likely future increase of service robots in domestic environments, it can be argued that anyone who experiences a robot in their own home on a continuous basis will become acclimatised (in some way) to living with a robot [9] and their responses will certainly change over time and be different to those of a robot-naive user.

An HRI study by Hüttenrauch and Severinson-Ecklundh [10] notably overcame many of these limitations (HRI exposure duration and ecological validity), where a service robot, CERO, was used for extended periods (during working hours, over a period of months) to provide fetch and carry services for a robot-experienced co-researcher who had some mobility impairment. It should be noted that by using an experienced robotics researcher as the HRI study participant, it was possible to overcome both the technical and logistical problems associated with running the study continuously over a very long period. As the participant was situated in her normal working environment, the study also enjoyed an ecologically valid (real-world) setting. Sabanovic et al. [11] have argued that "observational studies can be applied to human-robot social interactions in varying contexts and with differing tasks to quantitatively and qualitatively evaluate (and discover unanticipated aspects of) the social interaction." and stress the importance of an open human-inhabited setting for these studies.

In order to overcome these limitations of HRI (episode duration and continuous robot presence in an ecological valid environment), this work introduces a new methodological perspective. We call this approach Integrative Holistic Feedback Approach (IHFA), a definition in which:

1. "Holistic" refers to a co-habitation human-robot ecology, including smart home sensor grids in domestic environments;
2. "Integrative" refers to the integration of structured responses given by users into the design process of appearances and behaviours of robots;

3.   "Feedback" refers to the evaluative character of the users responses with regard to the appearance and behaviours of the robots

In the following sections the University of Hertfordshire Robot House (UHRH) is described and the IHFA methodology is presented in more detail, the various experiments conducted with the users are outlined and an overview of some of the findings is presented. In the final section the implications, outcomes and open questions identified by experimental study are discussed.

2   The Week Residence Experience

2.1   The University of Hertfordshire Robot House

The study was conducted at the UHRH, which is dedicated to HRI research in a realistic, domestic environment. It has the appearance of an ordinary British suburban, semi-detached house in a quiet residential street, with four fully-furnished bedrooms and a sizeable garden. The UHRH is also inhabited by different robots designed for robot companion research. It also has a large number of sensors, allowing the recording of a range of different user activities (see Fig.1.Left).

Fig. 1. Left: Robot House layout and sensor arrangement. Right: Experiment at the UHRH with a Care-O-Bot 3.

This set-up facilitates the realisation of interaction studies with human participants. These studies can be constrained experiments, in which the setting primarily is important due to configuration of the space around [7], but they can also be more open-ended, scenario based studies [12]. These studies can be thought of as a form of high-fidelity HRI prototyping [13], in which the user interacts with the robot in a fully interactive scenario. However, realistic studies in which participants move into the UHRH as residents have been difficult to carry out due to the nature of the space, the necessity to have trained personnel in place to respond to robot breakdowns as well as safety-related concerns.

2.2   Idea and Participants

In the HRI experiment presented in this paper the main participants (users) were two artists, Anna Dumitriu and Alex May, who were both robot-technology aware and who had an artistic interest in immersively experiencing robotic technology in a domestic environment. As professional artists, they saw it as an opportunity to create both installation and performance art using the facilities of the University of UHRH. To provide an ecologically valid environment for their experience, they stayed at the UHRH for one week.

During this period they experienced robotic technology as part of their ambient environment, while periodically playing the role of experimental participants in a series of exploratory experiments with different robot companions. These experiments involved gaze tracking during social interaction, the expression of non-verbal cues as indicator for intended actions, the use of a robot in remote communication and a sensor grid to log daily routines and analyse behavioural patterns. In order to do these experiments various robots were used. The robots were a Care-O-Bot 3 (COB) [14], the UH Sunflower Robot [15], KASPAR [16] and CHARLY [17]. Although these robots were initially introduced to the artists in constrained interaction episodes, the residency experiment quickly went beyond those constrained episodic interactions. The idea was for the participants to live as normally as possible within the UH Robot House: sleeping, eating and, more importantly, working there to create artistic responses to their experiences. The week concluded with a public event, which involved the display of digital and performance artworks created by the artists in collaboration with the researchers, and also show and tell demonstrations by the researchers.

Activities of researchers	Activities of participants
- presence during the day from 10am to 5pm - maintenance of robots when needed - preparation and support for experiments	- different studies concerned with expressiveness of Sunflower robot - eye gaze tracking study involving the Care-O-Bot 3 - communication via KASPAR robot - continued recording of activities in the house from 6pm to 8am

Fig. 2. Activities during the "Week in the Robot House" research

3   Initial Constrained Experiments

3.1   Eye Tracking

In order to understand how users perceive the appearance and behaviours of robots and where they are focusing their attention during interaction with them, it is useful to collect real time gaze data. Different human-robot interaction scenarios were performed with the participants wearing an eye gaze tracking device [18]. The scenarios included structured domestic tasks that required interaction with the COB (e.g. see Fig.1.Right: flower sorting).

The ASL Mobile Eye gaze tracking system was used to measure gaze direction within the field of vision of a scene camera mounted on the glasses, pointed straight ahead at what the wearer is facing. The system software logs gaze direction in image pixel coordinates, along with the video input from the scene and eye-directed cameras used by the tracking system [19]. This enabled the manual coding of which part of the COB each of the participants was looking at and for how long. This data will be used to design potential social behaviours for the COB.

3.2   Robot Mediated Communication

One of the more recent potential application areas being investigated for humanoid robots is the possibility of robot-mediated interviews [20]. During an experiment to explore more user-friendly interface options for non-technical users for the humanoid robot KASPAR [16] the artists communicated with each other via KASPAR and a live video link.

The person controlling KASPAR had a video link and could see and hear what was happening from the robots' perspective using a web-cam with an integrated microphone. The other user interacted with and spoke to the robot. KASPAR was controlled via a Java GUI and was pre-programmed with some basic body and facial gestures to aid communication. Body gestures included pointing (towards or away from robot), nodding and shaking of head. Facial options included sad, neutral, happy and very happy. Speaking through KASPAR was achieved with a microphone and voice conversion software. The voice conversion software had numerous options for different voices. Variations of high and low pitched human type voices were used which gave a masculine, feminine or childlike impression.

3.3   Robot Expressiveness

The COB expressiveness study aimed at actively involving the participants in creating robot behaviours for their own comfort and preferences, and thus interactively guide the development of the robots' behaviour for one of the ACCOMPANY Projects' scenarios [21]. The goal of this process was to improve the perceived social intelligence and consideration towards the person in the vicinity of the robot.

The study procedure involved the experimenter showing the participants the scenario in a step-by-step walk-through. At different stages of the scenario the participants were asked specific questions with regard to how the robot should perform the particular task and why (i.e. "Should the robot make any sound and if so, what kind of sound?"; "Should the robot display something on its colour LED display?"; "Where should the robot position itself?"; etc.).

4   Results

As the main evaluation method a series of structured interviews was performed. Each of the different researchers involved in the week long co-habitation experiment asked specific questions concerning their topics in different interviews. As a final stage, a general discussion with the participants was held. The interviews and the discussions were transcribed, analysed and is reported below:

4.1   Eye Tracking Systems

When asked about their overall impression of the ASL Mobile Eye device, the participants reported that wearing an eye tracker for extended periods of time was uncomfortable, because the device became a bit heavy, but nevertheless quite fun. According to them, wearing the eye-tracking device mostly did not change the way they looked at things and robots, but it did affect the way they looked at people. It made them more mindful while looking at other persons. "I was more worried about where I was looking at people", one of the participants said.

They also mentioned that walking around with the head mounted device made things in the environment a bit blurry and that it gave them the feeling something was in their field of vision.

The experiments with the eye-tracking device also led to the development of an art installation, which juxtaposed the participants' own robot HARR1 (Humanoid Art Research Robot 1) and users with the eye-tracking device during the concluding open house event. The idea behind this installation was to illustrate how the ASL Mobile Eye is used to examine at which parts of the robots' 'bodies' or 'heads' people look at and how HARR1 is perceived as looking back at people. HARR1 used an artificial vision system to scan the room and build up a picture of its location.

4.2   Robot Mediated Communication

At the end of the week the participants were asked for feedback on the remote communication system particularly with regard to 2 primary questions:

1. How user friendly the interface was for the person controlling the robot?
2. What type of voice would be most appropriate for the robot?

The participants felt that although the system was easy to use, it would have been easier to have a system that performed full body tracking with a Kinect sensor and that also replicated bodily gestures with the robot without the need for clicking and selecting gestures. From a purely remote presence perspective it would make sense to have a more seamless control mechanism to operate the robot without the need to explicitly select gestures on screen. The participants also reported that having a child-like voice accompanied the robot well, and that they did not feel the need to change it. In addition they mentioned that having a strong robotic voice would make it more difficult to understand the robot, which in an interview scenario would be problematic.

4.3   Robot Expressiveness

The data collected from the COB expressiveness study indicated that the participants preferred the robot to behave in their home in a consistent manner [22]. For example, both users supported the robot using its LED colour display as a simple context or task based status indicator. It could be used to signal both non-critical status (such as charging level etc.) and also more critical status (such as urgent attention required, i.e. a boiling kettle) and thus draw the users' attention to the current task of the robot and its associate potential level of hazard displayed (i.e. when using its arm in the vicinity of the participants).

The data also indicated that the COB should have a basic etiquette. It should for example always give way to allow users to go first and then follow after them. Its posture should indicate "after you", similar to an inviting gesture used when a home owner invites a guest into their house. During the experiment, the participants preferred the robot to remain in the living room with them, rather than moving away and coming back later (e.g. if the robot's responsibility is to remind them to drink).

4.4   Additional Findings

Privacy Issues at the Robot House

When asked whether their privacy at the UHRH was intruded upon by the robots, the participants reported that it was hard for them to separate the robots from the researchers working with them, because the robots were never really active without the researchers.

They commented that rather than being more used to robots after the co-habitation experiment, now they are more used to researchers. For them, the robots and researchers were in symbiosis, and the researchers appeared more like puppeteers to them – while most of the robots operated autonomously, they needed researchers to activate them and to monitor their behaviour.

Since both participants were familiar with technology and robots, their attitude towards robots – as they declared – did not really change after the week in the UHRH. Nevertheless, they reported that they had learned a lot about researchers and participating in experiments. They see this as really valuable for them to understand and to be able to share with the wider public.

They also described that they had the impression that KASPAR (even while inactive) still had a sort of feedback link to another room and the researcher on the other end was able to hear what they talked about. According to them, this made them wary about what they said in the presence of the KASPAR. On the other hand, the users were well aware that the UHRH was equipped with overhead cameras, microphones and a considerable amount of sensors, and that there was the real possibility of being recorded by these at any time as they actually could be remotely activated and accessed. Apparently, this awareness faded away over time, even though they were still well aware that the system could be activated remotely.

The users affirmed they were overall comfortable with the presence of the houses' sensor network system and overhead cameras. They even proposed the possibility to tweet live sensor information such as sending humorous messages like "someone has eaten a biscuit again", "toilet has been flushed" or "someone has left toilet seat up".

Idle Behaviour

The first reaction of the participants towards the researchers developing idle behaviour for the robots was to say: "Constant movements make noise." They considered that this would be potentially annoying, similar to the noise made by ticking clocks. Of course it depends of the sound of the noise and when it is made; "at night, the charging noise of the robot was annoying". Therefore, both participants agreed that it is best not to use sounds on the robots at all.

In order to avoid noise, they were asked if the use of lights would be any different and their answer was similar, "at night any small light can be very annoying". They came up with the idea that lights should be powered up only when the user comes near the robot, i.e. proximity related lights to indicate idle behaviour. "A single sort of light to indicate the robot is about to do something would be best, like amber colour". Note that the COB allows for customisable LED indicator colours for these behaviours.

5   Conclusions and Discussion

The Integrative Holistic Feedback Approach (IHFA) implemented in the described week long co-habitation experiment in the UHRH was a success in terms of insights from the users that can be fruitfully integrated in the HRI design process in progress at UH Adaptive Systems Research Group. Due to the continuous stay in the UHRH and the level of personal interest on the side of the participants, it was possible to repeatedly run an entire series of experiments. The integrative and informing atmosphere was maintained with the participants, and their level of technological awareness facilitated the IHFA implementation process.

The continuous long-term exposure also allowed the exploration of ambient aspects of living together with robots. It was very informing to hear the user accounts of the noises the robots made at night and the feel of the UHRH after things were calming down in the evening. It seems for example necessary to give more thought to the recharging noises a robot makes, or to what light or sound signals it should exhibit during the evening and night times, if any. Another interesting finding in this context is that the users would prefer to have the robot around them in the living room instead of having it come to them to inform or remind them, then move away between interactions

The participants described the relationship between the robots and the researchers as being in symbiosis. For them it was difficult to separate the robots from the researchers working with them. This was an unexpected outcome of the experiment. It converges with Suchmans' social situatedness theory [23]. In most demonstrations, and short-term experiments, participants usually see only the working robot behaving without much interference from researchers. Being actually present while the behaviours of the robot were developed and seeing the dependence of the technology on the human work was, according to the participants, a very valuable experience.

One example of an interesting specific outcome of the experiment was the description of the overall feeling of living in an environment with many sensors. The participants noted that they particularly felt watched by KASPAR. Being asked whether they felt the same way about the other sensors integrated in the UHRH or the COB, they stated in the case of the Robot House they got very quickly used to the situation and in the case of the COB it never even crossed their mind. We speculate that this effect occurred due to the way KASPAR was presented, namely its child-like appearance, realistic human-like features and its seemingly autonomous operation. The controlling computer and operating person was in a different room on the upper level of the house only for the specific interaction sessions

Finally we would like to remark that for the implementation of the IHFA methodology, it is important not only to realize a continuous and prolonged human-robot co-habitation in a smart home environment, but also to have knowledgeable and informed participants who have a personal interest to experience new technology.

We are aware that the feedback gained from two people may not generalize to other artists, let alone a wider population, so results gained in this study need to be critically reflected upon and compared with results from larger-scale user experiments and findings in the literature. The motivation of the participants, to gain experience with robots and smart home technology also differs significantly from scenarios where robots are meant to have specific assistive functions in concrete application areas. However, the present study gave us creative and novel insights and "food for thought" for our future work.

Acknowledgements. The research leading to these results has received funding from the [EU's] Seventh Framework Programme (FP7/2007-2013) under the grant agreement n°[287624].

References

[1] Heerink, M., Kröse, B., Evers, V., Wielinga, B.: Assessing acceptance of assistive social agent technology by older adults: the almere model. International Journal of Social Robotics 2(4), 361–375 (2010)

[2] Bickmore, T.W., Caruso, L., Clough-Gorr, K., Heeren, T.: It's just like you talk to a friend's relational agents for older adults. Interacting With Computers 17(6), 711–735 (2005)

[3] Sharkey, A., Sharkey, N.: Granny and the robots: ethical issues in robot care for the elderly. Ethics and Information Technology 14(1), 27–40 (2012)

[4] Díaz-Boladeras, M., Saez-Pons, J., Heerink, M., Angulo, C.: Emotional factors in robot-based assistive services for elderly at home. In: RO-MAN 2013, Gyeongju, Korea (2013)

[5] Duque, I., Dautenhahn, K., Koay, K.L., Willcock, I., Christianson, B.: Knowledge-driven user activity recognition for a Smart House. Development and validation of a generic and low-cost, resource-efficient system. In: ACHI 2013, Nice, France (2013)

[6] Kanda, T., Hirano, T., Eaton, D.: Interactive Robots as Social Partners and Peer Tutors for Children: A Field Trial. Human-Computer Interaction 19, 61–84 (2004)

[7] Walters, M.L., Oskoei, M.A., Syrdal, D.S., Dautenhahn, K.: A long-term Human-Robot Proxemic study. In: RO-MAN 2011, pp. 137–142 (2011)

[8] Leite, I., Martinho, C., Paiva, A.: Social Robots for Long-Term Interaction: A Survey. International Journal of Social Robotics 5(2), 291–308 (2013)

[9] Koay, K.L., Syrdal, D.S., Walters, M.L., Dautenhahn, K.: Living with Robots: Investigating the Habituation Effect in Participants' Preferences During a Longitudinal Human-Robot Interaction Study. In: RO-MAN 2007, pp. 564–569. IEEE, South Korea (2007)

[10] Hüttenrauch, H., Severinson-Ecklundh, K.: Fetch and Carry with CERO: Observations from a Long Term User Study with a Service Robot. In: RO-MAN 2002, Berlin, Germany, pp. 158–164 (2002)

[11] Sabanovic, S., Michalowski, M.P., Simmons, R.: Robots in the Wild: Observing Human-Robot Social Interaction Outside the Lab. In: AMC 2006, Istanbul, Turkey, pp. 756–761 (2006)

[12] Koay, K.L., Syrdal, D.S., Walters, M.L., Dautenhahn, K.: Five weeks in the robot house–Exploratory human-robot interaction trials in a domestic setting. In: ACHI 2009, Cancun, Mexico, pp. 219–226 (2009)

[13] Rudd, J., Stern, K., Isensee, S.: Low vs. high-fidelity prototyping debate. Interactions 3, 76–85 (1996)

[14] Parlitz, C., Hägele, M., Klein, P., Seifert, J., Dautenhahn, K.: Care-O-bot 3 - Rationale for human-robot interaction design. In: International Federation of Robotics u.a.: ISR 2008: 39th International Symposium on Robotics, Seoul, Korea, pp. 275–280 (2008)

[15] Koay, K.L., Lakatos, G., Syrdal, D.S., Gacsi, M., Bereczky, B., Dautenhahn, K., Miklosi, A., Walters, M.L.: Hey! There is someone at your door. A Hearing Robot using Visual Communication Signals of Hearing Dogs to Communicate Intent. In: Alife 2013, Singapore (2013)

[16] Dautenhahn, K., Nehaniv, C.L., Walters, M.L., Robins, B., Kose-Bagci, H., Assif, N., Blow, M.: KASPAR – a minimally expressive humanoid robot for human robot interaction research. Applied Bionics and Biomechanics 6, 369–397 (2009)

[17] Walters, M.L., Syrdal, D.S., Dautenhahn, K., Dumitriu, A., May, A., Christiansen, B., Koay, K.L.: My Familiar Robot Companion: Preferences and Perceptions of CHARLY, a Companion Humanoid Autonomous Robot for Living with You. In: Herrmann, G., Studley, M., Pearson, M., Conn, A., Melhuish, C., Witkowski, M., Kim, J.-H., Vadakkepat, P. (eds.) FIRA-TAROS 2012. LNCS, vol. 7429, pp. 300–312. Springer, Heidelberg (2012)

[18] Applied Science Laboratories. Mobile eye gaze tracking system,

[19] http://asleyetracking.com/

[20] Broz, F., Lehmann, H., Nehaniv, C.L., Dautenhahn, K.: Mutual Gaze, Personality, and Familiarity: Dual Eye-tracking During Conversation. In: RO-MAN 2012 (2012)

[21] Wood, L.J., Dautenhahn, K., Rainer, A., Robins, B., Lehmann, H., Syrdal, D.S.: Robot-Mediated Interviews-How Effective Is a Humanoid Robot as a Tool for Interviewing Young Children? PLOS ONE 8(3), e59448 (2013)

[22] Lehmann, H., Syrdal, D.S., Dautenhahn, K., Gelderblom, G.J., Bedaf, S., Amirabdollahian, F.: What can a robot do for you? - Evaluating the needs of the elderly in the UK. In: ACHI 2013, Nice, France, pp. 83–88 (2013)

[23] Koay, K.L., Walters, M.L., May, A., Dumitriu, A., Christianson, B., Burke, N., Dautenhahn, K.: Exploring Robot Etiquette: Refining a HRI home companion scenario based on feedback from two artists who lived with robots in the UH robot house. In: Herrmann, G., Pearson, M., Lenz, A. (eds.) ICSR 2013. LNCS (LNAI), vol. 8239, pp. 281–290. Springer, Heidelberg (2013)

[24] Suchman, L.: Human-machine reconfigurations: Plans and situated actions. Cambridge University Press (2007)

iCharibot: Design and Field Trials
of a Fundraising Robot

Miguel Sarabia, Tuan Le Mau, Harold Soh, Shuto Naruse, Crispian Poon,
Zhitian Liao, Kuen Cherng Tan, Zi Jian Lai, and Yiannis Demiris

Personal Robotics Lab
Department of Electrical and Electronic Engineering
Imperial College London, United Kingdom
{miguel.sarabia,y.demiris}@imperial.ac.uk

Abstract. In this work, we address the problem of increasing charitable donations through a novel, engaging fundraising robot: the Imperial
Charity Robot (iCharibot). To better understand how to engage passers-
by, we conducted a field trial in outdoor locations at a busy area in London, spread across 9 sessions of 40 minutes each. During our experiments,
iCharibot attracted 679 people and engaged with 386 individuals. Our
results show that interactivity led to longer user engagement with the
robot. Our data further suggests both saliency and interactivity led to
an increase in the total donation amount. These findings should prove
useful for future design of robotic fundraisers in particular and for social
robots in general.

1 Introduction

In 2011, an estimated £11bn were
raised for charities in the UK, with
almost 6 in every 10 adults contributing to charitable causes [5].
That said, charities still face challenges collecting sufficient funds in
our current harsh economic climate.
In fact, charitable giving fell 11%
in 2009 back to 2007 levels. In the
2011 UK Giving Report it was highlighted that "the operating environment for the sector remains tough
... and a number of organisations
are having to scale back or indeed
cease activity in light of financial difficulties" [5].

In this work, we consider raising
funds for charitable causes with an engaging donation-collection robot we call

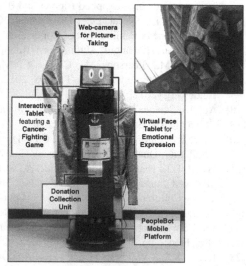

Fig. 1. Picture of iCharibot highlighting the
main components of the robot, and two participants who engaged with the it[1]

---

[1] A video of the system is available at: http://imperial.ac.uk/PersonalRobotics.

G. Herrmann et al. (Eds.): ICSR 2013, LNAI 8239, pp. 412–421, 2013.

the Imperial Charity Robot (iCharibot), shown in Fig. 1. iCharibot can express a range of facial expressions and has several abilities useful for social interaction; for example, waving hello and and offering passers-by to play a tablet game.

From a human-robot interaction (HRI) research perspective, we are interested in understanding the interaction mechanisms that would attract people to approach the robot and, hopefully, persuade them to donate. In other words, how should an engaging charity robot act and behave? In an effort to answer this question, we performed field trials spread across 9 sessions of 40 minutes each in three real-world locations in the South Kensington area of London; a busy locale, home to three museums, a concert hall and our university. During the 9 sessions, 386 people engaged with the robot and a total of £116.77 was raised. All funds collected by iCharibot were donated to Imperial Cancer Research, a subsidiary of Cancer Research UK.

Although there have been many robots interacting with the public in unconstrained environments [3,13]; the goal of such studies has traditionally been to verify whether a system is functional rather than to investigate robotic design principles. Further, iCharibot is uncommon amongst fundraising robots [7,8] in that it is the only one to have been deployed in a busy urban street.

This paper describes iCharibot's design and the results of our field trials. Specifically, we find that a more *sophisticated behaviour* (this concept is defined in Sect. 3.3) leads to greater user engagement duration and suggests a greater amount of donations.

2   Background

To the best of our knowledge, our work is preceded by only two other cases of fundraising robots. The first was Dona [7], which unlike iCharibot, was designed to be cat-like. As such, Dona had a very small foot-print and was designed to roam around its environment. From the point of view of interaction Dona bowed repeatedly when it encountered a person or an object. In contrast, iCharibot has a different roster of available actions, like waving its arms or reproducing synthesised speech.

The second fundraising robot was RoboBeggar, which took the appearance of a traditional Finnish begging statue [8]. RoboBeggar was similar in size to iCharibot and also had a touch screen for interaction but used a bank-card reader instead of a coinbox. The robot collected donations on behalf the Cancer Association of South-West Finland for two weeks at a shopping centre in Turku. A key difference between that experiment and ours was that RoboBeggar shared the role of fundraiser with other humans, whilst iCharibot was the only entity in charge of raising funds in our field trials. The authors do mention that donating to RoboBeggar was a lengthy process (mainly due to the use of a bank card); iCharibot attempted to avoid this, and much effort was spent on getting the robot to react to donations in a timely manner.

To better design iCharibot, we considered other robots that work in crowded, unconstrained environments, such as Rhino, the tour-guide robot [3]. Rhino was one of the earliest robots to be able to safely move in crowded environments and showed visitors around the Deutsches Museum Bonn. The robot operated for more 47 hours, and guided more than 2000 people; perhaps more importantly the authors remark that "Rhino's ability to react to people proved to be one of the most entertaining aspects, which contributed enormously to its popularity". The same team went on to build and design Minerva [13] another tour-guide robot that operated during two weeks at the National Museum of American History in Washington D.C. The robot ran for 94 hours and performed 620 tours. Crucially, Minerva was capable of showing emotional states through a mechanical face which satisfied and amused people more than Rhino's machine-like appearance. These two robots represent early examples of the importance of human-robot interaction design when dealing with the public. Furthermore, they also proved the feasibility of robotic field trials in crowded environments.

A more recent development is the Autonomous City Explorer (ACE) [1], which went from Technische Universität München's campus to Munich's city centre by asking for directions to 38 passers-by. Similar to iCharibot, ACE received commands from passers-by via a touch screen. It covered a distance of 1.5 km and took 5 hours to achieve its objective.

From these studies, we identified several guidelines that were relevant for iCharibot's design: a robot should be responsive to user actions, a touchscreen can be an intuitive and reliable interface and a robot should not threaten or inconvenience passers-by.

3   Design Considerations

iCharibot's main role was to raise money for charitable purposes and this goal guided our design decisions. A first prototype of the robot was tested with students from our university. The lessons learnt from these preliminary tests were incorporated in iCharibot (as we describe in Sect. 3.2). £48.83 were raised from these preliminary tests.

This section details our initial choices and modifications introduced after preliminary testing in terms of both hardware and software. It also introduces the different *behaviour modes* which will be our primary predictor variable.

3.1   Robot Appearance

As can be seen in Fig. 1, iCharibot's overall appearance is human-like in size with a cylindrical body. It was built using the PeopleBot platform. We ensured iCharibot did not mislead users by appearing to be more intelligent or capable than it really was [6].

To encourage interaction, a secondary Android tablet was fitted flat on the platform next to the robot's face. This interactive touch-interface allowed a user to play a cancer-fighting game. In addition, the robot had a three degree-of-freedom servo-operated mechanical right arm (for waving) and a transparent coin box (for donation collection). We posted the Imperial Cancer Research logo[2] below the coin collector surrounded by flashing blue LED strips to clearly and visibly announce iCharibot's fundraising purpose.

3.2 Robot Software and Hardware Design

The main computational unit for the system was a laptop PC mounted on the back of the iCharibot (on-board Robot PC in Fig. 2). This on-board PC provided physical connections via USB sockets for the Arduino microcontrollers that connected to the right arm and the coin detector. A second laptop was used to remotely-control the On-board Robot PC (Remote PC in Fig. 2).

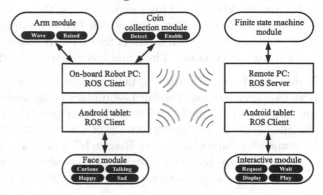

Fig. 2. iCharibot's modular software architecture. Rectangles correspond to hardware, while clear round-edged boxes represent modules. Black boxes denote the possible module states.

Figure 2 illustrates the iCharibot software architecture, developed using the Robot Operating System (ROS) [11]. In what follows, we describe each of these ROS modules.

Arm Module: our three-degree-of-freedom robot arm was based on a simple model of an anthropomorphic arm. The arm servos responded to control signals from an Arduino microcontroller which in turn received commands from ROS. The arm supported two states: *Wave* and *Raised*. The former was meant to attract attention. Originally, during preliminary trials, we used high-frequency waving (at 1Hz) to make the robot appear enthusiastic but this was slowed to 0.5Hz after initial tests revealed that it made iCharibot appear aggressive and prevented potential donors from approaching.

Face Module: iCharibot's face, which was displayed on the Android tablet, was graphically designed to evoke emotional connections between the robot and potential donors (following [2,15]). At the same time, we avoided showing a completely human face that may induce uneasy feelings (the uncanny-valley effect [10]). There were four different facial states: *Happy, Sad, Talking* and *Curious*, with switching between states controlled by a native Android application using the rosjava library.

Coin Collection Module: from preliminary tests, we discovered that the robot was expected to respond in a timely manner to donations; iCharibot initially used a coin sorter but we observed that this caused a donation-detection latency that led to the robot thanking the donor after she had left. This further caused

---

[2] Permission was granted to use Imperial Cancer Research's brand.

confusion to the next donor. Subsequently, we replaced the coin-sorter by a flat metal receptacle attached to a microswitch which was fitted underneath the coin collector to detect coins. This allowed for immediate detection of donations, triggering the appropriate robot action.

Interactive Module: This module ran on the secondary Android tablet and featured a cancer-fighting game which had been approved by Imperial Cancer Research. To play the game, users tapped their fingers on regions on the screens to "kill" cancer tumours. The game was won when all the tumours were removed.

Voice Module: iCharibot synthesised audio sentences using the Festival [4] text-to-speech engine. During preliminary tests, sentences such as "Hello, how are you?" proved ineffective at attracting people's attention. As such, we re-designed iCharibot's speech to prompt users to *take action* (eg. "Would you like to donate for Imperial Cancer Research?"). A synthetic robotic voice was used (instead of a pre-recorded human voice) to allow donors to quickly understand they were interacting with a robot, without raising user expectations [6].

Human-Detection Module: iCharibot was initially programmed to detect passers-by using PeopleBot's sonars. However, this detection mechanism did not work reliably during preliminary tests, which made the robot's behaviour unpredictable. To preserve the robot interactivity aspects, we decided to manually send a *human detected* signal to iCharibot from our remote PC when a person approached the robot. Apart from this, iCharibot was autonomous with its actions governed by the finite state machine module.

Finite State Machine Module: iCharibot's overall control algorithm was implemented using a finite state machine. The possible states of the robot organised into three distinct *behaviour modes*, which will be discussed next.

3.3   Behaviour Modes

The actions and abilities available to iCharibot during a given trial are determined by the behaviour mode. Importantly, behaviour mode is our main variable to design for greater engagement duration and donations. We devised the following modes (in order of increasing sophistication):

Baseline Behaviour. iCharibot stands still and ignores passers-by.

Salient Behaviour. iCharibot keeps waving its arm to ask for a donation but does not respond to the donors in any way when such donation is made. The tablet game is inactive in this mode.

Interactive Behaviour. iCharibot attempts to interact with potential donors by putting its arm down when a human approaches and asking her to play the cancer-fighting game. The game is designed to pause and iCharibot then asks for a donation. If a donation is received during the next 200 seconds, the robot will thank the donor, otherwise the robot will display an animated sad face. Notice that *interactive* behaviour is a superset of *salient* behaviour.

4   Experimental Procedure

In human philanthropy studies, there is evidence that active solicitation attracts more donations than passive solicitation [9]. Might this also be true for robots? Given that saliency and interactivity imply more active solicitations, we formulated the following hypotheses:

H_1. A user's *engagement duration* (τ) with iCharibot is longer in *interactive* mode than in *salient* mode.

H_2. A user's *engagement duration* (τ) with iCharibot is longer in *salient* mode than in *baseline* mode.

H_3. The total amount collected by iCharibot is higher in *interactive* mode than in *salient* mode.

H_4. The total amount collected by iCharibot is higher in *salient* mode than in *baseline* mode.

4.1   Interaction Types

Both H_1 and H_2 require a strict definition of what does it mean for a user to be *engaged*. The following categorisation of interactions between passers-by and the robot provides such definition as well as two other categories to avoid ambiguity during the field trials.

Attracted. The user made a definite stop and gazed at the robot.

Engaged. The user made a physical contact with iCharibot, either by touching parts of its body, arm or face; by playing the cancer-fighting game; or by making a donation. With this type of interaction we also recorded the *engagement duration* (τ).

Ignored. If a user was neither *attracted* nor *engaged*, then we classified the interaction as *ignored* (even if the user did take notice of the robot).

4.2   Engagement Duration

In our preliminary tests, we noticed that the robot was frequently approached by *groups*; this made keeping track of individual engagements more complicated. We further observed that each individual in a group generally remained interested throughout the time the group was engaged with the robot; either by continuing to gaze at the robot, moving around to look its different parts or talking about it with other group members. Moreover, since iCharibot is capable of multiple engagements, individuals sometimes played the cancer-fighting game together, touched the face or arms of the robot or donated coins simultaneously. To properly account for these cases, we defined the *engagement duration* (denoted as τ) as the difference between the time a user became *engaged* and either the time she left (if she was alone) or the whole group left (if she approached iCharibot with a group).

4.3   Field Trial Data Collection

Nine field trials were carried out during February 2012 at three locations along Exhibition Road in London, easily identified by the respective landmarks: outside the Science Museum, outside the Natural History Museum and close to South Kensington station. To even out any bias, each trial was carried out at a different time of the day (between 2pm and 6pm), with a different robot behaviour mode and at a different location.

Several pieces of information were recorded during the field trials. The most important was the interaction type of all passers-by that came close of iCharibot. For those participants who were categorised as engaged, the engagement duration (τ) was also logged. The high amount of foot traffic meant that no user surveys could be taken. Similarly, in order to ensure a swift donation process, no individual donations were recorded; note that even with a fast coin detector, individual donation tracking remains challenging as people donated concurrently. Instead of individual donations, we recorded the total funds raised per trial.

As the robot is almost fully autonomous, the duties of the researchers during experiments were limited to ensuring a safe operation and classification of interactions between passers-by and the robot. Consequently, the researchers stayed a few metres away from the robot. To aid with classification an Android application was developed which allowed for quick, convenient and concurrent categorisation of interactions. Thus a researcher only had to determine the type of interaction when a passer-by approached the robot. If the type of interaction was classified as *engaged* then the end of the interaction would also be logged (as per the definition of τ-duration introduced in Sect. 4.2). The classification task was always carried out by the same researcher for the sake of consistency.

5   Empirical Results, Analysis and Discussion

Throughout our nine 40-minute trials, we recorded a total of 9884 people who came within the iCharibot's vicinity, of which 679 people were *attracted* to the robot and 386 people were *engaged* with it.

Given our definitions of engagement duration (τ) and behaviour modes, we found the median τ-durations for the baseline, salient and interactive behaviours to be 18.96, 20.69 and 47.76 seconds respectively (boxplot shown in Fig. 3). To test our hypotheses, we first performed the non-parametric Kruskal–Wallis test and found the medians to be statistically different ($p \approx 4.9 \times 10^{-4}$). A further test on pairs using the Mann-Whitney U-test showed statistical difference between the *baseline* and *interactive* modes ($p \approx 1.7 \times 10^{-3}$), as well as the *salient* and the *interactive* modes ($p \approx 1.0 \times 10^{-3}$). However, the test did not show statistical difference between the *baseline* and *salient* behaviours. Looking closer at the data, the minimum and maximum τ-durations, discarding outliers, were $(1.00, 64.74)$ for baseline mode; $(0.56, 119.84)$ for salient mode; and $(0.86, 219.29)$ for interactive mode. These findings are in favour of hypothesis $\mathbf{H_1}$, but do not support hypothesis $\mathbf{H_2}$; that is, interactivity leads to longer

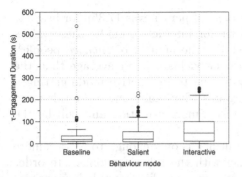

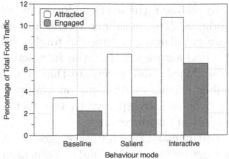

Fig. 3. Observed interaction durations (in seconds) for each behaviour mode. As hypothesised, interactivity led to an increase in the τ-duration. Results shown are statistically significant with $p < 0.01$.

Fig. 4. Classification of interactions by percentage over total foot traffic for each behaviour mode. Note data from different locations has been collated.

engagement duration, whereas saliency alone does not appear to be sufficient for longer engagements.

Figure 4 shows a comparison of the interaction types across the three behaviour modes of iCharibot as a bar graph. The total number of people who *ignored*, were *attracted* to and *engaged* with iCharibot were tallied and expressed as the percentage of the total foot traffic throughout the experiment period. The percentage of passers-by who ignored the robot consistently fell from 94.4% to 89.1% to 82.6%, making a step decrease of about 6% each. The percentage of passers-by who were *attracted* to iCharibot rose from 3.4% to 7.4% to 10.8% and of those who were *engaged* with iCharibot from 2.2% to 3.5% to 6.6%. Thus, as iCharibot's behaviour became more sophisticated, fewer passers-by *ignored* it and more of them were *attracted* to or became *engaged* with the robot.

Figure 5 shows the average amount of funds collected per trial (in pence sterling) over the total foot traffic per trial — note that this amount is not representative of the amount of money donated by any single person. As can be observed from the figure, the data suggests that the funds raised increased with more sophisticated iCharibot behaviours. The average donation per person was 0.39p for baseline mode, 0.84p for salient mode and 1.19p for interactive mode; amounts are pence sterling and were

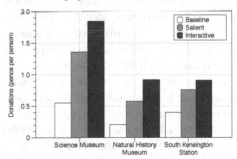

Fig. 5. Average donation amount per person, calculated as the money collected (in pence) over the total foot traffic per trial

calculated dividing the funds raised over the total flow. This represents an increase in average donations of around 0.4p as the sophistication of the robot grows, suggestive of H_3 and H_4. However, without individual donations, we cannot compute the statistical validity of these findings.

In addition, the average donation per *engaged* person was 17.35p for baseline mode, 24.21p for salient mode, and 18.09p for interactive mode; these values were calculated dividing the funds raised over the number of *engaged* people[3]. We find surprising that the interactive mode yields a lower average than the salient mode and speculate that it may be due to interactive mode getting a higher portion of its contribution from groups where only one person donated for everyone. Irrespectively, this fact deserves more attention and will be the subject of future research.

In summary, our results suggest that saliency is only enough to attract more passers-by but not to make them interact with the robot for longer. In order to achieve longer human-robot interactions, a more sophisticated behaviour was needed. In addition, we also observed promising evidence that a more sophisticated robot behaviour leads to a greater donation amount.

6   Conclusion and Further Work

In this article, we presented the design and realisation of iCharibot, a fundraising robot. We worked towards a robot that appeared friendly with cartoon-like facial expressions and ensured that it was responsive to donations. During experiments in a very busy area of London, 386 people *engaged* with iCharibot, and £75.94 sterling was raised for Imperial Cancer Research (plus an extra £48.83 in preliminary tests).

Overall, we found that increasing behaviour sophistication appealed to people, causing them to spend more time with iCharibot. Our results confirmed several intuitive notions: robot interactivity leads to a higher level of engagement while saliency alone does not (under the assumptions presented in Sect. 4.2). Moreover our data suggests that interactivity also leads to a higher amount of donations — we look forward to more tests with iCharibot, where individual donation amounts are recorded, to conclusively show this.

This work opens up several avenues for future research. From a technical standpoint, iCharibot can be improved in several ways. For example, sonars proved too unreliable for automatically detecting users and we are currently investigating the use of a Kinect sensor as a potential solution. From a HRI standpoint, interacting with iCharibot could be richer and more fluid. For example, communicating through a touch interface is effective and reliable but not as intuitive as speech. From an experimental perspective, we used a *conservative classification*; during the field-trials, many people glanced and smiled at the robot but were classified as *ignored*. Future work would go into refining these classifications in order to better capture the range of interactions that occurred. Another question left to be explored is whether interactivity is actually cost effective. The data suggests that interactivity yields higher donations at the expense of higher engagement time. This expense might not be without benefits since we have also observed a "group effect", where passers-by that walked in

---

[3] Note that, as per the definitions in 4.1, everyone who donates is classified as engaged, but not every engaged person donated.

groups were more likely to approach the robot ([12,14] report similar findings). Finding principled methods for balancing the multitude of factors that influence how engaging a robot is for a certain task remains a challenging and rewarding task that deserves further consideration.

References

1. Bauer, A., Klasing, K., Lidoris, G., Mühlbauer, Q., Rohrmüller, F., Sosnowski, S., Xu, T., Kühnlenz, K., Wollherr, D., Buss, M.: The autonomous city explorer: Towards natural human-robot interaction in urban environments. Intl. J. of Social Robot. 1(2), 127–140 (2009)
2. Breazeal, C.: Emotion and sociable humanoid robots. Intl. J. Human-Computer Stud. 59(1-2), 119–155 (2003)
3. Burgard, W., Cremers, A., Fox, D., Hähnel, D., Lakemeyer, G., Schulz, D., Steiner, W., Thrun, S.: Experiences with an interactive museum tour-guide robot. Artif. Intell. 114(1-2), 3–55 (1999)
4. Clark, R.A., Richmond, K., King, S.: Festival 2 – Build Your Own General Purpose Unit Selection Speech Synthesiser. In: Proc. of ISCA Speech Synthesis Workshop, pp. 173–178 (2004)
5. Dobbs, J., Jochum, V., Wilding, K., Lipscomb, L., Smith, M., Harrison, R.: UK Giving 2011: An overview of charitable giving in the UK (2011)
6. Goetz, J., Kiesler, S., Powers, A.: Matching robot appearance and behavior to tasks to improve human-robot cooperation. In: Proc. of IEEE RO-MAN, pp. 55–60 (2003)
7. Kim, M.S., Cha, B.K., Park, D.M., Lee, S.M., Kwak, S., Lee, M.K.: Dona: Urban Donation Motivating Robot. In: Proc. of ACM/IEEE HRI, pp. 159–160 (2010)
8. La Russa, G., Sutinen, E., Cronje, J.C.: When a robot turns into a totem: The RoboBeggar case. In: Soomro, S. (ed.) E-learning Experiences and Future, ch. 17, pp. 327–344. InTech (2010)
9. Lindskold, S., Forte, R.A., Haake, C.S., Schmidt, E.K.: The Effects of Directness of Face-to-Face Requests and Sex of Solicitor on Streetcorner Donations. J. of Social Psychology 101, 45–51 (1977)
10. Mori, M.: The uncanny valley. Energy 7(4), 33–35 (1970)
11. Quigley, M., Gerkey, B., Conley, K., Faust, J., Foote, T., Leibs, J., Berger, E., Wheeler, R., Ng, A.: ROS: an open-source Robot Operating System. In: Proc. of the Open Source Software Workshop at ICRA (2009)
12. Shiomi, M., Sakamoto, D., Kanda, T., Ishi, C.T., Ishiguro, H., Hagita, N.: Field Trial of a Networked Robot at a Train Station. Intl. J. of Social Robot. 3(1), 27–40 (2010)
13. Thrun, S., Bennewitz, M., Burgard, W., Cremers, A., Dellaert, F., Fox, D., Hahnel, D., Rosenberg, C., Roy, N., Schulte, J., et al.: MINERVA: A second-generation museum tour-guide robot. In: Proc. of IEEE ICRA, pp. 1999–2005 (1999)
14. Weiss, A., Bernhaupt, R., Tscheligi, M., Wollherr, D., Kuhnlenz, K., Buss, M.: A methodological variation for acceptance evaluation of Human-Robot Interaction in public places. In: Proc. of IEEE RO-MAN, pp. 713–718 (2008)
15. Yuasa, M., Saito, K., Mukawa, N.: Brain activity associated with graphic emoticons. The effect of abstract faces in communication over a computer network. Electrical Engineering in Japan 177(3), 36–45 (2011)

"It Don't Matter If You're Black or White"?

Effects of Robot Appearance and User Prejudice on Evaluations of a Newly Developed Robot Companion

Friederike Eyssel[1] and Steve Loughnan[2]

[1] Department of Psychology, Center of Excellence in Cognitive
Interaction Technology, University of Bielefeld
P.O. Box 100131, 33501 Bielefeld, Germany
[2] School of Psychological Sciences,
University of Melbourne, Australia
feyssel@cit-ec.uni-bielefeld.de

Abstract. Previous work shows that other people are evaluated more positively when they are perceived as part of the evaluator's ingroup. This phenomenon – ingroup bias – has been robustly documented in social psychological intergroup research. To test this effect to the domain of social robots, we conducted an experiment with 61 Caucasian (White) American participants who rated either a robot that resembled the participants' Caucasian ingroup or a social outgroup on *mind perception* (i.e., the attribution of *agency* and *experience*). First, we predicted that the ingroup robot would be evaluated more positively than the outgroup robot. Moreover, we assessed the Caucasian Americans' endorsement of anti-Black prejudice. Thus, second, we hypothesized interaction effects of *robot type* and *level of prejudice* on mind perception, meaning that effects of ingroup bias should be particularly pronounced under high (vs. low) levels of self-reported Anti- Black prejudice. Interestingly, we did not obtain main effects for *robot type* on mind perception. Contrary to our hypothesis, participants even seemed to attribute more *agency* and *experience* to the outgroup robot relative to the ingroup robot. As expected, no main effects for *levels of prejudice* on *mind perception* emerged. Importantly, however, we obtained the predicted interaction effects on the central dependent measures. Obviously, the interplay of design choices and user attitudes may bias anthropomorphic inferences about robot companions, posing a psychological barrier to human-robot companionship and pleasant human-robot interaction.

Keywords: Mind Perception, Social Categorization, Attitudes.

1   Introduction

Clearly, researchers in the domain of social robotics envision the emergence of robots as companions in everyday life. To some extent, this endeavour has already been realized if one takes into consideration the deployment of pet-like robots such as Paro (http://www.parorobots.com/index.asp) or AIBO (http://www.aiboworld.com/) as

G. Herrmann et al. (Eds.): ICSR 2013, LNAI 8239, pp. 422–431, 2013.
© Springer International Publishing Switzerland 2013

social companions in elderly care facilities. However, in the case of humanoid social companions, designers have to carefully consider *which* visual and functional features to implement in a newly developed system. Ideally, psychologists, in turn, have to identify whether, and in what way these design choices shape the perception and evaluation of the product by the potential end user. Obviously, design choices are crucial in determining the social perception of a newly developed robot before it launches onto the consumer market. To study this issue in more depth, we have thus examined the role of *visual cues* for a robot's alleged group membership, user attitudes, and the interaction of both aspects.

The present research builds heavily on the sparse existing literature on the role of social category cues in predicting anthropomorphic inferences about robots. Specifically, [1] have developed the Three-Factor Model of Anthropomorphism. Accordingly, the extent to which people anthropomorphize objects and nonhuman agents can be largely attributed to three key psychological factors: *sociality motivation*, *effectance motivation*, and *elicited agent knowledge*. Our research focuses specifically on the cognitive determinant of anthropomorphism - *elicited agent knowledge*. That is, people rely on their self-knowledge or their knowledge about "humans" in general when judging unfamiliar nonhuman entities, such as robots [1, 2]. We have argued elsewhere [3], however, that *elicited agent knowledge* would not be limited to knowledge structures related to the superordinate human category, but would also include subordinate, or other categories (i.e., nationality or specific prototypes of social robots). This reasoning is based on the notion that we easily and readily rely on age, gender, and ethnicity as core social categories when forming impressions of people and objects [4]. Only recently, we have extended the literature on social categorization and intergroup bias to social robots [3, 5, 6].

2   Related Work on Social Categorization and Its Consequences for the Perception of Robots

Robots may serve as companions at home, at the workplace or in any conceivable setting in the future. The question that naturally emerges from a design and marketing perspective, is related to the issue of how to configure such a robotic companion so that it is deemed an acceptable social interaction partner. Our previous research on the visual features of robot prototypes has already emphasized the effect of social category activation on social judgments regarding the respective technical system. For one, [5] have shown that people readily rely on *visual* gender cues in robots to base their judgments on. As predicted, participants even attributed gender stereotyped personality traits to a robot prototype that either appeared male or female in visual appearance. The transfer of gender stereotypes that are commonly used among humans to robot prototypes even extended to the perceived suitability of the gendered prototypes for the respective sex-typed application areas (e.g., managing the household vs. repairing technical gadgets). We also found that in same-sex dyads, male participants attributed more mind to a robot prototype that was using a male voice than to a prototype speaking with a female voice, whereas the opposite pattern was found in female user-female robot dyads [7]. The pattern of findings was also obtained regarding perceived feelings of psychological closeness, contact intentions, and the robot's sociability.

From human-human intergroup research, we know that a person's ethnic background determines attitudes and behavior toward the individual, often resulting in discrimination or so-called ingroup bias. Social psychologists have already researched this phenomenon for some decades showing that ingroup bias encompasses the preference for everything that is associated with the ingroup [8]. That is, people and even products are evaluated more positively on manifold dimensions if they are perceived as belonging to the ingroup [9, 10]. To date, however, there is sparse evidence for mental state attribution to robots and especially its sensitivity to intergroup boundaries.

In a recent experiment [3], we have therefore tested whether the alleged ingroup status of a robot would result in higher attributions of typically human traits to a robot, as has been shown in human-human intergroup contexts [11]. Thus, participants learned that they would have to rate an alleged German "ingroup" product versus a Turkish "outgroup" product that differed regarding name and location of production - all other aspects of the robot were held constant. Nevertheless, participants reproduced patterns of ingroup bias when making judgments of the robot: They attributed more humanity to the ingroup robot (e.g., warmth and mind) than to the outgroup robot. They even felt psychologically closer to the ingroup product and rated its design more favorably than the outgroup product, consistent with [9,10].

However, whereas [3] were able to demonstrate effects of ingroup bias on a wide range of measures, user attitudes regarding Turkish people as the relevant outgroup had not been taken into account in the sample of German participants. Thus, previous work lacked an important aspect because prejudice against Turkish people in Germany had not been measured. Interindividual differences in user attitudes, however, bear the potential of contributing to biased judgments and behavior [4]. From this follows that prejudiced attitudes, too, have an impact on social judgments, even if people do not openly admit to it. This might be due to the fact that in the course of societal and socio-political developments and changes, it has become a socio-cultural norm to endorse egalitarian, anti-racist, and anti-sexist values and beliefs. Or at least, it has become increasingly unacceptable to openly express negative affect or beliefs about out-groups. In the domain of racism research, it has been found that the expression of racism has taken more indirect and subtle forms in contemporary societies than in the past [12]. People refrain from displaying their racist attitudes and beliefs, their negative sentiments remain under the surface, yet they are still present.

In the present work, we investigated whether this reasoning would extend to the context of robots. Clearly, in the domain of intergroup research, prejudicial attitudes are oftentimes hidden in self-report questionnaire measures in order to avoid being perceived racist. This might also be an issue when it comes to evaluating a robot target that clearly represents a societal outgroup which is suffering from discrimination and stigmatization in contemporary society. We tested this by activating *elicited agent knowledge* in that we manipulated a robot's skin color. We reasoned that the activated *elicited agent knowledge* might not only affect judgments of the robot target, but simultaneously, it should trigger participants' motivation to appear unprejudiced and to respond in a socially desirable manner. To explore this assumption further, we asked Caucasian American participants to report their endorsement of prejudicial attitudes towards African Americans.. Moreover, in the current study, we extended the research by [3] by means of a) relying on a representative sample from the United States which did not include students only; b) by focusing on mind perception in particular; c) by examining the effects of more subtle *visual cues* for social category membership and their interplay with d) user attitudes, namely *prejudice* against African American people in the United States.. As with other interindividual difference measures (e.g., desirability for control, see [13]), intergroup attitudes and interacial prejudice might not only affect the perception of fellow humans, but even extend to nonhuman targets, such as robot companions. To address these open research issues, we conducted an online-experiment.

3 Method

3.1 Participants, Design, and Procedure

Initially, 90 participants took part in this experiment. However, because of the importance of this intergroup setting, we excluded participants who were either non-Caucasian Americans or who did not recognize the robots' ethnic origin correctly (e.g., perceiving it as Asian). Participants in the final sample were thus 61 Caucasian Americans (42 males, 19 females) with a mean age of 34.69 years ($SD = 11.89$; age range: 19 to 68 years).

They were recruited via Mechanical Turk and completed an online questionnaire that allegedly would help a robotics company to optimize the new companion robot "Sean" before launching it onto the market.

Participants were randomly assigned to one of two experimental conditions of a between-subjects design. They were asked to evaluate one of two *robot types* that either belonged to their Caucasian American ingroup or the African American outgroup.

Manipulation of Group Membership. Our goal was to examine the role of subtle visual cues in the evaluation of two potential future robot companions by Caucasian American participants. To induce *elicited agent knowledge* and subsequent differential social judgments, we used pictures of two *robot types* (ingroup vs. outgroup; [14]) as shown in Figure 1. The robot prototype was named SEAN in both conditions.

Fig. 1. Robot Types: Outgroup Robot (left), Ingroup Robot (right)

3.2   Measures

To assess participants' endorsement of the dependent variables, we used 7-point Likert scales ranging from 1 (*not at all*) to 7 (*very*) for *mind* perception. To assess prejudice, we used Likert scales from 1 (*strongly disagree*) to 5 (*strongly agree*). In both cases, high values indicate high agreement with the assessed dimensions. To compute indices for the central dependent measures, average scores were computed after conducting reliability analyses (Cronbach's α). Furthermore, we formed two groups of participants low ($n = 33$) and high in self-reported *prejudice* against African Americans ($n = 28$) based on a median split ($Md = 2.17$). Overall, participants reported a *prejudice* level of $M = 2.45$ ($SD = 0.73$).

Mind Perception. Mind perception can be interpreted as one form of humanity attribution [3, 7] to robots, indicating their anthropomorphisation. Gray, Gray and Wegner [15] proposed two dimensions of mind perception, namely, *agency* and *experience*. To assess *mind perception* of the robot prototypes, participants completed 12 items by responding to the following question: "To what extent do you think that SEAN is capable of: experiencing physical or emotional pain? / feeling afraid or fearful? / experiencing violent or uncontrolled anger? / having personality traits that make it unique from others?/ exercising self-restraint over desires, emotions, or impulses? / having experiences and being aware of things? / conveying thoughts or feelings to others? / longing or hoping for things? / making plans and working toward goal? / telling right from wrong and trying to do the right thing / understanding how others are feeling? / remembering things?".

These items were adapted from [15, see also 3, 7, 16] and they tap the *experience* dimension (e.g., feeling afraid or fearful; $\alpha = .66$) as well as the *agency* dimension (e.g., making plans and working toward goal; $\alpha = .73$). Internal consistency for the 12-item index of *mind perception* was $\alpha = .82$.

Self-reported Prejudice. *Modern racism* represents a relatively subtle form of prejudice that rules out problems of socially desirable response tendencies often associated with classic self-reports of prejudice [17, 18]. Persons high on this type of prejudice would deny that discrimination against African Americans in the United

States would still be a problem and accordingly, they would show resentments about actions or laws that could possibly favor the actually disliked out-group. In the present research, we measured participants' *level of prejudice* by using six items (α = .61). For example, "Over the past few years, the government and news media have shown more respect for Blacks than they deserve"; "Discrimination against Blacks is no longer a problem in the United States." or "Blacks are getting too demanding in their push for equal rights".

4   Results

To test the experimental hypotheses, we conducted multivariate analyses of variance (ANOVA) with *robot type* (ingroup vs. outgroup) and *level of prejudice* (low vs. high) as between-subjects factors on the key dependent measures *agency* and *experience*. Contrary to our predictions, we did not obtain significant main effects for *robot type*, neither on *agency* nor on *experience*, all $Fs < 1$, $ps > .76$. In line with our predictions, the ANOVA did not yield main effects for *level of prejudice* on the central dependent measures, all $Fs < 1.32$, $ps > .26$.

However, crucially, we obtained significant interaction effects of *robot type* and *level of prejudice*. The hypothesized interaction effects emerged for *agency*, $F(1, 59)$ = 6.13, $p = .02$, $\eta^2 = .10$, and for *experience*, $F(1, 57) = 6.04$, $p = .02$, $\eta^2 = .10$.

To illustrate, Figure 2 shows the pattern of means for *agency (left) and experience (right) attribution* as a function of *robot type* and *level of prejudice* against African Americans.

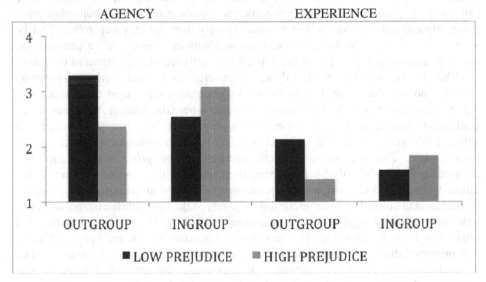

Fig. 2. Mean ratings of *agency* (left) and *experience* (right) attribution as a function of *robot type* (ingroup vs. outgroup) and *prejudice level* (low vs. high)

To investigate the interactions further, independent sample t-tests were conducted. Results revealed that participants low vs. high in prejudice did not differ in *agency* attribution to the ingroup robot ($M_{low\ prejudice}$ = 2.54; $SD_{low\ prejudice}$ = 0.91; $M_{high\ prejudice}$ = 3.08; $SD_{high\ prejudice}$ = 0.91; $t(25)$ = -1.53, p =.14). However, one-tailed t-tests showed that participants high in prejudice attributed significantly less *agency* to an outgroup robot ((M = 2.36; SD = 1.17) than participants low in prejudice (M = 3.29; SD = 1.39; $t(32)$ = 2.08, p =.05).

Secondly, we explored the pattern of means for *experience* attribution to the ingroup versus the outgroup robot as a function of *level of prejudice*. Once more, t-tests were conducted, and revealed the same pattern of means as were obtained for perceived robot *agency*: That is, participants low vs. high in prejudice did not differentially attribute *experience* to an ingroup robot ($M_{low\ prejudice}$ = 1.57; $SD_{low\ prejudice}$ = 0.54; $M_{high\ prejudice}$ = 1.83; $SD_{high\ prejudice}$ = 0.95; $t(25)$ = -0.92, p =.37). On the other hand, participants high in prejudice attributed significantly less *experience* to an outgroup robot ((M = 1.40; SD = 0.48) than participants low in prejudice (M = 2.13; SD = 1.03; $t(24.76)$ = 2.72, p =.01).

5   Discussion

Our research further tested the notion that processes of social categorization do not only apply to humans, but even generalize to robot companions.

Previous work by [3] has shown that when a robot was presented in terms of an ingroup member versus as an outgroup member by merely manipulating its name and country of production, preferential evaluations of the ingroup prototype were observed: As predicted, on all dimensions, participants rated the ingroup robot more favorably and anthropomorphized it more strongly than the outgroup robot. That is, given the very same background information about the robot (i. e., a picture), the mere manipulation of group membership affected subsequent evaluations of the robot.

Whereas the previous work [3] was conducted in German-Turkish intergroup context and was based on data of German participants, we sought to replicate and extend these findings in the US American context where the African American versus Caucasian American intergroup boundaries might yet be more pronounced. To achieve this goal, we used a different and less subtle manipulation of robot group membership. That is, in the current research, robot group membership was manipulated *visually,* only by presenting participants with a picture of a robot that either looked like a Caucasian ingroup member or showed afrocentric features, given its darker skin color. We were interested in effects of this less subtle manipulation in a new, but societally highly relevant intergroup context. To do so, we examined mind attribution to both robot types in a sample of Caucasian Americans. Specifically, we had predicted that Caucasian participants would attribute more mental capacities (i.e., as measured by perceived *agency* and *experience* and the total score) to an alleged ingroup robot compared to a robot that resembled African American appearance. Interestingly, however, and contrary to our predictions, we did not obtain main effects for *robot type* on any of the three measured dimension that tap inner life, vitality and cognitive capacities in a robot. From this follows that - whereas participants did not openly and explicitly differentiate between the ingroup and the outgroup robot in their ratings of the targets' *agency, experience,* and overall *mental capacities* - they might have done so because of social desirability concerns and their desire to appear egalitarian and unprejudiced [16-17].

Consequently, to shed more light on this assumption, we assessed individual differences in levels of endorsed anti-African American prejudice in our sample of Caucasian participants. Indeed, our data indicate that prejudice was moderately high in this group of participants. More interestingly, however, we were able to clearly show that participants' level of anti-African American prejudice affected the degree to which they attributed *agency, experience,* and *mind perception* as a whole to the *robot types.*

As predicted, participants' prejudiced attitudes did not affect their ratings on the three dimensions when an ingroup robot was in the focus of evaluation. However, when participants rated an outgroup robot companion, they attributed significantly less *agency* and *experience* to the robot. As such, this finding is novel and disturbing, as it illustrates unambiguously that design choices regarding the appearance of a newly developed robot prototype might affect subjective evaluations of the product (see also [3,10,11, 20-23] and behavior towards it, particularly so, when unknown variables such as participants' interindividual differences in endorsed prejudice also come into play. Future research should therefore not only investigate demographic variables and constructs that are seemingly of key relevance for robotics-related research questions (e.g., interindividual difference measures associated with previous experience with human-robot interaction or technology, or anthropomorphism [1,2]). Rather, future research in social robotics should also take into account interindividual difference variables that appear less focal at the first glance. Importantly, the present results clearly document that taking into account user prejudice against Black Americans made a difference in the perception of two robot prototypes: When examining the effects of *robot type* in an isolated way, no effects whatsoever emerged. This is contrary to our previous results as shown in [3] and [6], where we did not assess further interindividual difference measures and where group membership was unrelated to actual social categories. Our present results demonstrate clearly, however, that only in interaction with prejudiced attitudes towards the outgroup did the predicted patterns of ingroup bias emerge. This might be due to the political climate in contemporary societies and the aspect of social desirable responding which minimized participants' likelihood to openly admit a worse evaluation of a product that is obviously part of a socially relevant outgroup.

Certainly, more empirical evidence is needed to substantiate these findings, ideally using a yet wider range of dependent measures, controlling for social desirable response tendencies, and by way of generalizing the results to other intergroup contexts. For example, it might be worthwhile to conduct a study using the same stimulus materials but examining evaluations of the robot targets in a sample of African American participants. Furthermore, it would be interesting to make use of a broader range of indirect, reaction-time based measures of anthropomorphism and prejudice alike, see [6]. As demonstrated in previous work in robotics, processes that have been thoroughly studied in the realm of social psychological intergroup research are likewise observed in the context of robot evaluations [e. g., 3, 5]. Clearly, we use rules of thumb when differentiating among "us" and "them" – not only in the domain of human-human impression formation [4, 12, 17-19], but also when judging robots or other technical gadgets [5,6, 14, 21-23]. At the same time, we seem – just like in the intergroup context – to be inclined to refrain from "discriminating" against targets who appear different – even when these are robots. However, when controlling for prejudiced attitudes, those endorsing high levels of prejudice show the expected "discrimination" of outgroup robots or products, consequently evaluating the ingroup product more favorably [9,10, 20].

In summary, the present results emphasize the importance of taking into account elicited agent knowledge which obviously can be activated by minimal changes in visual appearance and design of a robot. Our data show, however, that design choices affect the users' mental models of the robot, particularly when users' bring their predispositions to the lab or into any other contexts where companion robots will be deployed in the future. Our findings speak to the fact that research in human-robot interaction and in social robotics can and should more stringently take into account product and user features, such as attitudes. As our findings nicely illustrate, considering the attitudes a user brings into a given situation can substantially clarify our understanding of experimental results and the underlying psychological processes.

References

1. Epley, N., Waytz, A., Cacioppo, J.T.: On Seeing Human: A three-factor theory of anthropomorphism. Psychological Review 114, 864–886 (2007)
2. Epley, N., Waytz, A., Akalis, S., Cacioppo, J.T.: When we need a human: Motivational determinants of anthropomorphism. Social Cognition 26, 143–155 (2008)
3. Eyssel, F., Kuchenbrandt, D.: Social categorization of social robots: Anthropomorphism as a function of robot group membership. British Journal of Social Psychology 51, 724–731 (2012)
4. Fiske, S.T.: Stereotyping, prejudice, and discrimination. In: Gilbert, D.T., Fiske, S.T., Lindzey, G. (eds.) Handbook of Social Psychology, pp. 357–411. McGraw-Hill, New York (1998)
5. Eyssel, F., Hegel, F.: (S)he's got the look: Gender-stereotyping of social robots. Journal of Applied Social Psychology 42, 2213–2230 (2012)
6. Kuchenbrandt, D., Eyssel, F., Bobinger, S., Neufeld, M.: When a robot's group membership matters: Anthropomorphization of robots as a function of social categorization. International Journal of Social Robotics 5, 409–417 (2013)
7. Eyssel, F., Kuchenbrandt, D., Hegel, F., de Ruiter, L.: Activating elicited agent knowledge: How robot and user features shape the perception of social robots. In: Proceedings of the 21st IEEE International Symposium in Robot and Human Interactive Communication, pp. 851–857 (2012)
8. Gaertner, S.L., Dovidio, J.F.: Reducing intergroup bias: the common ingroup identity model. Psychology Press, New York (2000)
9. Escalas, E.J., Bettman, J.R.: Self-construal, reference groups, and brand meaning. Journal of Consumer Research 32, 378–388 (2005)

10. White, K., Dahl, D.W.: Are all out-groups created equal? Consumer identity and dissociative influence. Journal of Consumer Research 34, 525–535 (2007)
11. Haslam, N.: Dehumanization: An integrative review. Personality and Social Psychology Review 10, 252–264 (2006)
12. Swim, J.K., Aikin, K.J., Hall, W.S., Hunter, B.A.: Sexism and racism: Old-fashioned and modern prejudices. Journal of Personality and Social Psychology 68, 199–214 (1995)
13. Eyssel, F., Kuchenbrandt, D.: Manipulating anthropomorphic inferences about NAO: The role of situational and dispositional aspects of effectance motivation. In: Proceedings of the 20th IEEE International Symposium in Robot and Human Interactive Communication, pp. 467–472 (2011)
14. Hegel, F., Eyssel, F., Wrede, B.: The social robot Flobi: Key Concepts of industrial design. In: Proceedings of the 19th IEEE International Symposium in Robot and Human Interactive Communication, pp. 120–125 (2010)
15. Gray, H.M., Gray, K., Wegner, D.M.: Dimensions of mind perception. Science 315, 619 (2007)
16. Eyssel, F., Reich, N.: Loneliness makes my heart grow fonder (of robots)? On the effects of loneliness on psychological anthropomorphism. In: Proceedings of the 8th ACM/IEEE Conference on Human-Robot Interaction (HRI 2013), pp. 121–122 (2013)
17. McConahay, J.B.: Modern racism, ambivalence, and the Modern Racism Scale. In: Dovidio, J.F., Gaertner, S.L. (eds.) Prejudice, Discrimination, and Racism, pp. 91–125. Academic Press, Orlando (1986)
18. Sears, D.O.: Symbolic racism. In: Katz, P., Taylor, D. (eds.) Eliminating Racism: Profiles in Controversy, pp. 53–84. Plenum Press, New York (1988)
19. Devine, P.G.: Stereotypes and prejudice: Their automatic and controlled components. Journal of Personality and Social Psychology 56, 5–18 (1989)
20. Ferguson, C.K., Kelley, H.H.: Significant factors in overevaluation of own-group's product. The Journal of Abnormal and Social Psychology 69, 223–228 (1964)
21. Powers, A., Kiesler, S.: The advisor robot: Tracing people's mental model from a robot's physical attributes. In: Proceedings of the Conference on Human-Robot Interaction, pp. 218–225 (2006)
22. Powers, A., Kramer, A.D.I., Lim, S., Kuo, J., Lee, S.-L., Kiesler, S.: Eliciting information from people with a gendered humanoid robot. In: Proceedings of the 14th IEEE International Workshop on Robot and Human Interactive Communication, pp. 158–163 (2005)
23. Reeves, B., Nass, C.: The media equation: How people treat computers, television, and new media like real people and places. Cambridge University Press, New York (1996)

A Humanoid Robot Companion
for Wheelchair Users

Miguel Sarabia and Yiannis Demiris

Personal Robotics Lab
Department of Electrical and Electronic Engineering
Imperial College London, United Kingdom
{miguel.sarabia,y.demiris}@imperial.ac.uk

Abstract. In this paper we integrate a humanoid robot with a powered wheelchair with the aim of lowering the cognitive requirements needed for powered mobility. We propose two roles for this companion: pointing out obstacles and giving directions. We show that children enjoyed driving with the humanoid companion by their side during a field-trial in an uncontrolled environment. Moreover, we present the results of a driving experiment for adults where the companion acted as a driving aid and conclude that participants preferred the humanoid companion to a simulated companion. Our results suggest that people will welcome a humanoid companion for their wheelchairs.

1 Introduction

Driving a wheelchair is a cognitively challenging task [7]. Users need to model the behaviour of the wheelchair, predict how it is going to behave and be spatially aware of their surroundings. Could there be a way of reducing the cognitive requirements driving a wheelchair demands? We hypothesise that a humanoid robot may be able to do so. This way, we aim to lower a wheelchair's entry barrier for people with cognitive disabilities and children whose cognitive faculties may not be fully developed.

Fig. 1. Child driving our paediatric wheelchair with Nao as companion during the Imperial Festival[2]

To that end we have added Nao, a small humanoid, to a paediatric wheelchair so that it can act as a companion for mobility-impaired persons[1]. In particular, we present two possible roles for Nao. In the first one, Nao points out the location of obstacles, explaining why the smart wheelchair may not be moving in the expected direction. For the second one, we set it to act as a driving aid giving directional instructions and compare it to more traditional driving aids (such as voice and on-screen arrows).

---

[1] This work was partially funded by the EU FP7 ALIZ-E project (248116).
[2] Videos of the system are available at: http://imperial.ac.uk/PersonalRobotics.

G. Herrmann et al. (Eds.): ICSR 2013, LNAI 8239, pp. 432–441, 2013.

We are also interested in understanding whether there are any benefits to having a physical robot as a companion rather than a simulated one. Our results will show that, at least for adults, participants much preferred the physical robot over the simulated one, even if the performance differences were inconclusive.

2   Related Work

In 2008, it was estimated that 61% to 91% of wheelchair users may benefit from a smart powered wheelchair at some point during their lives [12]. There is a wealth of research in the field of smart powered wheelchair reviewed in [11].

The need for powered mobility may be even more pronounced for children. Indeed, the Rehabilitation Engineering & Assistive Technology Society of North America supports the use of paediatric powered mobility as soon as the child possesses the necessary cognitive, sensorimotor and coping abilities [10]. According to the same study, the use of smart powered wheelchairs "enhances independence, improves psychosocial development and enables children to become productive and independent members of society".

It has been shown that a robotic wheelchair with haptic guidance can be used to teach children to drive [5]. This system, based on a game of "robot-tag", was used by 22 able-bodied children and a 8 year old with cerebral palsy. ARTY (Assistive Robotic Transport for Youngsters) is a paediatric wheelchair with a focus on safety. It was tested by 8 able-bodied children and a 5 year old with physical and cognitive disabilities [13]. Note both wheelchairs were transferred from able-bodied to cognitively impaired children. Further evidence of the potential of smart wheelchairs comes from [8] where 4 children with cerebral palsy aged 11 to 16 successfully navigated around their school using a touch-screen.

To the best of our knowledge there is little published research in companions for robotic wheelchair users. Thus, we review two European Framework Programme projects on robotic companions. IROMEC (Interactive Robotic Social Mediators as Companions) showcases the benefits of a robotic companion able to play with developmentally-impaired children [6]. CompanionAble, on the other hand, seeks to assist people in their homes and it is targeted for the elderly, particularly those who suffer from mild cognitive impairment [4].

Additionally, in 2007 the car manufacturer Nissan, added a PaPeRo robot to their Pivo 2 concept car to provide driving assistance. Whilst this was well reported on the press, we were unable to find any relevant scholarly material.

Finally, there has been research into the differences between a simulated robot and a physical one. In particular, it has been shown that people rate interactions with an actual robot more positively than with one seen through a live video display [1].

3   System Description

Our system has two distinct subsystems: ARTY and Aldebaran's Nao (Fig. 2). ARTY is composed of an Ottobock children's powered wheelchair, three laser

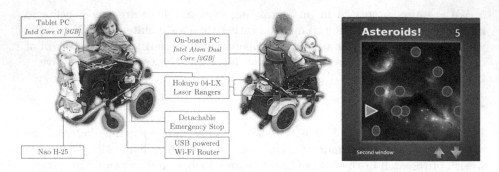

Fig. 2. Hardware components of our system   **Fig. 3.** Asteroids game

rangers, an on-board PC, a tablet PC and a wireless router. ARTY is an improved version of the one presented in [13]. Nao is a 60cm tall humanoid robot with 25 degrees of freedom which hangs from the front of the wheelchair.

Figure 4 summarises the main software components of our system, all of which are written atop ROS (the Robot Operating System) [9]. We reuse many software components readily available online and in what follows we describe the most relevant ones. Laser Combiner[3] takes the readings of the three laser rangers and combines them into a single coherent message [13]. Laser Scan Matcher[4] takes in this combined message and interpolates the wheelchair odometry using an iterative closest point algorithm [3]. AMCL[5] receives both the interpolated

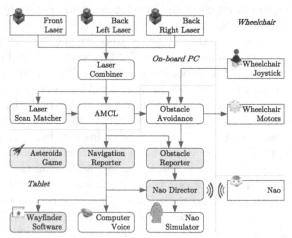

Fig. 4. Software components of our system. Square edges represent hardware, round edges represent software nodes. Shaded nodes are contributions of this article. See text for authorship of other nodes.

odometry and the combined laser message to localise the wheelchair on a pre-existing map. The Obstacle Avoidance[2] module moderates the input joystick signal according to the wheelchair proximity to obstacles using a dynamic window approach [13], thus avoiding potential collisions.

---

[3] Nodes written by Harold Soh, available from
http://imperial.ac.uk/PersonalRobotics.

[4] Node written by Ivan Dryanovski and William Morris, available from
http://www.ros.org/wiki/laser_scan_matcher.

[5] Node written by Brian Gerkey and Andrew Howards, available from
http://www.ros.org/wiki/amcl.

Let us now introduce the nodes written specifically for our system. Navigation Reporter raises an alert whenever the user has deviated from a pre-recorded path on the map and indicates the direction the user should follow at the junctions of said path. Every time a new pose is generated by the AMCL node, Navigation Reporter tries to find its closest match in the list of poses that compose the pre-defined path. To achieve this, the program evaluates all path poses within ± 1.5 metres of the last known path pose and chooses the one with highest score. The scoring function is defined as: score $= \exp\{\Delta/\alpha \times \ln 2\} + \exp\{\Gamma/\beta \times \ln 2\}$ where Δ represents the euclidean distance between the current wheelchair pose and the candidate path pose, Γ is the normalised angle difference between the current wheelchair pose and the candidate path pose; α and β are adjusting variables to control what distance and angles difference yield half score (we set $\alpha = 0.4$m and $\beta = \pi/4$ rad respectively, these values were determined empirically). If the score of all path poses considered was lower than a threshold (set to 1.3) no pose was chosen. This formulation is robust to errors in AMCL and can deal with paths that go over the same point repeatedly (thanks to the ± 1.5m sliding window).

Obstacle Reporter simply takes the information from the Obstacle Avoidance node and raises an alert if the user is driving towards an obstacle. To prevent Nao from becoming too repetitive, both Obstacle Reporter and Navigation Reporter suppress similar alerts that occur in a short period of time.

The asteroids game (Fig. 3) was developed as a secondary task for the path driving experiments (introduced in Sect. 5). The objective is to move the space-ship (triangle in the figure) away from the incoming asteroids (circles in the figure) using the up and down arrows.

3.1 Driving Aids: Pointing Out Obstacles and Giving Directions

Here we introduce the nodes in charge of communicating wheelchair alerts to the user. We start with Nao Director, which coordinates all of Nao's movements and speech. It instructs Nao to execute wake-up and power-off animations and provides a background behaviour for Nao — randomly looking around and blinking. If the node receives any alert from either Navigation Reporter or Obstacle Reporter it will stop the background behaviour and command Nao to indicate the direction the user has to follow (in case of a navigation alert) or the location of the obstacle (in case of an obstacle alert).

Nao Director can command both the actual Nao or an on-screen simulated Nao. The underlying code is the same. However the simulated Nao cannot emit sounds and instead writes them on the display. To minimise the differences between the simulated Nao and the physical one, Nao Director did not provide new information through voice, instead giving vague messages (eg. "we have to go that way"). One has to look at the arms of the robot or the simulator in order to understand its instructions.

We also implemented two other, more traditional, driving aids. The first one is the Wayfinder Software, a simple computer window to show navigation alerts. Upon receiving an alert, an arrow indicating the direction the user should follow is shown on-screen for 3 seconds. The second driving aid was the Computer

Voice, a program which would speak out loud the instructions received from Navigation Reporter using ROS text-to-speech facilities. Example utterances are: "go straight" and "drive left". The sound of the ROS computer voice was easily distinguishable from that of Nao.

4   Preliminary Field Trials

During our university's open-day in May 2013, we conducted trials of our wheelchair plus companion system with several able-bodied children. Since we had very little control over the environment (it was estimated that 10,000 people attended the event) the main focus of the exercise was to check whether children appreciated driving with Nao as a companion. Note Navigation Reporter was disabled and Nao only reported the location of obstacles.

We asked 14 participants, who had driven the wheelchair for about 5 minutes each, to fill in a questionnaire with three questions in five-point Likert-scale[6]. The responses came from 9 boys and 5 girls aged 4 to 12. Some of the younger participants were helped by their parents when filling in the questionnaire. Children were generally very positive about the field-trials: all 14 children strongly agreed they had enjoyed the wheelchair, 12 strongly agreed Nao had helped them not to crash and 11 participants strongly agreed they liked having Nao by their side when driving. The participants also had the opportunity to write comments about the experiment. All comments were positive; for example: *"I want him as a pet or brother"*, *"Nao is really helpful"*, *"I think HE IS AWESOME!"*.

Although there was an inherent pressure on the children to evaluate the wheelchair positively (it is hard to criticise the toy one has just played with), we consider the enthusiastic comments to be a proof that, at least in the short term, children really enjoy having Nao as a companion.

5   Path Driving Experiments

In this section we explore the effects of having a humanoid companion as a driving aid. We also investigate whether there was any advantage to having a physical robot rather than a simulator.

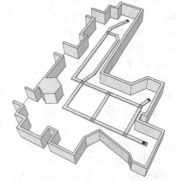

5.1   Experimental Set-up

Participants were tasked with navigating through a pre-defined path. The basic make-up of each path was the same, all shared the starting point and driving segments, the difference came from

Fig. 5. Path-driving environment with 6 junctions and 3 finishing points (in black)

the turns to be performed at each junction and the finishing point. There were

---

[6] To make the questionnaire friendlier we used smileys instead of the typical 'strongly disagree' to 'strongly agree' scale.

a total of 6 junctions and 3 finishing points (Fig. 5). All paths were designed so it would take around three minutes to complete them.

The participants did not know the exact directions to follow on the course, instead relying on different driving aids to guide them through the pre-defined paths. Specifically, driving aids told users to turn left, right or keep going straight at each junction. If participants did not follow the instructions the system would remember the last point where they had been on-course and instructed them to retrace their steps until they reached a known point (*lost* instruction).

To simulate real wheelchair driving conditions where a user may be distracted by some other task (eg. by having a conversation) and following the example set in [2], we devised a secondary task: the asteroids game introduced in Sect. 3.

Participants were asked to complete four paths. Each time, they were aided by one of the driving aids already described in Sect. 3.1[7]. At the end of each trial, they were asked to complete a questionnaire regarding the driving aid they had just used. Additionally, at the end of the experiment they were requested to chose their preferred driving aid.

Throughout these experiments Obstacle Reporter was disabled and Nao only acted as a driving aid. This was done as a control measure since neither of the traditional driving aids could communicate obstacle alerts. All data interchanged by the different ROS nodes of the system, including scores on the asteroids game, was recorded for subsequent analysis of user performance.

5.2 Performance Metrics and Questionnaires

The following metrics are used to compare the different driving aids:

Average driving speed computed as the sum of the euclidean distances between the poses reported by ACML over the time the trial lasted. A higher value indicates better overall performance in the task.

Impacts defined as number of impacts in the secondary task. A high score indicates participants attention was occupied with driving. Accordingly, a lower value implicates better overall performance.

Time lost computed as sum of the durations between a user receiving a *lost* instruction and the user receiving any other instruction, which only happens when the participant has driven back to the junction where she became lost. A lower value is suggestive of better overall performance.

We also collected subjective metrics through the use of questionnaires. The questions are listed below (note system was replaced by the actual driving aid):

- "I found driving the wheelchair whilst playing the game *difficult*".
- "I felt *safe* in the wheelchair".
- "I *understood* the instructions the system was trying to convey".
- "The system gave me *accurate* instructions".

---

[7] The order each participant used every driving aid was determined by choosing a random entry in a table with all permutations of driving aids.

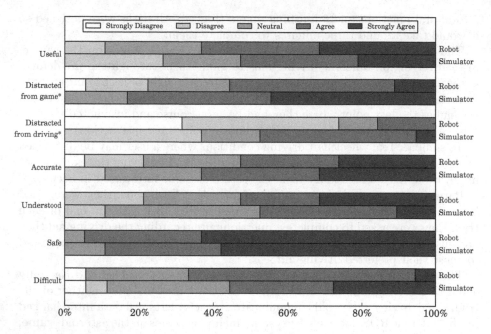

Fig. 6. Questionnaire responses for Simulator and Nao considering all attempts. Categories appended with * are statistically significant. See Sect. 5.2 for actual questions.

- "The system *distracted* me *from driving*".
- "The system *distracted* me *from* playing the *game*".
- "I found the system to be a *useful* driving aid".

5.3   Results

20 people (11 female and 9 male) aged between 21 and 38 completed the experiment. 40% of the participants declared to have worked with robots before. Due to a misaligned laser scanner we had to discard one of the trials as the AMCL module failed to localise properly.

Even though participants had a few minutes practice with the wheelchair before the actual experiment started, we found most of the learning occurred during the first attempt. This is evident from Fig. 7 where the deviation from the average performance is greatest in the first attempt for all three metrics. Therefore, we report on data distilled from *all* attempts as well as from the *last three* attempts.

Figure 6 shows the questionnaire results considering all attempts ($N = 19$). Safety and accuracy have equivalent scores for both the simulator and the robot. When asked about difficulty, understanding or usefulness, ratings are generally higher for the Nao robot but the differences do not reach statistical significance[8].

---

[8] We used the Wilcoxon signed-rank matched pairs test which is a non-parametric, within-subjects and two-tailed test.

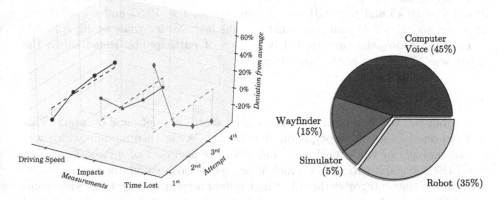

Fig. 7. Deviation from average across all trials in driving performance. Results from different driving aids have been aggregated.

Fig. 8. Preferred driving aids

In contrast, the robot has significantly lower ratings for distracting from driving ($p = 0.007$, $W = 11$) and distracting from game ($p = 0.032$, $W = 14.5$).

It is noteworthy that when considering only the last three attempts ($N = 9$)[9] both distracted from driving and distracted from game were no longer statistically significant, though they still favoured the robot. Difficulty and understanding continued to have higher, non-significant, ratings for the robot, whereas safety and accuracy had very similar ratings. Importantly, 89% agreed or strongly agreed that the robot was useful compared to only 33% for the simulator; this was found to be statistically significant ($p = 0.031$, $W = 21$).

With respect to the data recorded, we will indicate the median and the interquartile ranges (in square brackets) for driving speed (v), impacts (i) and time lost (τ). Note do not assume normality. Our data shows that when considering all trials ($N = 19$) the median driving speed was $\tilde{v}_{\text{robot}} = 0.19$ m/s [0.10 m/s] and $\tilde{v}_{\text{sim}} = 0.21$ m/s [0.11 m/s] for the Nao robot and the Nao simulator. Similarly, the median number of impacts was $\tilde{i}_{\text{robot}} = 69$ [45.75] and $\tilde{i}_{\text{sim}} = 64$ [60.5]; the time lost was $\tilde{\tau}_{\text{robot}} = 21.91$ s [82.69 s], $\tilde{\tau}_{\text{sim}} = 8.27$ s [72.53 s]. If only the last three trials are considered ($N = 9$) we find that $\tilde{v}_{\text{robot}} = 0.24$ m/s [0.08 m/s], $\tilde{v}_{\text{sim}} = 0.22$ m/s [0.07 m/s]; $\tilde{i}_{\text{robot}} = 52$ [55], $\tilde{i}_{\text{sim}} = 56$ [43]; $\tilde{\tau}_{\text{robot}} = 0$ s [15.86 s], $\tilde{\tau}_{\text{sim}} = 30.73$ s [57.49 s]. None of these differences were found to be statistically significant.

We were also interested in whether either the Nao robot or the Nao simulator would yield better performance than the control driving aids. Our data shows that the wayfinder software had the best median performance metrics ($\tilde{v}_{\text{wayfinder}} = 0.24$ m/s [0.06 m/s], $\tilde{i}_{\text{wayfinder}} = 49$ [43.25], $\tilde{\tau}_{\text{wayfinder}} = 5.57$ s [30.64 s]), followed by the computer voice ($\tilde{v}_{\text{voice}} = 0.22$ m/s [0.08 m/s], $\tilde{i}_{\text{voice}} = 52$ [43.25], $\tilde{\tau}_{\text{voice}} = 7.23$ s [23.93 s]), followed by the Nao simulator and Nao robot (see above for medians and interquartile ranges). Performing Friedman tests reveals a sta-

---

[9] This corresponds to the number of people who did not have either the Nao robot or the Nao simulator on the first attempt.

tistically significant effect of the driving aid on all three performance metrics. We found $T = 10.45$ and $p = 0.015$ for driving speed; $T = 12.97$ and $p = 0.005$ for impacts; and $T = 9.41$ and $p = 0.024$ for time lost. Notice that, again, $N = 19$.

Figure 8 shows the preferred driving aids of participants. Remarkably, the robot was 30% ahead of the Nao simulator.

5.4 Discussion

The robot and the simulator had very similar scores in safety and accuracy. This was expected as safety depends on the Obstacle Avoidance module which was active throughout the experiment. Likewise, the accuracy of instructions relies upon the Navigation Reporter which is shared by both simulator and robot.

In most other categories the robot had a slight advantage in ratings when compared to the simulator, though most of these differences did not reach statistical significance. The exceptions were distraction and usefulness.

Interestingly, the robot scored significantly lower in the distracted from game category than the simulator. This is surprising since the simulator shared the screen with the game and participants did not have to look away. Moreover, even if the accuracy and the actual task performance was similar across both driving aids, users much preferred the robot over the simulator (30% difference when it came to the favourite driving aid). Might the lack of embodiment cause the simulator to be more distracting and less useful? Further research is needed to clarify this question.

Although the simulator presented higher driving speed, lower number of impacts and less time lost, none of the results were statistically significant. Moreover when considering the last three attempts the robot had a slightly higher driving speed and a considerably lower time lost, suggesting more time may be required to habituate to Nao.

Taking into account the traditional driving aids it is clear that the robot and the simulator were not as effective as the control driving aids. More work is needed to ensure these robotic driving aids catch up with or even improve upon their traditional counterparts.

To be specific, we identified two issues which might explain why the performance of Nao was lower than that of the control driving aids. Firstly, Nao sometimes took too long to give instructions due to unpredictable network latency, which confused participants. Secondly, many participants were disappointed and mentioned in the comments that Nao never gave instructions by voice. By exploiting multi-modality Nao may become a more effective driving aid.

6   Conclusion and Further Work

We have investigated the potential of a humanoid companion for wheelchairs. Children were highly positive about having a robot accompanying them during an uncontrolled field trial. A physical robot was also found by users to be more useful and less distracting than a simulated one. This was the case even when there was no statistically significant indication of performance improvements.

In the future, we look forward to formulating new companion roles (besides the roles of obstacle reporter and driving aid we presented here) where embodiment is not only advantageous but a requirement. Having shown that able-bodied users appreciate a humanoid companion, we will be testing to verify how well our findings carry over to disabled children and people with cognitive disabilities. This is an area where robotic companions have a particular potential to greatly improve the quality of life of wheelchair users as well as lowering the entry barrier to powered mobility, allowing people that currently cannot drive a wheelchair to do so.

References

1. Bainbridge, W.A., Hart, J.W., Kim, E.S., Scassellati, B.: The Benefits of Interactions with Physically Present Robots over Video-Displayed Agents. Intl. J. Social Robot. 3(1), 41–52 (2010)
2. Carlson, T., Demiris, Y.: Collaborative control for a robotic wheelchair: evaluation of performance, attention, and workload. IEEE Trans. Syst., Man, Cybern., Part B 42(3), 876–888 (2012)
3. Censi, A.: An ICP variant using a point-to-line metric. In: Proc. of IEEE ICRA, pp. 19–25 (2008)
4. Gross, H.M., Schroeter, C., Mueller, S., Volkhardt, M., Einhorn, E., Bley, A., Langner, T., Martin, C., Merten, M.: I'll keep an eye on you: Home robot companion for elderly people with cognitive impairment. In: Proc. of IEEE SMC, pp. 2481–2488 (2011)
5. Marchal-Crespo, L., Furumasu, J., Reinkensmeyer, D.J.: A robotic wheelchair trainer: design overview and a feasibility study. J Neuroeng. Rehabil. 7(40) (2010)
6. Marti, P., Giusti, L.: A Robot Companion for Inclusive Games: a user-centred design perspective. In: Proc. of IEEE ICRA, pp. 4348–4353 (2010)
7. Massengale, S., Folden, D., McConnell, P., Stratton, L., Whitehead, V.: Effect of visual perception, visual function, cognition, and personality on power wheelchair use in adults. Assist. Technol. 17(2), 108–121 (2005)
8. Montesano, L., Díaz, M., Bhaskar, S., Minguez, J.: Towards an intelligent wheelchair system for users with cerebral palsy. IEEE Trans. on Neural Syst. Rehabil. Eng. 18(2), 193–202 (2010)
9. Quigley, M., Gerkey, B.P., Conley, K., Faust, J., Foote, T., Leibs, J., Berger, E., Wheeler, R., Ng, A.: ROS: an open-source Robot Operating System. In: Proc. Open Source Software Workshop at ICRA (2009)
10. Rosen, L., Arva, J., Furumasu, J., Harris, M., Lange, M.L., McCarthy, E., Kermoian, R., Pinkerton, H., Plummer, T., Roos, J., Sabet, A., Vander Schaaf, P., Wonsettler, T.: RESNA position on the application of power wheelchairs for pediatric users. Assist. Technol. 21(4), 218–225 (2009)
11. Simpson, R.C.: Smart wheelchairs: A literature review. J Rehabil. Res. Dev. 42(4), 423–438 (2005)
12. Simpson, R.C., LoPresti, E.F., Cooper, R.A.: How many people would benefit from a smart wheelchair? J. Rehabil. Res. Dev. 45(1), 53–72 (2008)
13. Soh, H., Demiris, Y.: Towards Early Mobility Independence: An Intelligent Paediatric Wheelchair with Case Studies. In: Proc. IROS Workshop on Progress, Challenges and Future Perspectives in Navigation and Manipulation Assistance for Robot Wheelchairs, Vilamoura, Portugal (2012)

Tuning Cost Functions for Social Navigation

David V. Lu[1], Daniel B. Allan[2], and William D. Smart[3]

[1] Washington University in St. Louis, St. Louis, MO 63130, USA
davidlu@wustl.edu
[2] Johns Hopkins University, Baltimore, MD 21218, USA
[3] Oregon State University, Corvallis, OR 97331, USA

Abstract. Human-Robot Interaction literature frequently uses Gaussian distributions within navigation costmaps to model proxemic constraints around humans. While it has proven to be effective in several cases, this approach is often hard to tune to get the desired behavior, often because of unforeseen interactions between different elements in the costmap. There is, as far as we are aware, no general strategy in the literature for how to predictably use this approach.

In this paper, we describe how the parameters for the soft constraints can affect the robot's planned paths, and what constraints on the parameters can be introduced in order to achieve certain behaviors. In particular, we show the complex interactions between the Gaussian's parameters and elements of the path planning algorithms, and how undesirable behavior can result from configurations exceeding certain ratios. There properties are explored using mathematical models of the paths and two sets of tests: the first using simulated costmaps, and the second using live data in conjunction with the ROS Navigation algorithms.

1 Introduction

Navigation is one of the fundamental tasks in mobile robotics. For robots with reasonable dynamics and operating speeds in indoor environments, efficient collision-free navigation is considered a solved problem from a practical standpoint. However, when humans are introduced into the environment, we must treat them differently than static obstacles by respecting their social norms, thus causing navigation to become more difficult.

Most path-planners in use today use a discretized costmap, where values in the costmap cells correspond to the "badness" of the robot being in that position. The path-planner then generates a path from the start position to the end that has the minimum accumulated cost. This will cause the robot, in many cases, to closely approach obstacles. While this is fine for furniture, it is not socially acceptable with people.

Using the full range of values in the costmap is vital for representing many social constraints for navigation algorithms. They represent general preferences or guidelines rather than hard and fast rules. As Kirby et al. [6] observed, "Human social conventions are tendencies, rather than strict rules," This approach discourages a subset of paths without disallowing them outright.

G. Herrmann et al. (Eds.): ICSR 2013, LNAI 8239, pp. 442–451, 2013.

In practice, creating the desired behavior around people is surprisingly difficult. While previous researchers have tuned their parameters to create working configurations, there exists no general guide for how to do this to effect a specific change in robot behavior. Furthermore, as we show below, the resulting behavior is not always intuitive from a consideration of the individual costmap elements.

2 Related Work

Methods for modeling social preferences in costmaps have existed for some time. Most uses of soft constraints in costmaps have been for representing a person's personal space. Dautenhahn et al. [4] constructed recommendations for planning motions where humans would be comfortable based on live HRI trials, taking proximity, visibility and hidden zones into consideration. These were formulated into a costmap system by Sisbot et al. [13], creating a Gaussian based "human aware motion planner." Kirby et al. [6] used an algorithm that in addition to minimizing path distance and avoiding obstacles, modeled proxemics and behaviors like passing on the right into the costmap, also using Gaussians. Work by Svenstrup et al. [14] created even more complicated models of personal space, integrating a mixture of four different constraints modeled as Gaussians. [15] expanded this work to maneuver among a field of multiple people while moving toward a goal. Soft constraints are occasionally used for other fields like autonomous vehicles. Ferguson and Likhachev [5] used large constant valued areas to favor driving on the right side of the road and to avoid curbs. Among the latest work in this field, Mainprice et al. [9] have expanded their original model to a three dimensional costmap in order to control positioning during hand off tasks, taking safety, visibility and the human's arm comfort into consideration. Scandolo and Fraichard [12] have also created a complex model that included proxemic, visibility and motion models as Gaussians, and "interaction areas" as constants. Lu and Smart [8] also have modified robot behavior using intermediate costmap values to improve the efficiency of human task completion.

Most of the obstacles added to the costmaps follow Gaussian distributions or constant values (with the minor exception of the representation of the intimate personal space in the work by Scandolo and Fraichard [12]). Little to no discussion is given about how the authors of the previous works found the parameters that worked best with their system.

The problem of designing costmaps that result in human-like behavior bears some similarity to inverse reinforcement learning (IRL) [11, 1, 2]. In IRL, the goal is to induce an immediate reward function based on example behaviors that will result in a similar final policy. In our setting, this would involve recording a number of human trajectories, determining a suitable set of features to serve as a state space (in a similar manner, perhaps, to [7]), then applying IRL algorithms to learn the functions that would create the local costmap based on these features. While an approach like this offers the promise of automatic costmap construction, it depends critically on the example trajectories and on a good feature set being selected.

3   Problem Statement

Each planned path depends on two separate components: the costmap and the planning algorithm. The costmap is represented by a two-dimensional grid, where each grid cell has a value $f(x,y)$. Values above some predefined threshold are designated as "lethal" values which will result in collisions. Standard algorithms such as Dijkstra's and A* are typically used as the path planner. However, if the total cost of a path is defined as the cost of the cells the path traverses alone, then the resulting path will be a very long path that avoids any cost. To avoid this scenario, wavefront planners are often used, which add a constant value, P, to each cell traversed to create a gradient from start to finish[3].

Formally, we define $C(p)$ as the total cost of the path p (including both the costmap costs and the path planning costs). Finding the best path involves minimizing the cost over all possible paths.

$$\min_{\forall \text{path} p} C(p) = \min_{\forall \text{path} p} \sum_{(x,y)\in p} \Big[f(x,y) + P \Big] \tag{1}$$

This definition for path cost assumes that each step in the path moves to another grid cell exactly 1 unit away, implying each cell is connected to its four immediate neighbors. We use this assumption throughout this paper, although it is possible to generalize it, for example, to also include diagonal moves.

For purposes of this paper, let us further refine the problem to reduce the number of cases we must consider. First, without loss of generality, let us consider paths that go from $(-n, 0)$ to $(n, 0)$. We further assume that there are no lethal cells in our costmap, since path planning algorithms already do a fine job of avoiding these.

In addition to the actual planning problem, there is also the parameter tuning problem. We would like to be able to design robot behavior from a high level and not need to fiddle with the parameters endlessly to find the perfect balance for the desired behavior. One goal of this is to find functions in which the parameter space is either intuitive or limited to only the best possible values.

The interplay between the values in the costmap and the path planning constant turns out to be crucial for determining the course of the planned path. The duality of optimizing for path length or for path cost results in a continuum of different paths that could be considered optimal depending on the weighting of the two sides.

4   Mathematical Properties of Gaussian Obstacles

Our analysis focuses on the frequently-employed two-dimensional Gaussian distribution, defined as

$$f(x,y) = A \exp\left(-\frac{x^2 + y^2}{2\sigma^2} \right) \tag{2}$$

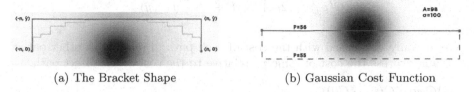

(a) The Bracket Shape   (b) Gaussian Cost Function

Fig. 1. Types of Paths - In 1a, an optimal bracket-shaped path in blue and a suboptimal path in yellow. In 1b, the discontinuity of optimal paths between $P = 55$ and $P = 56$.

The cost in each cell under the Gaussian depends on both the amplitude, A, and the variance, σ. One key feature of the Gaussian function (and all other monotonically decreasing functions) is that the optimal path is always bracket-shaped: from $(-n, 0)$ to $(-n, \hat{y})$ to $(n, \hat{y})$ to $(n, 0)$ for some $\hat{y}$. The proof of this is elided for space, but centers on the fact that all paths that reach $y = \hat{y}$ will have the same length (as seen in Figure 1a), and the bracket-shaped paths are the farthest paths of that length from the obstacle. Note that a direct path qualifies as a bracket with $\hat{y} = 0$.

The only variation in the paths is how far away from the obstacle they are, i.e. their $\hat{y}$ values. In this section we seek a relationship between the model parameters P, A and σ and the resulting distance of closest approach, $\hat{y}$. In order to cause the optimal path have to a smaller value of $\hat{y}$, the obvious strategies are to decrease the costs (decrease A or σ) or to increase the path constant P. This is generally true, but there exist certain conditions in which incrementally changing the parameters in this way will result in a drastically different path.

For instance, one would expect that increasing P incrementally would result in a gradual decline in $\hat{y}$. However, the resulting change sometimes is a discontinuous jump. Consider the paths in Figure 1b. With $A = 98$ and $\sigma = 100$, when P is increased from 55 to 56, one would expect $\hat{y}$ to undergo a small decrease. Instead, the path jumps suddenly through the origin ($\hat{y} = 0$). There is, in fact, no value of P which causes the optimal path to occur between these two paths.

4.1 Theory

Our goal is to show why there are certain parameters that result in one of three behaviors: **(1)** Minimum path cost is at $\hat{y} = 0$ **(2)** Minimum path cost is at finite $\hat{y} > 0$ **(3)** Minimum path cost is infinitely far away. The cost of a bracket path in terms of the model parameters and some choice of $\hat{y}$ is

$$C(p_{\hat{y}}) = C(\text{segment 1}) + C(\text{segment 2}) + C(\text{segment 3})$$

$$= \left[\hat{y}P + \sum_{i=0}^{\hat{y}-1} f(-n, i)\right] + \left[2nP + \sum_{x=-n}^{n} f(x, \hat{y})\right] + \left[\hat{y}P + \sum_{i=0}^{\hat{y}-1} f(n, i)\right].$$

Now let us assume that we begin far from the obstacle, so $n \gg \sigma$. In this limit, the cost due to the obstacle on segments 1 and 3 is very small compared to the baseline cost and the cost along segment 2.

$$C(p_{\hat{y}}) \approx (2n + 2\hat{y})P + \sum_{x=-n}^{n} f(x, \hat{y}) \tag{3}$$

We are only concerned with the cost of each path in relation to other paths, so we will express the cost of some $p_{\hat{y}}$ relative to the cost of the direct path.

$$\begin{aligned}
\Delta C(\hat{y}) &= C(\hat{y}) - C(0) \\
&= \left[(2n + 2\hat{y})P + \sum_{x=-n}^{n} f(x, \hat{y}) \right] - \left[(2n + 2(0))P + \sum_{x=-n}^{n} f(x, 0) \right] \\
&= 2P\hat{y} + \sum_{x=-n}^{n} \left[f(x, \hat{y}) - f(x, 0) \right]
\end{aligned}$$

When $\Delta C(\hat{y}) < 0$ for some $\hat{y}$, the direct path is not optimal.
With a Gaussian obstacle as f,

$$\Delta C(\hat{y}) = 2P\hat{y} + \sum_{x=-n}^{n} \left[A \exp\left(-\frac{x^2 + \hat{y}^2}{2\sigma^2} \right) - A \exp\left(-\frac{x^2}{2\sigma^2} \right) \right]$$

We have already assumed $n \gg \sigma$, the tails of the Gaussian will contribute negligibly, so we can approximate the sum over $x \in [-n, n]$ as the sum over all x, which has a simple solution. We use the following approximation.

$$\sum_{x=-n}^{n} A \exp\left(-\frac{x^2}{2\sigma^2} \right) \exp\left(-\frac{y^2}{2\sigma^2} \right) \approx \sum_{x=-\infty}^{\infty} A \exp\left(-\frac{x^2}{2\sigma^2} \right) \exp\left(-\frac{y^2}{2\sigma^2} \right)$$

$$= A\sigma\sqrt{2\pi} \exp\left(-\frac{y^2}{2\sigma^2} \right)$$

Finally, we have a closed-form expression for the cost of bracket path $p_{\hat{y}}$ compared to the direct path.

$$\Delta C(\hat{y}) = 2P\hat{y} + A\sigma\sqrt{2\pi} \left[\exp\left(-\frac{\hat{y}^2}{2\sigma^2} \right) - 1 \right] \tag{4}$$

Now, we locate the $\hat{y}$ that minimizes $\Delta C(\hat{y})$.

$$\frac{d\Delta C}{d\hat{y}} = 2P - \frac{\hat{y}}{\sigma} A\sqrt{2\pi} \exp\left(-\frac{\hat{y}^2}{2\sigma^2} \right) = 0 \tag{5}$$

$$\frac{P}{A} = \sqrt{\frac{\pi}{2}} \frac{\hat{y}}{\sigma} \exp\left(-\frac{\hat{y}^2}{2\sigma^2} \right) \tag{6}$$

The solution for $\hat{y}$ is related to the Lambert W-function [1], which cannot be written in closed form. Depending on the cost ratio P/A, it admits 0, 1 or 2 solutions. We can make several observations.

---

[1] See http://en.wikipedia.org/wiki/Lambert_W_function

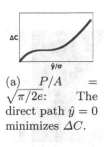

(a) $P/A = \sqrt{\pi/2e}$: The direct path $\hat{y} = 0$ minimizes ΔC.

(b) $P/A = 0.57$: A new global minimum is established at $\hat{y} > 0$.

(c) $P/A < 0.57$. $\hat{y}$ moves farther from the origin, and the minimum grows deeper.

(d) $P = 0$, and the global minimum is at $\hat{y} \to \infty$.

Fig. 2. Cost ΔC relative to the cost of the direct path, as a function of ratio of the baseline cost-per-step P and the amplitude A of a Gaussian obstacle. The minimum of this function determines where the optimal path is.

1. Eq. 6 peaks when $\hat{y}/\sigma = 1$, attaining $P/A = \sqrt{\pi/2e}$. If we set P and A such that $P/A > \sqrt{\pi/2e} \approx 0.760$, there is no solution to Eq. 6, and the only minimum of ΔC occurs on the boundary at $\hat{y} = 0$ like in Figure 2a. The direct path is optimal.
2. For values of P/A less than .760, the inflection point at $\hat{y}/\sigma = 1$ decreases as well, creating a local minimum. This minimum remains a local minimum until $P/A \approx 0.57$ (Figure 2b), a point we determined numerically where the minimum becomes the global minimum. For values even less than 0.57 (Figure 2c), the direct path is no longer optimal and the center of the obstacle is avoided.
3. When $P = 0$, $\Delta C(\hat{y})$ has a minimum at infinity (Figure 2d). This results in the path being as far away from the center of the obstacle as possible.

Thus, we have shown, with some simplifications, the mathematical underpinning for the relationship between the different parameters and why certain configurations lead to the three behaviors/values of $\hat{y}$ discussed at the beginning of this section.

5   Results from Simulation

To further explore the relationships between the parameters, we ran path planning algorithms over simulated costmaps as described in the Problem Statement section. Instead of showing all of the paths for each configuration, we represent the resultant $\hat{y}$ values in heatmaps as seen in Figure 3. As is evident from Equation 6, P and A are inversely related, a fact we could have surmised through dimensional analysis. As long as the ratio $P{:}A$ remains constant, the value of $\hat{y}$ also remains constant (plots not shown). This means we can explore the entire parameter space as a two dimensional heat map relating that ratio to σ, which we did by varying the value of P.

The three distinct behaviors are seen in the different colorations. Configurations represented in black result in paths that move straight through the obstacle

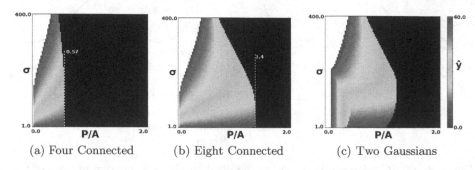

(a) Four Connected   (b) Eight Connected   (c) Two Gaussians

Fig. 3. Heatmaps - The value of $\hat{y}$ as a function of P/A and σ. Black values represent paths where $\hat{y} = 0$; white values represent paths where $\hat{y}$ is infinitely far away. Redder values indicate higher values of $\hat{y}$ and bluer values represent lower values.

($\hat{y} = 0$). On the opposite side of the spectrum, white configurations represent paths that are as far away as possible, i.e. the optimal $\hat{y}$ is infinitely large. The intermediate hues represent the finite values of $\hat{y} > 0$, with red values being the farthest away and blue values being the closest (smallest $\hat{y}$).

Let us begin the discussion with Figures 3a and 3b. Figure 3a was created using the von Neumann neighborhood (i.e. four connected) and corresponds with the math from the previous section. Figure 3b utilized the Moore neighborhood (i.e. eight connected), and thus the properties from the previous section need no necessarily apply. However, as evident in the two figures, both the von Neumann and Moore neighborhoods operate similarly, although with slightly different scaling. This lends substantiative support to our hypothesis that the properties operate similarly regardless of the precise path planning implementation used.

The next thing to note is the relationship between σ and $\hat{y}$ for a given P/A. For any given finite positive value of $\hat{y}$ we can decrease the variance to get a smaller $\hat{y}$. This leads us to our first general observation about the parameter space. **Decreasing the variance on a Gaussian will always lead to paths closer to the obstacle.** Similarly, we can assert that **Lowering the ratio $P : A$ will always increase $\hat{y}$ from a positive value to a greater value.**

However the inverse of these statements is not true. Increasing the variance will sometimes increase the distance from the obstacle, but can also lead to decreasing the distance down to $\hat{y} = 0$. **Increasing the variance on a Gaussian will only sometimes result in paths further from the obstacle.** Increasing $P : A$ will always lead to to decreased $\hat{y}$ values, however, at some point, those values jump to $\hat{y} = 0$. This discontinuity means that **For a given σ, some values of $\hat{y}$ cannot be expressed.** If this were not the case, the right-hand edge of the colored areas in the heatmaps would be blue for the lowest $\hat{y}$. Similarly, some values of $\hat{y}$ cannot be expressed for a given ratio $P : A$. These two limitations lead to the most important observation about the parameter space: **Finding paths with the full range of $\hat{y}$ values cannot be achieved by tuning just one parameter.**

This point is reiterated by the fact that there are some values of $P : A$ that do not have any values of σ resulting in $\hat{y} > 0$. Our simulated experiments also validated that these dead zones occur when $P : A > 0.57$ when the grid is four-connected, which can be seen in Figure 3a and since these values result in no solution to Equation 6. Values of $P/A < 0.57$ generally result in non-zero $\hat{y}$, until the variance gets sufficiently large, which is where our initial assumption that $n >> \sigma$ breaks down, resulting in unpredictable behavior. Based on the results shown in Figure 3b, we have further estimated that for eight connected neighborhoods, the dead zone starts with $P/A > 1.4$.

6 Results Using ROS Navigation

Much of this work is motivated by our experiences with the ROS navigation stack [10]. Initially, it did not have a way to directly input non-lethal obstacles, which made it difficult to model people's personal space. In a branched version of the software, we removed this limitation[8]. In the costmaps, the amplitudes can range from $[0, 254]$ with no limits on the variance. However, initially we were unaware that the default path planning constant was set so that $P = 50$, meaning that our options for A were limited, and we found it difficult to tune the parameters to get the passing distances we desired.

To further validate the principles in this paper, we used ROS navigation stack with real sensor data from Willow Garage's PR2. Paths were planned around a live human using a modified wavefront planner that performs interpolated gradient descent to determine a smooth path, i.e. not constrained to either the von Neumann or Moore neighborhoods. The results are shown in Figure 4 with three different values of A. As A increases, the value of $\hat{y}$ increases too, but then when A is very large, it contracts back down to $\hat{y} \approx 0$, as close to the person as possible. The full heatmaps are not included for space.

We found that even using the less restrictive planner, the same principles apply, although the constants have different values. Since costmaps and the discretized paths they produce are all approximations of the same continuous

Fig. 4. Results using ROS Navigation, $\sigma = 1.75$, and $P/A = \{3/3, 3/60, 3/235\}$

space, it is logical that the different grid connectivities would have similar results; however, it is nice to have the technical confirmation to back up the ideas. Not only is the solution more general, but it has the added benefit that the resulting smoother paths are more legible to people observing the robot.

7   Discussion

As robots require more and more social behaviors, it will be necessary to precisely design systems where the robot will navigate in a particular way. With this in mind, it is worth considering the placid assumption that the Gaussian costmap addition is the ideal tool for the job. In particular, the existence of the large dead zone where small changes to the parameters will have no effect on the path and the discontinuities when configuring just one parameter, together present numerous challenges to tackle when configuring a system. Additional and more complicated cost functions may alleviate some of the problems. Figure 3c shows the heatmap when using the sum of *two* Gaussian functions. This particular example creates greater range for $\hat{y}$ for small variances, but has its own problems with discontinuities. This is clearly not the only other option, and we advise that designers of new cost functions use the techniques in this paper to test their new functions.

The choice of costmap approach is vitally important for social navigation. The Gaussian adjustment explored in this paper are certainly better than no costmap modifications. However, in tight environments like narrow hallways (as we encountered previously[8]), the optimal $\hat{y}$ for a given set of parameters may result in an invalid path that collides with other obstacles. Limiting the $\hat{y}$ value means only considering the smallest values of $\hat{y}/\sigma$ in Figure 2b, and thus the robot may end up taking the most direct path, moving uncomfortably close to the person. In accordance with the study of proxemics, a robot which mostly drives far away, but sometimes moves into a person's intimate personal space seems much worse than a robot that always drove a medium distance away. This led to our decision not to use Gaussians in that experiment.

As a final note, social robots need to take many metrics into consideration to find the paths that will be best suited for interacting with humans. We do not claim that the sole definition of "best" path relies upon just path length and closest distance to an obstacle. Finding the best cost functions, and planners that optimize a wide variety of metrics, will likely see much fruitful work in the coming years.

References

[1] Abbeel, P., Ng, A.Y.: Apprenticeship learning via inverse reinforcement learning. In: Proceedings of the 21st International Conference on Machine Learning (ICML) (2004)
[2] Abbeel, P., Dolgov, D., Ng, A.Y., Thrun, S.: Apprenticeship learning for motion planning with application to parking lot navigation. In: Proceedings of the IEEE/RSJ International Conference on Intelligent Robots and Systems (IROS), pp. 1083–1090 (2008)

[3] Choset, H., Lynch, K.M., Hutchinson, S., Kantor, G., Burgard, W., Kavraki, L.E., Thrun, S.: Principles of robot motion: theory, algorithms, and implementations. MIT Press (2005)

[4] Dautenhahn, K., Walters, M., Woods, S., Koay, K.L., Nehaniv, C.L., Sisbot, A., Alami, R., Siméon, T.: How May I Serve You?: A Robot Companion Approaching a Seated Person in a Helping Context. In: Proceedings of the 1st ACM SIGCHI/SIGART Conference on Human-Robot Interaction, pp. 172–179. ACM (2006)

[5] Ferguson, D., Likhachev, M.: Efficiently using cost maps for planning complex maneuvers. Lab Papers (GRASP), p. 20 (2008)

[6] Kirby, R., Simmons, R., Forlizzi, J.: COMPANION: A Constraint-Optimizing Method for Person-Acceptable Navigation. In: Proc. RO-MAN 2009, Toyama, Japan, pp. 607–612 (2009)

[7] Levine, S., Popović, Z., Koltun, V.: Feature construction for inverse reinforcement learning. In: Advances in Neural Information Processing Systems (NIPS), pp. 1342–1350 (2010)

[8] Lu, D.V., Smart, W.D.: Towards More Efficient Navigation for Robots and Humans. In: IROS (2013)

[9] Mainprice, J., Akin Sisbot, E., Jaillet, L., Cortés, J., Alami, R., Siméon, T.: Planning human-aware motions using a sampling-based costmap planner. In: 2011 IEEE International Conference on Robotics and Automation (ICRA), pp. 5012–5017. IEEE (2011)

[10] Marder-Eppstein, E., Berger, E., Foote, T., Gerkey, B., Konolige, K.: The office marathon: Robust navigation in an indoor office environment. In: 2010 IEEE International Conference on Robotics and Automation (ICRA), pp. 300–307. IEEE (2010)

[11] Ng, A.Y., Russell, S.J.: Algorithms for inverse reinforcement learning. In: Proceedings of the 17th International Conference on Machine Learning (ICML), pp. 663–670 (2000)

[12] Scandolo, L., Fraichard, T.: An anthropomorphic navigation scheme for dynamic scenarios. In: 2011 IEEE International Conference on Robotics and Automation (ICRA), pp. 809–814. IEEE (2011)

[13] Sisbot, E.A., Marin-Urias, L.F., Alami, R., Simeon, T.: A human aware mobile robot motion planner. IEEE Transactions on Robotics 23(5), 874–883 (2007)

[14] Svenstrup, M., Tranberg, S., Andersen, H.J., Bak, T.: Pose estimation and adaptive robot behaviour for human-robot interaction. In: IEEE International Conference on Robotics and Automation, ICRA 2009, pp. 3571–3576. IEEE (2009)

[15] Svenstrup, M., Bak, T., Andersen, H.J.: Trajectory planning for robots in dynamic human environments. In: 2010 IEEE/RSJ International Conference on Intelligent Robots and Systems (IROS), pp. 4293–4298. IEEE (2010)

Child-Robot Interaction:
Perspectives and Challenges

Tony Belpaeme[1], Paul Baxter[1], Joachim de Greeff[1], James Kennedy[1],
Robin Read[1], Rosemarijn Looije[2], Mark Neerincx[2], Ilaria Baroni[3],
and Mattia Coti Zelati[3,*]

[1] Plymouth University, United Kingdom
[2] Organization for Applied Scientific Research, The Netherlands
[3] Fondazione Centro San Raffaele, Italy

Abstract. Child-Robot Interaction (cHRI) is a promising point of entry into the rich challenge that social HRI is. Starting from three years of experiences gained in a cHRI research project, this paper offers a view on the opportunities offered by letting robots interact with children rather than with adults and having the interaction in real-world circumstances rather than lab settings. It identifies the main challenges which face the field of cHRI: the technical challenges, while tremendous, might be overcome by moving away from the classical perspective of seeing social cognition as residing inside an agent, to seeing social cognition as a continuous and self-correcting interaction between two agents.

1 Introduction

Within the field of Human-Robot Interaction (HRI) interaction between children and robots takes up a special place. Child-Robot Interaction (often abbreviated as cHRI) is different from interaction between adults and robots in that children have got a different, immature cognitive development. Children typically do not see a robot as a mechatronic device running a computer program, but attribute characteristics to the robot which are typically expected to be attributed to living systems. This has been observed in both adults and children [1] and that anthropomorphisation is already strong at the age of 3 [2], and possible at even younger ages. Furthermore, it would seem that children anthropomorphise more than adults do; or at least are more eager to maintain the illusion that the robot has life-like characteristics [3]. This is most probably amplified by that fact that children, and indeed a wide range of young animals, engage in play [4]. *Pretend play* and *anthropomorphisation* seem relevant to the ability of children to engage with robots and treat them as life-like agents. Pretend play and anthropomorphisation are not typically observed in great apes, and seem to have evolved uniquely in humans. It is believed that both have a neuropsychological basis, and that they are integral to the development of sociocognitive and linguistic skills, while at the same time being active during a short period in the

* This work is supported by the EU Integrated Project ALIZ-E (FP7-ICT-248116).

G. Herrmann et al. (Eds.): ICSR 2013, LNAI 8239, pp. 452–459, 2013.

pre-school and primary school years of a child, i.e. before between the ages of 3 and 11 [5]. This propensity for social play spills over into technology: toys and specifically robots are readily treated as being alive and having "beliefs, desires and intentions" [6].

While, as roboticists, we know little about the neurological and psychological underpinnings of what makes social human-robot interaction work (however, see [7]), this does not stop us from actively using the human propensity to interact with robots on a social level. Through prototyping and rigorous evaluations, the field has uncovered that certain design decisions (for example, using a humanoid form rather than a zoomorphic form) or behaviours (for example, providing timely reactive responses to the user's actions) work better than others. In the same way, we have discovered that human-robot interaction works particularly well with younger users [8].

Child-Robot Interaction might be fundamentally different from Adult-Robot Interaction: children are not just small adults. Their neurophysical, physical and mental development are ongoing, and this might create entirely different conditions for HRI to operate in. For example, children are still developing their language skills and while doing so they often make linguistic errors, but also seem to be oblivious for errors made by adults [9]. As such, certain linguistic errors produced by a robot might also go unnoticed .

This paper wishes to take stock of the current state-of-the-art in child-robot interaction by identifying opportunities for cHRI and the obstacles that are still in the way of deploying cHRI out of the lab and in the real world on a large scale. The authors are all involved in various aspects of cHRI, from designing perceptual systems to evaluating cHRI in the wild, and have more than three years experience in designing, creating and evaluating child-robot interaction. In the next section we not only identify challenges, but also offer suggestion as to how these can be overcome.

2   Opportunities

Next to the aforementioned propensity for children to readily engage with social robots, there are a number of factors which make the study of social cHRI particularly attractive.

2.1   Hunger for Applications

As cHRI provides a relatively easy entry point into social HRI, there are a large number of applications where cHRI can have immediate and measurable impact. One is education: a robot can provide personalised, cheap and virtually tireless tutoring. Through the social rapport that a robot can build, tutoring has the potential of being very effective. The tutoring experiences that the robot offers are on a par with, and often exceed, the effectiveness of computer-based tutoring systems [10,11]. This effect is likely due to the embodied nature of the interaction, which is not achieved when interacting with a screen-based device. A robot can

Fig. 1. A young cerebral stroke patient engaging in robot-led physiotherapy

adapt its interaction and the level of its tutoring to the child's learning, using, for example, a Zone of Proximal Development approach (see Vygotsky [12]).

Another field where a robot can be effective is healthcare. Animal assisted therapy (AAT) is an often used method to improve the well-being of children during a stay in hospital. Unfortunately, AAT is expensive and not available to all young patients for hygiene reasons. However, robots provide an attractive alternative for this: Robot Assisted Therapy (RAT) is a fast growing sub-field of HRI. While robots can be used to just provide comfort and companionship, they can also take up a role in education and therapy. Robots have been shown to be effective for children diagnosed with diabetes [13,8], but have also been successfully trialled for children suffering from trauma requiring extensive physiotherapy (see Figure 1). Other domains in which RAT holds promise is respiratory diseases, such as asthma or cystic fibrosis, or cancer. The benefit the robot offers comes in a variety of forms, at the most basic level the robot provides much-needed diversion, but it can also be used as a companion and educator, while at the same time offering the ill child an experience which offers much-needed social status.

In the same sphere, we wish to highlight the use of robots for Autistic Spectrum Disorders (ASD). Children with ASD often respond well to robots, for reasons not yet fully understood [14]. People with Autism Spectrum Disorder have good systemising skills along with a preference for predictable rule based systems [15], which might explain why they feel comfortable working with computers and robots. but possibly. For ASD therapy (and indeed other robot assisted therapy) the robot can offer automated diagnosis, progress monitoring, semi-automated running of therapeutic programmes, as well as serving as an intermediary interaction medium between a human therapist and the child.

Finally, the edutainment industry is keeping a close watch on developments in cHRI. Low-cost robots which capitalise on elements of cHRI have the potential of having significant economic impact. While developments here are more sensitive to fashions and correct retail timing, academic research into cHRI still has a currently underestimated role to play.

2.2 Next Generation Cognitive Systems

The old views of cognitive systems being the product of solely the brain, or the brain interacting with the body, are giving way to a new view of cognition. In this new perspective cognition is no longer seen as just embodied, but is the product of interaction between two or more embodied brains (for some early musings see [16]). Indeed, social interaction is necessary for maturing cognition: without constant and extensive social interaction, the young child cannot grow into a socially and cognitively functioning adult. But this is only one side of the story: cognition does not happen in a social vacuum. Some of it does, such as object manipulation or locomotion, but most cognitive functions are social in nature: the manipulation of concepts and reasoning, language and multi-modal interaction all are inherently social.

cHRI provides an attractive point of entry into this new view of cognition (sometimes dubbed Cognition 3.0, 1.0 being the brain as seat of cognition, 2.0 being embodiment as focus of cognition). This view offers one very rich vein: the fact that cognitive interaction is the product of two agents, where it is not necessary for one agent to be fully cognitive. Instead social cognition is the product of high-resolution interaction between two or more agents, and when one agent provides impoverished interaction, gaps are often and readily filled by the cognition of the other agents. Take as illustrative example a conversation: spoken conversation does not follow the patterns seen in written conversation, instead spoken conversation is rife with hesitancies, disfluencies and interruptions. However, these do not negatively impact on the interaction or the intelligibility. The reason for this is that the conversation –and interaction in general– is the product of two agents, whose combined cognition caries the interaction forward.

This has tremendous potential for social HRI: the technical capabilities of robots fall far short of human cognitive capabilities. Nevertheless, given that social cognition is the product of two agents, any failures in the robot's cognition can potentially be covered by the human. This is never more obvious than in child-robot interaction. For example, the fact that a robot has no visual perception might go undetected for the entire duration of the interaction. Just the belief by the child that the robot can "see" is enough for the lacunas in the interaction (which are obvious to us engineers) to go undetected [17].

3   Challenges

Despite the positive notes in the previous section, there are a number of challenges which the field needs to address before cHRI can be considered a success.

3.1   Technical Challenges

As with most AI systems, perception remains a bottleneck. This is certainly so in HRI and cHRI: users often expect the robot to have the same perceptual modalities as the user has, and it are these modalities that have proven to be very hard to realise artificially. Artificial visual perception, for example, has after 50 years of steady improvement only reached a fraction human capability. Functionality such as face detection, face recognition, object recognition, figure ground separation and human behaviour understanding have come a long way and are currently of a standard where they can be used in restricted scenarios [18]. However, open-ended interaction in real-world environments still is not possible. The same goes for Natural Language Understanding: while speech recognition has come in leaps and bounds, child speech recognition is still under-performing. This has a knock-on effect on Natural Language Understanding, forcing human-robot interaction scenarios which use language to either resort to restricted interactions or to revert to Wizard of Oz control.

Another obstacle to autonomous cHRI (and HRI in general) is action selection: the problem of what to do next in response to current and past sensor input. Users have well-defined expectations of what the robot's response should be, but it has proven to be technically very challenging to select a correct response in open and unconstrained environments. Action selection mechanisms, and by extension cognitive architectures, are currently falling short when used in social interaction domains. New and radically different approaches will be needed to break through this barrier (see for example [19]).

However, as mentioned in the previous section, lacks in artificial processing and in generating appropriate responses, often go undetected by young users. This is a blessing for social robot builders and can be used to good effect. For example, the lack of visual perception can be disguised by still providing the robot with behaviours which make it appear as if it is seeing. Just the addition of two eyes to a robot head often suffices to "trick" the user into believing that the robot sees. Together with alternative technologies, such as RFID tags worn by the user for user identification and a laser range scanner for detection objects near the robot, this can be used to generate the illusion that the robot has visual perception.

3.2   Evaluation

Evaluating the effectiveness of cHRI has always been more problematic than with adults. While adults can be probed using questionnaires and self-reflection, children have the tendency to try to please the experimenter, rather than answer truthfully to survey questions. Likert scales always come back with extreme responses, and children often try to second guess the desirable response. As such alternative methods are required. A number of interaction metrics can be monitored: duration of interaction, proxemics, structure of utterances, biometrics, compliance with robot suggestions and instructions. However, these are often only circumstantial to the goal of cHRI. Often the interaction serves to offer consolation, or to educate a child or to encourage, and interaction metrics – while sometimes correlating with the goal of the interaction– cannot be trusted to inform us about the outcome of the interaction.

As an experimenter we need to measure if the desired outcome really is met. In some cases this is relatively straightforward: if the robot has a role in education, then a balanced pre and post test can be devised to measure what the contribution of the robot is to knowledge gain. However, other outcomes, such as the robot providing consolation, are much harder to objectively measure and might require very large sample sizes before returning significant results.

In conclusion, while other disciplines in robotics have relatively easy benchmarks to measure performance against, interaction with people introduces a level of noise which the field still has not fully addressed.

3.3   Expectations

When building social robots, it is important to set the right expectations. Not only in the young people interacting with the robot, but also in the parents, medical staff and teachers who use the robots. And it is the latter group which often proves problematic. In our efforts in creating social robot interaction, we have had to spend considerable effort talking to and demonstrating robots to adults in whose care children were. Children have preconceptions of what the robot can and cannot do, but are not noticeably troubled by unmet expectations. Adults, however, are different and as a robot builder you will need to invest considerable time and effort adjusting expectations. However, the investment will be very much worth it.

The authors found focus groups and regular meetings to be useful to identify a set of achievable goals for social robots in medical settings, which are both rewarding for us as researchers and for the medical staff we collaborate with. However, during this we learnt the hard way not to trust our own opinions. There is a temptation to think that, as an HRI expert, one knows what the user wants. This is seldom the case.

4   Conclusion

This paper did not contain a technical contribution, nor did it present results from evaluating technical systems or running user studies[1]. Most, perhaps all,

---

[1] However, the opinion and observations expressed in this paper are the result of 3 years and over 800 person months of experience in building and evaluating Child-Robot Interaction in the FP7 ALIZ-E project.

of the observations in the paper kick in open doors, but we feel the points made are worth making and worth communicating to our colleagues. In science and engineering too often we make mistakes and start redundant work which could have been avoided had we only engaged in a conversation about the challenges and potential solutions which others identified before us through hard graft.

We wish to reiterate that the interaction between children and robots is potentially very different from the interaction between adults and robots due to children's neurophysical and mental development being ongoing. At the same time, the view of cognition is being extended: we suggest that cognition is no longer the domain of the individual, but the product of a fine-grained interaction between agents, be they people or people and machines. This does not lead to new research questions, but also offers new opportunities for cHRI and HRI in general.

Acknowledgements. We would like to thank both anonymous reviewers for their critical and helpful suggestions.

References

1. Reeves, B., Nass, C.: The Media Equation: How People Treat Computers, Television, and New Media like Real People and Places, 2nd edn. Cambridge University Press (1998)
2. Berry, D.S., Springer, K.: Structure, motion, and preschoolers' perceptions of social causality. Ecological Psychology 5(4), 273–283 (1993)
3. Turkle, S., Breazeal, C.L., Dasté, O., Scassellati, B.: Encounters with kismet and cog: children respond to relational artifacts. In: IEEE-RAS/RSJ International Conference on Humanoid Robots, Los Angeles, CA (2004)
4. Fagen, R.: Animal play behav. Oxford University Press (1981)
5. Smith, P.K.: Children and Play: Understanding Children's Worlds. Wiley-Blackwell (2010)
6. Rao, A.S., Georgeff, M.P.: BDI agents: From theory to practice. In: Proceedings of the First International Conference on Multiagent Systems. AAAI Press (1995)
7. Rosenthal-von der Pütten, A., Schulte, F., Eimler, S., Sobieraj, S., Hoffmann, L., Maderwald, S., Brand, M., Krämer, N.: Neural correlates of empathy towards robots. In: Proceedings of the 8th ACM/IEEE International Conference on Human-Robot Interaction (HRI 2013), Tokyo, Japan (2013)
8. Belpaeme, T., Baxter, P., Read, R., Wood, R., Cuayáhuitl, H., Kiefer, B., Racioppa, S., Kruijff-Korbayová, I., Athanasopoulos, G., Enescu, V., Looije, R., Neerincx, M., Demiris, Y., Ros-Espinoza, R., Beck, A., Canamero, L., Hiolle, A., Lewis, M., Baroni, I., Nalin, M., Cosi, P., Paci, G., Tesser, F., Sommavilla, G., Humbert, R.: Multimodal child-robot interaction: Building social bonds. Journal of Human-Robot Interaction 1(2), 33–53 (2013)
9. Clark, E.V.: First Language Acquisition, 2nd edn. Cambridge University Press (2009)
10. Janssen, J.B., van der Wal, C.C., Neerincx, M.A., Looije, R.: Motivating children to learn arithmetic with an adaptive robot game. In: Mutlu, B., Bartneck, C., Ham, J., Evers, V., Kanda, T. (eds.) ICSR 2011. LNCS, vol. 7072, pp. 153–162. Springer, Heidelberg (2011)

11. Nalin, M., Baroni, I., Kruijff-Korbayová, I., Cañamero, L., Lewis, M., Beck, A., Cuayáhuitl, H., Sanna, A.: Children's adaptation in multi-session interaction with a humanoid robot. In: Proceedings of the Ro-Man Conference, Paris, France (2012)
12. Vygostky, L.S.: Play and its role in the mental development of the child. Soviet Psychology 5(3), 6–18 (1967)
13. Blanson Henkemans, O., Hoondert, V., Groot, F., Looije, R., Alpay, L., Neerincx, M.A.: "I just have diabetes": Children's need for diabetes selfmanagement support and how a social robot can accomodate. Patient Intelligence (2012) (accepted)
14. Dautenhahn, K., Werry, I.: Towards interactive robots in autism therapy: Background, motivation and challenges. Pragmatics & Cognition 12(1), 1–35 (2004)
15. Baron-Cohen, S.: The Essential Difference: men, women and the extreme male brain. Penguin/Basic Books (2003)
16. Seabra Lopes, L., Belpaeme, T., Cowley, S.: Beyond the individual: New insights on language, cognition and robots. Connection Science 20(4), 231–237 (2008)
17. Nalin, M., Bergamini, L., Giusti, A., Baroni, I., Sanna, A.: Children's perception of a robotic companion in a mildly constrained setting: How children within age 8-11 perceive a robotic companion. In: Proceedings of the Children and Robots Workshop at the IEEE/ACM International Conference on Human-Robot Interaction (HRI 2011), Lausanne, Switserland (2011)
18. Kruijff-Korbayová, I., Athanasopoulos, G., Beck, A., Cosi, P., Cuayáhuitl, H., Dekens, T., Enescu, V., Hiolle, A., Kiefer, B., Sahli, H., Schröder, M., Sommavilla, G., Tesser, F., Verhelst, W.: An event-based conversational system for the nao robot. In: Proceedings of IWSDS 2011: Workshop on Paralinguistic Information and its Integration in Spoken Dialogue Systems, pp. 125–132. Springer (2011)
19. Baxter, P., Wood, R., Morse, A., Belpaeme, T.: Memory-centred architectures: Perspectives on human-level cognitive competencies. In: 2011 AAAI Fall Symposium Series - Advances in Cognitive Systems, pp. 26–33 (2011)

Training a Robot via Human Feedback: A Case Study

W. Bradley Knox[1], Peter Stone[2], and Cynthia Breazeal[1]

[1] Massachusetts Institute of Technology
Media Lab, 20 Ames Street, Cambridge, MA USA
bradknox@mit.edu, cynthiab@media.mit.edu
[2] University of Texas at Austin
Dept. of Computer Science, Austin, TX USA
pstone@cs.utexas.edu

Abstract. We present a case study of applying a framework for learning from numeric human feedback—TAMER—to a physically embodied robot. In doing so, we also provide the first demonstration of the ability to train multiple behaviors by such feedback without algorithmic modifications and of a robot learning from free-form human-generated feedback without any further guidance or evaluative feedback. We describe transparency challenges specific to a physically embodied robot learning from human feedback and adjustments that address these challenges.

1 Introduction

As robots increasingly collaborate with people and otherwise operate in their vicinity, it will be crucial to develop methods that allow technically unskilled users to teach and customize behavior to their liking. In this paper we focus on teaching a robot by feedback signals of approval and disapproval generated by live human trainers, the technique of *interactive shaping*. These signals map to numeric values, which we call "human reward".

In comparison to learning from demonstration [1], teaching by such feedback has several potential advantages. An agent can display its learned behavior while being taught by feedback, increasing responsiveness to teaching and ensuring that teaching is focused on the task states experienced when the agent behaves according to its learned policy. A feedback interface can be independent of the task domain. And we speculate that feedback requires less expertise than control and places less cognitive load on the trainer. Further, a reward signal is relatively simple in comparison a control signal; given this simplicity, teaching

Fig. 1. A training session with the MDS robot Nexi. The artifact used for trainer interaction can be seen on the floor immediately behind Nexi, and the trainer holds a presentation remote by which reward is delivered.

by human-generated reward is a promising technique for improving the effectiveness of low-bandwidth myolectric and EEG-based interfaces, which are being developed to enable handicapped users to control various robotic devices.

G. Herrmann et al. (Eds.): ICSR 2013, LNAI 8239, pp. 460–470, 2013.

In this paper, we reductively study the problem of learning from live human feedback in isolation, without the added advantages of learning from demonstration or similar methods, to increase our ability to draw insight about this specific style of teaching and learning. This paper demonstrates for the first time that TAMER [6]—a framework for learning from human reward—can be successfully applied on a physically embodied robot. We detail our application of TAMER to enable the training of interactive navigation behaviors on the Mobile-Dexterous-Social (MDS) robot "Nexi". Figure 1 shows a snapshot of a training session. In this domain, Nexi senses the relative location of an artifact that the trainer can move, and the robot chooses at intervals to turn left or right, to move forward, or to stay still. Specifically, the choice is dependent on the artifact's relative location and is made according to what the robot has learned from the feedback signals provided by the human trainer. The artifact can be moved by the human, permitting the task domain itself—not just training—to be interactive. The evaluation in this paper is limited with respect to who trains the agent (the first author). However, multiple target behaviors are trained, giving the evaluation a different dimension of breadth than previous TAMER experiments, which focused on the speed or effectiveness of training to maximize a predetermined performance metric [9,11,8].

Though a few past projects have considered this problem of learning from human reward [4,21,20,16,18,13,9], only two of these implemented their solution for a robotic agent. In one such project [13], the agent learned partially in simulation and from hard-coded reward, demonstrations, and human reward. In another [18], the human trainer, an author of that study, followed a predetermined algorithm of giving positive reward for desired actions and negative reward otherwise. This paper describes the first successful teaching of a robot purely by free-form human reward. One contribution of this paper is the description of how a system for learning from human reward—TAMER—was applied to a physically embodied robot. A second contribution is explicitly demonstrating that different behaviors can be trained by changing only the reward provided to the agent (and trainer interaction with its environment). Isbell et al. [4] showed the potential for such personalization by human reward in a virtual online environment, but it has not previously been demonstrated for robots or for TAMER.

2   Background on TAMER

TAMER (for Training an Agent Manually via Evaluative Reinforcement) is a solution to the problem of how an agent can learn to perform a sequential task given only real-valued feedback on its behavior from a human trainer. This problem is defined formally by Knox [6]. The human feedback—"human reward"—is delivered through push buttons, spoken word, or any other easy-to-learn interface. The human's feedback is the *only* source of feedback or evaluation that the agent receives. However, TAMER and other methods for learning from human reward can be useful even when other evaluative information is available, as has been shown previously [21,5,17,11]. The TAMER algorithm described below has additionally been extended to learn in continuous action spaces through an actor-critic algorithm [22] and to provide additional information to the trainer—either action confidence or summaries of past performance—creating changes in the quantity of reward instances given and in learned performance [14].

Motivation and Philosophy of TAMER.   The TAMER framework is designed around two insights. First, when a human trainer evaluates some behavior, she considers the long-term impact of that behavior, so her feedback signal contains her full judgment of the desirability of the targeted behavior. Second, a human trainer's feedback is only delayed by how long it takes to make and then communicate an evaluation. TAMER assumes that trainers' feedback is focused on recent behavior; as a consequence, human reward is considered a trivially delayed, full judgment on the desirability of behavior. Following the insights above and TAMER's assumption of behavior-focused feedback, TAMER avoids the credit assignment problem inherent in reinforcement learning. It instead treats human reward as fully informative about the quality of recent actions from their corresponding states.

Mechanics of TAMER.   The TAMER framework consists of three modules, as illustrated in Figure 2: credit assignment to create labels from delayed reward signals for training samples; supervised learning from those samples to model human reward; and action selection using the human reward model. The three modules are described below.

TAMER models a hypothetical human reward function, $R_H : S \times A \to \mathbb{R}$, that predicts numeric reward based on the current state and action values (and thus is Markovian). This modeling, corresponding to the "supervised learner" box in Figure 2, uses a regression algorithm chosen by the agent designer; we call the model $\hat{R}_H$. Learning samples for modeling are constructed from experienced state-action pairs and the real-valued human reward credited to each pair as outlined below.

The TAMER algorithm used in this paper (the "full" algorithm with "delay-weighted aggregate reward" described in detail by Knox [6]), addresses the small delay in providing feedback by spreading each human reward signal among multiple recent state-action pairs, contributing to the label of each pair's resultant sample for learning $\hat{R}_H$. These samples, each with a state-action pair as input and a post-assignment reward label as the output, are shown as the product of the "credit assigner" box in Figure 2. Each sample's share of a reward signal is calculated from an estimated probability density function for the delay in reward delivery, f_{delay}.

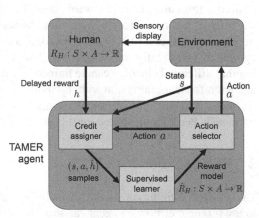

Fig. 2. An information-flow diagram illustrating the TAMER framework

To choose actions at some state s (the "action selector" box of Figure 2), a TAMER agent directly exploits the learned model $\hat{R}_H$ and its predictions of expected reward. When acting greedily, a TAMER agent chooses the action $a = argmax_a[\hat{R}_H(s,a)]$. This is equivalent to performing reinforcement with a discount factor of 0, where reward acquired from future actions is not considered in action selection (i.e., action selection is *myopic*). In practice, almost all TAMER agents thus far have been greedy, since the trainer can punish the agent to make it try something different, reducing the need for other forms of exploration.

Putting TAMER in Context.   Although reinforcement learning was inspired by models of animal learning [19], it has seldom been applied to reward created from non-expert humans. We and others concerned with the problem of learning from human reward (sometimes called interactive shaping) seek to understand how reinforcement learning can be adapted to learn from reward generated by a live human trainer, a goal that may be critical to the usability of reinforcement learning by non-experts. TAMER, along with other work on interactive shaping, makes progress towards a second major form of teaching, one that will complement but not supplant learning from demonstration (LfD). In contrast to LfD, interactive shaping is a young approach. A recent survey of LfD *for robots* cites more than 100 papers [1]; this paper describes the second project to involve training of robots exclusively from human reward (and the first from purely *free-form* reward).

In comparison to past methods for learning from human reward, TAMER differs in three important ways: (1) TAMER addresses delays in human evaluation through credit assignment, (2) TAMER learns a model of human reward ($\hat{R}_H$), and (3) at each time step, TAMER *myopically* chooses the action that is predicted to directly elicit the maximum reward ($argmax_a \hat{R}_H(s, a)$), eschewing consideration of the action's effect on future state. Accordingly, other algorithms for learning from human reward [4,21,20,16,18,13] do not directly account for delay, do not model human reward explicitly, and are not fully myopic (i.e., they employ discount factors greater than 0).

However, nearly *all* previous approaches for learning from human-generated reward are *relatively* myopic, with abnormally high rates of discounting. Myopia creates certain limitations, including the need for the trainer to communicate what behavior is correct in any context (e.g., going left at a certain corner); a non-myopic algorithm instead would permit communication of correct outcomes (e.g., reaching a goal or failure state), lessening the communication load on trainers (while ideally still allowing behavior-based feedback, which people seem inclined to give). However, the myopic trend in past work was only recently identified and justified by Knox and Stone [10], who built upon this understanding to create the first successful algorithm to learn non-myopically from human reward [12]. Along with their success in a 30-state grid world, they also showed that their non-myopic approach needs further human-motivated improvements to scale to more complex tasks.

Complementing this continuing research into non-myopic approaches, this paper focuses on applying an established and widely successful myopic approach to a robotic task, showing that TAMER can be used flexibly to teach a range of behaviors and drawing lessons from its application. TAMER has been implemented successfully in a number of simulation domains commonly used in reinforcement learning research: mountain car [9], balancing cart pole [11], Tetris [9], 3 vs. 2 keep-away soccer [17], and a grid-world task [10]. In comparison, other interactive-shaping approaches have been applied in at most two domains.

3   The MDS Robot Nexi

A main contribution of this paper is the application of TAMER to a physical robot, shown in Figure 3a. The Mobile-Dexterous-Social robot platform is designed for research at the intersection of mobility, manipulation, and human-robot interaction [2]. The mobile base of the MDS platform has 2 degrees of freedom, with two powered wheels and one unpowered, stability-adding wheel. The robot estimates its environmental state through a Vicon Motion Capture system that determines the 3-dimensional locations and orientations of the robot and the training artifact; in estimating its own position and orientation, the robot employs both the Vicon data and information from its wheel encoders. In addition to the Vicon system, the robot has a number of other sensing capabilities that are not employed in this work.

4   TAMER Algorithm for Interactive Robot Navigation

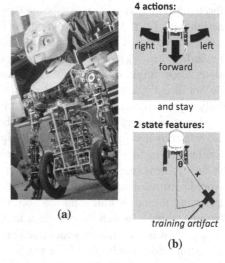

We implemented the full TAMER algorithm as described generally in Section 2 and in detail by Knox [6], using the delay-weighted aggregate reward credit assignment system described therein.

From the robot's estimation of the position and orientation of itself and the training artifact, two features are extracted and used as input to $\hat{R}_H$ along with the action. The first feature is the distance in meters from the robot to the training artifact, and the second is the angle in radians from the robot's position and orientation to the artifact. Figure 3b shows these state features and the four possible actions: turn left, turn right, move forward, or stay still.

In implementing a version of TAMER that learns interactive navigational behaviors, we specified the following components. $\hat{R}_H$ is modeled by the k-nearest neighbors algo-

Fig. 3. (a) The MDS robot Nexi. (b) Nexi's action and state spaces, as presented to TAMER.

rithm. More detail is given later in this section. The *training interface* is a presentation remote that can be held in the trainer's hand. Two buttons map to positive and negative reward, giving values of $+1$ and -1 respectively. Also, an additional button on the remote toggles the training mode on and off. When toggled on, TAMER chooses actions *and learns from feedback on those actions*; when off, TAMER does not learn further but does demonstrate learned behavior (see Knox [6] for details about toggling training). Another button both turns training off and forces the robot to stay still. This safety function is intended to avoid collisions with objects in the environment. The probability density function f_{delay}, which is used by TAMER's credit assignment module and describes the probability of a certain delay in feedback from the state-action pair it targets, is a Uniform(-0.8 seconds, -0.2 seconds) distribution, as has been employed in past work [6].[1]

The duration of time steps varies by the action chosen (for reasons discussed in Section 5). Moving forward and staying each last 1.5 seconds; turns occur for 2.5 seconds. When moving forward, Nexi attempts to move at 0.075 meters per second, and Nexi seeks to turn at 0.15 radians per second. Since changes in intended velocity— translational or rotational—require a period of acceleration, the degree of movement during a time step was affected by whether the same action had occurred in the previous time step.

$\hat{R}_H$ is modeled using k-nearest neighbors with a separate sub-model per action (i.e., there is no generalization between actions), as shown in Algorithm 1. The number of neighbors k is dynamically set to the floor of the square root of the number of samples gathered for the corresponding action, growing k with the sample size to counteract the lessening generalization caused by an increase in samples and to reduce the impact of any one experienced state-action pair, reducing potential erratic behavior caused by mistaken feedback. The distance metric is the Euclidean distance given the 2-dimensional feature vectors of the queried state and the neighboring sample's state. In calculating the distance, each vector element v is normalized within [0, 1] by $(v - v_{min})/(v_{max} - v_{min})$, where v_{max} and v_{min} are respectively the maximum and minimum values observed across training samples in the dimension of v.

To help prevent one or a few highly negative rewards during early learning from making Nexi avoid the targeted action completely, we bias $\hat{R}_H$ toward values of zero. This biasing is achieved by reducing the value of each neighbor by a factor determined by its distance d from the queried state, with larger distances resulting in larger reductions. The bias factor is calculated as the maximum of linear and hyperbolic decay functions as shown in line 11 of Algorithm 1.

Lastly, when multiple actions have the same predicted reward from the current state, such ties are broken by repeating the previous action. This approach lessens the number of action changes, which is intended to reduce early feedback error caused by ambiguously timed changes in actions (as discussed in Section 5). Accordingly, at the first time step, during which all actions are tied with a value of 0, a random action is chosen and repeated until non-zero feedback is given.

Algorithm 1. Inference by k-Nearest Neighbors

Given: Euclidean distance function d over state features
Input: Query q with state $q.s$ and action $q.a$, and a set of samples M_a for each action a. Each sample has state features, an action, and a reward label $\hat{h}$.

1:   $k \leftarrow floor(\sqrt{|M_{q.a}|})$
2:   **if** $k = 0$ **then**
3:     $\hat{R}_H(q.s, q.a) \leftarrow 0$
4:   **else**
5:     $knn = \emptyset$
6:     $preds\_sum \leftarrow 0$
7:     **for** $i = 1$ to k **do**
8:       $nn \leftarrow argmin_{m \subset M_{q.a} \setminus knn} d(m, q)$
9:       $knn \leftarrow knn \cup \{nn\}$
10:      $dist \leftarrow d(nn, q)$
11:      $prediction_i \leftarrow nn.\hat{h} \times max(1 - (dist/2), 1/(1 + (5 \times dist)))$
12:      $preds\_sum \leftarrow preds\_sum + prediction_i$
13:     **end for**
14:     $\hat{R}_H(q.s, q.a) \leftarrow preds\_sum/k$
15:   **end if**
16:   **return** $\hat{R}_H(q.s, q.a)$

---

[1] Two minor credit-assignment parameters are not explained here but are nonetheless part of the full TAMER algorithm. For this instantiation, these are $\epsilon_p = 1$ and $c_{min} = 0.5$.

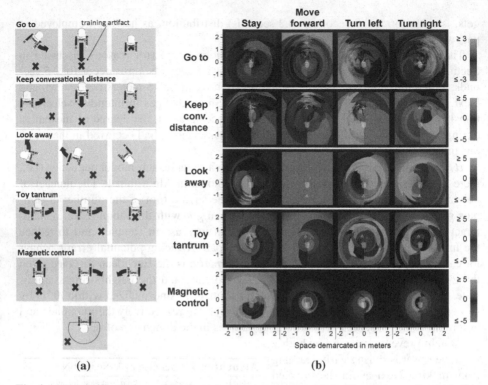

Fig. 4. (a) Iconic illustrations of the five interactive navigational behaviors that were taught to the MDS robot Nexi (described in Section 5). Each gray square represents a category of state space. The arrow indicates the desired action in such state; lack of an arrow corresponds to the stay action. (b) Heat maps showing the reward model that was learned at the end of each successful training session. Nexi is shown by a transparent birds-eye rendering of the robot, with Nexi facing the top of the page. The map colors communicate the value of the reward prediction for taking that action when the artifact is in the corresponding location relative to Nexi. A legend indicating the mapping between colors and prediction values for each behavior is given on the right. The small triangle, if visible, represents the location of the artifact at the end of training and subsequent testing of the behavior. (Note that in all cases the triangle is in a location that should make the robot stay still, a point of equilibrium.)

5   Results and Discussion

We now describe the results of training the robot and discuss challenges and lessons provided by implementing TAMER in this domain.

Behaviors Taught. Five different behaviors were independently taught by the first author, each of which is illustrated in Figure 4a:

- **Go to** – The robot turns to face the artifact, then moves forward, and stops before the artifact with little space between the two.
- **Keep conversational distance** – The robot goes to the artifact and stops at an approximate distance from the training artifact that two people would typically keep between each other during conversation (about 2 feet).

- **Look away** – The robot should turn away from the artifact, stopping when facing the opposite direction. The robot never moves forward.
- **Toy tantrum** – When the artifact is near the front of the robot, it does not move (as if the artifact is a toy that is in the robot's possession, satisfying the robot). Otherwise, the robot turns from side to side (as if in a tantrum to get the toy back). The robot never moves forward.
- **Magnetic control** – When the artifact is behind the robot, it acts as if the artifact repels it. The repulsion is akin to one end of a magnet repelling another magnet that faces it with the same pole. Specifically, when the artifact is near the center of the robot's back, the robot moves forward. If the artifact is behind its left shoulder, it turns right, moving that shoulder forward (and vice versa for the right shoulder). If the artifact is not near the robot's back, the robot does not move.

Videos of the successful training sessions—as well as some earlier, unsuccessful sessions—can be seen at `http://bradknox.net/nexi`. Figure 4b contains heat maps of the learned reward model for each behavior at the end of successful training.

Adjustments for Physical Embodiment. All of the videos were recorded during a one-day period of training and refinement of our implementation of the TAMER algorithm, during which we specifically adjusted action durations, the effects of chosen actions, and the communication of the robot's perceptions to the trainer. Nearly all sessions that ended unsuccessfully failed because of issues of *transparency*, which we addressed before or during this period. These transparency issues were mismatches between the state-action pair currently occurring and what the trainer believes to be occurring. The two main points of confusion and their solutions are described below.

The start and end of actions. As mentioned previously in Section 4, there can be a delay between the robot taking an action (e.g., turn right at 0.15 rad/s) and the robot visibly performing that action. This delay occurs specifically after any change in action. This offset between the robot's and the trainer's understandings of an action's duration (i.e., of a time step) can cause reward to be misattributed to the wrong action. The durations of each action—2.5 seconds for turns and 1.5 seconds otherwise—were chosen to ensure that the robot will carry out an action long enough that its visible duration can be targeted by the trainer.

The state of the training artifact. The position of the artifact, unlike that of the robot, was estimated from only Vicon data. When the artifact moved beyond the range of the infrared Vicon cameras, its position was no longer updated. The most common source of failed training sessions was a lack of awareness by the trainer of this loss of sensing. In response to this issue, an audible alarm was added that fired whenever the artifact could not be located, alerting the trainer that the robot's belief about the artifact is no longer changing.

The transparency issues above are illustrative of the types of challenges that are likely to occur with any physically embodied agent trained by human reward. Such issues are generally absent in simulation. In general, the designer of the learning environment and algorithm should seek to minimize cases in which the trainer gives feedback for a state-action pair that was perceived by the trainer but did not occur from the learning algorithm's perspective, causing misattributed feedback. Likewise, mismatches in perceived timing of state-action pairs could be problematic in any feedback-based learning system. Such challenges are related to but different from the correspondence problem in learning from demonstration [15,1], which is the problem of how to map from a demonstrator's state and action space to that of the emulating agent. Especially relevant is work by Crick et al. [3], which compares learning from human controllers who see a video feed of the robot's environment to learning from humans whose perceptions are

Table 1. Training times for each behavior

Target behavior	Active training time (in min.)	Total time (in min.)
Go to	27.3	38.5
Keep conv. dist.	9.5	11.4
Look away	5.9	7.9
Toy tantrum	4.7	6.9
Magnetic control	7.3	16.4

The middle column shows the cumulative duration of active training time, and the right column shows the time of the entire session, including time when agent learning is disabled.

matched to those of the robot, yielding a more limited sensory display. Their sensing-matched demonstrators performed worse at the task yet created learning samples that led to better performance.

Training Observations. The *go to* behavior was taught successfully early on, after which the aforementioned state transparency issues temporarily blocked further success. After the out-of-range alarm was added, the remaining four behaviors were taught successfully in consecutive training sessions. Table 1 shows the times required for training each behavior. Note that the latter four behaviors—which differ from *go to* training in that they were taught using an anecdotally superior strategy (for space considerations, described at http://bradknox.net/nexi), had the alarm, and benefitted from an additional half day of trainer experience—were taught in considerably less time.

6   Conclusion

In this paper, we described an application of TAMER to teach a physically embodied robot five different interactive navigational tasks. The feasibility of training these five behaviors constitutes the first focused demonstration of the possibility of using human reward to flexibly teach multiple robotic behaviors, and of TAMER to do so in any task domain.

Further work with numerous trainers and other task domains will be critical to establishing the generality of our findings. Additionally, in preliminary work, we have adapted TAMER to permit feedback on intended actions [7], for which we plan to use Nexi's emotive capabilities to signal intention. One expected advantage of such an approach is that unwanted, even harmful actions can be given negative reward before they occur, allowing the agent to learn what actions to avoid without ever taking them. We are also developing methods for non-myopic learning from human reward, which will permit reward that describes higher-level features of the task (e.g., goals) rather than only correct or incorrect behavior, reducing the training burden on humans, permitting more complex behavior to be taught in shorter training sessions.

Acknowledgments. This work has taken place in the Personal Robots Group (PRG) at MIT and the Learning Agents Research Group (LARG) at UT Austin. PRG research is supported in part by NSF (award 1138986, Collaborative Research: Socially Assistive Robots). LARG research is supported in part by NSF (IIS-0917122), ONR (N00014-09-1-0658), and the FHWA (DTFH61-07-H-00030). We thank Siggi Örn, Nick DePalma, and Adam Setapen for their generous support in operating the MDS robot.

References

1. Argall, B., Chernova, S., Veloso, M., Browning, B.: A survey of robot learning from demonstration. Robotics and Autonomous Systems 57(5), 469–483 (2009)
2. Breazeal, C., Siegel, M., Berlin, M., Gray, J., Grupen, R., Deegan, P., Weber, J., Narendran, K., McBean, J.: Mobile, dexterous, social robots for mobile manipulation and human-robot interaction. In: SIGGRAPH 2008: ACM SIGGRAPH 2008 New Tech Demos (2008)
3. Crick, C., Osentoski, S., Jay, G., Jenkins, O.C.: Human and robot perception in large-scale learning from demonstration. In: Proceedings of the 6th International Conference on Human-Robot Interaction, pp. 339–346. ACM (2011)
4. Ishell, C., Kearns, M., Singh, S., Shelton, C., Stone, P., Kormann, D.: Cobot in LambdaMOO: An Adaptive Social Statistics Agent. In: Proceedings of the 5th Annual International Conference on Autonomous Agents and Multiagent Systems (AAMAS) (2006)
5. Knox, W.B., Stone, P.: Combining manual feedback with subsequent MDP reward signals for reinforcement learning. In: Proceedings of the 9th Annual International Conference on Autonomous Agents and Multiagent Systems (AAMAS) (2010)
6. Knox, W.B.: Learning from Human-Generated Reward. PhD thesis, Department of Computer Science, The University of Texas at Austin (August 2012)
7. Knox, W.B., Breazeal, C., Stone, P.: Learning from feedback on actions past and intended. In: Proceedings of 7th ACM/IEEE International Conference on Human-Robot Interaction, Late-Breaking Reports Session (HRI 2012) (March 2012)
8. Knox, W.B., Glass, B.D., Love, B.C., Maddox, W.T., Stone, P.: How humans teach agents: A new experimental perspective. International Journal of Social Robotics, Special Issue on Robot Learning from Demonstration 4(4), 409–421 (2012)
9. Knox, W.B., Stone, P.: Interactively shaping agents via human reinforcement: The TAMER framework. In: The 5th International Conference on Knowledge Capture (September 2009)
10. Knox, W.B., Stone, P.: Reinforcement learning from human reward: Discounting in episodic tasks. In: 21st IEEE International Symposium on Robot and Human Interactive Communication (Ro-Man) (September 2012)
11. Knox, W.B., Stone, P.: Reinforcement learning with human and MDP reward. In: Proceedings of the 11th International Conference on Autonomous Agents and Multiagent Systems (AAMAS) (June 2012)
12. Knox, W.B., Stone, P.: Learning non-myopically from human-generated reward. In: International Conference on Intelligent User Interfaces (IUI) (March 2013)
13. León, A., Morales, E.F., Altamirano, L., Ruiz, J.R.: Teaching a robot to perform task through imitation and on-line feedback. In: San Martin, C., Kim, S.-W. (eds.) CIARP 2011. LNCS, vol. 7042, pp. 549–556. Springer, Heidelberg (2011)
14. Li, G., Hung, H., Whiteson, S., Knox, W.B.: Using informative behavior to increase engagement in the TAMER framework (May 2013)
15. Nehaniv, C.L., Dautenhahn, K.: 2 the correspondence problem. Imitation in Animals and Artifacts, 41 (2002)

16. Pilarski, P., Dawson, M., Degris, T., Fahimi, F., Carey, J., Sutton, R.: Online human training of a myoelectric prosthesis controller via actor-critic reinforcement learning. In: IEEE International Conference on Rehabilitation Robotics (ICORR), pp. 1–7. IEEE (2011)
17. Sridharan, M.: Augmented reinforcement learning for interaction with non-expert humans in agent domains. In: Proceedings of IEEE International Conference on Machine Learning Applications (2011)
18. Suay, H., Chernova, S.: Effect of human guidance and state space size on interactive reinforcement learning. In: 20th IEEE International Symposium on Robot and Human Interactive Communication (Ro-Man), pp. 1–6 (2011)
19. Sutton, R., Barto, A.: Reinforcement Learning: An Introduction. MIT Press (1998)
20. Tenorio-Gonzalez, A.C., Morales, E.F., Villaseñor-Pineda, L.: Dynamic reward shaping: Training a robot by voice. In: Kuri-Morales, A., Simari, G.R. (eds.) IBERAMIA 2010. LNCS, vol. 6433, pp. 483–492. Springer, Heidelberg (2010)
21. Thomaz, A., Breazeal, C.: Teachable robots: Understanding human teaching behavior to build more effective robot learners. Artificial Intelligence 172(6-7), 716–737 (2008)
22. Vien, N.A., Ertel, W.: Reinforcement learning combined with human feedback in continuous state and action spaces. In: 2012 IEEE International Conference on Development and Learning and Epigenetic Robotics (ICDL), pp. 1–6. IEEE (2012)

Real Time People Tracking in Crowded Environments with Range Measurements

Javier Correa, Jindong Liu, Member, IEEE,
and Guang-Zhong Yang, Fellow, IEEE

The Hamlyn Centre, Imperial College London, UK
{jc310,jliu4,gzy}@imperial.ac.uk

Abstract. Social and assistive robots have recognised benefit for future patient care and elderly management. For real-life applications, these robots often navigate within crowded environments. One of the basic requirements is to detect how people move within the scene and what is the general pattern of their dynamics. Laser range sensors have been applied for people tracking in many applications, as they are more precise, robust to lighting conditions and have broader field of view compared to colour or depth cameras. However, in crowded environments they are prone to environmental noise and can produce a high false positive rate for people detection. The purpose of this paper is to propose a robust method for tracking people in crowded environments based on a laser range sensor. The main contribution of the paper is the development of an enhanced Probability Hypothesis Density (PHD) filter for accurate tracking of multiple people in crowded environments. Different object detection modules are proposed for track initialisation and people tracking. This separation reduces the misdetection rate while increasing the tracking accuracy. Targets are initialised using a people detector module, which provides a good estimation of where people are located. Each person is then tracked using different object detection module with a high accuracy. The state of each person is then updated by the PHD filter. The proposed approach was tested with challenging datasets, showing an increase in performance using two metrics.

1 Introduction

With recent advances in assistive robots, such robots are expected to be increasingly used for a wide range of applications. Before these robots become an integral part of our daily life, many hurdles associated with both technical and social challenges need to be addressed. One of the key problems is to understand people dynamics and interactions while the robot navigates in crowded environment. Effective use of this information can make the robot behave more naturally, for example, follow basic social rules when navigating an environment. To obtain such information, the robot has to robustly detect and, most importantly, track people's movement in the environment.

G. Herrmann et al. (Eds.): ICSR 2013, LNAI 8239, pp. 471–480, 2013.
© Springer International Publishing Switzerland 2013

To track people from a mobile robot, a detection module and a tracking module are generally required. The detection module is typically in charge of detecting and localising people in the current sensor measurement. On the other hand, the tracking module keeps consistent estimation of people position and velocity over time. The basic approach to people tracking is to directly use the output from the detection module for tracking [1–3]. However, this approach relies on stable detections, which are rarely valid in crowded and cluttered environments. Thus far, a number of sensors have been applied for people detection and tracking. In recent years, depth and colour cameras have been used to do people tracking on mobile robots [4–7]. These sensors are discriminative, but with a narrow field of view, sensitive to lightning conditions and usually require higher computational resources to be analyse. In contrast, laser range sensors are robust to lightning conditions, provide a broader field of view and accurate distance measurements, all of which are beneficial for applications such as social navigation. For example, with a broader field of view, the robot can preemptively manoeuvre to avoid collisions and better identify approaching angles to a person.

To detect and track moving people, the difference between consecutive measurements and local minimum detection of the laser measurement was proposed by Schulz et al. [2]. However, this approach only works for moving people and is problematic in cluttered environments. Bellotto et al. [1] used predefined leg patterns to discriminate between people and other objects. They showed that some patterns were distinctive, but in general their detector can have higher values in both false positive rate and misdetection rate. In Arras et el. [8], a boosted people detector was proposed. This detector has good performance in the tests scenarios, in which high precision and recall were achieved. However, the best performance was only obtained by training the algorithm in per scenario basis, and in uncrowded environments.

In this work, a solution to people tracking in crowded environments by using a laser range sensor is proposed. The main contribution of this paper is to propose a tracking approach where the track initialisation and target tracking are computed using data from different modules. The proposed solution is based on a Probability Hypothesis Density (PHD) filter [9–11]. This filter has the property that the creation of new targets and tracking are completely separated processes. The PHD filter is usually initialised using constant starting positions or with a measurements-based prior [12] using the same tracking measurements. Any of these initialisations yield similar error to competing methods in crowded environments. To solve this problem, the proposed method applies different object detection modules for initialising targets and for tracking. The birth process of the PHD filter, which is used for track initialisation, is built entirely using the output from a people detector module. For spatial tracking, the output of a clustering module is used instead. Using these two complementary features reduces the misdetection rate and false positive rate.

The proposed approach has been tested in challenging crowded dataset. Each scenario of the dataset consists of recording the natural movement of people while the robot moves within the environment. Results show a decreased false positive rate and improved overall performance of the proposed approach.

2 Methodology

The proposed approach to people tracking can be conceptually divided into four modules. A schematic diagram of the proposed approach is shown in Figure 1. First, a People Detection Module is applied to detect people from a single measurement of the laser sensor. To this end, Bellotto's people legs detector was used [1]. A Clustering Module, which also makes use of the laser sensor measurement, is utilised to estimate clusters of points which can represent people in the environment. In the PHD filter, the output of the People Detection Module is used for Track Initialisation, while the output of the Clustering Module is used for the actual spatial tracking. Finally, the PHD filter is applied, naturally integrating the different modules outputs.

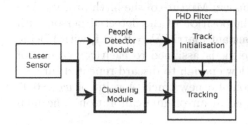

Fig. 1. Diagram of the proposed system. The Detection module is used to initialise tracks, while a Clustering Module is used for the spatial tracking. The bold lines highlight the separation between initialisation and tracking, and the new clustering component proposed in this work.

The PHD filter is a simplified solution to the Bayesian approach to multi-target tracking. This filter only propagates the first moment, or density, of the multi-target probability distribution, avoiding the computational complexity of the full multi-target Bayesian solution. The PHD filter has two main benefits when performing people tracking: it can efficiently deal with the data association problem and has a clear separation between track initialisation and tracking. The PHD recursion can be written as (using the notation of [11] and assuming no spawning):

$$v_{t|t-1}(\mathbf{x}) = \int p_{S,t}(\zeta) f_{t|t-1}(\mathbf{x}|\zeta) v_{t-1}(\zeta) d\zeta + \gamma_t(\mathbf{x}) \tag{1}$$

$$v_t(\mathbf{x}) = [1 - p_{D,t}(\mathbf{x})] v_{t|t-1}(\mathbf{x}) + \sum_{\mathbf{z} \in Z_t} \frac{p_{D,t}(\mathbf{x}) h_t(\mathbf{z}|\mathbf{x}) v_{t|t-1}(\mathbf{x})}{\kappa_t(\mathbf{z}) + \int p_{D,t}(\zeta) h_t(\mathbf{z}|\zeta) v_{t|t-1}(\zeta) d\zeta} \tag{2}$$

where:

- $v_{t|t-1}(\cdot)$ and $v_t(\cdot)$ are the prior and posterior densities at time t,
- $p_{S,t}(\cdot)$ the probability of a target surviving at time t,
- $f_{t|t-1}(\cdot|\cdot)$ the process update at time t,
- $\gamma_t(\cdot)$ the intensity of the birth process at time t,

- $p_{D,t}(\cdot)$ the probability of a target being detected at time t,
- $h_t(\cdot|\cdot)$ the observation function for time t,
- $\kappa_t(\cdot)$ the intensity of clutter process at time t.

The PHD recursion (Equations 1 and 2) can be solved by using Sequential Monte Carlo Method [10] or with a close form solution assuming a Gaussian Linear Mixture model [11]. The proposed method uses the latter representation.

In this work, the probability of a target surviving and detecting a target at time t are constant with values of 99% and 95% respectively. The clutter process $\kappa_t(\cdot)$ is set to a constant uniform process with value of 10^{-5}. Each person state $\mathbf{x} \in \mathbb{R}^4$, has two components: position and velocity (each one in $\mathbb{R}^2$). A constant velocity model with random acceleration is used to for the people's motion model, and finally, the measurement $\mathbf{z} \in \mathbb{R}^2$ is the measured position of a person.

Target initialisation in the PHD filter is achieved by the so called "birth process". This process is used to model how people appear in the environment and it is captured by the $\gamma_t(\cdot)$ component in Equation 1. To populate the birth process the output of the People Detector Module is used. Each detected person contributes with a component to the Gaussian Mixture of the birth process. The mean $\mathbf{m}_i$ of each component is set to the position of the detected person with no velocity. The covariance is set to a constant matrix $\Sigma_i = 0.3^2 \cdot \mathbf{I}$, with $\mathbf{I}$ being the 4x4 identity matrix. A constant weight is assigned to each component of the birth process. The weight has to be low enough to discard repeated targets, but high enough to not delay the creation of a new track. A value of $w_b = 0.05$ is used in this work and it was estimated in an off-line calibration. The final formula of the birth process is:

$$\gamma_t(\mathbf{x}) = w_b \sum_{i}^{N} \mathcal{N}(\mathbf{x}; \mathbf{m}_i, \Sigma_i), \qquad (3)$$

where N is the total number of detection at time t. The integration of the birth process into the PHD filter is done as proposed by Houssineau et al. [12].

The PHD filter does not naturally handle track's ID, and to evaluate the results an ID is required. To this end, a greedy heuristic is used assigning at time t the same ID as the target with the closest position in time $t - 1$.

To reduce the misdetection rate and maintain continuous tracking of people, stable features are needed. These features are extracted using the proposed Clustering Module. Figure 2 shows the details of this module. First, in the Conversions step the range measurements from the laser sensor are converted to

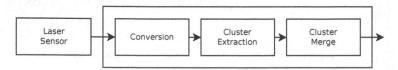

Fig. 2. Diagram showing the procedure to extract clusters corresponding to people from the laser sensor data

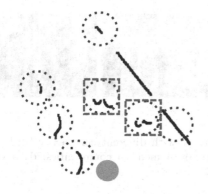

Fig. 3. The two types of features used. The black dots are the laser sensor output and the greet circle the position of the robot in this configuration. The dashed red circles shows the features corresponding to the position of the detection, while the dotted blue circles, the position of the relevant clusters.

Cartesian coordinates. The conversion allows to easily identify clusters of points by measuring the distance between them. Then, in the Cluster Extraction step, points are assigned to the same cluster if the distance between them is less than 15cm. To filter out clusters not corresponding to people, clusters with less than 4 points or greater than 100 points are discarded. After the Cluster Extraction step the legs of a person could still be in different clusters. To avoid this situation, the Cluster Merge step merges nearby clusters together. This final step is performed by measuring the distance between the centroids of each cluster. Clusters closer than 60cm are merged. Measuring the cluster's distances is only done pairwise, as at most two legs are going to be merged together. Using this heuristic manage merge two legs of the same person very accurately. The positions of the final clusters are the z variables used in Equation 2. The parameters used by this module were estimated experimentally.

In Figure 3 the distinction between the two used features is shown. The People Detector Module outputs the detected legs, shown in dashed red, while the output of the Clustering Module in blue. The features from the People Detector Module contribute with meaningful objects to be tracked, but increases the misdetection rate when a person is not detected. On the other hand, features from the Clustering Module, help with a stable spatial tracking, but increase the overall false positive rate. Using a combination of both features helps to lower both, misdetection rate and false positive rate. The combination of both features is done naturally by the PHD filter. In the deviation of this filter, it is assumed that each target generates one measurement only. This assumption makes that, when an already existing target is added to the birth process, the corresponding component is either merged with an existing component or removed when its weight drop below a threshold. Also, the extra number of measurements generated by the Clustering Module is modelled by the clutter component $\kappa_t(\cdot)$ of Equation 2.

3   Experimental Results

The proposed approach was tested in a set of challenging crowded scenarios. Each experiment is composed of a recording of the robot manually driven around

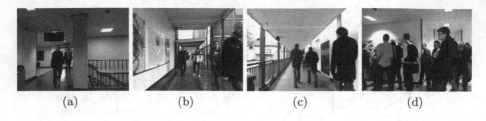

(a)   (b)   (c)   (d)

Fig. 4. Different snapshots of the used data set with different complexity levels. **a)** a scene of simple complexity. **b)** and **c)** scenes of medium complexities. **d)** a very complex scene.

different scenarios while the raw data of the sensors is being recorded. The robot uses a Pioneer P3DX base with an attached frame where the remaining hardware is mounted. The laser sensor is a Hokuyo UTM-30LX sensor. A colour camera is also recorded for references purposes but not used in the proposed work.

Four scenarios of five minute each were manually annotated. The annotation was carried out in the laser plane by visually identifying where people were located. Besides the spatial annotation, each recording was segmented with its perceived complexity: **simple**, **medium** and **complex**. This classification depends on how the scene looked from the colour camera. In Figure 4, four snapshots of the colour camera are shown. Figure 4(a) shows an example of a **simple** scenario, Figures 4(b) and 4(c) two different scenarios of **medium** complexity and 4(d) shows a **complex** scenario. Table 1 shows the total amount time recorded for each class.

In the first scene (*Long corridor*) the robot is located at the entrance of a building. After a few minutes, it starts moving towards a second building through a long, outdoors (but covered) corridor. The second scene (*Large hall*), the robots is standing in a large hall while people walk around it. Then the robot starts moving towards another side of the building. The third scene (*Entrance*) the robot is located in the entrance of one of buildings while people are coming in and out of the building. Finally, in the fourth scene (*Very crowded*) the robot is located in a corner of a very busy corridor.

Table 1. Total length in seconds of each class

	Simple	Medium	Complex
Static Robot	172.41	245.19	88.10
Moving Robot	437.46	140.05	116.01

To compare different approaches, the OSPA [13] metric and the CLEAR MOT index [14] were used. The parameters c and p of the OSPA metric were set to 1, allowing to subdivide this metric into two components (location and cardinality error). Lower values of the OSPA metric indicate lower error. The CLEAR MOT index is composed of two subindices, MOTA and MOTP. The cut-off distance for this index was set to 1m. Lower localisation error is obtained with lower values of the MOTP index, while higher MOTA values indicate more precise estimation of the people's configuration. The OSPA metric is averaged over all scenarios to illustrate the overall performance of the proposed approach, while the CLEAR MOT index is analysed per scenario basis.

The base comparison is done against [1], which implements a global Nearest Neighbour data association, an Unscented Kalman filter as the underlying tracking algorithm and uses the same people detector algorithm as the proposed approach. For comparisons, only the laser component of the base method was used. Also, comparisons are made against the Joint Probabilistic Data Association filter (JPDA) and the standard PHD filter using the people's legs detections as tracking measurements.

To make the comparisons, the recorded data was played back, and processed at real time by each method.

Figure 5 shows the average grouped OSPA results for different approaches. For all categories the proposed combination greatly outperforms other tracking approaches. It can be noted that, with increased complexity, the difference between different approaches decreases. Although initially the standard PHD filter behaves similarly as the other approaches, it has the worst performance on the complex scenario.

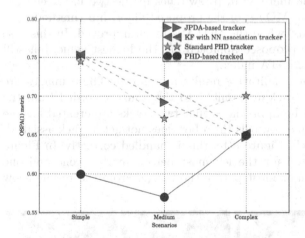

Fig. 5. Average OSPA(1) metric different methods used. The second method corresponds to tracking with Kalman Filters using the Nearest Neighbour association criteria.

The *Crowded* scenario was further analyse as it is the most challenging. In Table 2 the OSPA errors for the crowded scenarios are presented. It can be observed that the proposed method has an overall better OSPA metric than any other approach. This comes from a better cardinality component, which is the main expected result.

To further analyse the scenarios, the CLEAR MOT index is evaluated, as it is more informative on the sources of the errors. In Table 3 the CLEAR MOT indexes are shown for the different scenarios. In the first scenario shown in Table 3, the proposed approach has a clear performance advantage over the other

Table 2. OSPA metric for the very crowded scenario. (1) is the JPDA tracker with laser leg detections. (2) is the Nearest Neighbour tracker. (3) is the standard PHD filter with leg detections. (4) is the proposed approach with laser leg detections.

Method	OSPA(1)	OSPA(1) cardinality	OSPA(1) location
(1)	0.72	0.56	0.16
(2)	0.73	0.57	0.16
(3)	0.76	0.62	0.14
(4)	**0.68**	**0.46**	**0.21**

approaches in all metrics except in the misdetection rate compared to the PHD filter. For the *Large hall* scenario, although an overall improve in performance, an increase in the misdetection rate is observed. The two most challenging scenarios for the proposed approach are the *Entrance* and *Very crowded* scenarios. Both of these scenarios are the most crowded ones and a decrease in performance, compared to previous scenarios is me appreciated. This is correlated with results seen in the OSPA metric. For the *Entrance* scenario, the proposed method has the best MOTA and MOTP values, only surpassed on the misdetection rate by the standard PHD filter. The original PHD filter does not naturally handle tracks ID, which explain the increase in ID switches in the two last scenarios. In the *Crowded* scenario, the MOTP of the proposed algorithm (method 4) is comparable to other methods, while the MOTA is much improved. In this case the misdetection rate of the proposed approach has the highest value, but still manages to obtain the best MOTA index.

Finally, in Figure 6 some qualitative results are shown. These images are obtained by projecting the tracking results into the image recorded from the robot. In Figure 6(a) it can be appreciated that two tracks are created for one person, but the person moving towards the robot does not get a track assigned. Using the proposed approach, Figure 6(b), this is handled correctly. In Figure 6(c) two targets were created for the left most person in the scene and one extra false positive at the right most part of the image. From the results show

    (a)   (b)   (c)   (d)

Fig. 6. This figures show some qualitative results. Figures 6(a) and 6(c) show results for the JPDA based tracker. Figures 6(b) and 6(d) show results for the proposed approach. Figures 6(a) and 6(b) are extracted from the *Long corridor*, while Figures 6(c) and 6(d) are from the *Large Hall* scenario.

Table 3. MOTP and MOTA metrics for the different scenarios

Method	MOTP(m)	MOTA	Miss detec. rate	False pos. rate	ID Switches
Long corridor					
(1)	0.24	-66.26%	36.85%	126.34%	438
(2)	0.23	-72.98%	35.58%	133.99%	494
(3)	0.17	-93.00%	16.29%	173.29%	529
(4)	**0.17**	**23.12%**	**25.57%**	**48.33%**	**434**
Large hall					
(1)	0.18	-53.91%	20.36%	129.82%	785
(2)	0.18	-33.67%	16.78%	112.96%	716
(3)	0.16	-78.69%	16.39%	159.95%	492
(4)	**0.16**	**15.47%**	**23.93%**	**58.38%**	**465**
Entrance					
(1)	0.21	-54.64%	23.15%	127.83%	1058
(2)	0.19	-83.43%	23.45%	157.05%	304
(3)	0.16	-71.84%	16.03%	152.69%	905
(4)	**0.17**	**-0.59%**	**22.02%**	**75.50%**	**893**
Crowded					
(1)	0.19	-43.52%	17.99%	122.21%	747
(2)	0.19	-47.67%	17.78%	126.51%	752
(3)	0.22	-78.44%	17.81%	153.86%	1487
(4)	**0.22**	**-14.61%**	**24.45%**	**83.69%**	**1411**

previously, it is not strange to expect a greater number of false positive using this approach. On the other hand, Figure 6(d) shows that the proposed approach, manage to filter out the noise measurements, but has failed to maintain one track for the second leftmost person.

4 Conclusions

In this paper a framework for tracking people on a mobile robot in crowded environments by using a laser range sensor has been proposed. The solution is based on the PHD filter, making use of the clear separation between track initialisation and tracking. Track initialisation is carried out only using the People Detector Module output. Continuous tracking is achieved by using more stable but noisier measurements from the Clustering Module. The proposed approach was tested in different challenging crowded scenarios and different metrics are used to estimate its performance. Results show that the proposed method has an improved performance against previous methods according to misdetection rates and false positive rates. Moreover, it is shown that this approach manages to overcome some of the problems when using the PHD filter on its own. All the results shown in this paper are generated by real time implementation.

Complementing the laser with other types of sensor is an interesting extension to the proposed method. Another interesting research would be asses this method against a concrete task of human robot interaction.

References

1. Bellotto, N., Hu, H.: Multisensor-based human detection and tracking for mobile service robots. Trans. Sys. Man Cyber. Part B 39, 167–181 (2009)
2. Schulz, D., Burgard, W., Fox, D., Cremers, A.B.: People tracking with mobile robots using sample-based joint probabilistic data association filters. The International Journal of Robotics Research 22(2), 99–116 (2003)
3. Luber, M., Stork, J.A., Tipaldi, G.D., Arras, K.O.: People tracking with human motion predictions from social forces. In: 2010 IEEE International Conference on Robotics and Automation (ICRA), pp. 464–469 (May 2010)
4. McKeague, S., Liu, J., Yang, G.-Z.: Hand and body association in crowded environments for human-robot interaction. In: IEEE International Conference on Robotics and Automation (ICRA) (2013)
5. Mitzel, D., Leibe, B.: Real-time multi-person tracking with detector assisted structure propagation. In: IEEE International Conference on Computer Vision Workshops (ICCV Workshops), pp. 974–981 (November 2011)
6. Mitzel, D., Horbert, E., Ess, A., Leibe, B.: Multi-person tracking with sparse detection and continuous segmentation. In: Daniilidis, K., Maragos, P., Paragios, N. (eds.) ECCV 2010, Part I. LNCS, vol. 6311, pp. 397–410. Springer, Heidelberg (2010)
7. Munaro, M., Basso, F., Menegatti, E.: Tracking people within groups with RGB-D data. In: 2012 IEEE/RSJ International Conference on Intelligent Robots and Systems (IROS), pp. 2101–2107 (October 2012)
8. Arras, K.O., Mozos, O.M., Burgard, W.: Using boosted features for the detection of people in 2d range data. In: 2007 IEEE International Conference on Robotics and Automation, pp. 3402–3407 (April 2007)
9. Mahler, R.P.S.: Multitarget bayes filtering via first-order multitarget moments. IEEE Transactions on Aerospace and Electronic Systems 39(4), 1152–1178 (2003)
10. Vo, B.-N., Singh, S., Doucet, A.: Sequential monte carlo implementation of the phd filter for multi-target tracking. In: Proceedings of the Sixth International Conference of Information Fusion, vol. 2, pp. 792–799 (2003)
11. Vo, B.-N., Ma, W.-K.: The gaussian mixture probability hypothesis density filter. IEEE Transactions on Signal Processing 54(11), 4091–4104 (2006)
12. Houssineau, J., Laneuville, D.: PHD filter with diffuse spatial prior on the birth process with applications to GM-PHD filter. In: 2010 13th Conference on Information Fusion (FUSION), pp. 1–8 (July 2010)
13. Schuhmacher, D., Vo, B.-T., Vo, B.-N.: A consistent metric for performance evaluation of multi-object filters. IEEE Transactions on Signal Processing 56(8), 3447–3457 (2008)
14. Bernardin, K., Stiefelhagen, R.: Evaluating multiple object tracking performance: The CLEAR MOT metrics. J. Image Video Process (January 2008)

Who, How, Where: Using Exemplars to Learn Social Concepts

Alan R. Wagner[1] and Jigar Doshi[2]

[1] Georgia Institute of Technology Research Institute, Atlanta GA
[2] Georgia Insitute of Technology, Dept. of Electrical & Computer Engineering, Atlanta, GA
alan.wagner@gtri.gatech.edu, jdoshi8@gatech.edu

Abstract. This article introduces exemplars as a method of stereotype learning by social robot. An exemplar is a highly specific representation of an interaction with a particular person. Methods that allow a robot to create prototypes from its set of exemplars are presented. Using these techniques, we develop the situation prototype, a representation that captures information about an environment, the people typically found in an environment, and their actions. We show that this representation can be used by a robot to reason about several important social questions.

Keywords: exemplars, stereotype, prototype, social learning, user modeling.

1 Introduction

If robots are to ever act and interact in a wide variety of social contexts and with a wide variety of people, then these robots will need computational techniques that allow them to adapt and learn from their social experiences. Social robots will need to reason about who will be found at a particular location, where a particular type of person can be found, and how people will act in a given context. Consider, as a motivating example, a robot tasked with searching for a teenager in a crowded mall. Having a model that includes the interests and disinterests of a typical teenager will allow the robot to rule out large portions of the search space and make the problem more tractable by focusing on areas likely to result in locating the missing youth. As a further example, consider a robot looking for a human partner with a particular skill set, such as those of an EMT. Can a robot use knowledge gained by interacting with different people to predict a likely place to find people with a specific set of skills? In this paper we develop techniques that allow a robot to reason and predict facets of social interaction such as these.

Our research has focused on the development of a principled, formal framework for social action selection that allows a robot to represent and reason about its interactions with a human partner [1]. This framework operates by modeling the robot's interactive partner and using its model of the partner to predict the partner's preferences and guide its social interaction. The research presented here contributes a key extension of our framework into the domains of concept learning and mental

G. Herrmann et al. (Eds.): ICSR 2013, LNAI 8239, pp. 481–490, 2013.

simulation. In previous work we developed a method for creating stereotypes based on a prototype model of concept learning [2]. This work indicated that prototype-based stereotypes could be used to bootstrap the process of learning about a new person, used to preselect perceptual features important for recognition, and predict a person's appearance from their behavior.

Nevertheless, prototype-based stereotypes have important limitations such as their lack of context and details. These deficits have motivated our exploration of exemplars as an alternative model of social concept learning. The research presented here focuses on the development of methods designed to create exemplars from a robot's sensor data, create prototypes from a set of exemplars, and produce exemplars and prototypes which are context specific. We demonstrate that our approach can be used by a simulated robot to make predictions about several important aspects of a social environment, such as what types of people the robot will find in a given context and how people will act.

This article makes two key contributions. First, it begins to explore the development of categories related to context and the person's actions from a robot's interactive experience. This contribution has the potential to allow a social robot to reason about where certain categories of people can be found and the types of behavior these people exhibit within a social environment. Second, we begin to develop the representational and theoretical underpinnings that will allow a robot to utilize the exemplar model of stereotyping. Exemplars are an important potential representation because they act as an extensive source for generating predictions and inferences about the human partner's interactive behavior. For this research, exemplars are used to predict who is interacting, how they will act, where an interaction will occur. Further, the system we present is not given category labels or even the number of categories a priori. Rather the system learns these characteristics from its experience.

This article begins with a review of related work. Next, exemplar stereotypes are introduced followed by our techniques for creating representations that include context specific information. Experiments using this system as well as results are presented next. We conclude with a discussion of the results, their limitations, and provide directions for ongoing and future work.

2   Related Work

Stereotypes and stereotyping have long been a topic of investigation for psychologists. Schneider provides a good review of the existing work [3]. Numerous definitions of the term stereotype exist. Edwards defines a stereotype as a perceptual stimulus which arouses standardized preconceptions that, in turn, influence one's response to the stimulus [4]. Smith and Zarate describe three general classes of stereotype models: attribute-based, schematic-based, and exemplars [5].

With respect to computer science, the inclusion of techniques for stereotyping is not new. Human Computer Interaction (HCI) researchers have long used categories and stereotypes of users to influence aspects of user interface design [6]. The multi-agent systems community has also explored the use of stereotypes. Ballim and Wilks use stereotypes to generate belief models of agents [7].

Investigations of stereotyping with respect to robots are comparatively scarce. Fong et al. used predefined categories of users in conjunction with a human-robot collaboration task [8]. These categories influenced the robot's dialogue, actions, the information presented to the human, and the types of control afforded to the user. Duffy presents a framework for social embodiment in mobile autonomous systems [9]. His framework includes methods for representing and reasoning about stereotypes. He notes that stereotypes serve the purpose of bootstrapping the evaluation of another agent and that the perceptual features of the agent being stereotyped are an important representational consideration.

3   Exemplar Stereotypes

The exemplar approach to stereotyping emphasizes concrete rather than abstract representations [5]. Exemplars were developed as a means for reflecting the specificity and detail often associated with a person's cognitive representations. Exemplar researchers tend to stress that these memories are stored as highly detailed mental images and that the inclusion of these details serves as an additional source of information for making inferences. For instance, if one is asked to think of a child the mental image that results tends to be a specific memory trace related to a child that the person has recently or often interacted with. Exemplar categorization is achieved by comparing new instances of the perception of a person to stored representations of previously encountered individuals. Different dimensions of comparison can be weighted more heavily to reflect the robot's social motivation or objective [3].

Table 1. Features and values used in experiments

Partner Feature	Possible Values
Gender	Male, female
Age	Baby, child, teen, middle-aged, senior
Skin color	Pale white, white, dark white, brown, black
Skin texture	Smooth, wrinkled
Height	short, medium, tall
Hair style	bald, short, medium, long, bald
Hair color	brown, black, white, gray, red, blonde
Weight	thin, medium, heavy
Shirt style	collared t-shirt, t-shirt, long sleeve, collared long-sleeve, sweater, coat, no sleeves
Shirt/pant color	black, white, green, blue, red, yellow, brown, gray, multi-color
Glasses/facial hair	yes, no
Pant style	long, shorts, jeans, skirt, pajamas
Shoe style	tennis, sandal, high-heels, flip-flop, boot, slipper, no-shoes; dress
Actions	Eating, sitting, talking, drinking, walking, listening, watching, playing, writing, reading, coloring, standing, building, answering, running, singing, kicking, sleeping, teaching, hitting
Context Features	Chalkboard, clock, TV, book, wheelchair, food, fork, spoon, desk, chair, computer, pencil, wipe board, paper, toy, window, game, plants, table, art, plate, toilet, sand, cart, bicycle, car, ladder, bed, fence, glass

For a social robot, exemplars should ideally be represented as a collection of images or video segments capturing the robot's interaction with a person. In order for newly created exemplars to be compared to previously encountered exemplars, a process must exist during which the robot uses perceptual recognition and classification techniques to capture feature/value pairs related to the exemplar. These feature/value pairs are stored as a set and serve to discretely represent the exemplar. We term this intermediate representation an attributized exemplar. Table 1 shows our list of feature/value pair used during this experiment.

The creation of the attributized exemplar allows for the use of set distance measures, such as the Jaccard index, to be used with the representation. The Jaccard index,

$$j\left(m^A, m^B\right) = \frac{\left|m^A \cup m^B\right| - \left|m^A \cap m^B\right|}{\left|m^A \cup m^B\right|} \tag{1}$$

gauges the distance from one exemplar to another. The Jaccard index is used in conjunction with agglomerative or "bottom-up" clustering to determine if two exemplars are sufficiently similar to warrant their merging. Clustering affords a means for creating prototypes from exemplars. Agglomerative clustering operates by initially assuming each exemplar as a unique cluster and then merging exemplars which are sufficiently close to one another. The centroid that results from merging these clusters serves as a prototype representing a general category of individuals. As the distance required to merge decreases from 1 to 0, the generality of the category and concept being represented increases (Fig. 1).

From Exemplars to Prototypes

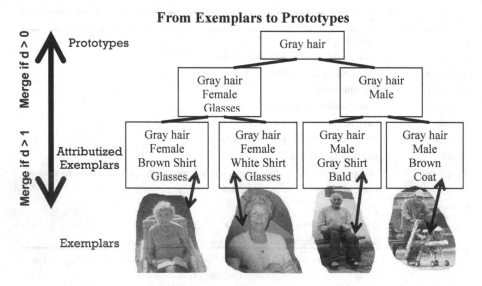

Fig. 1. The figure above depicts the transformation from exemplar to attributized exemplar to prototype. Exemplars are converted to attributized exemplars through a series of feature detectors. An attributized exemplar represents an exemplar as a set of feature value pairs. Agglomerative clustering results in increasingly general prototypes depending on the distance required to merge individual clusters.

3.1   Situation Prototype Representation

Consider, as a motivating example, a robot entering an elementary school classroom (Fig. 2). The people that the robot would likely encounter would typically be children. Perceptually these children would be shorter than adults, tend to have smooth skin, little facial hair, and wearing tennis shoes. Further, the people in this environment would utilize a specific set of actions such as raising their hand before speaking. Finally, the environment itself would include specific objects such as desks, bulletin boards, and books. A representation, which we call a situation prototype, consisting of a set of context features and a prototype of the type of person found in the environment captures the type of people in an environment, but also their behaviors and the features describing the context itself.. Context features relate to either salient objects or characteristics of the environment. In a classroom, for example, desks, chairs, books, bulletin boards, etc. could serve as context features.

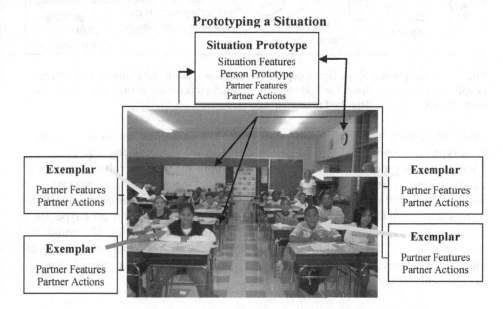

Fig. 2. The figure depicts the creation of a situation prototype from image information. Context features in the form of objects, such as clocks, desks, etc. are captured from the environment. Exemplars are created from the people in the environment. The exemplars are then used to create a person prototype consisting of prototypical partner features and actions. The lower portion of the figure depicts the process used to answer a question about the social environment. For example, the question "where do people look like this?" is examined by creating a list of partner features representing the person in question, for each context cluster the distance is calculated to determine the most similar target, and finally the cluster's context features are returned.

To create a situation prototype, first exemplars and attributized exemplars are created from several or all of the people found in the environment. The exemplars that are created capture both the person's perceptual features (feature/value pairs related to what the person looks like) and the action or actions they are performing in the

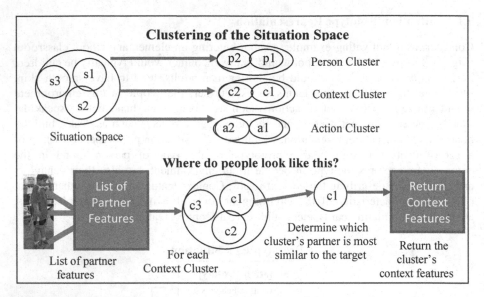

Fig. 3. The top portion of the figure depicts the situation space being clustered with respect to people, contexts and actions. The bottom portion then depicts the use of these clusters to reason about the six questions described in section 3.2.

environment (see Table 1). Next, these individual exemplars are used to create a prototype representing the typical person found in the environment. The process of creating a prototype from a set of exemplars within the scene involves determining which partner features and/or actions should be included in the prototype. A feature/action voting scheme in which individual features or actions receive votes from each exemplar in the scene was used to create the person prototype for a situation. Finally, context features are collected and added to a set. The result is a representation of the situation which includes information about what a typical member of the environment looks like, acts like, and the objects or characteristics of the environment itself (Fig. 2).

3.2   Six Questions

Given a robot with a collection of situation prototypes representing the individuals, contexts, and actions encountered in specific situations, we hypothesized that one could cluster over this situation space to generate centroids reflecting the concepts of who, how, and where people, actions, and environments could be found. More specifically, clustering over the situation space with respect to the person prototype portion of the representation generates centroids reflecting the types of people encountered irrespective of the environment. Clustering with respect to the context features, on the other hand generates concepts specific to particular environments. Finally clustering with respect to the action sets reflects the behaviors that occur independent of the environment. The top portion of Fig. 3 depicts this clustering of the situation space.

The process described thus far, (1) creating situation prototypes that reflect a social context (Fig. 2), and (2) clustering the resulting situation space with respect to people, context and actions (Fig. 3 top), results in concepts describing categories of people, places and behavior. As a final step, one can measure the distance from a particular exemplar to each context cluster (for example), select the closest cluster, and use this information to determine which context features are most likely to be associated with this type of person (e.g. Fig. 3 bottom). In other words, a robot can use the system to answer questions like, "where do people look like this?" where "this" refers to an exemplar indicating a person's appearance. Overall, we hypothesized that the system could be used by a robot to reason about the following six questions:

1. **Where do people look like this?** Given a set of partner features representing a person predict a set of context features detailing the type of environment in which people with this appearance are likely to be found.

2. **Where do people act like this?** Given a set of actions, predict a set of context features describing the environment in which these actions occur.

3. **What do people look like that act like this?** Given a set of actions predict a set of partner features describing the appearance of people that perform these actions.

4. **What do people look like here?** Given a set of context features predict a set of partner features describing the appearance of people in this environment.

5. **How do people act here?** Given a set of context features predict a set of actions likely to be performed in this context.

6. **How do people that look like this act?** Given a set of partner features predict a set of actions likely to be performed by people that look like this.

A procedure that allows a robot to generate predictions related to these six questions could afford significant advantages. For instance, such a system could afford a robot operating in a search and rescue mission to use information about the victims being found to predict its current location. For example, if many of the victims are elderly then the search location may be a nursing home. Alternatively, if the robot knows its current context than it can use this procedure to predict how to search based on predictions about the type of people found in that context. In order to test the success of this procedure an empirical evaluation was performed.

4   Empirical Evaluation

Our empirical evaluation examined whether and to what extent situation prototypes could be created and used to answer the preceding six questions. Three different contexts were used: nursing homes, elementary school classrooms, and restaurants. We believed that clustering situation prototypes with respect to the context features based on several classrooms, nursing homes, and restaurants would create a centroid representing a classroom that is distinct from a nursing home and restaurant. The resulting centroid would itself be a situation prototype and as such contain prototype information about the appearance of the person typically found in this environment and their behavior.

Situation prototypes were generated from images. Thirty photos from each environment were obtained by searching Google images for the key words "classroom", "restaurant" and "nursing home". Images that depicted the environment in its natural state were selected (e.g. images with people posing or obvious mismatches were rejected).

Amazon.com's Mechanical Turks were used create attributized exemplars from the images. The Amazon Mechanical Turk is a crowdsourcing Internet marketplace where the *Requesters* pay a pre-determined amount to the *Workers* for satisfactorily completing the Human Intelligence Task (HIT). Our study used only workers who were categorization masters and had at least a 90% acceptance rate. Images were prepared by marking a person in the image that the worker answered questions about. We marked people in the image whose faces were obviously visible and could have been easily detected by simple face recognition algorithms. To ensure consistency throughout the dataset, only 4 faces per image were selected. Workers were paid 50 cents for successfully completing a survey describing the person's features, actions, and context features.

The HITs that workers were asked to complete included of an image taken from a classroom, a nursing home, or a restaurant. In the scene an arrow pointed to a person about whom information is sought. For each marked individual in the image, questions were asked about the person's physical appearance and attire. Workers were also asked to select between 1-3 actions from a pre-defined set of 20 actions that the person was likely performing in the scene. Finally, questions were asked about the context such as the floor type, lighting and about the objects present in the room. From a pre-defined list of 30 objects, workers selected a minimum of three and a maximum of five objects. For each marked individual we collected independent surveys from three different workers. Table 1 lists the set of partner features, actions, and context features collected. All surveys that 1) had any answers that matched more than one possible value or did not match any possible values or 2) had a missing value for any feature (i.e. no response) were rejected. Any answers that were clearly typographical errors, had mismatched cases, and missing or additional special characters were accepted. Actions and context objects were selected from a dropdown menu. All entries that did not match the minimum quantity requirement (one for actions and three for context) were rejected.

Duplicate surveys were used to identify erroneous or random responses by the workers. The data relating to the perceptual features of a person depicted in an image, their actions, and characteristics of the context was aggregated by assigning votes to each of the HITs. Feature values with more than 50% agreement were accepted.

5   Experiments

Experiments were conducted in simulation and used the data generated by the workers. For each question from section 3.2, sixty people were randomly selected from the data. Depending on the question, either the features describing the person's appearance, the person's set of actions, or the features describing the context were used as input. Next, the distance from this input to each of the clusters was calculated. Finally the cluster with the smallest distance was used to predict the features in question. For example, the question, "where do people look like this?" takes as input a set of partner features describing a person's appearance (see Fig. 3 bottom). These features are compared to the person prototypes in each context cluster. The context features from the cluster with the smallest distance were used to predict the environment in which the person would likely be found. This prediction is then compared with the actual data to determine the correctness of the prediction.

Two types of controls were used to for comparison. A mode control used the most common person prototype, context, and action set to predict the answers to the questions posed in section 3.2. A random control used a partner prototype created from random features, a random action set, and random context features.

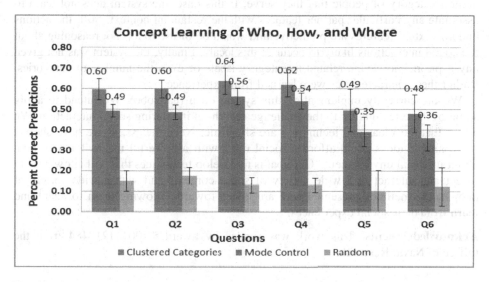

Fig. 4. The results from the experiment are depicted above. The bar to the left (blue) depicts the experimental condition. The center and rightmost bars depict the control conditions. The graph depicts results for each of the six questions from section 3.2.

The results from this experiment are depicted in Fig. 4. The experimental procedure is depicted by the bars on the left in blue. The middle bars depict the mode control and the random control is on the right. The experimental procedure performed significantly better than the both controls on Q1-Q4 ($p<0.04$ for all), yet did not perform better than the mode control on Q5 and Q6. Question 5 examines the type of actions performed in a context and question 6 examines the actions performed by a particular type of person. The lack of significance for these questions likely stems from the fact that the use of single images made distinguishing different actions from one another very difficult for the workers. Because of this, clustering the situation prototypes with respect to the action set lead to only a single cluster. We strongly believe that the use of movies or actual robots capable of detecting motion would result in better predictions of the partner's actions.

6   Conclusions

We have presented a system that uses exemplars in conjunction with prototypes to create a representation that captures the typical person and actions in an environment.

A technique that clusters this situation prototype representation and matches a person's appearance, actions, and context to a cluster centroid was described as a method affording a means for reasoning about important social questions.

The results presented make assumptions that would need to be addressed in order to use the system on a robot. First, clearly the use of Mechanical Turk workers would need to be replaced with the robot's own perceptual system. We have developed several classifiers toward this goal. Further, the procedure learns the category of people in context if such a category exists. Restaurants, for example, do not have a natural category of people that they serve. In this case, the system does not learn to associate any particular partner features with the restaurant context. Still, the actions and contextual features are predictive and would be of use to a robot reasoning about the social interactions likely to occur at this locale. Finally, the system was not given any a priori knowledge related to category labels or even the number of categories. Rather these characteristics were learned from experience.

We are currently implementing this system on a real robot. We intend to apply some of these techniques to the challenge of turn-taking during social interaction. We admit that the ideas and techniques are simplistic. Nevertheless, as we have shown, these simple ideas could afford a social robot with a powerful means for reasoning about its social environment. Our goal is to develop techniques that will allow a robot to act and interact in a wide variety of social contexts and with a wide variety of people, the methods presented here are a step towards allowing them to adapt and learn from their social experiences.

Acknowledgements. This work was funded by award #N00141210484 from the Office of Naval Research.

References

1. Wagner, A.R.: Creating and Using Matrix Representations of Social Interaction. In: The 4th ACM/IEEE International Conference on Human-Robot Interaction, San Diego CA (2009)
2. Wagner, A.R.: Using Cluster-based Stereotyping to Foster Human-Robot Cooperation. In: Proceedings of IEEE International Conference on Intelligent Robots and Systems (IROS 2012), Villamura, Portugal (2012)
3. Schneider, D.J.: The Psychology of Stereotyping. The Guilford Press, New York (2004)
4. Edwards, A.: Studies of Stereotypes: I. The directionality and uniformity of responses to stereotypes. J. of Soc. Psy. 12, 357–366 (1940)
5. Smith, E.R., Zarate, M.A.: Exemplar-Based Model of Social Judgement. Psy. Rev. (1992)
6. Rich, E.: User Modeling via Stereotypes. Cognitive Science 3(197), 329–354 (1979)
7. Ballim, A., Wilks, Y.: Beliefs, stereotypes, and dynamic agent modeling. User Modeling and User-Adapted Interaction 1(1), 33–65 (1991)
8. Fong, T., Thorpe, C., Baur, C.: Collaboration, dialogue, and human–robot interaction. In: Proceedings of the International Symposium on Robotics Research (2001)
9. Duffy, B.: Robots Social Embodiment in Autonomous Mobile Robotics. International Journal of Advanced Robotic Systems 1(3) (2004)

An Asynchronous RGB-D Sensor Fusion Framework Using Monte-Carlo Methods for Hand Tracking on a Mobile Robot in Crowded Environments

Stephen McKeague, Jindong Liu, Member, IEEE,
and Guang-Zhong Yang, Fellow, IEEE

The Hamlyn Centre, Imperial College London, UK
{sjm05,jliu4,gzy}@imperial.ac.uk

Abstract. Gesture recognition for human-robot interaction is a prerequisite for many social robotic tasks. One of the main technical difficulties is hand tracking in crowded and dynamic environments. Many existing methods have only been shown to work in clutter-free settings.

This paper proposes a sensor fusion based hand tracking algorithm for crowded environments. It is shown to significantly improve the accuracy of existing hand detectors, based on depth and RGB information. The main novelties of the proposed method include: a) a Monte-Carlo RGB update process to reduce false positives; b) online skin colour learning to cope with varying skin colour, clothing and illumination conditions; c) an asynchronous update method to integrate depth and RGB information for real-time applications. Tracking performance is evaluated in a number of controlled scenarios and crowded environments. All datasets used in this work have been made publicly available.

1 Introduction

Social robots are set to play an important role in our lives, with applications ranging from navigation to home-based companionship tasks. To command and control these robots to interact with people and the environment, effective human-robot interaction (HRI) must be implemented.

Gesture recognition is an important aspect of HRI, providing a means for an untrained human to issue instructions to a robot. It has been an active area of research for many years. However, limited work has been successfully translated from lab-based research into real-world applications. Real-world environments pose many challenges, such as crowds and changing dynamics of moving objects.

The introduction of depth cameras has initiated a host of new solutions to these problems. In particular, body part detection has seen much improvement [1–4]. However, in the recent ChaLearn Gesture Challenge [5], no top ranking method tracked individual body parts. Clearly there is a need for robust hand tracking solutions capable of performing in real-world environments.

G. Herrmann et al. (Eds.): ICSR 2013, LNAI 8239, pp. 491–500, 2013.

Hand tracking methods can be divided into two categories [6]: appearance-based [1–4] and model-based methods [7]. Appearance-based methods are generally faster, but only recognise a discrete number of hand poses. They represent a good choice for HRI due to real-time requirements. However, problems such as occlusions and computational cost are exacerbated in crowded and dynamic environments. In order to compensate for misdetections and reduced frame rates, this paper proposes a tracking algorithm that augments the depth detector with the RGB information available from RGB-D cameras.

Several appearance-based methods for hand detection in depth images have been proposed in recent research. In the body-part detector used by Microsoft's Xbox Kinect [1], each pixel is classified based on depth differences of neighbouring pixels using a random forest classifier. An alternate keypoint-based approach has been described by Plagemann et al. [3]. Keypoints are calculated based on finding the maximum geodesic distances from nodes in a surface mesh. Chen et al. [8] and Bretzner et al. [9] describe two algorithms for tracking a rigid hand in RGB images, using a single particle filter. Sigalas [10] uses a similar approach to track a subject's arms in RGB images, to perform full gesture recognition.

Oikonomidis et al. [7] recently presented an RGB-D, model-based approach, to track an articulated hand. The method generates a hand-model solution by comparing the observed depth map to a depth map rendered from an initial hand hypothesis. Wang and Popović [11] described an alternate method for tracking an articulated coloured glove using RGB images. A prior image database of hand poses is created for the coloured glove. The optimal pose for an input image is then given by its closest match in the image database.

This work builds upon a previous paper [2], which describes a framework for hand and body association in crowded environments using a depth camera. Hands are detected with a low number of false positives, and tracked using a Kalman filter. Similarly to [3], the method runs at around fifteen frames per second. Whilst the proposed sensor fusion tracker can work with any depth-based hand detector, these are the two main methods considered.

The proposed sensor-fusion algorithm has several novelties. First, it uses RGB information in order to provide position updates when the underlying depth-based detector fails. The method ensures accurate position updates at a 30Hz rate, regardless of the speed of the underlying depth detector. Unlike most RGB hand trackers the algorithm dynamically learns per-subject skin colour. This ensures that the system works simultaneously with people of different races, and is robust to illumination conditions and worn clothing. A novel asynchronous sensor update method is used to combine both input sources together.

2   Method

A schematic system structure of the sensor fusion algorithm is displayed in Figure 1. Rounded boxes denote functional components of the algorithm. Square boxes represent inputs and outputs. Boxes within the dashed line are components of the proposed tracker. Boxes outside represent the algorithm's two inputs: an

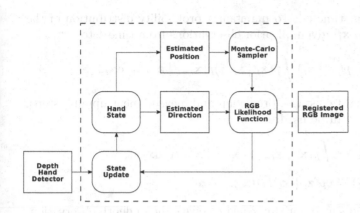

Fig. 1. System structure diagram of the sensor fusion algorithm. Rounded boxes denote functional components. Square boxes represent inputs and outputs. Boxes within the dashed line are components of the proposed tracker. Boxes outside represent the algorithm's two inputs.

RGB image registered to a corresponding depth image, and a hand detector that uses the depth image to return a single 3D position and direction for each hand.

The proposed tracker is comprised of four main components. The hand state stores a hand's position and direction estimates. Monte-Carlo sampling is used to analyse information local to a hand. An RGB likelihood function is applied to weigh the probability of samples belonging to the hand. A state update method asynchronously combines depth and RGB measurement updates to maintain a temporally consistent hand state.

The choice of state representation is partly based on our previous paper [2]. As the most natural likelihood function for a depth-based hand detector is Gaussian, the sensor fusion algorithm stores the hand state using a Kalman filter framework with a Brownian noise motion model. This is more computationally efficient than a non-parametric approach, such as a particle filter. A hand state, $\mathbf{s}$, is modelled with a 3D position, $\mathbf{x} = [x, y, z]^T$, and unit vector angle, $\vec{\mathbf{x}} = [ux, uy, uz]^T$: $\mathbf{s} = [x, y, z, ux, uy, uz]^T$.

The Kalman filter can provide state predictions at any time according to its motion model. Due to the unpredictable nature of hand motion during HRI gestures, the most appropriate motion model is Brownian noise. However, this will not provide a better estimate of the hand's real position than the most recent hand detection in the depth image. RGB information from the current frame can give a better state estimate than Brownian noise. Therefore, the algorithm has two aims: use RGB information to compensate for hand misdetections in depth images, and better predict hand motion with a 30Hz RGB update process.

2.1   Monte Carlo RGB Measurement Update

The proposed tracking algorithm adapts the well-known Bayes filter [12] to fuse measurement updates from two sources with different characteristics. The first is a hand detector using depth images, that provides a single value hand estimate with high precision but lower speed. The second is a hand likelihood function over the whole RGB image that is computed quickly, but with higher false positives.

The goal of the filter at time t is to maintain a probability distribution of the posterior state position $\mathbf{x}_t$, given all prior observations from time 1 to t, $\mathbf{z}_{1:t}$.

$$p(\mathbf{x}_t \mid \mathbf{z}_{1:t}) \propto p(\mathbf{z}_t \mid \mathbf{x}_t) \int p(\mathbf{x}_t \mid \mathbf{x}_{t-1}) p(\mathbf{x}_{t-1} \mid \mathbf{z}_{1:t-1}) \, d\mathbf{x}_{t-1}.$$

To facilitate the usual two stage, prediction and measurement update, steps, the following notation shall be used:

$$\overline{bel}(\mathbf{x}_t) = \int p(\mathbf{x}_t \mid \mathbf{x}_{t-1}) p(\mathbf{x}_{t-1} \mid \mathbf{z}_{1:t-1}) \, d\mathbf{x}_{t-1},$$

$$bel(\mathbf{x}_t) = \eta p(\mathbf{z}_t \mid \mathbf{x}_t) \overline{bel}(\mathbf{x}_t).$$

At time t the sensor fusion algorithm could receive either a depth observation, $\mathbf{z}_t^{depth}$, or an RGB observation, $\mathbf{z}_t^{RGB}$. Due to their independence:

$$p(\mathbf{z}_{1:t}) = p(\mathbf{z}_{1:t}^{depth}) p(\mathbf{z}_{1:t}^{RGB}).$$

$p(\mathbf{z}_t^{depth} \mid \mathbf{x}_t)$ is the depth measurement probability at time t given our latest predicted state, $\mathbf{x}_t$. As detailed before, this is a Gaussian probability density function (PDF) over all $\mathbf{x}_t$. The first challenge in the proposed sensor fusion algorithm is to model the RGB measurement probability at time t, $p(\mathbf{z}_t^{RGB} \mid \mathbf{x}_t)$.

The task of robustly detecting hands in RGB images is much more difficult than using depth images. A common method is to perform per-pixel skin detection [13], and then choose a somewhat arbitrary blob size. Blobs above this size would be considered hands, resulting in many false positives. Additionally, misdetections would result from subjects with darker skin colours. Partial occlusions and varying sleeve lengths would cause much frame to frame instability in calculating a single point to represent the detected hand.

By taking these disadvantages into consideration, we have avoided modelling $p(\mathbf{z}_t^{RGB} \mid \mathbf{x}_t)$ as the usual Gaussian PDF. Given an initial state estimate by the depth detector, there is a high likelihood that the hand will be nearby in a subsequent time period. By directly analysing adjacent surroundings for a skin coloured region, we could get an efficient and reliable estimate of how far the hand has moved. Sampling from the latest prediction PDF, $\overline{bel}(\mathbf{x}_t)$, provides a good way of evaluating this nearby information.

As noted in Figure 1, every RGB image is registered to a corresponding depth image. Thus each pixel will have six components, (x, y, z, r, g, b). Each sample, m of M, will have two defined properties. The first is a 3D position, $\mathbf{x}_t^{[m]} = [x, y, z]$, drawn from the latest prediction PDF. If the sample does not lie within a threshold, ϵ, of its nearest registered RGB pixel, $\mathbf{c}_t^{[m]}$, then it is discarded. The second sample property is its weight, $w_t^{[m]}$. This is based on the skin colour probability of the closest pixel.

$$\text{sample } \mathbf{x}_t^{[m]} \sim \overline{bel}(\mathbf{x}_t), \qquad\qquad \text{if } \|\mathbf{x}_t^{[m]} - \mathbf{c}_t^{[m]}\| < \epsilon,$$

$$w_t^{[m]} = p(\mathbf{z}_t^{RGB} \mid \mathbf{c}_t^{[m]}), \qquad\qquad m = 1 \ldots M.$$

Provided the sample weights are normalised, the complete sample set represents the non-parametric posterior PDF, $p(\mathbf{x}_t \mid \mathbf{z}_t^{RGB}, \mathbf{z}_{1:t-1})$. As mentioned previously, it is desirable to keep measurement updates within the Kalman filter framework. As well as computational advantages, it allows the covariance of the posterior PDF to be solely controlled by the more reliable depth detector.

Due to the larger number of false positives in the RGB likelihood function, the covariance of the posterior PDF after an RGB update, $p(\mathbf{x}_t \mid \mathbf{z}_t^{RGB}, \mathbf{z}_{1:t-1})$, will vary drastically from the covariance after a depth update, $p(\mathbf{x}_t \mid \mathbf{z}_t^{depth}, \mathbf{z}_{1:t-1})$. This will result in a widely differing RGB sample distribution in successive frames. Controlling covariance using only depth detections would have the desirable effect of spreading RGB samples over a wider area, the more depth misdetections a tracker experiences. RGB updates can then be neatly used to give a more accurate hand position estimate in-between depth updates.

For RGB measurement updates, the mean of $p(\mathbf{x}_t \mid \mathbf{z}_t^{RGB}, \mathbf{z}_{1:t-1})$ can be calculated as the weighted sum of samples. The mean position of the tracked state's posterior PDF, $E(bel(\mathbf{x}_t))$, should then be set to this value. The samples are then discarded and redrawn in subsequent RGB frames.

$$\hat{w}_t^{[m]} = \left[\sum_{n=0}^{M} w_t^{[n]}\right]^{-1} w_t^{[m]},$$

$$E(bel(\mathbf{x}_t)) = \sum_{m=0}^{M} \mathbf{x}_t^{[m]} \hat{w}_t^{[m]}.$$

2.2 RGB Likelihood Function

Having described how to incorporate the RGB measurement probability, $p(\mathbf{z}^{RGB} \mid \mathbf{x})$, into the sensor fusion algorithm, the task remains of defining the likelihood function that generates it.

Each hand tracker is assigned an initially empty skin colour histogram. To aid illumination invariance, only the hue and saturation are stored [14]. Any time a tracker performs a depth measurement update, the subsequent registered RGB image is used to update its skin colour histogram. At this time, denoted as t, the mean position of the latest prediction PDF, $E(\overline{bel}(\mathbf{x}_t))$, provides the current best estimate of the tracked hand. Its nearest registered RGB pixel is found, $\mathbf{c}_t$. If the distance between these points is below a threshold, ϵ, then the tracker's histogram is updated with the hand's skin colour information.

To isolate skin coloured pixels, all points within a set geodesic distance, max_d, to $\mathbf{c}_t$ are analysed. The geodesic distance between two pixels in the registered RGB frame is equal to the shortest cumulative distance between them, whilst traversing neighbouring pixels on the same surface [2]. The hue and saturation values of the segmented pixels are then used to update the histogram of the appropriate tracker. Figure 2 shows the hand segmentation results of this method. It allows the system to adaptively learn a particular subject's skin colour, and provides invariance to varying illumination conditions.

Fig. 2. Image showing a segmented hand from the geodesic distance analysis. The hue and saturation values of the segmented pixels will be used to update the appropriate tracker's skin colour histogram. This provides the proposed algorithm with illumination invariance and the ability to model various colours of skin.

Using only skin colour analysis, the mean of the RGB likelihood function lies near the palm. However the main depth detectors considered [2,3] provide hand detections around the fingertips. This difference in estimated hand position could cause the tracked state's posterior distribution, $bel(\mathbf{x})$, to unstably fluctuate.

To centre the mean of the RGB likelihood function on the fingertips, two additional likelihood functions are introduced. The first is a distance transform, where the likelihood, p^d, of a registered pixel, $\mathbf{v}$, depends on its 3D distance to the closest edge, $edge(\mathbf{v})$. The second likelihood function is an angular weighting, where likelihood, p^a, depends on a pixel's distance to the hand's direction vector. This vector is the mean direction of the prediction PDF, $E(\overline{bel}(\overrightarrow{\mathbf{x}}))$, from the mean position, $E(\overline{bel}(\mathbf{x}))$.

$$p^d(\mathbf{v}) = e^{-\alpha edge(\mathbf{v})^2},$$
$$p^a(\mathbf{v}) = e^{-\beta|(\mathbf{v}-E(\overline{bel}(\mathbf{x})))\cdot E(\overline{bel}(\overrightarrow{\mathbf{x}}))|^2},$$

where α and β are scaling parameters, controlling the spread of the probability around areas of maximum likelihood.

The combined likelihood function for pixel $\mathbf{v}$, $p(\mathbf{v})$, is simply the product of the three individual likelihood functions. $p^c(\mathbf{v})$ shall represent the skin probability of pixel $\mathbf{v}$, obtained from the current tracker's skin colour histogram. The mean of this combined likelihood function will lie over the fingertips, as desired:

$$p(\mathbf{v}) = p^c(\mathbf{v})p^d(\mathbf{v})p^a(\mathbf{v}).$$

2.3   Asynchronous Measurement Update

With the sensor fusion algorithm defined, the last problem remains to be solved: the processing time for depth detections is longer than RGB detections. Assume that from an image at time t_0, a tracker is created from a depth detection. Figure 3 illustrates the problem graphically. Images are generated at intervals of Δ_t, depth detections take Δ_d to process and RGB detections take Δ_{RGB}. Performing a standard depth update at $t_1 + \Delta_d$ will cause RGB update information from images at t_2 and t_3 to be lost, as $\Delta_{RGB} < \Delta_t$ and $\Delta_d > \Delta_t + \Delta_{RGB}$.

To process the depth detection from Figure 3 in a temporally consistent manner, the tracker's state should first be reverted to what is was at t_1. To accomplish

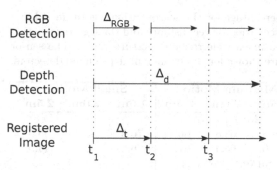

RGB
Detection

Depth
Detection

Registered
Image

Fig. 3. Illustration of the asynchronous measurement problem. Images are generated at intervals of Δ_t, depth detections take Δ_d to process and RGB detections take Δ_{RGB}. Performing a standard depth update at $t_1 + \Delta_d$ will cause RGB update information from images at t_2 and t_3 to be lost, as $\Delta_{RGB} < \Delta_t$ and $\Delta_d > \Delta_t + \Delta_{RGB}$.

this, a history of RGB update information since the last depth update must be maintained. This history should include the image used for the update and its time. In the current example, the most recent depth measurement was at t_0. The history would thus include images from times: t_1, t_2, t_3.

After the new depth update, the tracker's state would be in an estimated position for time t_1, despite the current time being $t_3 + \Delta_{RGB} < t < t_4$. To re-include RGB information from images at t_2 and t_3, RGB updates should be iteratively re-performed using the relevant images in the history. This would put the tracker's state in the most accurate possible estimate for the current time.

3   Results

Four gestures were defined for use in the experiments: A wave for attracting a robot's attention, a subtle push to indicate that interaction has finished, a subtle follow me motion, and a subtle raised hand for stopping the robot's movement. Using the tri-phase gesture model [15], a gesture can be split into three stages: preparation, stroke and retraction. As the preparation and retraction stages contribute little information to the gesture recognition process [10], only ground truth annotations from the stroke phase are considered.

All datasets were recorded with a Microsoft Kinect at a resolution of 640 by 480 pixels, and have been made publicly available[1]. Recordings were manually annotated with 2D ground truth, for automatic results generation. Ignoring a position's z-coordinate, a tracked hand is considered detected if it lies within 0.1m of the ground truth.

Fifty samples are used during the RGB update process. The maximum distance for the closest registered RGB pixel to a point in space, ϵ, is set to 0.15m. The skin colour histogram contains 30 hue bins and 32 saturation bins. Finally, the probability scaling parameters, α and β, are set to 0.5 and 750 respectively.

To ensure that the algorithm's performance improvements are independent of the underlying depth hand detector, three different methods are compared: the hand detector from [2], the shape context hand detector on which it was based, and the method from [3]. For each method, the results of the sensor fusion algorithm are compared with the performance of a Kalman filter.

---

[1] http://www.imperial.ac.uk/hamlyn/eo/gesturedataset

Table 1. Table showing detection percentage for the sensor fusion algorithm (S) and Kalman filter (K) in nine controlled scenarios. Three distance and three gesture motion range categories are analysed, each using a wide range of gesture angles. The sensor fusion algorithm's improved results are shown for three different depth hand detectors.

	Large Motion			Medium Motion			Small Motion		
	1.5m	2.0m	2.5m	1.5m	2.0m	2.5m	1.5m	2.0m	2.5m
				McKEAGUE					
S	79.7	70.2	62.2	76.9	68.8	66.8	76.5	65.9	56.0
K	66.9	53.3	48.7	77.0	63.4	67.4	76.2	65.1	44.9
				SHAPE CONTEXT					
S	79.8	68.0	48.8	74.7	60.9	47.6	71.7	52.5	21.6
K	67.6	50.7	36.3	64.4	57.1	47.8	59.9	48.2	23.7
				PLAGEMANN					
S	77.2	64.9	36.9	27.5	36.5	18.6	18.5	39.6	11.1
K	41.6	41.7	29.0	23.4	21.1	10.4	10.2	26.1	13.2

3.1   Controlled Scenarios

Three parameters governing hand tracking difficulty were determined: angle and distance to camera, and range of gesture motion. Nine scenarios were recorded whilst varying these parameters. Each is around a minute long and contains a single subject gesturing towards the camera, at a wide range of angles. The camera distances analysed are: 1.5m, 2.0m and 2.5m. Gesture motion ranges are: a large motion range using only waves, a medium motion range using half waves and half subtle gestures, and a small motion range using only subtle gestures.

Detection results comparing the sensor fusion algorithm to the Kalman filter for each of these categories is displayed in Table 1. The performance of the sensor fusion algorithm greatly surpasses the Kalman filter in nearly all scenarios. This increase applies to all evaluated depth detectors. Indeed, some scenarios exhibit a 20% detection increase when using the sensor fusion algorithm.

3.2   Crowded Environments

Crowded and dynamic environments are extremely challenging for hand trackers. The sensor fusion algorithm's performance was evaluated in three such publicly available scenarios [2]. Each scenario is over a minute long and contains various subjects gesturing towards the robot.

Figure 4 displays several results from the crowded environments, using the sensor fusion tracker and the depth detector from [2]. They highlight situations that would cause the most problems for traditional RGB hand trackers. 4(a) shows correct hand tracking from a subject with darker skin colour. 4(b) and 4(c) illustrate the algorithm's invariance to different clothing styles. A lot of specular reflection can be observed in 4(d), further justifying the importance of the algorithm's learned skin colour histogram.

Formal detection results for the crowded environments are displayed in Table 2. Again, the performance of the sensor fusion algorithm greatly surpasses the

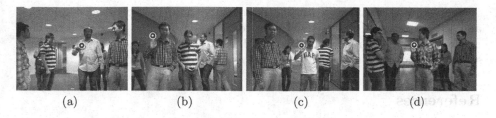

(a)   (b)   (c)   (d)

Fig. 4. Example results showing the performance of the sensor fusion algorithm in a range of crowded environments. Tracked hands are displayed with a white and blue circle. The RGB detection process is shown to work well despite different subject of different skin colour (a), clothing styles (b, c), and illumination conditions (d).

	Scenario 1	Scenario 2	Scenario 3
	MCKEAGUE		
[S]	88.4	78.8	89.2
[K]	87.4	76.7	90.2
	SHAPE CONTEXT		
[S]	80.4	63.0	65.3
[K]	70.5	57.1	53.9
	PLAGEMANN		
[S]	58.0	44.6	18.5
[K]	34.3	34.0	14.2

Table 2. Table showing the greater detection percentage of the sensor fusion algorithm (S) over Kalman filter (K) in three crowded environments. Each scenario has gesturing subjects with varying skin colour and clothing styles, under varying illumination conditions.

Kalman filter, for all but one of the results. The biggest increases are shown when the underlying hand detector doesn't perform well. The sensor fusion algorithm effectively uses RGB information to compensate for depth detector misdetections, thus maintaining a higher tracking accuracy than the Kalman filter.

4   Conclusions

In this paper, we have presented an RGB-D sensor fusion algorithm for tracking hands in crowded environments. The algorithm solves two problems with current hand detection methods. The first is misdetections caused by occlusions, sensor noise, and the highly articulate nature of the hand. This problem is exacerbated in crowded and dynamic environments. The second problem is the computational complexity of depth hand detectors, which the algorithm solves by giving accurate hand position estimates at a 30Hz rate.

Three main novelties were presented. A Monte-Carlo RGB update process was devised to compensate for the large number of false positives produced by traditional skin colour detection. A method of online skin colour learning was proposed, so that the algorithm works with a wide range of skin colours, clothing styles and illumination conditions. Finally, an asynchronous measurement update step was developed, to integrate depth and RGB detections at different computational speeds into the tracking algorithm.

Tracking performance was evaluated in a large number of controlled scenarios, and several crowded and dynamic environments. All tested datasets have been made publicly available. The improved performance of the algorithm over a standard tracking method was demonstrated, for three different hand detectors.

References

1. Shotton, J., Fitzgibbon, A., Cook, M., Sharp, T., Finocchio, M., Moore, R., Kipman, A., Blake, A.: Real-time human pose recognition in parts from single depth images. In: IEEE Conference on Computer Vision and Pattern Recognition (CVPR), pp. 1297–1304 (2011)
2. McKeague, S., Liu, J., Yang, G.-Z.: Hand and body association in crowded environments for human-robot interaction. In: IEEE International Conference on Robotics and Automation (ICRA) (2013)
3. Plagemann, C., Ganapathi, V., Koller, D., Thrun, S.: Real-time identification and localization of body parts from depth images. In: IEEE International Conference on Robotics and Automation (ICRA), pp. 3108–3113 (2010)
4. Li, Z., Kulic, D.: Local shape context based real-time endpoint body part detection and identification from depth images. In: Canadian Conference on Computer and Robot Vision, pp. 219–226 (2011)
5. Guyon, I., Athitsos, V., Jangyodsuk, P., Hamner, B., Escalante, H.J.: Chalearn gesture challenge: Design and first results. In: IEEE Conference on Computer Vision and Pattern Recognition Workshops (CVPRW), pp. 1–6 (2012)
6. Donoser, M., Bischof, H.: Real time appearance based hand tracking. In: International Conference on Pattern Recognition (ICPR), pp. 1–4 (2008)
7. Oikonomidis, I., Kyriazis, N., Argyros, A.: Efficient model-based 3d tracking of hand articulations using kinect. In: British Machine Vision Conference (BMVC), pp. 101.1–101.11 (2011)
8. Chen, B., Huang, C., Tseng, T., Fu, L.: Robust head and hands tracking with occlusion handling for human machine interaction. In: IEEE/RSJ International Conference on Intelligent Robots and Systems (IROS), pp. 2141–2146 (2012)
9. Bretzner, L., Laptev, I., Lindeberg, T.: Hand gesture recognition using multi-scale colour features, hierarchical models and particle filtering. In: IEEE International Conference on Automatic Face and Gesture Recognition, pp. 423–428 (2002)
10. Sigalas, M., Baltzakis, H., Trahanias, P.: Gesture recognition based on arm tracking for human-robot interaction. In: IEEE/RSJ International Conference on Intelligent Robots and Systems (IROS), pp. 5424–5429 (2010)
11. Wang, R., Popović, J.: Real-time hand-tracking with a color glove. ACM Trans. Graph. 28(3), 63:1–63:8 (2009)
12. Thrun, S., Burgard, W., Fox, D.: Probabilistic Robotics (Intelligent Robotics and Autonomous Agents). The MIT Press (2005)
13. Kovac, J., Peer, P., Solina, F.: Human skin color clustering for face detection. In: EUROCON. Computer as a Tool, vol. 2, pp. 144–148 (2003)
14. Stern, H., Efros, B.: Adaptive color space switching for face tracking in multi-colored lighting environments. In: IEEE International Conference on Automatic Face and Gesture Recognition, pp. 236–241 (2002)
15. Mitra, S., Acharya, T.: Gesture recognition: A survey. IEEE Transactions on Systems, Man, and Cybernetics 37, 311–324 (2007)

Using Spatial Semantic and Pragmatic Fields to Interpret Natural Language Pick-and-Place Instructions for a Mobile Service Robot

Juan Fasola and Maja J. Matarić

Interaction Lab, Computer Science Department, Viterbi School of Engineering,
University of Southern California, Los Angeles, CA, USA
{fasola,mataric}@usc.edu

Abstract. We present a methodology for enabling mobile service robots to follow natural language instructions for object pick-and-place tasks from non-expert users, with and without user-specified constraints, and with a particular focus on spatial language understanding. Our approach is capable of addressing both the semantic and pragmatic properties of object movement-oriented natural language instructions, and in particular, proposes a novel computational field representation for the incorporation of spatial pragmatic constraints in mobile manipulation task planning. The design and implementation details of our methodology are also presented, including the grammar utilized and our procedure for pruning multiple candidate parses based on context. The paper concludes with an evaluation of our approach implemented on a simulated mobile robot operating in both 2D and 3D home environments.

1 Introduction

Spatially-oriented tasks such as fetching and moving objects are commonly referenced among the tasks most desired by non-expert users for household service robots to perform [3]. In accomplishing these types of tasks, autonomous service robots will need to be capable of interacting with and learning from non-expert users in a manner that is both natural and practical for the users. In particular, these robots will need to be capable of understanding natural language instructions in order to learn new tasks and receive guidance and feedback on task execution.

Spatial language plays an important role in instruction-based natural language communication, and is especially relevant for object pick-and-place tasks whose purpose lies in achieving desired spatial relations between specified figure and reference objects [12]. Spatial relations, both dynamic and static, expressed in language are often expressed by prepositions [1]. Therefore, the ability for robots to understand and differentiate between spatial prepositions in spoken language is crucial for their interaction with the user to be successful. However, in performing mobile manipulation tasks, service robots must consider not only the semantics of spatial relations within the natural language instruction, but also the *pragmatics* of the task itself. These consist in the implicit (unvoiced) constraints/specifications that accompany the

G. Herrmann et al. (Eds.): ICSR 2013, LNAI 8239, pp. 501–510, 2013.

spoken instructions and further specify the meaning the speaker intends to convey; which can come from context, prior knowledge, norms, etc.

In this paper, we extend upon our previous work and present a general methodology for enabling mobile service robots to follow natural language instructions for object pick-and-place tasks from non-expert users, with and without user-specified constraints. Our approach is capable of addressing both the semantic and pragmatic properties of natural language instructions, and in particular, proposes a novel computational field representation for the incorporation of spatial pragmatic constraints in mobile manipulation task planning. Furthermore, we describe extensions to our spatial language understanding framework, including a revised grammar for the inclusion of figure objects, and a procedure for pruning multiple candidate parses based on context. We conclude the paper with an evaluation of our approach implemented on a simulated mobile robot in both 2D and 3D home environments to demonstrate the generalizability and usefulness of our approach towards real world applications.

2   Related Work

Previous work that has investigated the use and representation of spatial language in human-robot interaction (HRI) includes the work of Skubic et al. [9], who demonstrated a robot capable of understanding and relaying static spatial relations in natural language instruction and production tasks. Computational field models of static relations have also been used in systems for object pick-and-place tasks on a tabletop [8], and for visually situated dialogue [11]. These works implemented pre-defined models of spatial relations, however, researchers have also designed systems capable of learning these types of static spatial relations automatically from training data (e.g., [10]).

Towards enabling HRI-based object pick-and-place instruction understanding and task execution, various approaches have previously been explored. Representative work includes approaches tested on mobile robots with grippers (e.g., [13]), and on humanoid robots with articulated arms (e.g., [14]). Approaches capable of interpreting explicit spatial relations in spoken language instructions include the work of Sisbot et al. [15], who designed a perspective-taking approach to situational awareness using discretized representations of spatial relations, and Brenner et al. [16] who alternatively employed a continuous potential field model. Our approach extends upon this related work by modeling both static and dynamic spatial relations, and by incorporating the implicit pragmatics of the instruction for robot task planning.

Recent work has also investigated approaches to interpreting natural language instructions involving dynamic spatial relations (DSRs). Tellex et al. [6] developed a probabilistic graphical model to infer object pick-and-place tasks for execution by a forklift robot from natural language commands. Kollar et al. [7] employed a Bayesian approach for interpreting route directions on a mobile robot. In both of these works there was no explicit definition of the spatial relations used, static or otherwise, and instead they were learned from labeled training data. However, these approaches typically require the system designer to provide an extensive training data set of natural language input for each new application context, without taking advantage of the domain-independent nature of spatial prepositions. In addition, as the meanings of spatial relations are learned via task demonstration, the semantics and pragmatics of the instructed tasks are not clearly separable, thus limiting the ability of the robot to generalize the learned task execution to new contexts with different pragmatic constraints. Our methodology employs domain-generalizable spatial relations as primitives, and probabilistic reasoning for the grounding, semantic, and pragmatic interpretation of phrases, thereby allowing for context-based instruction understanding and user modifiable robot execution paths.

3   Spatial Language-Based HRI Framework

We have developed a framework for human-robot interaction that incorporates our methodology for representing spatial language, and DSRs in particular, to enable natural language-based interaction with non-expert users. Fundamentally, our approach encodes spatial language within the robot *a priori* as primitives. Static spatial relation primitives are represented using the *semantic field* model proposed by O'Keefe [2], where the semantic fields of static prepositions, parameterized by figure and reference objects, assign weight values to points in the environment depending on how accurately they capture the meaning of the preposition (e.g., for the static spatial preposition 'near', points closer to an object have higher weight).

The software framework contains five system modules that enable the interpretation of natural language instructions, from speech or text-based input, and translation into robot action execution. They are: the syntactic parser, noun phrase (NP) grounding, semantic interpretation, planning, and action modules. In this section we will provide a brief overview of the framework design and module functionality. For a complete description of the software modules, we refer the reader to [4, 5].

The syntactic parser represents the entry point of our software architecture, as it responsible for parsing the user-given natural language instruction into a format that the remaining modules can interpret. After the syntax of the instruction has been determined, the parse tree is passed on to the grounding module which attempts to associate parsed NPs with known objects in the world, grounding hierarchical NPs probabilistically using semantic fields. All groundings are then passed on to the semantic interpreter for final instruction meaning association. The semantic interpretation module utilizes a Bayesian approach to infer the semantics of the given instruction probabilistically using a database of learned mappings from input observations to instruction meanings. The four observation inputs include: the verb and preposition utilized, and the associated grounding types observed for the figure and reference objects. The resulting semantic output of the module includes: the command type, the DSR type, and the static spatial relation (if available). The command type is domain-specific, and may include commands such as: robot movement, object movement, learned tasks, etc. While the output specification was designed to represent the instruction of spatial tasks, the inference procedure utilized is general and can easily be modified to accommodate the requirements of the specific application domain.

4   Parsing Spatial Language Instructions with Figure Objects

4.1   Grammar Extension for Figure Objects

The syntactic parser of our human-robot interaction framework has previously been shown to successfully interpret natural language English instructions involving spatial language [4, 5]. However, while our previous parser was able to interpret a variety of spatial language instructions, including those with hierarchical noun phrases, the grammar utilized was only able to capture spatial relationships between an implicit figure object (i.e., the robot) with regards to reference objects specified in a single noun phrase.

To account for the interpretation of explicit figure objects in the instruction semantics, in addition to the previously interpreted reference objects, the phrase structure grammar of our syntactic parser was extended to accept directive sentences with two noun phrase parameters. The extended constituency rules, which define valid sentences (S), noun phrases (NP), and terminal noun phrases (N'), are presented in Table 1.

Table 1. Grammar Constituency Rules for English Directives using Spatial Language

$S \rightarrow V \ (P_s \,|\, P_p)^* \ NP$	$N' \rightarrow (Det) \ A^* \ N+$	$NP \rightarrow N' \ P_s \ NP$
$S \rightarrow V \ NP \ (P_s \,|\, P_p)^* \ NP$	$NP \rightarrow N'$	$NP \rightarrow NP \ and \ NP$

Part-of-Speech (POS) Tags: V = Verb, P_s = Static Preposition, P_p = Path Preposition, N = Noun, A = Adjective, Det = Determiner

The syntactic parser of our framework extracts part-of-speech (POS) tags for all words in the natural language input using the Stanford NLP Parser, with the exception of prepositions, which are instead identified using a manually constructed lexicon. Spatial prepositions in the lexicon are divided into two categories: static (e.g., near, in, on) and path prepositions (e.g., to, from, through), with POS tags of P_s and P_p, respectively. The two categories are not mutually exclusive, as some spatial prepositions are members of both categories (e.g., around). This categorization serves to facilitate the identification of both correct and incorrect preposition usage within noun phrases, as represented by our constituency rules (e.g., "Give me [the ball *by* the couch]" vs. "Give me [the ball *to* the couch]").

4.2   Pruning Multiple Parses of a Single Instruction

Static prepositions are often used to express path relations in natural language directive instructions, in substitution of semantically related path prepositions (e.g., the use of "in" instead of "into"; "on" instead of "onto"). This characteristic of natural language often results in the generation of multiple candidate parses for the given directive instruction, each with differing semantics (typically). In these cases, the optimal parse (i.e., the most likely interpretation of the instruction) is determined by evaluating each candidate parse according to both: 1) the resulting parse semantics, and 2) the context of the current environment.

Table 2. Possible Flags Raised during Parse Pruning

Flag Type	Description
No Ground	No object/label association found
Low Probable Ground	Object/label association found but with low probability for semantic match of prepositional phrase
Multiple Ground	Multiple candidate groundings for NP
Parameter Count	Missing parameter for command
Parameter Type	Figure and/or reference object type mismatch for command and/or prepositional phrase

Fig. 1. Apartment environment with cup locations and *on*(bookcase) semantic field shown

As an example, under the grammar described above, the phrase "Put the cup on the bookcase *into* the kitchen" has only a single valid parse. In contrast, the phrase "Put the cup on the bookcase *in* the kitchen" has three possible valid parses, listed below:

(1) [$_V$ Put] [$_{NP}$ the cup] [$_{Ps}$ on] [$_{NP}$ the bookcase in the kitchen]
(2) [$_V$ Put] [$_{NP}$ the cup on the bookcase] [$_{Ps}$ in] [$_{NP}$ the kitchen]
(3) [$_V$ Put] [$_{NP}$ the cup on the bookcase in the kitchen]

In evaluating each candidate parse, our methodology first attempts to ground the NPs of the parse with known objects in the world, and if successful, proceeds to infer the semantics of the instruction as a whole (command, DSR, and static relation), as described in Section 3. If an error is encountered during the grounding process, or during parameter validation after semantic interpretation of the parse is completed, a flag is raised. Consequently, candidate parses that raise flags are weighted as less likely than parses that do not raise flags. Example flags for grounding and command parameter errors, along with their descriptions, are listed in Table 2.

If among the candidates a single parse emerges without errors, it is considered to be the optimal parse and is subsequently used for robot task planning and execution. If however, multiple parses are found equally likely, or if all parses raise flags, the user is needed to provide additional clarification before robot task planning can occur.

To illustrate the parse pruning procedure further, consider the three candidate parses listed above in the context of the environment shown in Fig. 1. Candidates (1) and (3) both fail due to NP grounding errors (low probable ground flag), as the environment does not contain a bookcase in the kitchen. Parse candidate (2) succeeds without errors, as there is a single cup sitting on the bookcase as determined by the probabilistic semantic field grounding procedure (i.e., no multiple ground errors, even with three known cups in the environment), a single ground match for "the kitchen", and two correctly typed figure (Mobile object) and reference object (Room) parameters for the inferred command of object movement (as determined by the semantic interpreter). Thus, in this example, parse (2) is chosen as the optimal parse to be used for subsequent robot task planning and execution.

5   Object Pick-and-Place Movement Planning

Once the optimal parse is determined and the results of the semantic interpreter indicate a user instruction of object movement, the robot must first plan to gain possession of the specified figure object, and if applicable, proceed to plan an appropriate placement for the object meeting the requirements of the specified spatial relations of the instruction. In this section we will describe our methodology for accomplishing both of these subtasks using a combination of semantic and pragmatic fields, in addition to taking into account the inherent uncertainty in object grasping in real world scenarios with the use of domain-dependent probabilistic robot-object grasp fields.

5.1   Object Pick Up Planning with Grasp Fields

For mobile service robots with onboard manipulators, satisfying an object placement request typically involves first gaining possession of the object in question by picking it up. To address this task, our approach employs a domain-dependent robot-object grasp field centered on the figure object and computed for all points in the environment. This grasp field is analogous to a probability density function, wherein every point in the environment is assigned a weight value in the range [0, 1] that approximates the probability of success of grasping the object with the robot base positioned at that point. The field is domain-dependent as it may incorporate robot characteristics such as arm reach distance, as well as object attributes such as size, weight distribution, handles, etc. These types of fields can either be learned from a corpus of robot-object grasp attempts, or approximated with the use of a general-purpose proximity field based on the robot's grasp radius. Fig. 2 shows two example grasp fields for household objects; one proximity-based and the other specific to object orientation.

Once the grasp field of the figure object is computed for all points in the environment, the A* search algorithm is utilized to find a solution path for robot movement to the point of maximum grasp potential in the free space, where the robot can attempt to pick up the object using pre-defined grasp behaviors during task execution.

5.2   Object Placement Planning with Semantic Fields

To accomplish the task of object placement, our methodology plans a solution for robot action by first computing the semantic field corresponding to the spatial relation of the instruction (inferred during semantic interpretation) with respect to the reference object, over all points in the robot's workspace; where the workspace of the robot is defined as all points in the environment reachable by the robot's end effector.

Once the semantic field is computed, suitable placement points are identified utilizing the maximum field value recorded in the workspace (e.g., all points with weight values $\geq$ 90% of maximum). The point in the environment where the robot base should be positioned as a pre-condition for the object placement action is determined in two steps: 1) candidate robot base target points are identified using the brushfire algorithm, starting at the suitable placement points as initial positions and expanding until reaching the robot's maximum reach distance, and 2) A* search is run from the robot's current location until reaching the closest base target point.

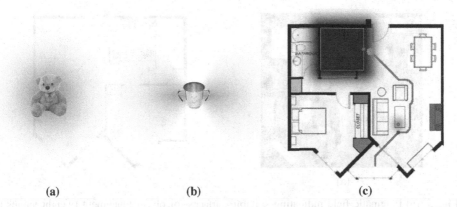

| (a) | (b) | (c) |

Fig. 2. (a) Proximity-based grasp field for teddy bear object; (b) Orientation-based grasp field for cup with handles; (c) Task solution for "Take the cup to the kitchen" with grasp field and semantic field shown at the pick (1) and place (2) locations, respectively

The robot movement plan returned by A* is then utilized during task execution for the robot to position itself before attempting to set the object down at the placement point associated with the chosen base target point. Fig. 2c displays an example solution plan for the object movement instruction "Take the cup to the kitchen", showing both the grasp field used during pick up planning, and the semantic field for *at*(the kitchen) computed during placement planning.

5.3   Pragmatic Fields for Object Placement Planning

In our approach, the use of semantic fields to guide object placement planning successfully enables appropriate placement of figure objects with respect to the spatial relations of the natural language instruction. However, this method only captures the explicit semantics of the instruction, without addressing the *pragmatics* of the task.

Our methodology allows for the incorporation of specific pragmatic constraints with the introduction of spatial *pragmatic fields*. These pragmatic fields are similar to semantic fields in that they assign weight values ($\mathbb{R}[0,1]$) to points in the environment depending on their appropriateness at meeting the goals of the specific spatial pragmatic constraint. As the underlying representations of pragmatic and semantic fields are the same, they can easily be combined for use in robot task planning.

To illustrate the usefulness of incorporating pragmatic fields during planning, consider the example task solution displayed in Fig. 2c for the instruction "Take the cup to the kitchen". Here, the robot sets the cup down on the floor at the entryway of the kitchen because, as the cup is located within the kitchen, the explicit semantics of the instruction are satisfied. However, this solution most likely fails to meet the expectations and intentions of the user's request, which are context-dependent but would likely indicate placement locations such as: on a counter top, in a cupboard, in the sink, etc. User preferences for such spatial locations, and many others (e.g., on surfaces, away from surface edges, away from obstacles, in drawers), can all be incorporated by multiplying their associated pragmatic fields together with the previously calculated semantic fields for final placement planning. Fig. 3a shows an example pragmatic field for object placement preference on surfaces (with weighted surfaces), and Fig. 3b illustrates the use of this field in our described robot task example.

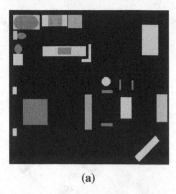
(a)

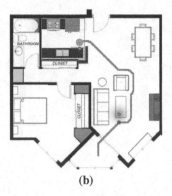
(b)

Fig. 3. (a) Pragmatic field indicating suitable surfaces for object placement (weight values in grayscale); (b) Task solution for "Take the cup to the kitchen" incorporating pragmatic constraints, with combined semantic/pragmatic field shown at object placement location

6   Evaluation

To evaluate the ability of our robot architecture to follow natural language directives involving object pick-and-place tasks, we conducted two separate test runs of the system testing the robot's ability to respond to multiple commands involving object relocation, provided as a sequence of instructions, with and without corresponding user-specified constraints. The test runs served to demonstrate the effectiveness of the semantic interpretation module in inferring the correct command specifications (command, DSR, static relation) given the natural language input, and to demonstrate the path generation capabilities of the system. Our testing domain consisted of a simulated mobile robot operating within a 2D map of a home environment.

A dataset of 189 labeled training examples (each containing a list of observations with correct command specifications), was utilized for the probabilistic inference procedure of the semantic interpretation module. This dataset included the use of 11 different DSRs, 10 separate static spatial relations, 5 commands, and 38 different verbs, each appearing multiple times and in novel combinations among the examples. The instruction sequence provided to the robot in the test runs, including the natural language constraints that were specified for each instruction, is listed in Table 3. The two test runs evaluated the same instructions, with the only difference being the addition of the constraints for the second run. The path generation results for the entire instruction sequence of both test runs are provided in Fig. 4.

The path generation procedure for the first test run utilized a combination of spatial semantic fields, and the pragmatic field for surface placement shown in Fig. 3a. For the second test run, the planning procedure remained the same as in the first, except for the addition of pragmatic fields corresponding to the spatial language constraints specified for each instruction. As evidenced by the resulting robot execution paths for the pick-and-place instruction sequence, our spatial language-based HRI framework was able to demonstrate its potential by successfully following the natural language directives, including under user-specified constraints, during each of the test runs performed for the purposes of system evaluation.

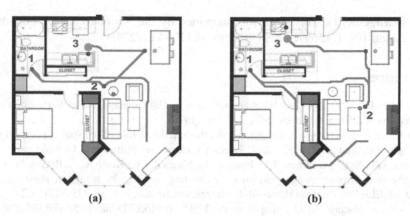

Fig. 4. Robot execution paths for test runs (a) Run #1 (no constraints) (b) Run #2 (constraints)

Table 3. Instruction Sequence for Test Runs with Corresponding Constraints

Instruction 1	Instruction 2	Instruction 3
Grab the medicine	Drop the medicine in the living room	Take the olive oil to the kitchen
Go inside my room	*Put the Tylenol between the couch and the tv*	*Place the oil near the stove*

To demonstrate the generalizability of our approach and its usefulness in practice with real robots in real environments, we have implemented our approach in the 3D Gazebo simulator under the ROS framework. Translation of the discretized plan returned by the planner to continuous robot motor commands (e.g., wheel velocities) was accomplished by providing a cost map based on the task solution path to the ROS navigation stack. The local planner employed is able to respond to dynamic obstacles not represented in the map (e.g., people, objects) which facilitates its usage in real world domains. Autonomous pick and place behaviors were also incorporated utilizing software packages available for the PR2 robot. Fig. 5 shows a snapshot of a successful run of the robot executing a pick-and-place task within the 3D household environment. Future work includes testing our approach on the actual PR2 with end users.

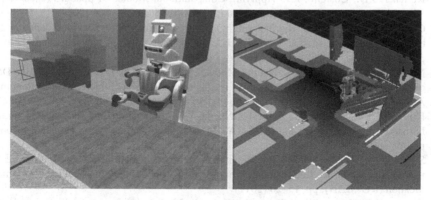

Fig. 5. PR2 robot executing task for "Put the Coke can on the coffee table" in 3D household environment using Gazebo simulator/ROS framework

Acknowledgements. This work was supported by the National Science Foundation under Grants IIS-0713697, CNS-0709296, and IIS-1117279.

References

1. Landau, B., Jackendoff, R.: What and where in spatial language and spatial cognition. Behavioral and Brain Sciences 16(2), 217–265 (1993)
2. O'Keefe, J.: Vector grammar, places, and the functional role of the spatial prepositions in English. In: van der Zee, E., Slack, J. (eds.). Oxford University Press, Oxford (2003)
3. Beer, J.M., Smarr, C., Chen, T.L., Prakash, A., Mitzner, T.L., Kemp, C.C., Rogers, W.A.: The domesticated robot: design guidelines for assisting older adults to age in place. In: Proc. ACM/IEEE Intl. Conf. on Human-Robot Interaction, Boston, MA, pp. 335–342 (2012)
4. Fasola, J., Matarić, M.J.: Using Semantic Fields to Model Dynamic Spatial Relations in a Robot Architecture for Natural Language Instruction of Service Robots. In: IEEE/RSJ International Conference on Intelligent Robots and Systems (IROS), Tokyo, Japan (2013)
5. Fasola, J., Matarić, M.J.: Modeling Dynamic Spatial Relations with Global Properties for Natural Language-Based Human-Robot Interaction. In: Proc. IEEE International Symposium on Robot and Human Interactive Communication (Ro-Man), Korea, August 26-29 (2013)
6. Tellex, S., Kollar, T., Dickerson, S., Walter, M.R., Banerjee, A.G., Teller, S., Roy, N.: Approaching the Symbol Grounding Problem with Probabilistic Graphical Models. AI Magazine 32(4), 64–76 (2011)
7. Kollar, T., Tellex, S., Roy, D., Roy, N.: Toward Understanding Natural Language Directions. In: Proc. ACM/IEEE Int'l Conf. on Human-Robot Interaction (HRI), pp. 259–266 (2010)
8. Sandamirskaya, Y., Lipinski, J., Iossifidis, I., Schöner, G.: Natural human-robot interaction through spatial language: A dynamic neural field approach. In: 19th IEEE International Symposium on Robot and Human Interactive Communication, pp. 600–607 (2010)
9. Skubic, M., Perzanowski, D., Blisard, S., Schultz, A., Adams, W., Bugajska, M., Brock, D.: Spatial Language for Human-Robot Dialogs. IEEE Transactions on SMC Part C, Special Issue on Human-Robot Interaction 34(2), 154–167 (2004)
10. Roy, D.K.: Learning visually grounded words and syntax for a scene description task. Computer Speech & Language 16(3-4), 353–385 (2002)
11. Kelleher, J.D., Costello, F.J.: Applying Computational Models of Spatial Prepositions to Visually Situated Dialog. Computational Linguistics 35(2), 271–306 (2009)
12. Carlson, L.A., Hill, P.L.: Formulating spatial descriptions across various dialogue contexts. In: Coventry, K., Tenbrink, T., Bateman, J. (eds.) Spatial Language and Dialogue, pp. 88–103. Oxford University Press Inc., New York (2009)
13. Nicolescu, M., Matarić, M.J.: Natural Methods for Robot Task Learning: Instructive Demonstration, Generalization and Practice. In: International Joint Conference on Autonomous Agents and Multiagent Systems (Agents), Melbourne, Australia, pp. 241–248 (2003)
14. Dominey, P.F., Metta, G., Nori, F., Natale, L.: Anticipation and initiative in human-humanoid interaction. In: 8th IEEE-RAS International Conference on Humanoid Robots, pp. 693–699 (2008)
15. Sisbot, E.A., Ros, R., Alami, R.: Situation Assessment for Human-Robot Interactive Object Manipulation. In: International Symposium in Robot and Human Interactive Communication (RO-MAN), pp. 15–20 (2011)
16. Brenner, M., Hawes, N., Kelleher, J., Wyatt, J.: Mediating between qualitative and quantitative representations for task-orientated human-robot interaction. In: Proc. of the 20th International Joint Conference on Artifical Intelligence (IJCAI), pp. 2072–2077 (2007)

Bodily Mood Expression: Recognize Moods from Functional Behaviors of Humanoid Robots

Junchao Xu[1], Joost Broekens[1], Koen Hindriks[1], and Mark A. Neerincx[1,2]

[1] Intelligent Systems, EEMCS, Delft University of Technology,
2628 CD, Delft, The Netherlands
{junchaoxu86,k.v.hindriks,joost.broekens}@gmail.com
[2] Human Factors, TNO, 3769 DE, Soesterberg, The Netherlands
mark.neerincx@tno.nl

Abstract. Our goal is to develop bodily mood expression that can be used during the execution of functional behaviors for humanoid social robots. Our model generates such expression by stylizing behaviors through modulating behavior parameters within functional bounds. We have applied this approach to two behaviors, waving and pointing, and obtained parameter settings corresponding to different moods and interrelations between parameters from a design experiment. This paper reports an evaluation of the parameter settings in a recognition experiment under three conditions: modulating all parameters, only important parameters, and only unimportant parameters. The results show that valence and arousal can be well recognized when the important parameters were modulated. Modulating only the unimportant parameters is promising to express weak moods. Speed parameters, repetition, and head-up-down were found to correlate with arousal, while speed parameters may correlate more with valence than arousal when they are slow.

Keywords: mood expression, nonverbal behavioral cues, body language, social robots, human robot interaction (HRI).

1 Introduction

Nonverbal expression of affect, as a key ability of social robots, helps humans to understand robots' internal states (e.g., emotions, moods, beliefs, and intentions) and improves the life-like quality of robots [1]. Besides facial expression, bodily expression is a major communication channel of affect. Experimental studies showed that people can recognize these expressions (e.g., [2], [3], [4], and [5]). Furthermore, bodily expression improved humans' recognition of robots' emotion ([2], [3]). In addition, bodily expression is important for robots that lack facial features (e.g., NAO and ASIMO). One way of constructing bodily expression is to build from scratch by "mimicking" humans' behaviors (static postures and dynamic movements). These bodily expressions are typically designed as explicit behaviors. They usually consist of body actions that express emotions deliberately. For example, raising both hands shows happiness [2]; arms akimbo shows anger [3]; covering eyes by hands shows

G. Herrmann et al. (Eds.): ICSR 2013, LNAI 8239, pp. 511–520, 2013.

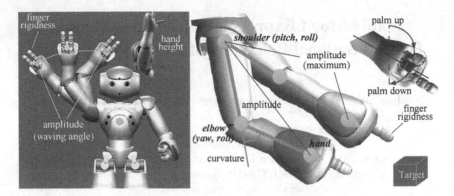

Fig. 1. The parameterized waving (left) and pointing (right) behaviors: our model contains three pose parameters of the arm shown in the figure, two pose parameters of head (head-vertical and head-horizontal), and four motion parameters containing motion-speed, decay-speed, repetition, and hold-time. More details can be found in [6].

fear [4]. However, these body actions rise and dissipate quickly and do not extend over time. Thus, we believe that this type of expression is suitable for expressing emotions, but not moods. Moreover, these body actions may interrupt functional behaviors. For example, a robot cannot express excitement while it is pointing to the object or person that makes it excited by raising both hands. Our work aims at integrating bodily expression of mood with functional behaviors, e.g., task execution, communicative gestures, walking, etc. To this end, we parameterized functional behaviors so that modulating parameters can generate affective cues. Hence, moods can be reflected from the same behavior executed in different "styles", rather than the behavior "contents" per se. As a result, mood can be expressed continuously over time, even when robots are executing tasks. Therefore, we believe that this method is suitable for mood expression. Moreover, bodily mood expression may enhance the affective interaction by prolonging it and providing more modalities.

We investigated our behavior model with a humanoid robot NAO, with interests in whether parameter modulation can be effectively applied to a robotic platform for showing mood. In particular, mood is expressed less explicitly through our approach. In addition, we studied high-DOF functional behaviors, allowing us to define more parameters that may enrich the mood expression. We are also interested how behavior parameters can be combined to show different moods. In previous work [6], our model has been applied to two functional behaviors, waving and pointing (Fig. 1), and we obtained general design principles about the relations between mood variables and behavior parameter modulation from a design experiment, in which participants were asked to design mood expression according to five levels of valence labeled by very-unhappy, unhappy, neutral, happy, and very-happy. In addition, the relative importance and the interrelations between parameters were investigated [7]. Table 1 summarizes the main findings. It is not clear whether people can recognize moods in the presence of behavior functions, since people may devote their attention to behavior functions. This paper reports the findings of a study on people's recognition of the mood expressions resulted from the design experiment, and whether the conclusions of the design experiment correspond to people's perceptions.

Table 1. The design principles and parameter importance

Waving	HandHeight	Finger	Amp	Rep	HoldTime	DecaySpd	MotSpd	HeadVer.	HeadHor.
Relation[1]	+		+	+	+		+	+	+
Import.[2]	2		5	4			3		1
Pointing	PalmDir.	Finger	Amp	Rep	HoldTime	DecaySpd	MotSpd	HeadVer.	HeadHor.
Relation[1]	+		+	*	*		+	+	+
Import.[2]			2				3		1

[1] */+ denotes significant correlations with valence; + denotes increase with valence;
[2] The number denotes the importance; small - important; unnumbered - unimportant.

Several studies addressed behavior parameter modulation. Wallbott [8] studied the emotional bodily movements and postures of actors/actresses. His study indicated that the body movement "qualities" can reflect emotions. Laban movement analysis [9] models body movements from different aspects, e.g., effort and shape. Chi et al. [11] developed EMOTE framework for synthesizing expressive gestures of virtual agents. An evaluation of effort elements showed that trained observers can recognize the displayed effort at a moderate rate, whereas this study also indicated that prominent effort elements may mask other elements when they are showed in combination. In contrast to EMOTE, which performs as a post process of pre-generated behaviors, Pelachaud et al. [9] modifies gestures before the computing the animation. They characterized behavior expressivity using six parameters: spatial, temporal, fluidity, power, overall activation, and repetition. Their model was applied to an embodied conversational agent for communicating cognitive and affective states through modulated gestures. Evaluation showed that spatial and temporal extents received high recognition rate, but power and fluidity quite low; abrupt and vigorous received high recognition rate but not for sluggish. To achieve a better concord between mood expression and behavior functions, our approach defines behavior parameters while defining the behavior functional profile, so behaviors are also modified first and then the robot joints are computed.

2   Experiment Design and Hypotheses

The recognition experiment first evaluated whether participants can differentiate the five valence levels from modulated behaviors of the design experiment [6]. Second, we tested whether people's recognition is different when modulating different parameter (sub)sets according to the relative importance [7]: 1) all parameters (APS); 2) only important parameters (IPS), which are numbered in Table 1; and 3) only unimportant parameters (UPS), which are unnumbered. We expect that modulating only the IPS parameters can still express moods without reducing the recognition rate considerably. Although statistical results and participants' ranks showed that the importance of the UPS parameters was low, but participants did modify them during the design. Thus, we suspect that the UPS parameters can express "weak" moods, which are more implicit and less intense, so we tested whether modulating only the UPS parameters can still express moods. Moreover, the behavior naturalness is one of participants' design criteria in the design experiment. Thus, we suspect that modulating all the parameters may result in more natural behaviors than modulating only IPS parameters. Hence, the behavior naturalness was assessed in the recognition experiment. Therefore, our hypotheses are formulated as follows:

H1. People can distinguish different valence levels from modulated behaviors when all behavior parameters (APS) are modulated. The relationship between parameter settings and perceived valence levels is consistent with the relationship found in the design experiment;

H2. People can perceive different levels of valence when only important parameters (IPS) are modulated; People can still recognize the valence when only modulating unimportant parameters (UPS), but the recognition rate is worse than the APS and IPS conditions;

H3. The behaviors generated by modulating all the parameters (APS) are perceived more natural than the ones generated by modulating only the important parameters (IPS).

The test settings (videos can be found at our website[1]) for the recognition experiment is based on the average setting obtained from the design experiment [6]. An average setting may be not the best design due to the inconsistency and unnaturalness caused by mixture of different individual designs [15]. In our case, the diversity on the arousal dimension is averaged out: for the negative valence, most participants designed sadness (low arousal), but a part of participants designed anger (high arousal). Therefore, we corrected the weighted settings in terms of consistency and naturalness within the boundary of the design principles found in the design experiment, and added anger to recover the diversity on the arousal dimension.

Besides, we tested whether people can perceive arousal from the test settings, since the participants of the design experiment did consider the arousal dimension as just mentioned. We also studied whether parameter sets influence the recognition of arousal. Note that the important parameters (IPS) were obtained from the task where participants were asked to design mood expression only according to the valence. The importance may be only or more in regard with the valence. Thus, whether the parameter sets influence the perception of arousal was unclear.

Paired comparison was used to test how well people perceived valence and arousal from behaviors under APS, IPS and UPS conditions (H1, H2). This method provides more precise results in interval scales than a direct scaling, because it transforms the scaling task, which is difficult for humans, to a comparison task [12, 13]. Participants were asked to compare (not paired comparison) the naturalness of generated behaviors corresponding to each mood under the IPS and APS conditions respectively (H3). The notions of valence and arousal were explained to participants before the experiment using categorical emotion labels and SAM manikins. Naturalness was explained mainly in terms of natural interaction. Participants were provided a user interface for inputting answers and proceeding with the experiment. Two grey NAO robots were used to perform behaviors modified by two moods simultaneously to reduce participants' cognitive workload. Waving and pointing were arranged in a

---

[1] http://ii.tudelft.nl/~junchao/mood_expr_recog.html

Table 2. The correlation (Pearson r) between parameters and valence or arousal

Waving	HandHeight	Finger	Amp	Rep	HoldTime	Decay	MotSpd	HeadVer.	HeadHor.
Valence	0.889	0.936	0.966	0.858		0.848	0.848	0.950	
Arousal				0.653	-0.976	0.977	0.977	0.797	
Pointing	PalmDir.	Finger	Amp	Rep	HoldTime	Decay	MotSpd	HeadVer.	HeadHor.
Valence	0.507		0.914	0.810	0.315	0.927	0.927	0.984	
Arousal			0.923			0.978	0.978	0.924	

counter-balanced order. For each behavior, the six moods were presented in pairs in a random order, and they were presented under APS, IPS, and UPS conditions successively. 26 participants (13 females and 13 males) were recruited from Delft University of Technology. The participants' ages ranged from 21 to 35 years (M = 28.6, SD = 3.3). 13 participants are Chinese, and the other 13 are not. All the participants signed the informed consent form. A pre-experiment questionnaire confirmed that the participants had little experience of designing robots or animated characters. Each participant received a gift as compensation for their time.

3   Analysis and Result

The method based on Thurstone model from [13] was used to analyze the paired comparison data. To see how well participants recognized the moods under the APS, IPS, and UPS conditions, only the mood factor was input into the analysis. For convenience, all results are combined and illustrated in Fig. 2. Assuming that valence and arousal are orthogonal [14], the tested moods are denoted in the valence-arousal space (Fig. 2). First, we interpret the recognition of valence from the five settings derived from the design experiment; second, we interpret the recognition of arousal; finally, we interpret the additional mood anger.

To analyze the recognition of valence (H1), we first looked at the results under APS condition (Fig. 2a, b). Regardless of anger (interpreted later), for both behaviors the valence of each pair of moods was significantly differentiated by participants except for unhappy and neutral pointing. This result shows that people can recognize the valence from the behavioral cues in general (H1). Pearson correlations between parameter values and the perceived valence scales (Fig. 2a, b) were computed. The results (Table 2) show that the relationship between parameters and perceived valence is generally consistent (H1) with the findings of the design experiment (Table 1).

Secondly, we interpret how participants' recognition under the IPS and UPS conditions differs from the APS condition (H2). To this end, we added the parameter set condition as a factor [13] to the paired comparison analysis. The overall result (Table 3) affirms that the parameter set condition influenced participants' perception significantly for both behaviors with regard to valence. In addition, we compared the parameter set conditions in pairs using the same method above. For both behaviors, there are no significant differences between the parameter set APS and IPS (Table 3), which suggests that modulating only the important parameters is capable of expressing valence almost equally well as modulating all the parameters. The generated scale of valence under IPS condition is

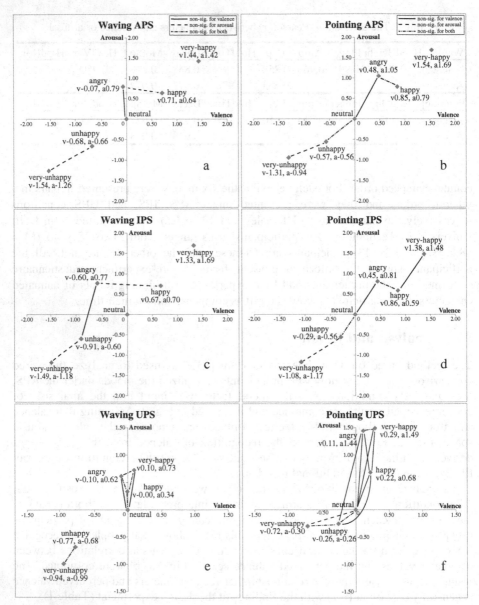

Fig. 2. The figure shows the position of each perceived mood in the valence-arousal space under the APS, IPS and UPS conditions for the waving and pointing behaviors. The valence or arousal of unconnected moods was significantly differentiated, while for the connected ones either valence or arousal or both was not significantly differentiated.

Table 3. Significant differences of recognition between parameter set conditions

		Overall	APS vs. IPS	APS vs. UPS	IPS vs. UPS
Waving	Valence	p < 0.001***	p = 0.205	p < 0.001***	p < 0.001***
	Arousal	p < 0.010*	p = 0.931	p = 0.026*	p = 0.001**
Pointing	Valence	p < 0.001***	p = 0.671	p < 0.001***	p < 0.001***
	Arousal	p = 0.006**	p = 0.879	p = 0.011*	p = 0.001**

*p<0.05, **p<0.01, ***p<0.001

similar with the APS condition (Fig. 2c, d). The only difference is that the happy and very-happy pointing were differentiated under the APS condition but not for the IPS condition. Possible reason is that repetition increased for very-happy under the APS condition, but not for the IPS condition, since repetition was rated unimportant in previous study. Further study is needed to address whether repetition is important to valence in different situations.

The recognition of valence under APS and IPS conditions is significantly better than UPS condition (Table 3). The high-arousal moods (anger, happy, and very-happy) and neutral were less successfully differentiated by participants for waving (Fig. 2e). Similar results were obtained for pointing (Fig. 2f). Besides, the unhappy and neutral pointing were not significantly differentiated. This suggests that none of the UPS parameters is sufficient to present the valence of high-arousal moods. However, as we hypothesized, some moods can still be recognized even without modulating important parameters. The valence of unhappy and very-unhappy waving was significantly differentiated from waving of neutral and high-arousal moods (Fig. 2e). The long hold-time, slow decay-speed, head turning away from both hand and the front distinguished the unhappy and very-unhappy. We exclude finger since few participants mentioned it in the post-questionnaire. For pointing, the valence of very-unhappy was significantly differentiated from other moods except unhappy. Thus, we conclude that the UPS parameters are promising for "weak" mood expressions for at least two valence levels: positive and negative.

Results also show that participants recognized arousal levels from behaviors. Under APS (Fig. 2a, b) and IPS conditions (Fig. 2c, d), the arousal of high-arousal moods and neutral was significantly differentiated for both behaviors, regardless of anger (integrated later). The arousal of low-arousal moods (unhappy and very-unhappy) was significantly differentiated from high-arousal moods and neutral for waving, whereas only very-unhappy was significantly differentiated form high-arousal moods and neutral for pointing. Statistically, there are no significant differences of perceived arousal between the APS and IPS conditions for both behaviors (Table 3), which suggests that the IPS parameters are capable to express arousal equally well as the APS parameters. However, the perceived arousal under UPS condition differs significantly from either the APS or IPS condition (Table 3). For waving (Fig. 2e), the arousal of very-happy and anger significantly differentiated from neutral, whereas other high-arousal moods were not. Possible reasons are that the zero hold-time and fast decay-speed of angry and very-happy waving made the overall movement fast and fluent, resulting in the perception of a high arousal. The arousal of high-arousal moods was better recognized for the pointing behavior than waving behavior. For pointing (Fig. 2f), the arousal of all high-arousal moods was significantly differentiated from neutral, and very-happy was differentiated from happy. Fast decay-speed and high repetition may account for this. This suggests that the decay-speed and repetition correlate more with arousal than valence. They were actually considered unimportant to valence.

The arousal of unhappy and very-unhappy waving was significantly differentiated from other moods (Fig. 2e), but unhappy and very-unhappy were not differentiated from each other. For pointing, the arousal of neither unhappy nor very-unhappy was significantly differentiated from neutral (Fig. 2f). In fact, the arousal between unhappy and very-unhappy was not significantly differentiated for both behaviors under all conditions, but their valence was significantly differentiated under APS and IPS conditions. The arousal-correlated parameters (e.g., speed, repetition) seem not able to render arousal for low-arousal moods. Back to the UPS condition (Fig. 2e, f), we found that the very slow decay-speed distinguished the valence of very-unhappy from neutral. It seems that the speed parameter like decay-speed may correlate more with valence when it is slow, whereas correlates more with arousal when it is fast.

The recognition of angry waving showed the promise of expressing anger through parameter modulation. The valence of anger was perceived as negative for all conditions (Fig. 2a, c, e), although it was not significantly differentiated from neutral under APS and UPS conditions. Surprisingly, the valence was better differentiated from neutral under the IPS condition (Fig. 2c). We considered that the longer hold-time under IPS condition caused the movement jerkier resulting in a more negative perception, whereas the zero hold-time and the faster decay-speed under APS condition made the movement smoother resulting in a relative more positive perception. Furthermore, the head turned away from the moving hand in the APS condition, which made the robot seem to avoid the eye-contact resulting in a feeling of fear, while fear has a more positive valence than anger [14]. The valence of anger was recognized better for waving than pointing, since it was recognized as positive for pointing under all conditions. Perhaps, the presence of arousal (by large amplitude, repeated movements, and fast speed) in angry pointing was dominant and masked the expression of negative valence, which led people to consider the mood as excitement.

As discussed before, the arousal of anger was recognized significantly higher than neutral and low-arousal moods for both behaviors under all conditions. Interestingly, the perceived arousal of angry pointing and waving under UPS condition was as high as very-happy (Fig. 2e, f), whereas in other conditions it is significantly lower than very-happy. Possible reason is that most parameters were set to the same value between these two moods under the UPS condition. However, the only element that made the arousal of the very-happy pointing under APS and IPS conditions higher than angry pointing is the high-raised head. This suggests that head-up-down correlates with arousal. According to the above discussion, we summarize the parameters that correlate with arousal in Table 2, where Pearson correlation was computed between parameter values and the perceived arousal scale.

Binomial tests were used to analyze whether behaviors under the APS condition was perceived more natural than the IPS condition (H3). Participants' choices between the APS and IPS conditions are not significantly above chance level for each mood and behavior. Thus, our study did not show that modulating UPS parameters improves the behavior naturalness. We also tested the effect of gender and culture (Chinese and Non-Chinese) by adding them as a factor into paired comparison analysis separately. The results do not show any significant differences between gender and culture conditions.

4   Discussion

The modulation of the important parameters expresses moods better than unimportant parameters. Most important parameters like hand-height, amplitude, motion-speed, and repetition are "global" parameters, which influence the overall movement. Changing these parameters will alter the movement appearance noticeably. Head position also has strong effect on affect expression [5], probably because the head is a special body part that people usually pay attention to during interaction. The unimportant parameters are "local" parameters that influence only a small region of the body parts (e.g., finger-rigidness, palm-up-down) or a short period (e.g., hold-time and decay-speed are temporally local) of the whole movement. Thus, they may not produce sufficient affective cues or people may not even notice them. Hence, behaviors with more "global" parameters may be more affectively versatile, For example, waving has higher expressivity than pointing. In fact, moods were recognized better through waving than pointing in general.

Interactions may exist between valence and arousal. According to Table 1 and Table 2, parameters like motion-speed, head-up-down and repetition of waving were found to correlate with both valence and arousal. In addition, a 5-point Likert scale (from 1: "extremely disagree" to 5: "extremely agree") post-experiment questionnaire suggests that the participants generally agreed on that valence and arousal are related. The mean rating is 3.85 (SD=0.88). Several studies also reported that valence and arousal are not orthogonal [16]. The interaction between valence and arousal should be taken into account when we design mood expressions.

Our model is possible to be generalized to other behaviors in terms of the relations between behavior parameters and mood variables. As in our model parameters are defined at the stage of constructing behavior functional profiles, parameters are dependent on behavior functions. Thus, the same parameters may have different meanings for different behaviors. Despite the differences, design principles may still hold. For example, although the amplitude is the swing angle for waving but the arm extension for pointing, larger amplitude corresponds with a positive mood for both behaviors. However, design principles may also be different for the same parameters. For example, the hold-time means smoothness for waving but persistence for pointing. Hence, shorter hold-time (smoother movement) corresponds with a positive mood for waving, whereas longer hold-time (more persistent) of pointing generally expresses a positive mood. We suggest designers pay attention to the meaning of a parameter for specific behaviors when modulating the parameter to express mood.

5   Conclusion and Future Work

This paper presents a study on people's recognition of humanoid robots' bodily mood expression through behavior parameter modulation. The results indicated that five valence levels can be expressed through parameter modulation for the two behaviors studied. Arousal can also be expressed with at least four levels. The important parameters that influence the behavior overall have a major effect on both valence and arousal. The unimportant parameters can express "weak" moods for at least two levels of valence and three levels of arousal for both behaviors, but no effect on naturalness of these parameters was observed. The speed parameters, repetition, and head-up-down were found to correlate with arousal. Speed parameters are capable to render arousal when they are fast, but not when they are slow. In the future, we will improve the angry pointing and study the relation between the pointing direction and mood expression. While mood expressions via parameter modulation can be recognized in an experimental setting, whether people can recognize them correctly, even notice, in real HRI scenarios still remains a question. We will apply the design principles into more behaviors used in HRI and address the question in the future.

References

1. Breazeal, C., Takanishi, A., Kobayashi, T.: Social robots that interact with people. In: Springer Handbook of Robotics, pp. 1349–1369. Springer, Berlin (2008)
2. Zecca, M., Mizoguchi, Y., Endo, K., Iida, F., Kawabata, Y., Endo, N., Itoh, K., Takanishi, A.: Whole body emotion expressions for kobian humanoid robot - preliminary experiments with different emotional patterns. In: RO-MAN, pp. 381–386. IEEE (2009)
3. Hirth, J., Schmitz, N., Berns, K.: Towards social robots: Designing an emotion-based architecture. Int. J. Social Robotics 3, 273–290 (2011)
4. Häring, M., Bee, N., Andre, E.: Creation and evaluation of emotion expression with body movement, sound and eye color for humanoid robots. In: RO-MAN, pp. 204–209. IEEE (August 2011)
5. Beck, A., Stevens, B., Bard, K.A., Cañamero, L.: Emotional body language displayed by artificial agents. ACM Trans. Interact. Intell. Syst. 2(1), 1–29 (2012)
6. Xu, J., Broekens, J., Hindriks, K., Neerincx, M.A.: Mood Expression through Parameterized Functional Behavior of Robots. In: ROMAN, Gyeongju, Republic of Korea (in press, 2013)
7. Xu, J., Broekens, J., Hindriks, K., Neerincx, M.A.: The Relative Importance and Interrelations between Behavior Parameters for Robots' Mood Expression. In: ACII, Geneva, Switzerland (in press, 2013)
8. Wallbott, H.: Bodily expression of emotion. European J. Social Psychology 28(6), 879–896 (1998)
9. Pelachaud, C.: Studies on gesture expressivity for a virtual agent. Speech Communication 51(7), 630–639 (2009)
10. von Laban, R., Ullmann, L.: The Mastery of Movement, 4th edn. Dance Books, Limited (2011)
11. Chi, D., Costa, M., Zhao, L., Badler, N.: The emote model for effort and shape. In: SIGGRAPH, pp. 173–182. ACM (2000)
12. Engeldrum, P.G.: Psychometric scaling: a toolkit for imaging systems development. Imcotek Press (2000)
13. Rajae-Joordens, R., Engel, J.: Paired comparisons in visual perception studies using small sample sizes. Displays (2005)
14. Russell, J.A.: A circumplex model of affect. Journal of Personality and Social Psychology 39, 1161–1178 (1980)
15. Bergmann, K., Kopp, S., Eyssel, F.: Individualized gesturing outperforms average gesturing – evaluating gesture production in virtual humans. In: Safonova, A. (ed.) IVA 2010. LNCS, vol. 6356, pp. 104–117. Springer, Heidelberg (2010)
16. Lang, P.J., Bradley, M.M., Cuthbert, B.N.: International Affective Picture System (IAPS): Technical Manual and Affective Ratings (1997)

Cooperative Robot Manipulator Control with Human 'pinning' for Robot Assistive Task Execution

Muhammad Nasiruddin Mahyuddin* and Guido Herrmann**

Bristol Robotic Laboratory and Department of Mechanical Engineering,
University of Bristol, Bristol, UK
{memnm,g.herrmann}@bristol.ac.uk

Abstract. This paper presents the use of a multi-agent controller in application to a human-robot cooperative task where the human is the lead and two robotic manipulators act as agents allowing for physical assistance of the human in a lifting task: the human aids in providing direction for two synchronized robot arms placing a tray with a water glass in a human-robot interaction experiment.

Novel adaptive multi-agent theory is exploited to achieve precise coordination between the arms, while being lead by the human. A novel finite-time adaptation scheme aids changing structures, such as the removal of the leading agent, so that consensus and synchronisation is retained. This is permitted by the decentralized control structure, where each agent is supported by an agent-specific controller and information exchanged between the agents is limited to position and velocity of each manipulator. Thus, the controller is robust to structural changes in the multi-agent network, e.g. the removal of the pinning agent.

Recent research on control systems for human-robot interaction has considered several areas such as compliant control approaches for humanoids [1–4]. Other work has also considered the learning of the impedance level [5] required in human robot interaction based on a cost function idea which presents the compromise for compliance or tracking [6]. This provides a more abstract, but ultimately practical tool for quantitative impedance level learning.

The creation of suitable trajectories for human-robot interaction is equally important, as it is affected by human expectations and behaviour. For instance, the control of humanoid robots to achieve human-like motion via online optimization [7, 8] has shown an approach of creating psychological confidence of a

---

* Muhammad Nasiruddin Mahyuddin is currently doing a PhD in Bristol Robotics Laboratory, Department of Mechanical Engineering, University of Bristol. His PhD is sponsored by University Sains Malaysia and Ministry of Higher Education of Malaysia.

** Guido Herrmann is a Reader in Dynamics and Control with the Department of Mechanical Engineering and Bristol Robotics Laboratory, University of Bristol, BS8 1TR.

G. Herrmann et al. (Eds.): ICSR 2013, LNAI 8239, pp. 521–530, 2013.

human interacting with a robot. This is based on the fact that human motion tends to follow certain paths, patterns and constraints to produce fairly consistent motion patterns [9]. There is evidence that some form of optimisation is taking place during the planning and/or execution of motion [10]. Other workers have looked at the aspect to estimate human intention, i.e. a neural network control algorithm estimates to an extent human motion intention so that a robot manipulator supports the human in a human robot interaction task [11].

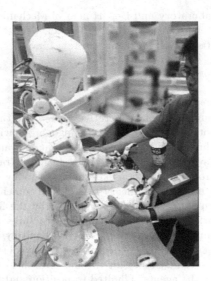

Fig. 1. BERT II system used in the cooperative human-robot interaction

These ideas are complemented here in this paper, where we look at multiple robot manipulators supporting a lifting or placement task in a human-robot experiment (Figure 1), where both robot manipulators have to be well balanced and provide well-coordinated precision in safe human-robot interaction. This is here achieved via cooperative control approaches, where the human takes the lead in the human-robot cooperative task, i.e. pins the robot arm to follow a desired trajectory via physical interaction with the robot, while the robot arm controllers are the cooperative agents following in a synchronized manner. The other advantage is that the overall controller structure is decentralized, i.e. only required position, velocity information is exchanged between the decentralized controller.

This type of control scheme is different to the work done in the eighties as evident in [12],[13],[14] and [15]. They had mainly a master/slave structure [16] or were of a centralised type [17],[18]. A master/slave scheme does not provide the general criterion for choosing the master or the slave, which is to the paper here of importance. The work here is therefore also not suited to centralised control or some of the model-based decentralised adaptive controller for multiple manipulators in [19], [20] and [21].

The flexibility achieved from cooperative control, i.e. multi-agent theory [22–24], is key to this work. Thus, multi-agent theory provides a framework for a distributed leader-follower control of an interconnected multi-manipulator system. The controller architecture is structured in a distributed form since, each manipulator's controller only uses information of its connected manipulators or so-called its 'neighbours'. In this sense, the human takes lead in the cooperative, human-robot interaction control scheme, while when needed the achieved precise consensus between the robot manipulators is retained once the human releases pinning. Thus, this permits that a human can guide a robot in a coordinated lifting task when needed.

The suggested controller for this human-robot interaction task is based on a recently developed finite adaptation scheme which is incorporated in a multi-agent control scheme: A pair of humanoid arms being subjected to a delicate cooperative task, such as carrying a serving tray with a full glass of water, benefits as both robot arm motions are synchronised accurately with the aids from the dynamic estimation of the decentralized controllers.

The paper is structured as follows. The theoretical background of the multi-agent controller is explained in Sections 1-4. Thus, Section 1 provides some background about the multi-agent system, the humanoid BERT II, and the respective underlying graph theory. Section 2 provides the basic set-up of the decentralized controller. The adaptive scheme is presented in Section 3, while the underlying analytical result is given in Section 4. Finally, the experimental results are discussed in Section 5.

1   System Definition and Problem Formulation

Bristol Robotic Elumotion II or BERT II two robot arm system can be modeled as an interconnected N (in this case, $N = 2$) manipulator system composed of m-link. The communication exchange between the two arms across the Control Area Network (CAN) bus can be modeled by a graph which is inspired by the multi-agent theory.

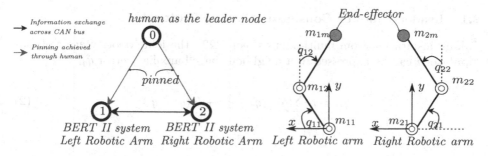

Fig. 2. (a) Two nodes with a human leader pinning and (b) the two manipulators

1.1 Leader - Agent Communication Structure

A general directed graph $\mathcal{G} = (\mathbb{V}, \mathbb{E})$ with nonempty finite set of N nodes $\mathbb{V} = \{v_1, \ldots, v_i, \ldots, v_N\}$ is considered. Node i represents the ith manipulator. It is also assumed that the graph is strongly connected and fixed. The graph consists of directed edges or arcs $\mathbb{E} \subseteq \mathbb{V} \times \mathbb{V}$ with no repeated edges and no self loops $(v_i, v_i) \notin \mathbb{E}, \forall i$. Denote the connectivity matrix as $A = [a_{ij}]$ with $a_{ij} > 0$ if there is a directed edge $(v_j, v_i) \in \mathbb{E}$, while $a_{ij} = 0$ otherwise. It is assumed that the integrity of communication between ith and jth manipulators is the same i.e., $a_{ij} = a_{ji}$ such that A is symmetric matrix. The set of neighbours of a node v_i is $N_i = \{v_j : (v_j, v_i) \in \mathbb{E}\}$, i.e. the set of nodes with arcs incoming to v_i. The in-degree matrix is a diagonal matrix $D = [d_i]$ with $d_i = \sum_{j \in N_i} a_{ij}$ the weighted in-degree node i (i.e. ith row sum of A). Define the graph Laplacian matrix as $L = D - A$, which has all row sums equal to zero. Define $d_i^0 = \sum_j a_{ji}$, the weighted out-degree of node i, that is the ith column sum of the connectivity matrix, A. Figure 2 shows the topology of the interconnected nodes represented by the digraph. There is a directed path from v_i to v_j for all distinct nodes $v_i, v_j \in \mathbb{V}$. The connectivity matrix A and L are irreducible [25, 26].

The manipulator-leader communication is again directed from leader to manipulator only, which is identified through the pinning gain $b_i \geq 0$. Thus, in case, a manipulator i, $(0 < i \leq N)$, is pinned, then $b_i > 0$. Note that there is at least one manipulator for which $b_i > 0$. The leader node carries the index 0, e.g. Figure 2. Thus, $b_i \neq 0$ if and only if there exists an arc from the leader node to the ith node in $\mathcal{G}$. According to [27], the manipulator i dynamics equation can be given as,

$$M_i(q_i)\ddot{q}_i + V_i(q_i, \dot{q}_i)\dot{q}_i + G_i(q_i) = \tau_i \tag{1}$$

where $q_i, \dot{q}_i, \ddot{q}_i \in \mathbb{R}^m$ denote the joint position, velocity, and acceleration vectors of manipulator i, respectively; $M_i(q_i) \in \mathbb{R}^{m \times m}$ is the ith manipulator's inertia matrix; $V_i(q_i, \dot{q}_i) \in \mathbb{R}^{m \times m}$ is the centripetal-Coriolis matrix of the ith manipulator; $G_i(q_i) \in \mathbb{R}^m$ is the ith manipulator's gravitational vector; $\tau_i \in \mathbb{R}^m$ represents the torque input vector of manipulator i.

2 Distributed Motion Synchronisation Controller Design

2.1 Leader-Follower Consensus Control Protocol

Taking inspiration from multi-agent theory [22], the joint velocity error of manipulator i can be expressed in a neighbour-based auxiliary error $\dot{q}_{di}$

$$\dot{q}_{di} = \mu_i \sum_{j=1}^{N} a_{ij}(q_j - q_i) + b_i(q_0 - q_i) \tag{2}$$

$q_0 \in \mathbb{R}^m$ is the human node leader's joint position served to be followed by the manipulator trajectory. The term l_{ij} is the corresponding element of the Laplacian matrix L. Consensus control in (2) can be also expressed in terms of overall network expression as

$$\dot{q}_d = -(I_N \otimes \mu)\left[(L + B) \otimes I_m\right](q - \bar{q}_0) \qquad (3)$$

where $I_N \in \mathbb{R}^{N \times N}, I_m \in \mathbb{R}^{m \times m}$ are the identity matrices, $\bar{q}_0 = 1_N \otimes q_0 \in \mathbb{R}^{Nm}$ (Noting that $1_N = (1, 1, \cdots, 1_N)^T \in \mathbb{R}^N$) and $\mu \in \mathbb{R}^m$ is the control gain for each manipulator i to be designed.

2.2 Closed-Loop Dynamic of the Cooperative Manipulator

The following control torque input τ_i is proposed for the ith manipulator

$$\tau_i = \phi_i(q_i, \dot{q}_i, \dot{q}_{di}, \ddot{q}_{di})\hat{\Theta}_i - \beta_i e_i - k_i \varepsilon_i \qquad (4)$$

where $\tilde{e}_i = \dot{q}_i - \dot{q}_{di}$. The control gains β_i and k_i are to be designed in Section 4. ε_i is a vector containing sliding variables for each joint m of each manipulator i defined as $\varepsilon_i = \frac{e_i}{\|e_i\|} \in \mathbb{R}^m$.

The closed-loop dynamic equation of the corresponding cooperative manipulator can be written as

$$M_i(q_i)\dot{e}_i + V_i(q_i, \dot{q}_i)e_i = \phi_i(q_i, \dot{q}_i, \dot{q}_{di}, \ddot{q}_{di})\hat{\Theta}_i - \beta_i e_i - k_i \varepsilon_i$$
$$-\phi_i(q_i, \dot{q}_i, \dot{q}_{di}, \ddot{q}_{di})\Theta_i - M_i(q_i)\ddot{q}_0$$
$$-V_i(q_i, \dot{q}_i)\dot{q}_0 \qquad (5)$$

3 Adaptation Algorithm to Assist the Cooperative Task

Following the same construct of the adaptation algorithm as in [28],[29] and avoiding lengthy derivation, the adaptation law featuring finite-time adaptive algorithm suited to assist in the human robot cooperative task is presented. The parameter estimation algorithm comprises of a synchronisation error e_i and the switched parameter R_i for each agent manipulator i:

$$\dot{\hat{\Theta}}_i = -\Gamma_i \phi_i(q_i, \dot{q}_i, \dot{q}_{di}, \ddot{q}_{di})e_i - \Gamma_i R_i \qquad (6)$$

$$R_i = \omega_{1_i} \frac{W_i(t)\hat{\Theta}_i - N(t)_i}{\|W_i(t)\hat{\Theta}_i - N_i(t)\|} + \omega_{2_i}(W_i(t)\hat{\Theta}_i - N(t)_i),$$
$$i = 1, \ldots, N. \qquad (7)$$

where W and N can be generated by the following differential equation,

$$\dot{W}_i(t) = -k_{FF_i}W_i(t) + k_{FF_i}\phi_{f_i}^T(q_i, \dot{q}_i)\phi_{f_i}(q_i, \dot{q}_i),$$
$$W_i(0) = wI_l, \qquad (8)$$
$$\dot{N}_i(t) = -k_{FF_i}N_i(t) + k_{FF_i}\phi_{f_i}^T(q_i, \dot{q}_i)\tau_{f_i}, \qquad (9)$$
$$N_i(0) = 0$$

and $k_{FF_i} \in \mathbb{R}^+$, can be interpreted as a forgetting factor. where ω_{1_i} and ω_{2_i} are positive scalars which are to be chosen large enough in the Lyapunov based design to achieve robust stability. Γ_i is a diagonal positive definite matrix.

4   Analysis

Theorem 1 *Consider the two arm manipulator system with dynamics defined by (1), if the cooperative controller is designed by (4), then both manipulators will move in synchronised* mirrored *motion following the human node leader's trajectory.* $\diamondsuit$

Proof Following the same proof in [28], the following Lyapunov function is proposed

$$\mathcal{V} = \mathcal{V}_1 + \mathcal{V}_2 + \mathcal{V}_3 \tag{10}$$

where

$$\mathcal{V}_1 = \frac{1}{2}(q - \bar{q}_0)^T \left[(L + B) \otimes I_m\right] (q - \bar{q}_0)^T \tag{11}$$

$$\mathcal{V}_2 = \frac{1}{2}\sum_{i=1}^{N} \left(e_i^T M_i(q_i)e_i\right) \tag{12}$$

$$\mathcal{V}_3 = \frac{1}{2}\sum_{i=1}^{N} \left(\tilde{N}_i^T W_i^{-1} \Gamma_i^{-1} W_i^{-1} \tilde{N}_i\right) \tag{13}$$

where $\tilde{N}_i(t) = N_i(t) - W_i(t)\hat{\Theta}_i = W_i(t)\tilde{\Theta}_i - e^{-k_{FF}t_i}w_i\Theta_i$. $q = (q_1^T, q_i^T, \cdots, q_N^T)^T \in \mathbb{R}^{Nm}$, $\Gamma = diag(\gamma_1, \gamma_i, \cdots, \gamma_l) \in \mathbb{R}^{l \times l}$. In particular, finite time convergence of the parameter relevant function V_3 can be shown after a number of omitted derivations:

$$\dot{V}_3 \leq -\sum_{i=1}^{N}\omega_{1b_i}\|\tilde{N}_i\|\|W_i^{-1}\| - \sum_{i=1}^{N}\omega_{2_i}\|\tilde{N}_i\|^2\|W_i^{-1}\| \tag{14}$$

$$\leq -\bar{\omega}_2 \left[\varpi_{1b}\sqrt{V_3} + \varpi_2 V_3\right] \tag{15}$$

where ω_{1b_i} is a scalar resulting from $\omega_{1i} = \omega_{1ai} + \omega_{1bi}$ (so that $\omega_{1ai} > \Gamma_i^{-1}\xi_i$), $\bar{\omega}_2 = \sum_{i=1}^{N}\omega_{2i}, \varpi_{1b} = \sum_{i=1}^{N}\frac{1}{\bar{\sigma}(\Gamma_i^{-1/2})}, \varpi_2 = \sum_{i=1}^{N}\frac{1}{\bar{\sigma}(\Gamma_i^{-1})\bar{\sigma}(W_i^{-1})}$. Finite-time convergence of the parameter estimates to its true value can enhance the human assisting cooperative control task. Further elaboration on the finite-time convergence can be found in the author's previous work in [28][29]. ∎

5   Experimental Results

An experiment is conducted by having a BERT II system holding a tray with a water cup. As shown in Figure 3, the person (which acts as the leader node) moves both arms in synchrony to achieve a cooperative, robot (mutually) assistive task between the human and the robot. The robot's role in this context can be seen as physically lifting the tray, while the human provides the guidance for the position of the tray. The advantage of the multi-agent controller is that it keeps the two robot arms in synchrony, i.e. the tray with the water remains balanced: the human positions the tray into a desired position by light pressure. According to the consensus protocol (2), this desired position is the *consensus equilibrium*. The interaction between the human and the robot can be modeled by a graph (Figure 3c), showing also the option for removal of the human pinning.

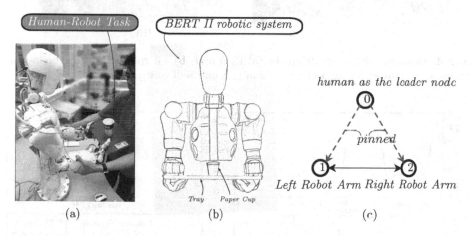

Fig. 3. (a) Human initially engaged with the robot arms giving the input for the desired assisted task. (b) The front view of BERT II in action holding the serving tray with a paper cup. (c) The human-BERT II interaction can be modeled by the graph.

Figure 4 shows the hands-off operation whereby the human input previously catered for leader-following is removed from being in contact with the BERT II's arm. Both of the robot's arm maintained the same position and gesture without affecting the integrity of the paper cup on the serving tray. This is supported by the fast decentralized control scheme with finite-time adaptation, which permits fast structural changes.

Figure 5 shows that the two arms remain in synchrony at all times, where the control torques only indicate at what point in time the human experimentalist removed pinning from one of the arms.

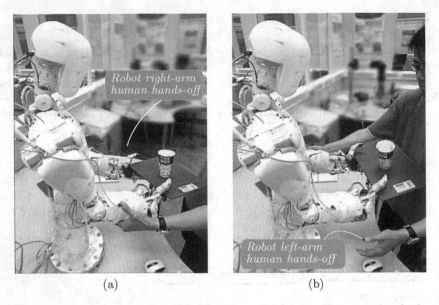

(a)   (b)

Fig. 4. Deliberate hands-off from the (a) right and (b) left robotic arm by the human to show the tray carrying a cup is still in tact and well controlled

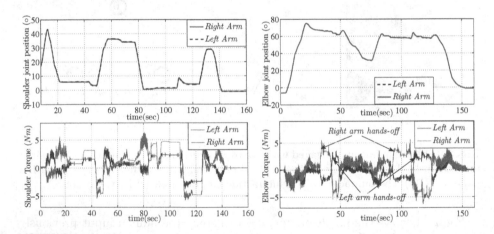

Fig. 5. The trajectories of the (a) shoulder and (b) elbow joint angles for both robot arm. The corresponding robot arms' shoulder and elbow torque can be shown in (c) and (d) respectively.

6   Conclusion

This paper presented the use of multi-agent control theory in application to human-robot assistive task. The multi-agent controller showed particular advantage for a highly synchronised motion between multiple robot manipulators where careful balance is needed as for instance in a lifting exercise. The human is merely used for guidance, while the decentralized controller easily compensates for structural changes to retain balance, i.e. consensus between the robotic arms. This can be highly important in cooperative human-robot tasks for sick, elderly or even in industrial context. Thus, it is important to note that the human retains for this type of control scheme lead and control in a direct human-robot interaction task, while the robot is physically active. The control scheme requires a minor amount of sensor systems.

References

1. Albu-Schäffer, A., Ott, C., Hirzinger, G.: A unified passivity-based control framework for position, torque and impedance control of flexible joint robots. The International Journal of Robotics Research 26(1), 23–39 (2007)
2. Chen, Z., Lii, N.Y., Wimboeck, T., Fan, S., Jin, M., Borst, C.H., Liu, H.: Experimental study on impedance control for the five-finger dexterous robot hand dlr-hit ii. In: 2010 IEEE/RSJ International Conference on Intelligent Robots and Systems (IROS), pp. 5867–5874. IEEE (2010)
3. Braun, D.J., Petit, F., Huber, F., Haddadin, S., van der Smagt, P., Albu-Schaffer, A., Vijayakumar, S.: Optimal torque and stiffness control in compliantly actuated robots. In: 2012 IEEE/RSJ International Conference on Intelligent Robots and Systems (IROS), pp. 2801–2808 (2012)
4. Khan, S.G., Herrmann, G., Pipe, T., Melhuish, C., Spiers, A.: Safe adaptive compliance control of a humanoid robotic arm with anti-windup compensation and posture control. International Journal of Social Robotics 2(3), 305–319 (2010)
5. Cheah, C.C., Wang, D.: Learning impedance control for robotic manipulators. IEEE Transactions on Robotics and Automation 14(3), 452–465 (1998)
6. Li, Y., Sam Ge, S., Yang, C.: Learning impedance control for physical robot-environment interaction. International Journal of Control 85(2), 182–193 (2012)
7. Demircan, E., Besier, T., Menon, S., Khatib, O.: Human motion reconstruction and synthesis of human skills. In: Advances in Robot Kinematics: Motion in Man and Machine, pp. 283–292. Springer (2010)
8. Spiers, A., Herrmann, G., Melhuish, C., Pipe, T., Lenz, A.: Robotic implementation of realistic reaching motion using a sliding mode/Operational space controller. In: Kim, J.-H., et al. (eds.) FIRA 2009. LNCS, vol. 5744, pp. 230–238. Springer, Heidelberg (2009)
9. Lacquaniti, F., Soechting, J.: Coordination of arm and wrist motion during a reaching task. The Journal of Neuroscience 2(4), 399–408 (1982)
10. Flash, T., Hogan, N.: The coordination of arm movements: An experimentally confirmed mathematical model. The Journal of Neuroscience 5(7), 1688–1703 (1985)
11. Ge, S.S., Li, Y., He, H.: Neural-network-based human intention estimation for physical human-robot interaction. In: 2011 8th International Conference on Ubiquitous Robots and Ambient Intelligence (URAI), pp. 390–395 (2011)
12. Koivo, A.: Adaptive position-velocity-force control of two manipulators. In: 1985 24th IEEE Conference on Decision and Control, vol. 4, pp. 1529–1532. IEEE (December 1985)
13. Hemami, A.: Kinematics of two-arm robots. IEEE Journal on Robotics and Automation 2(4), 225–228 (1986)

14. Uchiyama, M., Dauchez, P.: Hybrid Position/Force control for coordination of a two-arm robot. In: Proceedings of the IEEE International Conference on Robotics and Automation, Raleigh, North Carolina, U.S.A., Number c, pp. 1242–1247 (1987)
15. Tarn, T., Bejczy, A., Yun, X.: Design of dynamic control of two cooperating robot arms: Closed chain formulation. Institute of Electrical and Electronics Engineers (1987)
16. Zheng, Y., Luh, J.S.: Control of two coordinated robots in motion. In: 1985 24th IEEE Conference on Decision and Control, pp. 1761–1766. IEEE (December 1985)
17. Yoshikawa, T., Sugie, T., Tanaka, M.: Dynamic hybrid position/Force control of robot manipulators–Controller design and experiment. In: Proceedings of the 1987 IEEE International Conference on Robotics and Automation, pp. 2005–2010 (1987)
18. Jean, J.H., Fu, L.C.: An adaptive control scheme for coordinated multimanipulator systems. IEEE Transactions on Robotics and Automation 9(2), 226–231 (1993)
19. Slotine, J.J.: Adaptive manipulator control: A case study. IEEE Transactions on Automatic Control 33(11), 995–1003 (1988)
20. Naniwa, T., Arimoto, S., Vega, V.P.: A model-based adaptive control scheme for coordinated control of multiple manipulators. In: Proceedings of IEEE/RSJ International Conference on Intelligent Robots and Systems (IROS 1994), pp. 695–702. IEEE (1994)
21. Liu, Y.H., Arimoto, S., Parra-Vega, V., Kitagaki, K.: Decentralized adaptive control of multiple manipulators in co-operations. International Journal of Control 67(5), 649–674 (1997)
22. Cheng, L., Hou, Z.G., Tan, M.: Decentralized adaptive consensus control for multimanipulator system with uncertain dynamics. In: 2008 IEEE International Conference on Systems, Man and Cybernetics, pp. 2712–2717 (October 2008)
23. Hong, Y., Hu, J., Gao, L.: Tracking control for multi-agent consensus with an active leader and variable topology. Automatica 42(7), 1177–1182 (2006)
24. Chen, G., Lewis, F.: Distributed adaptive controller design for the unknown networked lagrangian systems. In: 2010 49th IEEE Conference on Decision and Control (CDC), pp. 6698–6703 (2008)
25. Das, A., Lewis, F.L.: Cooperative adaptive control for synchronization of second-order systems with unknown nonlinearities. International Journal of Robust and Nonlinear Control (2010)
26. Qu, Z.: Cooperative Control of Dynamical Systems. Springer, Heidelberg (2009)
27. Lewis, F.L., Dawson, D.M., Abdallah, C.T.: Robot Manipulator Control: Theory and Practice, 2nd edn. Marcel Dekker (2004)
28. Mahyuddin, M.N., Herrmann, G.: Distributed motion synchronisation control of humanoid arms. In: Omar, K., et al. (eds.) FIRA 2013. CCIS, vol. 376, pp. 21–35. Springer, Heidelberg (2013)
29. Mahyuddin, M.N., Herrmann, G., Khan, S.: A novel adaptive control algorithm in application to a humanoid robot arm. In: Herrmann, G., Studley, M., Pearson, M., Conn, A., Melhuish, C., Witkowski, M., Kim, J.-H., Vadakkepat, P. (eds.) FIRA-TAROS 2012. LNCS, vol. 7429, pp. 25–36. Springer, Heidelberg (2012)

Effects of Politeness and Interaction Context on Perception and Experience of HRI

Maha Salem[1], Micheline Ziadee[2], and Majd Sakr[3]

Carnegie Mellon University in Qatar, Doha, Qatar
MahaSalem@web.de, michelinekz@gmail.com, msakr@qatar.cmu.edu

Abstract. Politeness is believed to facilitate communication in human interaction, as it can minimize the potential for conflict and confrontation. Regarding the role of politeness strategies for human-robot interaction, conflicting findings are presented in the literature. Thus, we conducted a between-participants experimental study with a receptionist robot to gain a deeper understanding of how politeness on the one hand, and the type of interaction itself on the other hand, might affect and shape user experience and evaluation of HRI. Our findings suggest that the interaction context has a greater impact on participants' perception of the robot in HRI than the use – or lack – of politeness strategies.

Keywords: Natural Language Dialogue, Politeness, Communication Style.

1 Introduction

Interactive service robots are increasingly employed to provide people with information in various environments, thereby acting as educational tutors, museum guides or receptionists. To succeed in the role of a social actor that is intended to interact with humans, such robots should therefore be equipped with an appropriate level of communicative capabilities and social competencies. Being very skilled social actors, humans maintain the effectiveness of their social communication practices by means of numerous strategies, e.g. to elicit cooperation and establish common ground between interlocutors. Politeness represents one such strategy and is believed to facilitate interaction, as it minimizes the potential for conflict and confrontation that is inherent in human interchange [8].

In their seminal work on politeness, Brown and Levinson [2] developed their theory around the notion of "face". As face represents a public image that we seek to protect, any acts that undermine that image are considered face-threatening acts (FTAs). The role of politeness is to mitigate such acts and minimize their threat, which is achieved through four types of strategies employed in communication: *bald on record* (i.e. no attempts to minimize threats to the hearer's face), *positive politeness* (i.e. a type of metaphorical extension of intimacy, to imply common ground or shared desires), *negative politeness* (i.e. to minimize the imposition that a FTA unavoidably effects) and *off-record* (i.e. making it impossible to determine one sole, clear communicative intention). The choice of

G. Herrmann et al. (Eds.): ICSR 2013, LNAI 8239, pp. 531–541, 2013.

strategy depends on factors like the social distance and power relations between the interaction partners as well as the extent to which the act is face-threatening.

While politeness is important for human communication, it is unclear which role it might play in human-robot interaction (HRI). The present work thus aims to shed light on how the use of politeness strategies may affect social perceptions of a robot and HRI. For this, we conducted an experiment in which we employed a receptionist robot intended for use at Carnegie Mellon University in Qatar as an interaction partner. Based on our findings, we discuss general implications of our work with regard to future research directions in the field of HRI.

2   Background and Related Work

Research in human-computer interaction showed that people tend to respond to and treat media and computer systems in social ways, similar to real people [12], and that people act politely towards machines [10]. In effect, Reeves and Nass, the authors of *The Media Equation*, claim that "the biggest reason for making machines that are polite is that people are polite to machines. Everyone expects reciprocity, and everyone will be disappointed if it's absent" [12]. As a result, the natural human tendency to relate to computers as social actors is often exploited for the design of interactive sociable agents and robots.

Politeness has been of special interest to researchers in agent-based teaching. According to Johnson et al. [7], "real tutors use politeness as a means for respecting the student's social face, and for indirectly fostering his intrinsic motivation". Consequently, they developed interaction tactics for animated pedagogical agents based on the model by Brown and Levinson. To evaluate the impact of politeness in learning settings, Wang et al. [15] conducted an experiment in which students completed a learning task with the help of an on-screen agent. The agent produced polite or direct suggestions in response to the learners' queries. Overall the polite agent had a positive effect on the students' learning outcomes compared to the direct agent. This 'politeness effect' was stronger with students who needed more help and hence had a richer interaction with the agent.

While a number of studies have explored politeness strategies for animated agents and confirmed their effectiveness in human-agent interaction, only few researchers in the field of HRI have addressed this topic so far. Moreover, previous politeness studies typically focused on a single interaction with a specific (e.g. learning) task. For example, Nomura and Saeki [11] present a study in which participants interacted with a small-sized humanoid robot giving them instructions to perform a task. During interaction, the robot performed different types of motion to express varying politeness levels, while generated speech was identical across conditions. The results did not show a beneficial effect of the robot's polite motions on human task performance, but suggested that human perception of the robot was enhanced. However, the authors concede that their findings are limited to a specific interaction task and robot. In a pre-study of human dialogues for direction retrievals, Weiss et al [16] assumed that politeness would have a major influence on the success of the dialog. However, this assumption was not supported and was therefore not further investigated in the main study.

Given the contradicting findings in the wider human-machine interaction literature, we lack a comprehensive understanding of how people perceive robots that employ politeness strategies and of the role of verbal and nonverbal cues in this process. It is further unclear whether such strategies are beneficial in specific interaction contexts only, e.g. teaching scenarios, or whether they can be recommended as general design guidelines as done by the Media Equation Theory. The present paper is thus driven by the following research questions. First, does the level of politeness expressed by a robot influence human perception and acceptance of the robot, task performance and the overall HRI experience? And second, what role does the context play in which the interaction takes place, i.e. do humans prefer a polite robot in one situation more than in another?

3   Method

We conducted an experiment to gain a deeper understanding of how politeness and interaction context might impact and shape user experience and evaluation of HRI. Hypotheses, experimental design, procedure, measures and information on participants are described in the following.

3.1   Hypotheses

Based on findings from related work on politeness strategies in human-machine interaction (e.g. [12,15]), we developed two main hypotheses for our experiment:

1. *Effect of condition.* Manipulation of the robot's politeness strategy will affect
   (a) participants' perception of the robot and subjective assessment of HRI.
   (b) participants' performance at recalling the route directions presented by the robot (objective assessment of HRI).
2. *Effect of HRI context.* The change of HRI context from goal-oriented interaction (*Task 1*) to open dialogue (*Task 2*) will affect participants' perception of the robot and subjective assessment of HRI.

3.2   Experimental Design

We conducted a between-participants study in which participants interacted with the Hala receptionist robot [14], which consists of a stationary anthropomorphic torso with a screen mounted on a pan-tilt unit (see **Fig. 1**). We manipulated the *politeness level* of the robot's verbal behaviors in two experimental conditions and measured how this manipulation might impact participants' task performance, perception of the robot and HRI experience in general. Our manipulations were informed by the theoretical framework provided by Brown and Levinson [2], as well as by results from previous experiments using the Hala platform, particulary [9]. Specifically, our 'polite' condition was designed to follow the principle of *positive politeness*, while the control condition was inspired by the *bald on record* strategy. To additionally verify their effectiveness, verbal stimuli

Fig. 1. Hala receptionist robot and the experimental setting

Fig. 2. The robot's facial expression set to neutral

for the two conditions were evaluated prior to running the experiment by means of an online questionnaire. It comprised the original audio files to be uttered by the robot, organized in 32 pairs of utterances which could be replayed with a mouse click. 13 independent raters were asked to identify the polite version of each pair, resulting in an overall inter-rater agreement of 84.61 %.

The robot's face, providing the main channel for the robot's expression of non-verbal behaviors such as emotions, was identical across conditions and set to a neutral facial expression as shown in Fig. 2. Visemes were generated to accompany the robot's speech output, for which the Acapela text-to-speech system [1] was used. The decision to refrain from the use of gaze and emotional display via facial expressions was motivated by the idea to focus on the verbal interaction level to avoid effects caused by the interplay of the two modalities.

The experiment comprised two consecutive tasks, each representing a different HRI context, and are described in the following.

1. Direction-giving Task. In the first interaction task, participants were given a list with names of three locations in a fictitious building and were instructed to ask the robot for directions to these places. The robot's verbal utterances varied across experimental conditions for dialogue acts of greeting, direction-giving as well as handling failure and disagreement when giving directions. At the beginning of the interaction, the robot greeted the participant either with the phrase "Hello, how may I help you?" in the polite condition group or only with the word "Hello" in the control group. Participants were instructed to then ask for directions to Dr. Miller's office, which consisted of five steps (e.g. "turn right", "walk straight until you reach a fountain") and varied across the two conditions based on the use of politeness markers such as "please" or "you will want to turn right" as opposed to "turn right". Since we used participants' ability to recall these directions to measure their task performance, the directions were pretested to be feasible but not trivial to remember. For the second location on the list, participants had to ask the robot for directions to the bookshop, which triggered the robot's failure handling behavior. In the polite condition the robot responded with "Sorry, I don't know where it is, because I am new" in contrast to the response "I have absolutely no idea" in the control condition. Finally, for the third location, participants had to ask for the cafeteria and were instructed to imagine that they misunderstood the robot after being given the directions by asking to clarify if they should "turn left?". This triggered the robot's disagreement handling behavior, resulting in the response "Yes, *turn right*" in the polite condition to avoid explicit disagreement as opposed to the response "*No*, turn right" in the control condition. The robot's response to the participant's final utterance (typically a variation of "thank you" or "goodbye") did not differ across conditions and was either "you're welcome" or "goodbye'.

2. Chitchat Task. For the second interaction, participants were instructed to have an open conversation with the robot, but they were informed that the robot's knowledge was still under development and thus limited. In reality, the restriction of the robot's knowledge base was a design choice made for this experiment to allow for a comparable interaction experience across participants. Following a short greeting sequence ("Hello. It's nice to meet you! How are you?" in the polite condition versus "Hello. How are you?" in the control condition), participants could ask the robot any questions they liked. If the knowledge base did not provide an answer to a question asked, the robot responded by outlining some topics available for discussion (e.g. "Unfortunately, I have no reply to this question. How about talking about Qatar, Carnegie Mellon University, my job or my hobbies, instead?" in the polite condition versus "I have no reply to this question. We can talk about Qatar, Carnegie Mellon University, my job, and my hobbies, instead" in the control condition). As with the first task, the robot's verbal utterances varied across the two conditions based on the use (or lack) of politeness markers such as "please" when asking the participant for something and "I'm very sorry" or "unfortunately" when handling failure to respond.

We refrained from randomizing the order of the two interactions tasks, since the unstructured nature of the chitchat task could result in participants having different initial perceptions of the robot depending on how the interaction proceeded and what dialogue content was triggered. This, in turn, could affect participants' assessments of the robot after the more structured direction-giving task.

3.3   Experimental Procedure

The experiment was conducted in a closed office room on the campus of Carnegie Mellon University in Qatar. The interaction area was separated from the table at which participants filled out the subsequent questionnaires by means of a movable wall (see Fig. 1). Participants were tested individually. They were greeted by the experimenter outside the experiment room where they received a brief description of the experimental process and were asked to review and sign a consent form. After consenting, they were asked to fill out a short questionnaire that recorded their demographic background and previous experience with robots and technology. Participants were then introduced to the first interaction, the *direction-giving task*. Following this task, participants were asked to sit at a table with a laptop in the same room (see Fig. 1) and to fill out a questionnaire evaluating the robot and their HRI experience. Upon completion, participants received instructions for the second interaction, the *chitchat task*. After the second interaction, they were asked to fill out another questionnaire on the laptop to document their interaction experience.

To ensure minimal variability in the experimental procedure while allowing for interaction based on natural speech, which is not currently supported by the system, the robot was controlled using a Wizard-Of-Oz technique [5] during the study. The experimenter initiated the robot's verbal interaction behavior from a selection of pre-determined utterances. The ordering of the utterance sequence remained identical for each experimental run within the same condition group for the first task (*direction-giving*), but differed for the second task (*chitchat*) depending on the participants' input. The entire interaction was recorded using a microphone and a voice recorder, while the experimenter listened to and controlled the interaction from an adjacent room.

The second questionnaire was followed by an interview in which the investigator asked participants to describe and comment on their experience in response to open-ended questions. After the interview, participants were carefully debriefed about the purpose of the experiment and received a tartlet as a thank-you before being dismissed. The total experiment time was approximately 30 minutes, out of which the participants interacted with the robot for about 7 to 10 minutes.

3.4   Dependent Measures

Dependent variables comprised a number of **subjective measures** based on two questionnaires (one after each interaction task). These were used to tap various dimensions of HRI including the participants' perception of the robot, their general HRI experience and their enjoyment of and involvement in the interaction tasks. Five-point Likert scales were used for all scales; the measures are described in more detail in the following and are summarized in Table 1.

Politeness We measured perceived politeness of the robot based on two items: 'the robot is polite' as well as the reversely coded ratings of 'the robot is rude'. This measure was mainly intended as a manipulation check to verify that our manipulations were effective.

Competency Participants were asked to report how they rated the following competency-related attributes with regard to the robot: 'competent', 'helpful', 'effective', 'cooperative', 'organized' and 'intelligent'.

Extroversion The degree of extroversion regarding the robot's personality was assessed using four traits: 'extrovert', 'outgoing', 'sociable', and 'fun-loving'.

Perceived Warmth We measured perceptions of the robot's interpersonal warmth, a core dimension of human social cognition [4]. Warmth was assessed using the reversely coded ratings of the six items 'cold', 'hardhearted', 'shallow', 'aggressive', 'rude' and 'impatient'. Since these items were borrowed from measures of typically human traits proposed by Haslam et al [6], the warmth index can be further used as an indicator of anthropomorphization in HRI (see [13]).

Shared Reality We further administered three items ('How close do you feel to the robot', 'How much do you think you have in common with the robot?', and

Table 1. Dependent measures, respective questionnaire items and scales used to evaluate participants' perception of the robot and HRI experience

Index/Measure (Cronbach's α):	Questionnaire Items:	Scale:
Politeness ($\alpha = .753$)	polite, rude_rev	1 = strongly disagree, 5 = strongly agree
Competency ($\alpha = .881$)	competent, helpful, effective, cooperative, organized, intelligent	1 = strongly disagree, 5 = strongly agree
Extroversion ($\alpha = .894$)	extrovert, outgoing, sociable, fun-loving	1 = strongly disagree, 5 = strongly agree
Perceived Warmth ($\alpha = .815$)	cold_rev, hardhearted_rev, shallow_rev, aggressive_rev, rude_rev, impatient_rev	1 = strongly disagree, 5 = strongly agree
Shared Reality ($\alpha = .766$)	'How close do you feel to the robot?' 'How much do you think you have in common with the robot?' 'How pleasant was the interaction with the robot for you?'	1 = not at all, 5 = very much
Future Contact Intentions	'Would you like to interact with the robot again?'	1 = not at all, 5 = very much

'How pleasant was the interaction with the robot for you?') to assess participants' degree of shared reality with the robot. The shared reality index taps perceptions of similarity and experienced psychological closeness to the robot [3]. Moreover, it covers aspects of human-robot acceptance, because participants had to indicate how pleasant the interaction with the robot was for them.

Future Contact Intentions Finally, using a single item to measure participants' future contact intentions, we asked participants to indicate to what extent they would like to interact with the robot again.

In addition, assessment of *task performance* based on participants' ability to recall the directions to Dr. Miller's office was used as an **objective measure** (with values 0 = incorrect and 1= correct).

3.5   Participation

A total of 44 participants (20 female, 24 male) took part in the experiment, ranging in age from 22 to 70 years ($M = 39.45$ years, $SD = 12.19$). All participants were English native speakers and were recruited at Carnegie Mellon University in Qatar. Based on five-point Likert scale ratings, participants were identified as having negligible experience with robots ($M = 1.89$, $SD = 0.97$) and advanced skills regarding technology and computer use ($M = 4.27$, $SD = 0.82$). Based on the same scale, their video game use was found to be moderate ($M = 2.91$, $SD = 1.12$). Participants were randomly assigned to one of two experimental conditions that manipulated the politeness level of the robot's verbal behaviors, while maintaining gender- and age-balanced distributions.

4 Results

First, reliability analyses (Cronbach's α) were conducted to check for internal consistency where applicable. The indices proved reliable with values for Cronbach's α ranging between .75 and .89 for all dependent measures (see Table 1). Consequently, participants' responses to the respective items were averaged to form indices of *politeness, competency, extroversion, perceived warmth* and *shared reality*. We then conducted repeated-measures ANOVA to see if differences between measurements after Task 1 and 2 were significant across conditions, while further exploring potential effects using the factors 'condition' and 'gender'.

Results confirmed that the manipulation of *politeness* level for the robot's verbal utterances had a significant effect on participants' ratings of how polite they found the robot, $F(1,40) = 10.48$, $p = 0.002$. There was a significant interaction effect between the level of politeness and the gender of the participant: female participants rated the robot as less polite after the first task than male participants; however, they gave it higher ratings for perceived politeness after the second task compared to their male counterparts whose ratings dropped after the second interaction round, $F(1,40) = 4.77$, $p = 0.035$ (see Fig. 3a).

Our analysis showed a main effect of HRI context on the robot's perceived *competency*: participants found the robot more competent after the first interaction task involving direction-giving than after the second task comprising the chitchat interaction, $F(1,40) = 6.49$, $p = 0.015$.

There was a significant main effect of HRI context on participants' ratings of the robot's level of *extroversion*: participants perceived the robot as more extrovert after the chitchat interaction than after the direction-giving task, $F(1,40) = 16.81$, $p < 0.001$.

Our analysis showed a main effect of HRI context on participants' ratings of *perceived warmth*: participants reported greater perceived warmth after meeting the robot for the chitchat interaction than after the first task involving direction-giving, $F(1,40) = 4.84$, $p = 0.034$. We found a significant interaction between HRI context and gender for the perceived warmth measure: female participants perceived less warmth towards the robot after the first task than male participants, but gave it higher ratings for perceived warmth after the second task compared to their male counterparts whose ratings dropped after the chitchat interaction, $F(1,40) = 9.59$, $p = 0.004$ (Fig. 3b). A marginal interaction was found between HRI context and politeness level for participants' ratings of perceived warmth, with a greater increase in ratings after the second interaction for participants in the control condition compared to those in the polite condition, $F(1,40) = 3.24$, $p = 0.079$ (Fig. 3c). Following up on this finding, independent t-tests conducted for this measure showed that after the first interaction, participants in the polite group reported significantly greater perceived warmth with regard to the robot ($M = 3.83$, $SD = 0.54$) than those in the control group ($M = 3.36$, $SD = 0.79$), $t(42) = -2.34$, $p = 0.024$. A comparison of participants' ratings after the second interaction yielded no significant differences.

We found a significant main effect of HRI context on participants' reported experience of *shared reality*: participants experienced greater shared reality with

a)
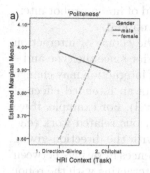

b)
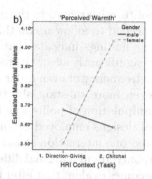

c)
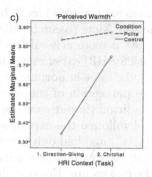

Fig. 3. Interaction effects between a) HRI context and gender regarding ratings of *politeness*; b) HRI context and gender regarding ratings of *perceived warmth*; c) HRI context and experimental condition regarding ratings of *perceived warmth*

the robot after the second task involving the chitchat encounter than after the direction-giving task, $F(1,40) = 4.26$, $p = 0.046$.

Finally, our analysis showed a main effect of HRI context on participants' *future contact intentions* with regard to the robot: participants expressed a greater desire to interact again with the robot after the direction-giving interaction than after the chitchat interaction, $F(1,40) = 5.58$, $p = 0.023$.

Apart from the manipulation check variable measuring the robot's politeness level as perceived by the participants and the *perceived warmth* index measured after the first interaction, our analysis showed no effect of condition on the remaining dependent variables. Importantly, with regard to our objective measure, we found no effect of condition on participants' *task performance* when asked to recall the route directions to Dr. Miller's office, $U = 220.0$, $Z = -0.64$, $p = n.s.$

5   Discussion and Conclusion

The results indicate that the manipulation of politeness level – although perceived by the participants – had no major impact on participants' perception of the robot and overall HRI experience following the two interactions, nor on their task performance. This finding rejects our first hypothesis which predicted an effect of experimental condition. In contrast, the change of interaction context from the goal-oriented Task 1 to the open dialogue of Task 2 had a significant impact on all other dependent measures besides the manipulation check variable. This is in line with our second hypothesis which predicted an effect of HRI context. Although these findings may partly be attributed to the general repeated encounter with the robot, e.g. based on a familiarization effect, and our experimental design did not include a randomized order of the interaction tasks to further investigate our observations, they have some general implications.

Our results contradict previous findings from human-agent research (e.g. [7,15]) regarding the effectiveness of politeness strategies for HRI. In addition, they challenge the general applicability of the Media Equation Theory, which claims that users expect machines to be polite, with regard to the field of human-robot interaction. More research will be needed to follow up on this notion.

At a more general level, our findings indicate that the choice of interaction task for HRI experiments may significantly affect the results, since the type and – maybe more importantly – the frequency or even order of interaction may change the perception of the robot even more substantially than an intended effect of condition (in our case the manipulation of politeness level). For example, if we had followed the experimental designs employed by previous related work (e.g. [11,16]) and had only conducted one interaction round with the direction-giving task, we would have, similarly, only found marginal differences in ratings between the two groups. However, it becomes evident that after interacting with the robot *again* and in a different context (e.g. less goal-oriented), perception and ratings of the robot changed immensely and outweighed the expected effect of condition.

This could raise a question regarding the validity of some HRI studies that comprise only a single, typically short, encounter with a specific robot in which an effect of condition may or may not be shown: what if the same manipulation of robot behavior had been used as a measure in a different interaction scenario?

In view of these controversies, our findings could have some wider implications for future HRI studies, since they emphasize the importance of long-term studies with robots, or at least ensuring multiple encounters, which are still rather rare in HRI research. Therefore, we hope that our present work will stimulate the discussion of better design guidelines for HRI experiments in the future.

Acknowledgement. This publication was made possible by NPRP grant 09-1113-1-171 from the Qatar National Research Fund (a member of Qatar Foundation). Statements herein are solely the responsibility of the authors.

References

1. http://www.acapela-group.com/ (accessed July 2013)
2. Brown, P., Levinson, S.C.: Politeness: Some Universals in Language Usage. Studies in Interactional Sociolinguistics. Cambridge University Press (1987)
3. Echterhoff, G., Higgins, E.T., Levine, J.M.: Shared reality: Experiencing commonality with others' inner states about the world. Perspectives on Psychological Science 4, 496–521 (2009)
4. Fiske, S.T., Cuddy, J.C., Glick, P.: Universal dimensions of social cognition: Warmth and competence. Trends in Cognitive Science 11(2), 77–82 (2006)
5. Fraser, N.M., Gilbert, G.N.: Simulating speech systems. Computer Speech & Language 5(1), 81–99 (1991)
6. Haslam, N., Bain, P., Loughnan, S., Kashima, Y.: Attributing and denying humanness to others. European Review of Social Psychology 19, 55–85 (2008)
7. Johnson, W.L., Rizzo, P., Bosma, W., Kole, S., Ghijsen, M., van Welbergen, H.: Generating socially appropriate tutorial dialog. In: André, E., Dybkjær, L., Minker, W., Heisterkamp, P. (eds.) ADS 2004. LNCS (LNAI), vol. 3068, pp. 254–264. Springer, Heidelberg (2004)
8. Lakoff, R.T.: Talking Power: The Politics of Language in Our Lives. Basic Books, New York (1990)

9. Makatchev, M., Simmons, R.G., Sakr, M., Ziadee, M.: Expressing ethnicity through behaviors of a robot character. In: ACM/IEEE International Conference on Human-Robot Interaction (HRI 2013), pp. 357–364 (2013)
10. Nass, C., Moon, Y., Carney, P.: Are people polite to computers? Responses to computer-based interviewing systems. Journal of Applied Social Psychology 29(5), 1093–1109 (1999)
11. Nomura, T., Saeki, K.: Effects of polite behaviors expressed by robots: A case study in japan. In: Proceedings of the 2009 IEEE/WIC/ACM International Joint Conference on Web Intelligence and Intelligent Agent Technology (WI-IAT 2009), vol. 2, pp. 108–114. IEEE Computer Society, Washington, DC (2009)
12. Reeves, B., Nass, C.: The media equation: How people treat computers, television, and new media like real people and places. Cambridge University Press (1996)
13. Salem, M., Eyssel, F., Rohlfing, K., Kopp, S., Joublin, F.: To err is human(-like): Effects of robot gesture on perceived anthropomorphism and likability. International Journal of Social Robotics, 1–11 (2013)
14. Simmons, R., Makatchev, M., Kirby, R., Lee, M., Fanaswala, I., Browning, B., Forlizzi, J., Sakr, M.: Believable robot characters. AI Magazine 32(4), 39–52 (2011)
15. Wang, N., Johnson, W.L., Mayer, R.E., Rizzo, P., Shaw, E., Collins, H.: The politeness effect: Pedagogical agents and learning outcomes. Int. J. Hum.-Comput. Stud. 66(2), 98–112 (2008)
16. Weiss, A., Mirnig, N., Forster, F.: What users expect of a proactive navigation robot. In: Proceedings of the Workshop on the Role of Expectations in Intuitive Human-Robot Interaction at the ACM/IEEE International Conference on Human-Robot Interaction (HRI 2011), pp. 36–40 (2011)

Facial Expressions and Gestures to Convey Emotions with a Humanoid Robot

Sandra Costa, Filomena Soares, and Cristina Santos

Algoritmi Centre, University of Minho
Guimaraes, Portugal

Abstract. This paper presents the results of a perceptual study with ZECA (Zeno Engaging Children with Autism), a robot able to display facial expressions. ZECA is a robotic tool used to study human-robot interactions with children with Autism Spectrum Disorder. This study describes the first steps towards this goal. Facial expressions and gestures conveying emotions such as sadness, happiness, or surprise are displayed by the robot. The design of the facial expressions based on action units is presented. The participants answered a questionnaire intended to verify if these expressions with or without gestures were recognized as such in the corresponding video. Results show that participants were successfully able to recognize the emotion featured in the corresponding video, and the gestures were a valuable addition to the recognition.

1 Introduction

The Robotica-Autismo research project aims to use findings in social assistive robotics in children with autism spectrum disorder (ASD). It is expected that the use of robots promotes the development of social interaction skills [1]. Research on human-robot interaction (HRI) has shown that robots improve participation and prompt new social behaviours within this target group [1]. ASD is defined as a global developmental disorder, which usually manifests during the first three years of life [2]. The current criteria (DSM-V) characterizes ASD by a severe deficit in social communication and restrict and repetitive behaviour patterns, activities and interests [3].

The focus of this research project is to improve the social exchanges between a child with ASD and a partner. The tool used by the researchers to foster social interactions, communication and emotion recognition is a humanoid robot with the capability of displaying facial expressions called ZECA (Zeno Engaging Children with Autism), produced by Hanson Robotics [4].

In particular, the aim is to use affective robotics defined as robotic systems that rely on detection, synthesis and production of affective and emotional behaviours to enable visual recognition of emotions. This recognition is defined as the ability to visually identify the socio-emotional processes such as emotions. In this article, we present the evaluation of a perceptual study of facial expressions and gestures displayed by ZECA. This study involved two different groups: one composed by typically developing children and another with adults. With this

G. Herrmann et al. (Eds.): ICSR 2013, LNAI 8239, pp. 542–551, 2013.

preliminary study, we expect to validate the produced facial expressions and gestures.

2   State of Art

There are already a few research projects that have devoted their attention to the specific theme of the design of emotional expressions in robots for HRI. However, only a small number of projects focuses specifically on the use of robots with children with ASD, testing the recognition of facial expressions and emotions. We are going to focus on those projects, having in mind the specificity of the ASD.

In their most recent work, the WikiTherapist project [5] aims to promote embodied interaction through interfaces able to develop an understanding of the intent of the emerging natural movement of the human body. The movements are then simulated on robots as part of games that aim to improve the social interaction skills of children with ASD. The authors describe a framework which enables the expression and interpretation of patterns of emotional movement.

The humanoid robot FACE [6] was built to allow children with ASD to handle expressive and emotional information, being able to express and convey emotions, and emulate empathetic behaviours. This platform integrates therapeutic adaptive information derived from sensors used on children and in the surrounding environment. With control and data processing algorithms, the expressions and movements of FACE are modeled to harmonize with the feelings of the user.

The robot Bandit project [7] aims at facilitating HRI in a natural way, increasing interactions of the child. A sound speaker was mounted on the robot to display pre-recorded vocalizations communicating the emotional state of the robot (e.g., happy or confused). They found that more verbal children showed more interest, but also had more expectations for the social capabilities of the humanoid robot.

In [8] KASPAR is used in tactile interaction scenarios with children with ASD. It responds according to different types of touches showing happiness or sadness. The goal of this feedback is to automatically produce a response to the childrens tactile interaction, teaching appropriate physical social engagement, reinforcing suitable behaviours when using touch to interact with another agent. Despite having no significant differences when evaluating tactile interaction, more than 90% of the times the children touched the robot gently.

The robot Kismet [9] produces a variety of social cues from visual and auditory channels, and delivers social signals to people with whom it interacts through gaze direction, facial expressions, body postures and vocalizations. In their research, the authors studied how children reacted to the robot when it generated a variety of social gestures and facial expressions in response to stimuli from the child. The result consisted on an ongoing turn-taking between robot and human aimed at maintaining the robot's drives within homeostatic bounds.

3   Methodology

The final goal of the project is to use ZECA to label emotional behaviours to children with ASD. Several questions arise: can I teach a "happy state" with ZECA smiling or it is better to accomplish this state with a "happy gesture"? Is ZECA really happy? In order to answer these questions, a preliminary study was undertaken focused on displaying ZECA's emotional states (happy, sad, surprised, among others) defined only with facial expressions and facial expressions with gestures. As mentioned before, this is only a preliminary study, and the question on the basis of this research is "How can affective robotics contribute to develop visual emotion recognition in children with ASD?". This paper constitutes then a basic step to test this question. Two different pre-tests were performed to validate ZECA's expressions used in the perceptual study.

3.1   Pre-test 1

The basic emotions defined by Ekman [10] plus a neutral facial expression were chosen to be displayed on the robot: neutral, happiness, sadness, surprise, fear, anger, and disgust. The display of these facial expressions and the corresponding gestures were defined in experimental pre-tests in laboratory. A pre-study with a first version of the set of facial expressions set was done with seven adults. It was requested from these adults to classify each of the facial expressions, using a forced choice. This experiment was done individually on a computer, and through the observation of seven videos with the robot displaying these emotions. The participants had then to register their answers in a notebook.

3.2   Pre-test 2

Having in mind the results from pre-study 1, the Action Units (AU) used by Ekman [10] in the Facial Action Coding System (FACS) were studied to reach better and faster the desirable expressions. Ekman defined which action units (fundamental actions of individual muscles or groups of muscles) would be necessary to define the basic emotions, and using this information, the robot joints were defined to get the correct correspondence. The description of all the AU can be found in [10]. In some cases, due to some limitations in the degrees of freedom of the robot nor all AU, especially the ones correspondent to micro-expressions could be specified. This way the possible approximation based on observation was done.

Using FaceReader of Noldus [11], the set of facial expressions was evaluated. FaceReader is a software tool for automatic facial expression analysis. This software works in three steps: face finding - an accurate position of the face is found; face modeling - the active appearance model is used to synchronize an artificial face model, which describes the location of 500 key points as well as the texture of the face; and face classification - output is presented as six basic expressions and one neutral state [11]. ZECA has in its head eleven degrees of freedom (DOF): Neck Yaw, Neck Roll, Neck Pitch, Brows Pitch, Eyelids, Eyes Pitch,

Eye Left, Eye Right, Jaw, Smile Left, Smile Right. Due to the lack of DOF on the nose, cheeks and the absence of teeth, it was not possible with the software to identify the disgust facial expression. Therefore, it was decided to exclude this facial expression from the final perceptual study.

3.3   Perceptual Study

The authors intended to verify the reliability of the produced data and for this purpose, a perceptual study was designed. We wanted to provide a precise representation of the correct perceptions both from children and adults, using the videos from the Pre-Test 1 and 2, an online questionnaire was built to collect the answers of the sample. The questionnaire, available on a computer, was divided in two parts. In the first part, the users had to give information about their age and gender and make the matching between videos showing only different facial expressions and seven options ("I'm sad", "I'm happy", "I'm angry", "I'm scared", "I'm neutral", "I'm surprised", "I do not know"). In the second part of the questionnaire, the same facial expressions were complemented with gestures. It was decided to add gestures, as a component of non-verbal emotion expression, because they are believed to help people for interpreting the emotional state of another agent [12]. Experts (e.g., psychology, puppetry, drama) were not used to create gestures in this exploratory study because it was unclear what expertise was needed. Further, using engineers without formal experience in gesture creation reflects current practice.

The participants only had to choose the correct option which he/she considered appropriated for each video. Both videos and options were randomized. Before the final study, this questionnaire was tested, to find if technical details should be improved. Small changes like the size of the font, the buttons, among others were changed to cater for the difficulties of the younger children while dealing with computers specially regarding fine-motor coordination.

3.4   Participants

Two distinct groups participated in this study. These participants attended the perceptual study through a web page specifically created for this purpose.

Group A was constituted by typically developing children and the sample used in this study had 42 participants between 8 and 10 years old (M = 9.046; SD = 0.688). The test was performed in Primary School of Gualtar, with whom our team has a signed protocol. The experiment with this group was performed in a computer room of the school, with 11 computers with Internet connection. Each trial had an approximated duration of 30 minutes. First, the protocol was explained, and then the children performed the experiment, on their own computer. When they finished filling in the questionnaire, another child took his/her place on the same computer. Group B was composed by 61 adults aged between 18 and 59 years old (M = 32.393; SD = 9.730). Both groups were instructed to complete the questionnaire selecting the most appropriate correspondence for each video.

4   Results

The results of the two pre-tests and the perceptual study are presented in this section.

4.1   Pre-test 1

The results of the pre-test 1 are represented in fig. 1.

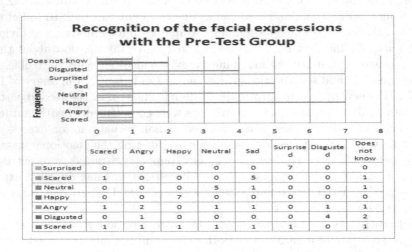

	Scared	Angry	Happy	Neutral	Sad	Surprised	Disgusted	Does not know
Surprised	0	0	0	0	0	7	0	0
Scared	1	0	0	0	5	0	0	1
Neutral	0	0	0	5	1	0	0	1
Happy	0	0	7	0	0	0	0	0
Angry	1	2	0	1	1	0	1	1
Disgusted	0	1	0	0	0	0	4	2
Scared	1	1	1	1	1	1	0	1

Fig. 1. The result of the recognition of the facial expressions on ZECA with the Pre-Test Group

Despite the good results on the corresponding facial expressions to happiness (100%), sadness (71.429%), surprise (100%), fear (71.429%), and neutral (71.429%), some more development needed to be done to the disgust (57.142%) and anger (28.571%) facial expressions.

4.2   Pre-test 2

Several images were subjected to analysis due to the limitations of the FaceReader software regarding the image quality. Still, the performance of the robot in the rest of the facial expressions was quite satisfactory and two examples can be observed in the following figures.

In fig. 2, the robot is displaying the sad face, and a great intensity is obtained from the raising of the inner brow. The Brow Lowered AU was not identified maybe because the distance between the eye and the brow is considerable higher compared to a human being. Regarding the Lip Corner Depressor AU, this was also not recognized by the software.

Fig. 2. Results from FaceReader software when analysing the sad face expression

4.3   Perceptual Study

The results of the questionnaires performed with Group A and Group B were quite encouraging. Fig. 3 and 4 show the results of Group A in the first and second part of the questionnaire, respectively. In the Fig. 3, we can see that only two of the facial expressions (scared and angry) had less than 50% of the correct recognition rate, but still above chance level. But the other facial expressions yielded more than 75% of recognition rate (Happy - 83.333%, Neutral - 85.714%, Sad - 97.619%, Surprised - 76.190%).

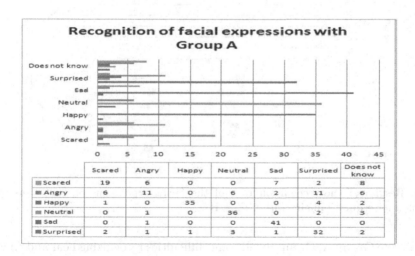

Fig. 3. Results of the recognition rate of the facial expressions on ZECA using a multiple-choice questionnaire with Group A

Adding the gesture to the facial expression helped the recognition of the associated emotion (fig. 4). The recognition of Fear improved from 45.238% to 73.81% and of Anger from 26.191% to 47.619%.

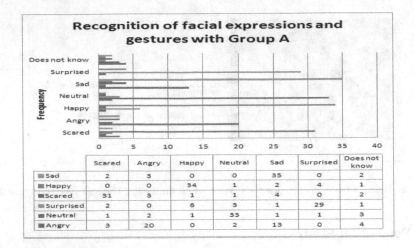

Fig. 4. Results of the recognition rate of the facial expressions and gestures on ZECA using a multiple-choice questionnaire with Group A

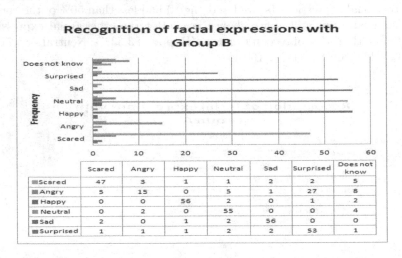

Fig. 5. Results of the recognition rate of the facial expressions on ZECA using a multiple-choice questionnaire with Group B

Similar to Group A, Group B had some difficulties recognizing Fear and Anger, as it can be seen in fig. 5. However, the recognition rates were considerably better, specially with the associated gestures (fig. 6). The proportions were as follows (without gestures - with gestures): Scared: 77.049% - 93.442%, Angry: 24.59% - 70.492%, Happy: 91.803% - 98.361%, Neutral: 90.164% - 91.803%, Sad: 91.803% - 88.525%, Surprised: 86.885% - 83.607%.

Using the data provided above, a Chi-Square test was performed with the goal to verify if there were differences between the observed and expected frequencies

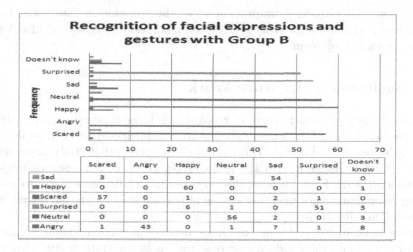

Fig. 6. Results of the recognition rate of the facial expressions and gestures on ZECA using a multiple-choice questionnaire with Group B

of the choices given by the participants on the perceptual study. The recognition rate of basic emotions from the designed movement behaviors in [5] were Sadness - 76%, Anger - 43%, Happiness - 5%, Nervous - 74%, Fear - 76%. We expected that at least 80% of the participants would be able to correctly recognize the emotions conveyed by the robot, our results show that for both Group A $(\chi(5, N = 42) = 22.97, p < .05)$ and B $(\chi(5, N = 61) = 27.16, p < .05)$, the observed and expected frequencies regarding the videos without gestures were different. However, when the gestures were added to the facial expressions display, the null hypotheses is not rejected (Group A - $\chi(5, N = 42) = 5.88, p < .05$, Group B - $\chi(5, N = 61) = 6.10, p < .05)$, meaning that the observed choices in the recognition of the emotions displayed by the robot were as expected.

5   Discussion

Before testing the facial expressions and gestures with the participants of the perceptual study, the Pre-Test was fundamental to verify, after the design in laboratory, which improvements had to be done. It was clear that the lack of a few physical characteristics on the robot prevented us to explicitly design with better quality some of the emotions. However, we considered the final result fairly satisfying, and indeed acceptable for further tests.

The use of the FaceReader software provide us with more information about which AU should be enhanced, and particularly which action units could give the hint to the correct identification of the emotion displayed by the robot. The results from the perceptual study clearly show that the facial expressions combined with the gestures displayed by ZECA can convey emotions normally displayed by human beings, and are also recognized by both children and adults.

Having observed an overall improvement of the recognition rate using facial expressions and gestures, further studies using emotional displays of ZECA will have these two components.

6   Conclusion and Future Work

The goal of this study was to produce recognizable facial expressions and gestures to be displayed on ZECA. These information was identified by both adults and typically developing children. The results of the perceptual study allow us to continue further research with the main target group of our research project: children with ASD. Even knowing the very specific impairments of children with ASD in recognizing emotions, this was an important step for the envisaged work with children with autism.

Future work includes the use of these displays in different game scenarios, in order to activate the children's representations of the state or situation, promoting empathy between the child and other partners. A collaboration with psychologists and experts in this domain is going to be made to carefully design a new set of experiments and tests with children with autism.

The creation of this empathy can be used to activate the perception of the external world [13]. Special attention should be given to the emotional body language displayed by the robot in order to be socially accepted. Also, the introduction of context which may help the improvement of recognition rate during human-robot games, leading to better outcome of the interactions [14].

Acknowledgment. The authors are grateful to the professionals, and the students of the Gualtar School, Braga, Portugal, for their participation in the project. The authors are also grateful to the Portuguese Foundation (FCT) for funding through the R&D project RIPD/ADA/109407/2009 and SFRH/BD/71600/2010. This work is also supported by a QREN initiative, from UEFEDER (Fundo Europeu de Desenvolvimento Regional) funds through the "Programa Operacional Factores de Competitividade - COMPETE" (FCOMP-01-0124-FEDER-022674).

References

1. Feil-Seifer, D., Mataric, M.: Defining socially assistive robotics. In: 9th International Conference on Rehabilitation Robotics, ICORR 2005, pp. 465–468. IEEE (2005)
2. Rutter, M.: Diagnosis and definition of childhood autism. Journal of Autism and Developmental Disorders 8(2), 139–161 (1978)
3. Association, A.: American Psychiatric Association. Diagnostic and Statistical Manual of Mental Disorders: DSM V®. American Psychiatric Association (2013)
4. Hanson, D., Baurmann, S., Riccio, T., Margolin, R., Dockins, T., Tavares, M., Carpenter, K.: Zeno: A cognitive character. In: AI Magazine, and Special Proc. of AAAI National Conference, Chicago (2009)

5. Barakova, E., Lourens, T.: Expressing and interpreting emotional movements in social games with robots. Personal and Ubiquitous Computing 14(5), 457–467 (2010)
6. Mazzei, D., Billeci, L., Armato, A., Lazzeri, N., Cisternino, A., Pioggia, G., Igliozzi, R., Muratori, F., Ahluwalia, A., De Rossi, D.: The face of autism. In: The 18th IEEE International Symposium on Robot and Human Interactive Communication, RO-MAN 2010, pp. 791–796 (2010)
7. Mundy, P.: Development of socially assistive robots for children with autism spectrum disorders
8. Costa, S., Lehmann, H., Robins, B., Dautenhahn, K., Soares, F.: Where is your nose?-developing body awareness skills among children with autism using a humanoid robot. In: The Sixth International Conference on Advances in Computer-Human Interactions, ACHI 2013, pp. 117–122 (2013)
9. Breazeal, C.: Sociable machines: Expressive social exchange between humans and robots. PhD thesis, Massachusetts Institute of Technology (2000)
10. Ekman, P., Rosenberg, E.: What the face reveals: Basic and applied studies of spontaneous expression using the Facial Action Coding System (FACS). Oxford University Press, USA (1998)
11. Den Uyl, M., Van Kuilenburg, H.: The facereader: Online facial expression recognition. Proc. Measuring Behaviour, 589–590 (2005)
12. Ekman, P.: Emotions revealed: Recognizing faces and feelings to improve communication and emotional life. Holt Paperbacks (2007)
13. Preston, S.D., De Waal, F.: Empathy: Its ultimate and proximate bases. Behavioral and Brain Sciences 25(1), 1–20 (2002)
14. Beck, A., Hiolle, A., Mazel, A., Cañamero, L.: Interpretation of emotional body language displayed by robots. In: Proceedings of the 3rd International Workshop on Affective Interaction in Natural Environments, pp. 37–42. ACM (2010)

PEPITA: A Design of Robot Pet Interface for Promoting Interaction

Eleuda Nuñez[1], Kyohei Uchida[2], and Kenji Suzuki[3]

[1] University of Tsukuba, Japan
eleuda@ai.iit.tsukuba.ac.jp
[2] University of Tsukuba, Japan
uchida@ai.iit.tsukuba.ac.jp
[3] University of Tsukuba / JST, Japan
kenji@ieee.org

Abstract. Sharing social information such as photos and experiences is becoming a part of everyday interaction. In this study, we introduce a robot pet, which is able to bring people together by promoting interaction both among humans who share the same place and those separated by distance. However, the use of communication technologies has led to concerns when people are immersed in virtual experiences and real-life social interactions decrease as a consequence. To avoid this, we propose a way of sharing interactions in the same space by using a projector and allowing distant communication via internet in order to share one basic kind of human communication such as hugs. For this first stage, we present the design and implementation of the proposed robot pet which is based on smartphone and a projector to display photos and the pet's internal states. In addition, an LED color study in a public space is also given as a set of design parameters.

Keywords: Robot design, haptic interaction, social robots, handheld projector.

1 Introduction

At the present time, the majority of applications for robots are limited to the industrial field, such as automobile, power generation and manufacturing. However, the next generation of robots will interact more with humans in social environments. Researchers now are putting considerable efforts on developing social robots that can be integrated and co-exist in the human society. Social robots will interact with people in daily life environments, and they could be found at home doing household chores and playing with children, at offices, on streets, in hospitals helping with therapies and even in schools.

Robot pets or animal-like partner are considered a social robot application from the therapeutic point of view. The main purpose of these robots is to be human companions, offering comfort and entertainment, and these artificial companions will be implemented in the human daily life in the near future.

G. Herrmann et al. (Eds.): ICSR 2013, LNAI 8239, pp. 552–561, 2013.

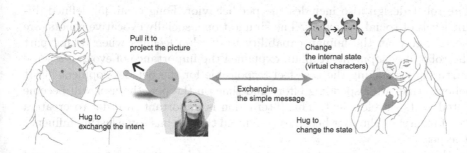

Fig. 1. Robot pet mediated communication in a distant place

The robotic pets can perfectly fit in environments where living pets are not allowed or the conditions are not good enough to keep living pets. The future robot pets are expected to provide the similar pleasure of petting a living pet.

Technology is increasing the possibilities of communication, where even if the persons are not sharing the same physical space, sharing social information becomes the key factor in human social interaction. We would like to propose a robot pet able to bring people together by promoting interaction both among humans who share the same space and those separated by distance. In addition to many popular ways of communication such as smartphones and social networks, we are interested in the idea of sharing "social events" by using a robotic device via internet. In this communication, a very simple message without any relevant or detailed information is exchanged among people who know each other well. Norman [1] defined the meaning of instant messages as a presence detector, or a way to know that someone "is there". He also explains that communication devices are kind of social facilitators, when even if the message content is imprecise, the emotional content is strong. In particular, we focus on "hugging interaction" as a simple message. Hugs are haptic interactions and a basic kind of communication; they transfer energy and give an emotional lift. They are a form of non-verbal communication.

However, the use of communication technologies has led to concerns when people are immersed in virtual experiences and real-life social interactions decrease as a consequence. This problem, which was further analyzed by Turkle [2], emphasizes on how the social interaction has been deteriorating with the inclusion of technologies. Even if we consider robots as a possible solution to change from the traditional screen medium, this strongly depends on which feature the robot exhibits. For this reason we aim to use a projector-based human-robot interaction. We assume that the interaction can be shared with more than one user at the same time and same place. The projected images or characters represent a powerful visual feedback, which can be used to show different kinds of information. Xiang et at.[3], and Karl et al.[4], explored a set of projector-based interaction concepts for multiusers that share the same space, and which motivates us to explore the potential application of projectors on human-robot interaction.

The robot design also includes the pet behavior. Fong et al. [5] defined different levels of social behaviors. The simplest one, socially evocative robots, are those that rely on the human capability to develop feelings when taking part in the robot development. Kaplan, explained the importance of evoking a sense of nurturing or making the user feel responsible for the pet development [6]. It is believed that after spending effort and time on the pet, the user will become invested in its development. This interaction is important in order to create a sense of reality, as the user becomes convinced that he has some lasting influence on the pet.

In this paper, we present the design and implementation of the proposed robot pet interface for promoting human-human social interaction as illustrated in figure 1. A smartphone and projector, both embedded in the robot, are used to display photos and pet's internal states. We first describe the robot pet design, and introduce the interaction scenario. An LED color study is conducted during a long-term exhibition. We will then demonstrate the performance and show an example of use with the developed robot pet. Finally we will present our conclusions and future work.

2   Robot Pet Design

2.1   Device Design

The external round appearance tries to emulate a seed or an egg that "contains a creature inside". A plastic sphere contains the internal circuit as well as a smartphone and a handheld projector. The microcontroller used here is an Arduino UNO, connected to a Samsung Galaxy Nexus by wireless communication using a Seeed Studio Wi-Fi shield. The smartphone is connected to a Pico projector compatible with Android devices, Samsung EAD-R10. The tail of the robot pet is a stretch sensor, based on the idea presented by Slyper [7], about using the structure and elastic properties of the object that contains the sensor. The tail is made of silicon and contains a magnet with a linear Hall effect sensor and an LED on the tip. A system overview can be seen on figure 2.

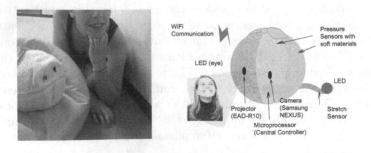

Fig. 2. PEPITA: System overview of the developed robot pet

This device features a pressure sensor to detect hug interaction. In this context, it is important to achieve a comfortable sensation and choose the materials accordingly. The pressure sensor is composed of conductive foam (Inventables Inc.), whose resistance values change when it is pressed, and one piece of conductive fabric on each side.

The structure of the sensor is important to distinguish hugs from other touch interactions. It is made up of 8 different pieces surrounding the plastic ball and the combined readings from all the sensors are used to detect a hug. The greatest goal of the pet robot design it is to be easily accepted and enjoyed by the user. Therefore, the design parameters aim to meet with three of the four definitions from Norman's [1] interpretation of a scientific study of pleasure and design:

1) Psysio-pleasure: pleasures of the body, involving the 5 senses. From the evaluation survey by H. Aghaebrahimi et al. [8], it was concluded that when it is deliberated about the principal causes of human's affection toward robots, haptic and motion/position feedback are the most important. This supports the principle of using hugging interaction as one of the main feedback on this device. Therefore, touch interaction plays an important role for companion robots [9].

2) Socio-pleasure: It's derived from interaction with others. By design, all communication technologies play important social roles, but sometimes the social pleasure can arise only by using the product. For example at home there are some spaces, like the living room or kitchen, for many spontaneous social interactions. Based on this idea, by using the projector, it is intended to provide this kind of socio-pleasure not only by the concept of detection presence, but also by the social interaction that can arise from people sharing the same space.

3) Psycho-pleasure: it is related to the user's psychological states when using the product. This can be evaluated by reactions, comments or impressions. Even though this device has not been tested with different users, it is intended to develop comfort and joy during the interaction with the robot; hence this definition also plays an important role as design parameter.

2.2   Interaction Scenarios

In this prototype, the user can interact with the pet by either hugging it or pulling the tail. The device will show three internal states based on hug interaction, and it will capture and project photos when the tail is pulled. The internal states change when the device is left without interaction for a certain time, with the purpose of engaging the user to interact with it. A schematic diagram with all possible interactions can be seen in figure 3. For better understanding, we assigned names, "happy", "sad" and "annoyed", to the three different internal states. The central controller will take all the pressure sensors readings and decide whether it is a hug or not and consequently modify the internal state. After that, it sends a message to the Android device indicating which state should be displayed on the projector. All the processing on the Android device is managed by an application developed for this purpose. A different message is sent to the

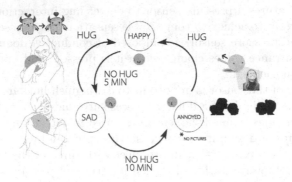

Fig. 3. State transition rule: PEPITA has currently three internal states which are changed by hugging

Android device when the tail is pulled activating the camera to take pictures. The feedback is provided by different LED colors, vibration, sounds and projections.

2.3   Color Lighting as a Feedback of Internal States

Aesthetic impression is related to the emotional response that results from the aesthetic attractiveness of the object, and it can be whether positive or negative [10]. From the data collected by Creusen and Schoormans [11], it was concluded that some of the features that influence aesthetic attractiveness are roundness, size and color. Following this it was clear that it is important to take into account users' perception of visual communication. Since this device shows different color pattern for each internal state, it is necessary to understand if there is an intrinsic meaning to them. This can be used, for example, to choose the color that is the most attractive, for the feedback of the state just after hugging the pet.

To simplify the task, we started with the three basic colors: Red, green and blue, and made an evaluation about users' preference. The study was done with a previous version of PEPITA. This communication device displayed one of these three colors each time it was rolled by the user, and transferred this action to a second device whose LED also lit up. These colors changed without any pattern or special control over them, and a study was done to determine if there was a preference for any of the colors.

2.4   Personalization

We aim to develop a sense of "uniqueness" for each pet, in other words, to allow each user to make a different pet. Although it is difficult to modify the physical features of the robot for each person, the following factors can be customized and personalized by each user:

- Skin: foam to surround the internal mechanics, making the pet huggable and comfortable. And also, a removable and colorful fabric skin that can be adapted to each user's preference.
- Internal states: The internal state changes are driven by the user-pet interaction. In this case we started with the simple concept of having from three different changing internal states based on the hugging interaction.
- Digital Avatar: The purpose is to graphically visualize the state of the pet. Many studies have demonstrated the complexity of making robotic facial expressions that can be easily understood by humans, but cartoon expressions are simple to perceive and comprehend.

The body of the robot pet is not animal-shaped, since robot pets are not meant to substitute living animals [12]. As many researchers concluded before, the use of imaginative, caricatured or funny representation of animals are useful and effective to accomplish robot acceptance; rather than using complex, realistic representations [8]. Simple and intuitive interfaces are preferred by elderly people [13], and can also be safe and attractive if they are to be manipulated by children. Following these ideas we propose the design and development of a robotic device with a soft and rounded body, which we consider the most appropriate for the intended interaction. A set of soft sensors was designed to serve this purpose, and a smartphone centered framework endows the pet with a variety of capabilities. All this features were implemented with the final purpose of evaluating the feasibility of promoting human-human interaction using a pet robot.

3 System Performance

3.1 LED Color Study

A preliminary experiment to investigate the colors and the perception people have of them was conducted in Melbourne, Australia as part of the International Biennial of Media Art. As it was proposed for PEPITA, this kind of

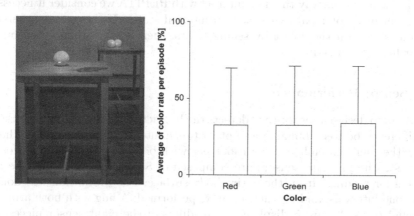

Fig. 4. Interaction study: color change during a long term exhibition

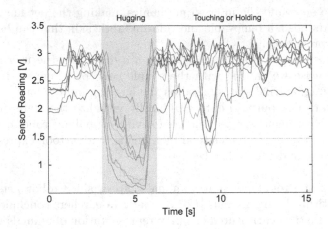

Fig. 5. Sensors performance

event-interaction and instant message-like communication is the main purpose of the proposed device. As it can be seen on figure 4, for this experiment multiple devices were connected each other via Wi-Fi, and LED color display and vibrations were given as a feedback every time a user interacts with the device. When a user rolled the device, it changes color and gives vibration feedback. At the same time, the color of all other devices changed to the same color as the user's device. A total of 3476 episodes were analyzed in 53 days, and the color data was obtained. The device changed colors between green, red and blue, and sent the LED color information to a server.

The results showed that there is no difference regarding the time when showing colors. Figure 4 shows that each color displayed almost the same rate of episodes. Because this device does not have any strategy to change the color, we conclude that the lack of meaning of the color feedback does not encourage users to change the device color to a particular one. For this reason, to use color feedback to elicit some influence in the way the user interact with PEPITA we consider it necessary to implement a color pattern based on internal states. If the user can perceive different states on the robot by seeing colors, then it will probably encourage him or her to change it.

3.2   Sensor Performance

We intend to design a sensor that detects the hug action and is also to understand the difference between hugging and other touch interactions. We assume that in a hug, the user will embrace the robot pet with both arms and hold it close to the chest. The pressure sensors are composed of 8 pieces surrounding the ball, allowing for readings from the entire body surface. The evaluation was done in ten second intervals, and two actions were performed: A hug with both arms and one hand press. Figure 5 displays the readings of the eight sensor pieces, and it can be seen how sensitive the sensors are, reacting even from manipulation.

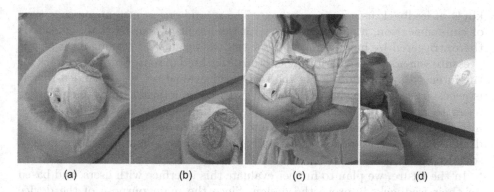

Fig. 6. PEPITA: (a) PEPITA on the chair while lighting up LED on the tip of tail. (b) PEPITA projects a virtual character to show its internal state. (c) Hugging PEPITA. (d) PEPITA projects a photo just taken by the PEPITA itself.

The first valley illustrates the hug action, and it shows how the sensor reading fall due to the foam compression. The difference in sensor values is due to the amount of force applied during the hug. The second valley shows how the signals change when the robot is pressed with one hand. It can be seen that only some of the sensor signals drop this time, because it is only being pressed from one side. In this study, we used a simple threshold method to detect hug actions.

3.3 Robot Preliminary Test

We conducted a preliminary test of PEPITA as illustrated in Figure 6. Two subjects participated in this study. We could test the internal states changed by hugging, and take and display photos by pulling the tail. In addition, we received feedback from the participants regarding the use of the interface. For example, we inquired whether it is comfortable when the device is hugged, or if the LED, sounds and vibrations were pleasant and how they perceived the appearance. Generally users reported that the pet was comfortable to hug and they were able to feel the vibration feedback as well as see the visual feedback from the projector and LED. The pet appearance was perceived as cute or interesting. This evaluation helped us to decide some future improvement for this device, as well as confirm that the device's elements carry out the expected actions.

4   Conclusions and Future Work

In this work, we presented the design and implementation of a robot pet interface that can be used to promote interaction. To achieve this, two ways of interaction using this device were proposed. This first implementation starts by projecting pictures through the robot's eyes and displaying avatars for each of the pet's internal states. We started exploring the versatility of projectors for different kinds of feedback using a smartphone. The second one is instant-message type of communication, a short message with the intent of sharing that someone is there, triggered by hug interaction. For this we began by designing a set of soft pressure sensors. In the performance study, this kind of sensor proved to work for the intended application. From the LED color evaluation, we concluded that it is necessary to implement a color strategy based on internal states to encourage users to interact with the device. In future implementations, we aim to use internet features to transfer the hug interaction from one device to another. The design parameters were chosen carefully following previous works in the area of social robotics.

In the future, we plan to further evaluate this interface with users, and based on their feedback, improve the design. Since the main purpose of the device is to promote interaction, the next step is to add internet features that allow, for example, the user to send the pictures of PEPITAs point of view, from one device to another. In this way, by using the hug interaction, users will be able to communicate with each other in a more emotional way. Further, we consider the possibility of sending pictures and some information about the interaction with the device, to a social network and share them with a larger group of people, and, in this way, exploit all the features of the Android device.

Finally, another possible evaluation to be done would be related to achieving a long term interaction with a robot pet that brings people together by providing different ways of communication and interaction. In this work we followed some of the guidelines presented by Leite et al. [14] in their comprehensive survey about desired social robots characteristics. At the end they provided a set of direction for future design of social robots, and even if our current proposal is only focused on the appearance, we plan to extend our work to other areas.

References

1. Norman, D.A.: Emotional Design: Why we love (or hate) everyday things (2004)
2. Turkle, S.: Alone Together: Why We Expect More from Technology and Less from Each Other. Basic Books (2011)
3. Cao, X., Forlines, C., Balakrishnan, R.: Multi-user interaction using handheld projectors. In: Proceedings of the 20th Annual ACM Symposium on User Interface Software and Technology, pp. 43–52. ACM (2007)
4. Willis, K.D.D., Poupyrev, I., Hudson, S.E., Mahler, M.: Sidebyside: Ad-hoc multi-user interaction with handheld projectors. In: Proceedings of the 24th Annual ACM Symposium on User Interface Software and Technology, pp. 431–440. ACM (2011)
5. Fong, T.W., Nourbakhsh, I., Dautenhahn, K.: A survey of socially interactive robots. Robotics and Autonomous Systems (2003)
6. Kaplan, F.: Free creatures: The role of uselessness in the design of artificial pets. In: Proceedings of the 1st Edutainement Workshop (2000)
7. Slyper, R., Poupyrev, I., Hodgins, J.: Sensing through structure: designing soft silicone sensors. In: Proceedings of the Fifth International Conference on Tangible, Embedded, and Embodied Interaction, pp. 213–220. ACM (2011)
8. Aghaebrahimi Samani, H., Cheok, A.D., Tharakan, M.J., Koh, J., Fernando, N.: A design process for lovotics. In: Lamers, M.H., Verbeek, F.J. (eds.) HRPR 2010. LNICST, vol. 59, pp. 118–125. Springer, Heidelberg (2011)

9. Stiehl, W.D., Lieberman, J., Breazeal, C., Basel, L., Lalla, L., Wolf, M.: Design of a therapeutic robotic companion for relational, affective touch. In: IEEE International Workshop on Robot and Human Interactive Communication, ROMAN 2005, pp. 408–415 (2005)
10. Sun, Q., Sridhar, N., O'Brien, M.: Consumer perception of product stimuli: An investigation into indian consumer psychology and its implications for new product development, process and strategy. In: D2B2: The 2nd Tsinghua International Design Management Symposium (2009)
11. Creusen, M.E.H., Schoormans, J.P.L.: The different roles of product appearance in consumer choice. Journal of Product Innovation Management 22(1), 63–81 (2005)
12. Vanderborght, B., Goris, K., Saldienand, J., Lefeber, D.: Probo, an Intelligent Huggable Robot for HRI Studies with Children. In: Human-Robot Interaction. InTech (2010)
13. Kidd, C.D., Taggart, W., Turkle, S.: A sociable robot to encourage social interaction among the elderly. In: Proceedings of the 2006 IEEE International Conference on Robotics and Automation, ICRA 2006, pp. 3972–3976 (2006)
14. Leite, I., Martinho, C., Paiva, A.: Social robots for long-term interaction: A survey. International Journal of Social Robotics 5(2), 291–308 (2013)

A Socially Assistive Robot to Support Physical Training of Older People – An End User Acceptance Study

Franz Werner and Daniela Krainer

CEIT Raltec, 2320 Schwechat, Austria
f.werner@ceit.at

Abstract. Training support and motivation to conduct physical training enhances the efficiency of training at home. A robotic trainer is proposed to support physical training by using its physical presence to provide exercise demonstration, feedback and motivation. A pre-pilot study with 14 potential end users was conducted to evaluate the technical system, user motivation and acceptance. Early results regarding the motivational abilities and the user acceptance towards the system are given.

Keywords: Human robot interaction, physical training, acceptance of socially assistive robots.

1 Introduction

Physical training is an important and often used measure for physical rehabilitation and prevention of physical deficiencies of older people. Applying strategies for motivating older users prior and during the training at home is important to ensure quantity and quality of the training execution and the efficiency of the physiotherapy. [1]

Within the nationally funded project "PhysicAAL"[1] user studies with a training system based on the socially assistive robot [2] "NAO" from "Aldebaran Robotics"[2] and the "Microsoft Kinect"[3] are currently being undertaken to assess the acceptance of human–robot interaction (HRI) for physical training support and the impact of the system on the user's motivation.

A prototype system was developed in cooperation with physiotherapists that simulates a human physiotherapeutic trainer by performing prescribed exercises in front of the user. To motivate users to take part in the training a human-human like interaction was resembled, including speech, gestures and mimics (simulated via LEDs in the face of the robot to show emotions [3]) for communication. (see also [4])

The Microsoft Kinect-Sensor with an RGB camera and depth sensor was chosen as input channel to analyse the movements of the users and react upon them. By facilitating motion analysis, non-optimal exercise execution was detected and commented on by the robot.

---

[1] http://physicaal.raltec.at (link contains
  video material showing the used human-robot-interaction)
[2] http://www.aldebaran-robotics.com
[3] http://www.microsoft.com/en-us/kinectforwindows

G. Herrmann et al. (Eds.): ICSR 2013, LNAI 8239, pp. 562–563, 2013.

2   Methods

To assess usability, acceptance and technical reliability of the system user studies with 12 potential end users are planned to be conducted in a simulated real-life test setting.

Within a pre-pilot study a group of 14 (n=14) seniors was invited to a gymnasium of the "senior citizen centre Schwechat" [5] for a 15 minute demonstration of the prototype robotic system. The group had an average age of 69 years and consisted of seven women and seven men. After an explanation of the system, physical exercises were conducted by the robotic trainer with the older users by mimicking the movements of the robotic assistant. A specifically tailored questionnaire was asked directly after the demonstration and a focus group session was conducted to generate qualitative results.

3   Results and Conclusion

Results from the pre-pilot show that users found the system highly entertaining (mean score 4.4 on a 5-point Likert scale) and rather motivating (mean score 3.8 on a 5-point Likert scale). When compared to similar training support systems such as video and paper based support, users preferred the video course over the robot trainer (6 video, 2 robot, 2 paper, 4 nothing) for regular training at home but 10 out of 13 would prefer to take the robotic system if offered the chance to test one of the systems for the duration of one month at home.

Our first results suggest that potential end users accept the idea of physical training with a socially assistive robotic trainer and find the robotic prototype entertaining and motivating. The fact that users preferred video supported training over the robotic solution but would opt for the robotic solution for testing suggests that the initial excitement of the new and innovative solution was high and longer trials are needed to fully assess the long-term acceptance of the system.

References

[1] World Health Organization, Global Recommendations on Physical Activity for Health (2010), http://whqlibdoc.who.int/publications/2010/9789241599979_eng.pdf
[2] Matarić, M.J., et al.: Defining socially assistive robotics. Journal of NeuroEngineering and Rehabilitation (2007)
[3] Torta, et al.: Attitude towards socially assistive robots in intelligent homes: Results from laboratory studies and field trials. Journal of Human-Robot Interaction (2012)
[4] Fasola, J., Matarić, M.J.: Using Socially Assistive Human-Robot Interaction to Motivate Physical Exercise for Older Adults. In: Kanade, T. (ed.) Proceedings of the IEEE - Special Issue on Quality of Life Technology (August 2012)
[5] Hlauschek, Panek, Zagler: Involvement of elderly citizens as potential end users of assistive technologies in the Living Lab Schwechat. In: Proceedings of the PETRA 2009, article no. 55 (2009)

How Do Companion-Type Service-Robots Fit into Daily Routines of the Elderly to Improve Quality of Life?

Katja Richter and Nicola Döring

TU Ilmenau, Institute of Media and Communication Science,
Department of Media Psychology and Media Design, Ehrenbergstr. 29, D-98693 Ilmenau
{katja.richter,nicola.doering}@tu-ilmenau.de

Abstract. Within an interdisciplinary research project companion-type Service-Robot is developed. The study investigates daily routines of the elderly in order to identify how the robot that shall live with the elderly in their home fits best into everyday life to support, elongate independent living and thus improve quality of life.

Keywords: quality of live, independent living, companion, service-robot, elderly, demographic change.

1 Relevance and Research Question

Demographic change challenges the health care system: while the proportion of the working age population is decreasing, the number of the elderly that need support and care is increasing. In addition, a trend reveals that more of the elderly attend living on their own at private homes [7]. But age-related loss of skills aggravates or even prevents that [10]. Robots offer numerous possibilities to elongate independent living and therefore improve quality of live [4], [6]. For living successfully as companions with the elderly, robots have to be able to perform a variety of tasks, interact flexible and adapted in non-standard situations and environments, learn, communicate and react socially and appropriate [8]. So far, robot-development is mostly technic-driven, available robots are predominantly prototypes [1]. The robot developed within the project SERROGA offers a range of companion-type services that intend to support the wellbeing of the elderly: helping in case of emergency, managing health care, structuring daily routines or simplifying communication and contacting. But for successful purpose individual needs and technological possibilities have to match [3]. Thus, usability-needs have to be guaranteed on the one hand - what is ensured by formative evaluation. On the other hand several studies conclude that robot use need to fit to individual daily routines of the elderly [3].

2 Methods and Data

Daily routines are quiet heterogeneous, are performed automatically, are difficult to control, unintentionally and unaware [5]. In order to explore these, respondents lead a

G. Herrmann et al. (Eds.): ICSR 2013, LNAI 8239, pp. 564–566, 2013.

semi-structured diary over the period of a week. Additionally, information about age, gender, relationship status and health status is recorded by a standardized question-naire. Subsequently semi-structured qualitative interviews are conducted to provide detailed understanding of how needs and life situation determine routines [2]. The investigated sample is restricted to elderly that are at least 60 years old, are retired and still live in their own homes without needing permanent professional care. So far 6 persons were interviewed: 4 female & 2 mal; 4 living alone & 2 living with partner; characterized by different health states. The collected data will be analyzed in order to identify behavioral patterns that can be condensed to personas. Therefore everyday activities will be analyzed regarding the time at which they take place, regularity, type of cooperation, degree of interactivity and the type of need that is satisfied concerning to Maslow's hierarchy of human needs [9].

3   Preliminary Results

The results proof that differences in daily routines and support requirements are re-lated to health and relationship status, basically. Thus, typical day course scenarios considering robot usage will be created for the following four personas: Person living alone with limitations on physical health concerning limited mobility, Person living alone with limitations on physical health concerning serious diabetes, Person living alone with limitations on mental health concerning depression and Person living in a partnership without serious limitations on physical or mental health, but the partner having de-mentia. By means of the personas, scenarios for robot use within daily routines will be developed as next step.

References

1. Bekey, G.A.: Current Trends in Robotics: Technology and Ethics. In: Lin, P., Abney, K., Bekey, G.A. (eds.) Robot Ethics. The Ethical and Social Implications of Robotics, pp. 17–34 (2012)
2. Bukov, A.: Individuelle Ressourcen als Determinanten sozialer Beteiligung im Alter. In: Backes, G.M., Clemens, W. (eds.) Lebenslagen im Alter. Gesellschaftliche Bedingungen und Grenzen, pp. 187–214. Leske + Budrich, Opladen (2000)
3. Decker, M.: Service robots in the mirror of reflective research. Poiesis und Praxis 9, 181–200 (2012)
4. Forlizzi, J., DiSalvo, C., Gemperle, F.: Assistive Robotics and an Ecology of Elders Living Independently in Their Homes. Human–Computer Interaction 19, 25–59 (2004)
5. Koch, T.: Macht der Gewohnheit? Der Einfluss der Habitualisierung auf die Fernsehnut-zung. VS Verlag, Wiesbaden (2010)
6. Meyer, S.: Mein Freund der Roboter. Servicerobotik für ältere Menschen – eine Antwort auf den demografischen Wandel? VDE, Berlin (2011)
7. Statistisches Bundesamt: Pflegestatistik 2009 – Pflege im Rahmen der Pflegeversicherung. 3. Bericht: Ländervergleich – ambulante Pflegedienste. Statistisches Bundesamt, Wiesba-den (2011)

8. Syrdal, D.S., Koay, K.L., Walters, K.L., Dautenhahn, K.: A personalized robot companion? The role of individual differences on spatial preferences in HRI scenarios. In: 16th IEEE International Conference on Robot and Human Interactive Communication, pp. 1143–1148 (2007)

9. Thielke, S., Harniss, M., Thompson, H., Patel, S., Demiris, G., Johnson, K.: Maslow's Hierarchy of Human Needs and the Adoption of Health-Related Technologies for Older Adults. Aging International 37(4), 470–488 (2011)

10. Thieme, F.: Alter(n) in der alternden Gesellschaft. Eine soziologische Einführung in die Wissenschaft vom Alter(n). VS Verlag für Sozialwissenschaften, Wiesbaden (2008)

Do Children Behave Differently with a Social Robot If with Peers?*

Paul Baxter, Joachim de Greeff, and Tony Belpaeme

Centre for Robotics and Neural Systems
Plymouth University, U.K.
paul.baxter@plymouth.ac.uk

Abstract. Having robots interact socially with children is an increasingly relevant goal. We conducted a study with 7-8 year-olds looking at how single and pairs of children interacted with an autonomously acting robot in the context of a collaborative game. A preliminary examination of the interactions reveals a notable difference in how the robot is viewed: in one-on-one interactions with a robot, children appear to be more forgiving of failures in robot social contingencies; however, these deficiencies are not tolerated to the same extent in multi-party interactions.

1 Social Robots for Children

Autonomous robotic agents require some competencies for social interaction if they are to interact with humans [1]. There are a number of domains in which such socially-capable robots for interacting with children are useful; the present work is part of an ongoing effort to provide children with social robotic companionship in a healthcare context [2]. One aspect of importance here is the manner in which children interact with social robots in naturalistic environments, and to what extent the children's behaviour towards the robot is dependant on the social context. This emphasis on non-laboratory environments for studying interaction facilitates the exhibition of naturalistic behaviour, but then also requires a rigorous behavioural evaluation of the social interaction involving the robot [3]. In this abstract, we present initial observations from a study involving children interacting with an autonomous social robot.

2 HRI Study: Setup and Observations

The study involved children in year 3 (7 to 8 year-olds) drawn from two schools. The children interacted with a robot in the context of a touchscreen-based sorting game, in which the robot participates autonomously [4], and gives feedback. Two conditions were used (figure 1): (1) a single child interacting with the robot ($n_s = 12$), and (2) a pair of children interacting with the robot ($n_p = 7$ pairs). For each interaction, the children involved were only instructed to play the game and told that the robot would join in; no instructions were given regarding how to interact with the robot. All interactions were recorded for further analysis.

* This work is funded by the EU FP7 ALIZ-E project (grant 248116).

G. Herrmann et al. (Eds.): ICSR 2013, LNAI 8239, pp. 567–568, 2013.

Preliminary observations of the interactions seem to indicate a clear difference in how the children treat the robot between the two conditions. In condition 1, the children typically engage in a turn-taking behaviour with the robot, using gaze as a cue: this indicates that the robot is potentially viewed as a social inter-actant. However, in the pair-robot interactions, this dynamic was fundamentally changed: the children would become socially interactive between themselves, typically treating the robot as a non-social other that could also play the game. The topic of conversation between the children was typically either on shortfalls in the socially contingent robot behaviours, or playing the sorting game but largely ignoring the robot's contribution.

Fig. 1. Example interactions from the two conditions: (left) single child interacting with the robot; (right) two children interacting with the robot

The observations presented above indicate a greater effect of number of human participants in an interaction with a robot than mere issues of robot sensory lim-itations [3]. It seems that the way the robot is perceived by the children changes according to the social context (i.e. different number of children) [1]. This could for example be due to the inadequacy of the robot to respond to complex social cues in as timely a manner as the two children would expect of one another, thus breaking the illusion of robot social competence, which would not necessarily oc-cur with just a single child present. Further analysis of the recorded interactions is required to formally characterise this effect. While solutions to these issues are typically approached from a robot centred view (i.e. the introduction of better sensing/modelling), the observation presented here indicates that it is the chil-dren's responses to the robot that is of greater importance regarding naturalistic social interaction: the social context can make or break the illusion of robot social competence even given the same control system.

References

1. Dautenhahn, K.: Socially intelligent robots: Dimensions of human-robot interaction. Philosophical Transactions of the Royal Society B 362, 679–704 (2007)
2. Belpaeme, T., et al.: Multimodal child-robot interaction: Building social bonds. Journal of Human-Robot Interaction 1(2), 33–53 (2012)
3. Sabanovic, S., et al.: Robots in the wild: Observing human-robot social interaction outside the lab. In: 9th Int. Workshop on Advanced Motion Control, pp. 596–601 (2006)
4. Baxter, P., et al.: A touchscreen-based sandtray to facilitate, mediate and contex-tualise human-robot social interaction. In: HRI 2012, pp. 105–106 (2012)

Is a Furry Pet More Engaging? Comparing the Effect of the Material on the Body Surface of Robot Pets

ManPing Zhao[1] and Angel P. del Pobil[1,2]

[1] Department of Interaction Science, Sungkyunkwan University, Seoul, South Korea
manping.310@gmail.com
[2] Robotic Intelligence Laboratory, Jaume-I University, Castelló, Spain
pobil@uji.es

Abstract. The development of attractive robot pets calls for features that the users find more and more appealing and animal-like. There are a number of design issues that need to be addressed: How can we design robot pets to make them look more real? How can we make robot pets more easy to use? How can we make them more interactive and their human-robot relationship more natural? These problems should be solved as soon as possible to satisfy the rising demand of this kind of robots. Though morphology is a primary design decision, the importance of the material covering the body surface has somehow being neglected. In this paper, we focus on the role of this material for robot pets since we believe it is a fundamental design factor. In our study we compare two different materials for the body surface of a pet robot and consider five categories (friendliness, enjoyment, attractive, presence and immersion).

1   Introduction

Social robots can play various social roles in our lives, these robots have abilities to communicate and interact with humans by following social behaviors [1]. Nowadays, robot pets as a kind of social robots are being used more and more widely. There are a number of recent examples of technological substitutes of pets, such as I-Cybie, AIBO, Paro and I-cat. Among them, AIBO and Paro are probably the two most popular ones used in robot-assisted therapy and educational areas as social companions. The challenge is to make robot pets more interactive so that they can play social roles in the users' lives. Progress depends not only on technological development, but also in their appearance; the success of both AIBO and Paro can be accounted for by the way they look and behave like real animals, including sounds, movement, and appearance. The importance of morphology has been acknowledged to the effect that different morphologies suggest different affordances to users, triggering a variety of cognitive heuristics and thereby shaping their interactions with the robots [2]; for example, AIBO was designed to resemble a real puppy. However, the interaction with pets is not only visual, but it heavily relies on the sense of touch. We claim that the material covering the body surface of a pet robot is as important as its morphology, though it has somehow being neglected. This can partly explain the success of Paro, which is covered with furry fabrics similar to that of a baby seal.

G. Herrmann et al. (Eds.): ICSR 2013, LNAI 8239, pp. 569–570, 2013.

In our research, we compare two different materials covering the body surfaces of two robots and study the effect on five users' qualities.

2   Method and Experiment Design

The participants were 40 students (20 females, 20 males), all of them from a university in Korea. They did not have any negative attitude towards pets, especially dogs. In this research, the robot pet Genibo produced in Korea (DASAROBOT) was chosen. It can act like a real pet and expresses its emotions, especially when users touch its head and back and flank sensors. 20 of the participants took an experiment with a plastic surface robot; the others took an experiment with a furry surface version of the robot.

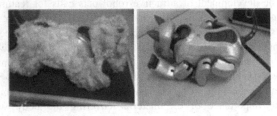

Fig. 1. Furry surface type vs. plastic surface type

2.1   Results and Conclusion

In our results, the friendliness category of the furry robot (Mean=4.621) is higher than the plastic robot (Mean=3.257). The mean value of the furry robot is also higher than the plastic robot in the enjoyment, social attractive, social presence, and immersion categories. In the enjoyment category, the furry robot (Mean=4.867) is higher than the plastic one (Mean=3.483), in the social attractive category, the furry robot (Mean=4.483), is higher than the plastic robot (Mean=2.800), in the social presence category, the furry robot (Mean=4.633) is higher than the plastic robot (Mean=3.575), in the using immersion category, the furry robot (Mean=5.010) is higher than the plastic robot (Mean=4.000). Our results show that the participants consistently evaluated the interaction with the furry robot dog as more positive than in the case of the plastic one in all the five categories. This supports our hypothesis regarding the role of the tactile modality in the interaction with robot pets and, consequently, the importance of the material covering the body surface.

References

1. Breazeal, C.: Designing Sociable Robots. MIT Press, Cambridge (2002)
2. del Pobil, A.P., Sundar, S.S.: Interaction science perspective on HRI: Designing robot morphology. In: Workshop at the 5th ACM/IEEE International Conference on Human-Robot Interaction (HRI 2010), Osaka, Japan (2010)

Can You Trust Your Robotic Assistant?

Farshid Amirabdollahian[1], Kerstin Dautenhahn[1], Clare Dixon[2],
Kerstin Eder[3,4], Michael Fisher[2], Kheng Lee Koay[1], Evgeni Magid[3,4],
Tony Pipe[3,5], Maha Salem[1], Joe Saunders[1], and Matt Webster[2]

[1] Adaptive Systems Research Group, U. Hertfordshire
adapsys.feis.herts.ac.uk
[2] Centre for Autonomous Systems Technology, U. Liverpool
www.liv.ac.uk/cast
[3] Bristol Robotics Laboratory
www.brl.ac.uk/vv
[4] Computer Science Dept., U. Bristol
www.bristol.ac.uk
[5] Engineering, Design and Mathematics Dept., U. West of England
www.uwe.ac.uk

Abstract. Robotic assistants are being developed to assist with a range
of tasks at work and home. Besides designing and developing such robotic
assistants, a key issue that needs to be addressed is showing that they are
both safe and trustworthy. We discuss our approach to this using formal
verification, simulation-based testing and formative user evaluation.

Keywords: Safety, trust, verification, service robotics, robotic assistants.

1 Introduction

Robotic Assistants are now being designed to help us at work and at home, e.g.,
in our everyday activities or in health-care scenarios. A robotic assistant can do
a variety of things, from simply fetching your glasses or helping you put together
flat-pack furniture, to running a bath, and even analysing your health needs. A
wide range of robotic assistants is currently under development within academia
and industry, from surgery and rehabilitation robots to flexible manufacturing
robots. These all indicate a rise in development and use of robotic technologies.

At the present time, the major challenge no longer lies in producing such
robotic helpers, but in demonstrating that they are *safe* and *trustworthy*. The
Trustworthy Robotic Assistants project[1] aims to address exactly this. Specifically, we propose a holistic approach to developing and providing a unified safety
framework for human–robot interaction (HRI). It combines our previous experiences with *formal verification* (e.g., [4]), *simulation-based testing* (e.g. [3]) and
formative user evaluation (e.g. [1,2]). Based on this combination of techniques,
we aim to tackle the holistic analysis of safety and trustworthiness in HRI.

---

[1] http://www.robosafe.org/

G. Herrmann et al. (Eds.): ICSR 2013, LNAI 8239, pp. 571–573, 2013.
© Springer International Publishing Switzerland 2013

2   Research Approach

In practice, no single verification/validation technique is adequate to cover all safety aspects of an HRI scenario. Therefore the analysis of trustworthiness in robotic assistants requires a combination of different techniques.

Formal verification can exhaustively analyse all of the robot's possible choices, but uses a vastly simplified environmental model. Simulation-based testing of human–robot interactions can be carried out in a fast, directed way and involves a much more realistic environmental model, but is essentially selective and does not take into account true human interaction. Formative user evaluation provides exactly this validation, constructing a comprehensive analysis from the human participant's point of view. Though non-trivial to achieve, this combined approach will be very powerful. Not only will analysis from one technique stimulate new explorations for the others, but each distinct technique actually remedies some of the deficiencies of the others. Therefore, this combination provides a new, strong, comprehensive, end-to-end verification and validation method for assessing safety in human-robot interactions.

We start with the user requirements, which are transformed into a system specification that formally describes the HRI framework and characterizes all components of the framework: the assumptions about the human behaviour, the robot and the environment, the HRI protocols, and the expected outcomes of interactions between these components. The key components of the specification, presented as explicit requirements, include the safety of the human and the environment. The system is designed according to the specification and will be implemented with real robots, based on ROS^2 platforms. Formal verification will verify that the design is correct with respect to the specification, and simulation-based methods will verify the system design against the implementation. To strengthen our approach, simulation-based methods will be applied to verify the implementation against the specification. Finally, the end-user evaluation will validate our implementation with regard to the initial user requirements.

We will be developing a unified safety framework for HRI. This framework can be used to generate evidence of the correctness of robotic assistants, allowing them to gain the trust of their users. The safety and trustworthiness components provided by the Trustworthy Robotic Assistants project will inform future safety standards and will eventually contribute to professional safety assessment.

Acknowledgements. The authors were partially supported by the Trustworthy Robotic Assistants[1] project funded by EPSRC grants EP/K006193/1, EP/K006-320/1, EP/K006509/1 and EP/K006223/1.

References

1. Dautenhahn, K.: Methodology and Themes of Human-Robot Interaction: A Growing Research Field. Int. J. of Advanced Robotic Systems 4(1), 103–108 (2007)

---

[2] Robot Operating System, http://www.ros.org/

2. Grigore, E.C., Eder, K., Pipe, A., Melhuish, C., Leonards, U.: Joint Action Understanding improves Robot-to-Human Object Handover. In: IEEE IROS (2013)
3. Magid, E., Tsubouchi, T., Koyanagi, E., Yoshida, Y., Tadokoro, T.: Controlled Balance Losing in Random Step Environment for Path Planning of a Teleoperated Crawler Type Vehicle. J. of Field Robotics 28(6), 932–949 (2011)
4. Stocker, R., Dennis, L., Dixon, C., Fisher, M.: Verifying Brahms Human-Robot Teamwork Models. In: del Cerro, L.F., Herzig, A., Mengin, J. (eds.) JELIA 2012. LNCS, vol. 7519, pp. 385–397. Springer, Heidelberg (2012)

A Robotic Social Reciprocity in Children
with Autism Spectrum Disorder

Giovanni Gerardo Muscolo[1,2,*], Carmine Tommaso Recchiuto[3], Giulia Campatelli[4],
and Rezia Molfino[2]

[1] Creative and Visionary Design Lab, Humanot s.r.l., Prato, Italy
[2] PMAR Lab, Dept. of Applied Mechanics and Machine Design, Scuola Politecnica,
University of Genova, Genova, Italy
muscolo@dimec.unige.it
[3] Electro-Informatic Lab, Humanot s.r.l., Prato, Italy
[4] IRCSS Fondazione Stella Maris, Calambrone, Italy

Abstract. The authors aim at deeply investigating the underlying mechanisms
of social reciprocity in children with autism spectrum disorders (ASD) by
designing and developing a new generation of humanoid robots able to interact
in an unstructured environment with children with ASD, stimulating their
reaction, giving and receiving objects and finally anticipating their actions.
During the interaction, an external sensors network (eye tracking, movement's
analysis, inertial measurement units) will measure all the fundamental
parameters for the models analysis. The research aims at the development of
new psychological and neuro-scientific models related to the communication
and the social reciprocity in children with ASD are expected to be developed.

Keywords: ASD, Humanoid robots, Social reciprocity, eye-tracking system.

1 Concept and Research Objectives

Social reciprocity involves an individual participating in long chains of back and forth
intentional interactions with the other. It includes verbal communication (language),
non-verbal communication (vocalization, affective communication) and gestures (face
expressions, body movements, spinal posture). Social reciprocity is a fundamental
capacity of development and it allows the individual to relate and communicate with
the other reciprocally and purposefully [1]. ASD children often may show aimless
behaviors, misreading of other person's cues, fragmented circles of purposeful
interaction and self-absorbed behavior as the proof of challenges in this ability [2].
These difficulties within reciprocal social interaction have been interpreted by some
as the consequence of an underlying deficit in Theory of Mind [3]. *Theory of Mind is
the ability to understand that others have intentions and desires that are different
from one's own.* Social reciprocity can also be conceptualized as a first step in reality
testing: throughout purposeful interactions with the other, the child learns that affect

---

* Corresponding author.

G. Herrmann et al. (Eds.): ICSR 2013, LNAI 8239, pp. 574–575, 2013.
© Springer International Publishing Switzerland 2013

behaviors (both positive as negative) have all their consequences in other's behavior. These cause-and-effect experiences teach a child that the world is a purposeful place. When cause-and-effect behavioral patterns do not occur, the most fundamental aspect of the sense of causality may be compromised. *These deficits in social reciprocity can have a lifelong negative effect on the social communication capacities and cognitive development of children with ASD. Humanoid robotics and ICT technologies are always more important in this research area* [4]. ICT can be fundamental not only as a "sensors network system", for the analysis of movements and of the gaze, but also as a therapeutic tool to be placed alongside the classic psychologist approach. The robotic technologies will be used for both purposes. *Advanced sensors systems will be used to monitor the behavioural patterns of children with ASD in an unstructured environment:* in particular, an *eye-tracking system in a completely free space* and *new non-invasive, miniaturized and wireless inertial measurement units* to be placed on the children will be developed during this project. The analysis of these data will help the *developing of new psychological models related to the communication ability and the social reciprocity schemes of children with ASD;* but they will also provide the foundations for *a new generation of humanoid robots able to interact with the children.* The humanoid robots will simulate and stimulate the social interaction with the subjects, by moving with them in the environment, making use of the analysis of the sensors data to react to their movements and to anticipate their action. Finally, the robots will also be able to be controlled by an external user that will make them execute the desired action: indicate a point, give an object to the children, and so on. Based on these observations and hypotheses, our vision *is to develop two humanoid robots for two different age classes of children with ASD; the robots will be able to walk, run and grasp objects both in an autonomous way and in an external controlled way, and interact with children anticipating their movements.* From a long-term perspective, the key characteristics of the project are: 1) Psychological models and gaze analysis in children with ASD; 2) Humanoid robotics as a therapeutic tool; 3) Sensing; 4) Customized behavioural control. The research will open new horizons in the field of the therapy for children with ASD and in the study of the psychological models underlying the social reciprocity in these children.

References

1. Klin, A., Saulnier, C., Sparrow, S., Cicchetti, D.V., Volkmar, F., Lord, C.: Social and communication abilities and disabilities in higher functioning individuals with autism spectrum disorders. Journal of Autism and Developmental Disorders 37, 788–793 (2007)
2. Colombi, C., Liebal, K., Tomasello, M., Young, G., Warneken, F., Rogers, S.: Examining correlates of cooperation in autism: Imitation, joint attention, and understanding intentions. Autism 13, 143–163 (2009)
3. Bauminger, N., Kasari, C.: Brief report: Theory of mind in high-functioning children with autism. Journal of Autism and Developmental Disorders 29, 81–86 (1999)
4. Shamsuddin, S., et al.: Initial response of autistic children in human-robot interaction therapy with humanoid robot NAO. In: 2012 IEEE 8th International Colloquium on Signal Processing and its Applications (CSPA). IEEE (2012)

Learning Complex Trajectories by Imitation Using Orthogonal Basis Functions and Template Matching

Mohsen Falahi, Mohammad Mashhadian, and Mohsen Tamiz

Neural and Cognitive Sciences Laboratory, Amirkabir University of Technology, Tehran, Iran
{mfalahi13,m-mashhadian,mohsen_tamiz}@aut.ac.ir

Abstract. In this paper a new method based on Orthogonal basis Functions and Template Matching (OFTM) for learning trajectory by imitation is introduced. In this method, the robot uses primitive movements including template and orthogonal basis trajectories, which are learnt by using Gaussian Mixture Model (GMM), to construct new given trajectories. To obtain this goal, the robot calculates the dissimilarity between the new trajectory and arbitrary templates, then the similar parts will be replaced by the template, and the rest of the new trajectory will be constructed by using the orthogonal learnt trajectories. The results show that our method is more accurate and requires less computation in comparison with learning the whole trajectory by GMM.

Keywords: Learning by imitation, template matching, orthogonal basis functions, primitive movements.

1 Introduction

As the importance of social robots increases, the focus on learning methods, especially learning by imitation as a natural and safe approach [1] has increased. There are several methods for learning a trajectory as the fundamental component in imitation learning systems [2] such as GMM [3] and HMM [4]. In this paper a new method based on OFTM is introduced to address this issue.

2 Proposed Method for Learning Trajectories by Imitation

In our algorithm, firstly, the demonstrator shows some arbitrary trajectories to the robot, and the robot learns them via GMM. These trajectories are learned as templates. Then, we demonstrate sine and cosine trajectories as the orthogonal basis functions to the robot. Now the robot's learning steps are completed. Hereinafter, when a new trajectory is given to the robot, it calculates the dissimilarity between templates and the given trajectory as a criterion of similarity by

$$E_k = \frac{1}{\tau_f}\sum_{i=1}^{\tau_f}\left(\frac{Tr(i+k)}{\left|\arg_{j\in[k,\tau_f+k]}\max Tr(j)\right|} - \frac{Tm(i)}{\left|\arg_{j\in[1,\tau_f]}\max Tm(j)\right|}\right)^2 \tag{1}$$

G. Herrmann et al. (Eds.): ICSR 2013, LNAI 8239, pp. 576–578, 2013.

Where k is the start point of the template which varies from zero to the length of the given trajectory, E_k denotes dissimilarity, $Tr(i)$ and $Tm(i)$ are the values of the given trajectory and the template at the time i, respectively. In cases that E_k is less than μ, the defined threshold, those parts will be eliminated from the trajectory and replaced by the corresponding templates. It is notable that dissimilarities are calculated with various periods (τ_f) and scales of the learned templates. Since sine and cosine functions are learnt as orthogonal functions, the robot applies Fourier transform to construct rest of the given trajectory. As Fourier transform is not able to reconstruct functions with high frequency changes, because of Gibbs phenomenon, and considering robot's frequency limitation in its movements, it is recommended to teach these kinds of trajectories by template.

The results of our tests show that OFTM is more accurate than GMM and needs less computation. As an example of our tests, in which a new trajectory is given to both methods, the accuracy of OFTM with μ=0.05 is more than GMM/GMR in reconstructing it (see Fig. 2). Moreover, the average time it takes to produce the result for OFTM and GMM/GMR is 0.84 and 13.09 seconds respectively, on a same computer.

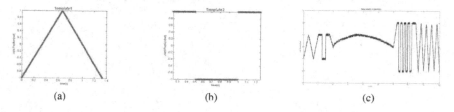

(a)   (h)   (c)

Fig. 1. (a) and (b) are two templates and (c) is a new given trajectory which the robot should learn

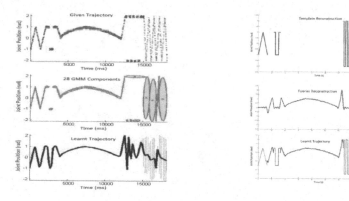

Fig. 2. (a) Using GMM to learn the new trajectory

Fig. 2. (b) Using OFTM to learn the new trajectory

References

1. Bandera, J.P., Rodriguez, J.A., Molina-Tanco, L., Bandera, A.: A Survey of Vision-Based Architecture for Robot Learning by Imitation. Int. Journal of Humanoid Robotics 9(1) (2012)
2. Aleotti, J., Caselli, S.: Robust Trajectory Learning and Approximation for Robot Programming by Demonstration. Int. Journal of Robotics and Autonomous Systems 54, 409–413 (2006)
3. Cho, S., Jo, S.: Kinesthetic Learning of Behaviors in a Humanoid Robot. In: 11th Int. Conf. on Control, Automation and Systems, Gyeonggi-do, Korea
4. Calinon, S., Sauser, E.L., Billard, A.G., Caldwell, D.G.: Evaluation of a Probabilistic approach to Learn and Reproduce Gestures by Imitation. In: 2010 IEEE Int. Conf. on Robotics and Automation, Anchorage, Alaska, USA (2010)

Does Physical Interaction Escalate?

Atsushi Takagi, Carlo Bagnato and Etienne Burdet

Imperial College of Science, Technology and Medicine, London, UK
{at1107,cb2410,eburdet}@imperial.ac.uk

Abstract. To improve human-robot interaction, we investigate how humans communicate sensory information. The study [1] has shown that when partners are instructed to pass the force they sensed the force monotonically increases, which might explain why fights can escalate. We investigated whether the escalation was due to competitive bias induced by the presence of the partner [2,3] rather than the attenuation of sensory percepts [4]. The experiment in [1] was repeated over a larger range of forces with a robotic device to separate the partners. Our results exhibited a different behavior to [1]: with low forces the forces iteratively increased and with large ones it decreased, such that partners converged in force and plateaued. This suggests the utility of robotic devices in removing unwanted competitive behavior.

Introduction and Method

Shergill *et al.* investigated the perception of, and ability to reproduce forces [1]. Participants placed their index finger on a molded support, sensed a force applied by a motor and reproduced the same force on their partner. To examine the effect of competitive bias [3], we replicated Shergill's study over a larger force range with a robotic device to separate partners.

A dual robotic interface [5] was used for this study. 4 pairs of subjects were instructed to sense a force using their index finger, then after a 5 s gap reproduced the same force against the robot with the same finger. A curtain was placed between the partners and noise isolation headphones were worn. Subjects were told that the robot was applying the force. The initial force was provided by the robot to a subject in a pair picked at random. This force was passed between the pair back and forth 6 times. 3 trials at 5 initial force levels between 2–10 N were tested, compared with 0.5–3 N in [1].

Results and Discussion

Figure 1(a) shows results from a representative pair. We found that all pairs converged in force and plateaued, initially escalating if the starting force was small and reducing when large. We examined subjects on how well they could reproduce forces as individuals. Figure 1(b) shows that the force was underestimated for small and overestimated for large forces. During sensing, the finger

G. Herrmann et al. (Eds.): ICSR 2013, LNAI 8239, pp. 579–580, 2013.

is squeezed between the mold and the interface, and force is measured by skin mechanoreceptors as pressure. When the force is reproduced by pushing against the robot, the individual applies a force such that the sensation on their finger pad is matched [6], which may explain the underestimation of force. The overestimation of small forces is still unclear. The combination of over- and underestimation leads to the convergence of force between partners.

Our results differ significantly with those of Shergill et al., which may suggest that their task had become competitive as subjects in [1] were facing each other like in games of chess or checkers. Such spatial arrangement is known to be a strong cue of confrontation [7]. Our study shows that a robotic device can remove competitive behavior in tasks where it is obstructive [3].

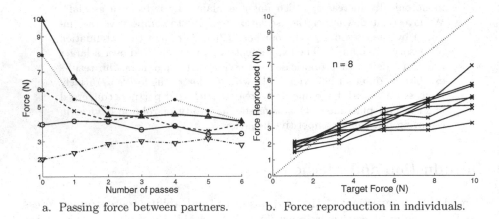

a. Passing force between partners.   b. Force reproduction in individuals.

Fig. 1. (a) Force convergence between partners (from a representative pair). (b) Over- and underestimation for small and large forces respectively in individuals.

References

1. Shergill, S.S., Bays, P.M., Frith, C.D., Wolpert, D.M.: Two eyes for an eye: The neuroscience of force escalation. Science 301(5630), 187 (2003)
2. Zajonc, R.B.: Social facilitation. Science 149(whole no. 3681), 269–274 (1965)
3. Rogoff, B.: The Cultural Nature of Human Development. Oxford University Press (January 2003)
4. Blakemore, S.J., Wolpert, D.M., Frith, C.D.: Central cancellation of self-produced tickle sensation. Nature Neuroscience 1(7), 635–640 (1998)
5. Melendez-Calderon, A., Bagutti, L., Pedrono, B., Burdet, E.: Hi5: A versatile dual-wrist device to study human-human interaction and bimanual control. In: Intelligent Robots and Systems (IROS), pp. 2578–2583 (September 2011)
6. Johansson, R.S., Flanagan, J.R.: Coding and use of tactile signals from the fingertips in object manipulation tasks. Nature Reviews Neuroscience 10(5), 345–359 (2009)
7. Kendon, A.: Conducting Interaction: Patterns of Behavior in Focused Encounters. CUP Archive (November 1990)

Being There: Humans and Robots in Public Spaces

Paul Bremner[1], Niki Trigoni[2], Ian Brown[2], Hatice Gunes[3], Chris Bevan[4],
Danae Stanton Fraser[4], and Mark Levine[5]

[1] Bristol Robotics Lab, University of the West of England, Bristol
[2] Department of Computer Science, University of Oxford
[3] School of Electronic Eng. & Computer Science, Queen Mary University of London
[4] CREATE Lab, Department of Psychology, University of Bath
[5] Department of Psychology, University of Exeter

Abstract. Here we present an introduction to the collaborative research project Being There: Humans and Robots in Public Spaces. In the project we will investigate human-human and human-robot interactions in a sensored public space, using automated affect and behavioural analysis.

Keywords: Human-Robot Interaction, Tele-Presence, Public Space.

1 Concept and Research Objectives

The principle virtue of public space is that it allows for strangers to meet. This in turn facilitates a number of public goods: it promotes communication; it allows people to experience difference and thus not to fear it; it lets us learn about the interests and needs of others, it helps us develop shared understandings and common purpose, and finally it allows us to practice and refine the emotions which allow successful citizenship. Research shows that there is significant importance in physical and co-present interpersonal contact to both the development of personal relationships [1] and to the reduction of prejudice between individuals from different communities [2].

An obvious barrier to participation in a physical public space is that of accessibility, whether due to geographical distance or infirmity. A possible solution to this issue is to allow remote participation is tele-operated robots. A robot used in this context would need to be able to convey the social presence of the remote user, allowing them to participate as if they were present. Our EPSRC Digital Personhood Project titled Being There: Humans and Robots in Public Spaces aims to develop a tele-presence user interface that is able to utilise both implicit (e.g., motion capture) and explicit controls to operate several robot platforms (Aldebaran Robotics' NAO, Engineered Arts' RoboThespian, ActivMedia Robotics' PeopleBot). Various aspects of tele-presence will be investigated including varying the appearance, behaviours and capabilities, of each robot platform, as well as the controls and data present on the user interface.

In order to provide quantifiable analysis of robot mediated social interactions in a public space, we will create a model public space to function as a living

G. Herrmann et al. (Eds.): ICSR 2013, LNAI 8239, pp. 581–582, 2013.

laboratory in real public venues. We will equip the space with the capacity to track micro-contacts between multiple individuals in real time, and concurrently capture and process information about the emotional and non-verbal communicative qualities of behaviours across time. Accurate, cm-level, location data for individuals in the space is needed to allow capture of micro-interactions between them. We propose to develop a localization system based on low frequency magnetic fields: small magnetic beacons will generate low frequency fields, which will be sensed by low-power miniature receivers carried by individuals and robots. Automatic, behavioural analysis for multi-participant and multi-group environments will be developed using data from physiological, visual and vocal sensory systems, in conjunction with information from the 3D location tracking.

We will utilise sensored environment to perform a programmatic series of behavioural and psychological studies. These studies will focus on social cohesion, pro-social behaviour and trust. A recognised key element of social cohesion and community relations is behavioural synchrony [3], and the space will be curated to focus on this; hence, we will use it as both a focus for the space, and as a key independent variable which we will manipulate for research purposes. We will be able to explore changes in the remote sensing of emotional tone and behaviour in the space, as well as physiological correlates of individual level changes. In addition to these implicit measures, we will also explore participants trust in the relationships between individuals and groups. We will conduct studies with and without the tele-operated robots; thus, we will build a firm understanding both of human behaviour in public spaces, and how robot mediated interactions differ from normal behaviours, and how best to minimise such differences. Additionally, the data gathered from the human interactions will be used in development of the tele-operation system, to determine desired robot behaviours, useful sensory data to feed back to the operator, and additional implicit robot controls.

Our assumption is that the human-robot interaction experiments we will be conducting will provide quantitative benchmark tests for socially interactive robot systems, with respect to a baseline of controlled human behaviours. By using multiple robot platforms, and varying their behaviours and appearance we will gain insight into the requirements for socially acceptable robots. Our hypothesis is that tele-presence will allow us to gain this understanding about socially acceptable robots without the inherent issues from using autonomous systems in a public space, while still being informative to the design of autonomous systems.

References

1. Hove, M.J., Risen, J.L.: It's all in the timing: Interpersonal synchrony increases affiliation. Social Cognition 27(6) (2009)
2. Pettigrew, T., Tropp, L.: When Groups Meet: The Dynamics of Intergroup Contact. Psychology Press (2011)
3. Wiltermuth, S.S., Heath, C.: Synchrony and Cooperation. Psychological Science 20(1), 1–5 (2009)

Signaling Robot Trustworthiness: Effects of Behavioral Cues as Warnings

Rik van den Brule[1,2], Gijsbert Bijlstra[2], Ron Dotsch[2], Daniël H.J. Wigboldus[2], and Pim Haselager[1]

[1] Donders Institute for Brain, Cognition and Behaviour, Radboud University Nijmegen, The Netherlands
r.vandenbrule@donders.ru.nl
[2] Behavioural Science Institute, Radboud University Nijmegen, The Netherlands

Abstract. By making use of existing social behavioral cues, domestic robots can express their uncertainty about their actions. Ideally, when a robot can warn that a mistake is likely, humans can take preemptive actions to ensure the successful completion of the robot's task. In the present research we show that a robot employing behavioral cues predictive of making mistakes is judged as more reliable and understandable than robots that do not use these cues predictively, and is trusted on par with a robot that does not employ behavioral cues.

1 Introduction

Domestic robots will face tasks in widely varying circumstances, ranging from an empty house to, for instance, a children's party. In certain situations a robot's functioning may require additional attention from those present. Minor interventions by humans might enable robots to carry out their task successfully. Therefore, a domestic robot may benefit from sending out warning signals when unexpected circumstances arise or a mistake becomes likely.

Current domestic robots already employ warning signals. However, an important drawback is that such signals are mostly used *after* a malfunction has occurred. Ideally, a robot would send online, proactive communication about uncertainty levels so that its owner can take preemptive action. Although such mistake-detection systems do not exist presently, it is possible to explore the feasibility of such systems by manipulating a robot's warning signals and the likelihood of mistakes following them in an experimental setting.

Robots can communicate uncertainty about their actions in many different ways. In this project, in which we borrow ideas from social psychology, we use nonverbal behavioral cues that occur naturally in human interaction. The advantage of employing existing behavioral cues to express uncertainty is that they are easily interpretable. Such attributes can influence a robots trustworthiness [1,2].

By utilizing uncertainty cues as a warning that a mistake is more likely to occur, the perceived reliability and understandability of a robot's can be increased.

G. Herrmann et al. (Eds.): ICSR 2013, LNAI 8239, pp. 583–584, 2013.

The robot's trustworthiness can be regulated by making the robot appear to be less trustworthy at the right times. In other words, the robots trustworthiness can be calibrated such that a users current level of trust is well mapped to the system's capabilities in the current circumstances [3].

2 Current Research

We tested whether the predictability of a robot's task performance with behavioral cues expressing uncertainty affects trustworthiness judgments in Human-Robot Interaction by means of a video study with a simulated Nao robot. The robot selected one of three soda cans by means of gaze behavior, after which it did or did not display a cue (the robot wipes its forehead with its hand). Participants then predicted whether the robot would point towards the selected can, or (accidently) push it off the table, after which they observed the robots action. After completing 60 of such trials, the participants rated the robot on a trustworthiness scale, as well as perceived reliability and perceived understandability scales based on [4].

82 participants were assigned to one of four conditions: predictive cues, no cues, random cues, or always cues. Importantly, only in the predictive condition the probability of a mistake given a behavioral cue was above chance level.

Analyses of variance (ANOVA) indicated that there was a significant effect of the presentation of the cue on trustworthiness, perceived reliability and perceived understandability, all $F(3, 78)$s > 5.96 , all ps $< .0010$. Post hoc Tukey's HSD tests revealed that participants in the predictive cues condition perceived the robot as more understandable and reliable compared to all other conditions, although for reliability the difference between no cues and predictive cues was marginally significant. There was a somewhat different pattern for trustworthiness. In the predictive condition, the robot was rated as significantly more trustworthy compared to the non-predictive conditions in which the behavioral cue was shown, but ratings did *not* differ significantly from the no cues condition. This suggests that a robot's trustworthiness is not based on a robot's behavior or predictability alone, but that the relationship between these attributes is more complex. In conclusion, signaling trustworthiness with behavioral warning cues is a promising, yet challenging, way to calibrate trust in Human Robot Interaction.

References

1. Hancock, P.A., Billings, D.R., Schaefer, K.E., Chen, J.Y.C., de Visser, E.J., Parasuraman, R.: A Meta-Analysis of Factors Affecting Trust in Human-Robot Interaction. Human Factors 53(5), 517–527 (2011)
2. Van den Brule, R., Dotsch, R., Bijlstra, G., Wigboldus, D.H.J., Haselager, W.F.G.: Human-robot trust: A multi-method approach (submitted)
3. Lee, J., See, K.: Trust in Automation: Designing for Appropriate Reliance. Human Factors 46(1), 50–80 (2004)
4. Madsen, M., Gregor, S.: Measuring Human-Computer Trust. In: Proceedings of the 11th Australasian Conference on Information Systems, pp. 6–8 (2000)

Activity Switching in Child-Robot Interaction: A Hospital Case Study*

Joachim de Greeff[1], Joris B. Janssen[2], Rosemarijn Looije[2], Tina Mioch[2], Laurence Alpay[2], Mark A. Neerincx[2,3], Paul Baxter[1], and Tony Belpaeme[1]

[1] Centre for Robotics and Neural Systems,
Plymouth University, United Kingdom
Joachim.deGreeff@plymouth.ac.uk
[2] TNO, Soesterberg, the Netherlands
[3] TU Delft, Delft, the Netherlands

Abstract. A case study is presented in which a socially capable robot played multiple activities during multiple sessions with hospitalised children. In this paper we show how an environment was created in which children engage with the robot and feel at liberty to express their preferences regarding which activities to play. Valuable lessons regarding robot behaviour, control and setup that allow for such a multi-activity child-robot interaction scenario were learned.

1 Introduction and Experimental Setup

The ALIZ-E project aims to develop robots that are capable of sustained interaction with diabetic children, over the course of multiple encounters in various conditions. In Child-Robot interaction, the capacity of a robot to adequately respond to the child is of great importance [1], however this is typically highly context dependant and can be hard to predict. To achieve natural interactions, the robot should seamlessly switch between multiple activities, in such a manner that the child feels encouraged to play with the robot and express his/her preference. In this switch behaviour, the robot should carefully combine the child's preference, the goals of the interaction and the preferred learning style [2].

An experiment was conducted in which a social robot (the Nao) played three game activities with children (see Fig. 1) in the WKZ child hospital (Utrecht, the Netherlands): A) Quiz: in which robot and child take turns asking each other questions presented on a see-saw tablet, B) Sandtray: a sorting game on a touchscreen, in which two interlocutors can play simultaneously without a specified structure, and C) Imitation: a physical activity in which two players remember, repeat, and extend a sequence of arm movements.

13 hospitalised children (aged 7-11) interacted on average for ± 30 minutes per session with the robot (either in the play room or in their hospital room);

---

* The authors wish to thank the children and parents, the employees of the WKZ hospital, Marie-Louise Denie, Liselotte Kroon and Bert Bierman for their contributions. This work is funded by the EU FP7 ALIZ-E project (grant 248116).

G. Herrmann et al. (Eds.): ICSR 2013, LNAI 8239, pp. 585–586, 2013.

Fig. 1. Three activities: Quiz (left), Sandtray (middle) and Imitation (right)

6 of them interacted with the robot for multiple sessions. Data was collected through camera recordings, game activity logs, a semi-structured interview with the child, and notes by care professionals on robot related talk from the child.

The robot's behaviour was prescripted; it acted mostly semi-autonomous, with an experimenter providing input of children's speech and behaviour through a Wizard of Oz setup. For these non-autonomous parts the experimenter followed a protocol. During the first session all three activities were played in a prescribed order; in subsequent sessions the children were left free to play the activity of their choosing.

2   Observations and Conclusion

Typically children were engaged with the robot, playing multiple activities and asking questions regarding its capabilities and preferences. Some (timid) children were happy for the robot to initiate another activity, others clearly vocalised their preferred activity; particularly in follow up interactions (e.g. participant 5, session 2: "We could also do this game", while pointing at a different game than the robot proposed). The children's varying approach towards switching activities can be seen as an indication that at least in some cases the robot's behaviour did induce the sought after environment in which children feel the liberty to express their preferences. Cases in which this did not happen might be due to individual differences in child behaviour and/or inadequacy on the robot's part.

This research explored multi-activity child-robot interaction in a hospital environment. Interview and video coding are part of ongoing, more formal analysis; this will provide further insights in the important dynamics regarding activity switching in child-robot interaction.

References

1. Belpaeme, T., et al.: Multimodal Child-Robot Interaction: Building Social Bonds. Journal of Human-Robot Interaction 1(2), 33–53 (2013)
2. Mayer, R.E., Massa, L.J.: Three facets of visual and verbal learners: Cognitive ability, cognitive style, and learning preference. Journal of Educational Psychology 95(4), 833 (2003)

Addressing Multiple Participants: A Museum Robot's Gaze Shapes Visitor Participation

Karola Pitsch, Raphaela Gehle, and Sebastian Wrede

Bielefeld University, CITEC & CoR-Lab

Abstract. Using videorecordings from a real-world field study with a museum guide robot, we show procedures by which the robot manages (i) to include and (ii) to disengage users in a multi-party situation.

1 Introduction

A museum guide robot is faced with the task of dealing with multiple visitors. It needs to detect who addresses the robot [3], understand the individual users' shifting states of participation, and make use of strategies for addressing multiple visitors [2]. In controlled laboratory studies, the effectiveness of a robot's gaze behavior for influencing the users' state of participation has been shown [2]. [1] demonstrate that a robot can distinguish which listener in a group would be willing to provide an answer to a question. However, when the robot is deployed in a real-world museum, the visitors happen to pass by and engage with the robot whenever they want to and can disattend and walk away at any moment [3]. Thus, the robot is faced with the task of initially securing the users' attention, maintaining their engagement and to deal with the heterogeneity of individuals who happen to stop at the robot's site at different moments in time. In this paper, we investigate the ways in which a museum guide robot's gaze strategies can shape - in a museum field trial - the users' engagement and state of participation. In particular, we are interested in the users' reactions, dynamic shifts of participation status and the micro-dynamics emerging between the visitors.

2 Robot System and Analytical Method

A humanoid robot (Nao) was deployed as guide in a German arts museum to offer information to uninformed visitors who happen to pass by. It used predefined talk and gestures and adjusted its head orientation dynamically to the nearest visitor's position. Analysis of the videorecorded interactions is informed by Conversation Analysis (CA) and investigates the interrelationship between the robot's and the visitors' actions and how users interpret the robot's conduct.

3 Analysis and Implications

Sequential micro-analysis of the HRI data reveals how visitors treat the robot's head orientation (gaze) as interactionally meaningful. Two cases are considered:

G. Herrmann et al. (Eds.): ICSR 2013, LNAI 8239, pp. 587–588, 2013.

From a 1:1 situation to engaging a group: In a first case (VP159), the robot manages to enlarge an initial 1:1 setting to include two further visitors. When a visitor (V1) enters the room and positions himself in front of the robot, it detects V1 and reacts by shifting its head in this direction. The robot greets and V1 reciprocates. Thus, a focused encounter has been established with V1 reacting to the robot's communicational offers. Then two further visitors (V2, V3) enter the room and V2 assumes a position that allows her a direct view to the robot. The system now turns its head to V2. V1 reacts to the robot's shifting head orientation by turning around, looking where the robot seems to be looking and takes a step backwards to the side. By adjusting the spatial configuration, he integrates V2 in the encounter with the robot. When Nao then asks a question and shifts its head to an intermediary position between V1 and V2, both participants respond (head shake, "no"). Thus, V2's participation status shifts from observing V1's engagement to becoming an active co-participant.

Disengaging an individual from a group: In a second case (VP075), two visitors enter the room. V1 walks straight to the robot and positions himself in front of it, the system detects V1 and directs its head towards him. V2 arrives a few seconds later, stops slightly behind V1, monitors the ensuing interaction between V1 and the robot and closely follows the robot's explanations. When the robot's head orientation shifts slightly to the opposite site from V2, V1 reacts by repositioning himself again in the robot's line of sight. V2, however, stops to follow the robot's actions, disengages and walks away (with V1 remaining).

In sum, to enable autonomous robot systems to systematically deal with such multi-party situations more fine-grained interactional coordination and reliable procedures for pro-actively engaging the user are required. Therefore, advanced perceptual skills, control architectures and novel interaction models need to be developed in concert with each other which allow for incremental processing.

Acknowlegement. The authors acknowlegde the financial support from CITEC (project 'Interactional Coordination and Incrementality in HRI'), the Volkswagen Foundation (Dilthey Fellowship 'Interaction & Space', K. Pitsch), and the CoR-Lab.

References

1. Kobayashi, Y., Shibata, T., Hoshi, Y., Kuno, Y., Okada, M., Yamazaki, K.: "I will ask you" choosing answerers by observing gaze responses using integrated sensors for museum guide robots. In: 2010 IEEE RO-MAN, pp. 652–657 (2010)
2. Mutlu, B., Shiwa, T., Kanda, T., Ishiguro, H., Hagita, N.: Footing in human-robot conversations: How robots might shape participant roles using gaze cues. In: HRI 2009, New York, NY, USA, pp. 61–68 (2009)
3. Pitsch, K., Kuzuoka, H., Suzuki, Y., Süssenbach, L., Luff, P., Heath, C.: "The first five seconds": Contingent stepwise entry into an interaction as a means to secure sustained engagement. In: IEEE International Symposium on Robot and Human Interactive Communication, pp. 985–991 (2009)

Towards a Robotic Sports Instructor
for High-Interaction Sports

Ingmar Berger, Britta Wrede, Franz Kummert and Ingo Lütkebohle

Applied Informatics Group, Bielefeld University, Bielefeld, Germany

1 Introduction

For many people, instructor-led sports courses are a particularly motivating form of physical exercise. Usually known as "spinning" or "body pump", they are offered in gyms as group courses. The instructor continously provides goals, guidance, and correction, leading to high motivation and effectiveness through interaction.

However, instructors are not always available, e.g., at home, or in isolated settings[1]. We have therefore developed a robotic system that can provide some training functions, targeting the "spinning" or indoor-cycling domain.

Compared to other sports assistance approaches, our robotic system is distinguished by: i) Frequent interaction (multiple times per minute), ii) multi-modal monitoring, and iii) use of the robot's embodiment for multi-modal instruction.

To our knowledge this is the first robotic system that targets human-robot interaction during exercising. Most systems focus either on automatic interpretation of data during exercises (like [3]) or provide feedback by visualising sensor data for professional athletes that can interpret the data directly (cf. [1]).

2 System

Our setup is based on an instrumented exercise bike[2] delivering speed, cadence and power output, a heart-rate monitor, two Kinect depth cameras for pose analysis, and a Nao robot. See Figure 1 for an overview.

An important aspect for realization is to cover enough typical trainer-trainee interactions for a meaningful training. From Human-Human-Studies, we could extract so called "movement patterns" that represent reconfigurable and re-usable situational behavior models [2]. A sequence of these constitutes a training session. Our current system consists of 36 different configurations of these patterns, with close to 300 individual exercise actions.

Fig. 1. Photo of the setup

---

[1] Our own target application are space missions.
[2] SRM Indoor Trainer, see http://www.srm.de

G. Herrmann et al. (Eds.): ICSR 2013, LNAI 8239, pp. 589–590, 2013.
© Springer International Publishing Switzerland 2013

A simple example of such a pattern is the "static movement" pattern (cf. Figure 2). It is used to instruct, and monitor, stationary exercises, such as posture ("stand up"), or cadence ("pedal with 60rpm").

The crucial aspect of the system is that it monitors the execution of the movements, and, can correct it, if necessary. We employ a multi-phase *repair* strategy: i) the athlete is notified about his deviation and requested to correct it; ii) an optional re-correction step follows; iii) in case the athlete reaches the target range of the deviating parameter, the robot finally praises him and the interaction ends successful, otherwise the repair fails.

In a first step we investigated the average number of successful repairs per day over all exercises and participants. With success rates between 40% and 70% and an average of around 60% successful repairs, the results show that athletes respond to the repairs – especially, since the physical exhaustion of the athletes must be considered as a potential cause for failed repairs.

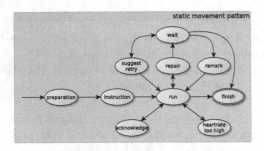

Fig. 2. Static Movement Pattern

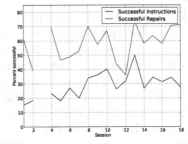

Fig. 3. Avg. successful rates/ day

3   Conclusions

We have presented a robot sports instructor for highly interactive sports exercises. For this early report, we have considered interactive exercise repairs, and found that they are regularly necessary, and that the proposed guidance approach is effective.

References

1. Baca, A., Kornfeind, P.: Rapid Feedback Systems for Elite Sports Training. IEEE Pervasive Computing 5(4), 70–76 (2006)
2. Berger, I., Süssenbach, L., Lütkebohle, I.: Requirements and interaction-design of a robotic sports-instructor prototype (submitted, 2013)
3. Novatchkov, H., Baca, A.: Machine learning methods for the automatic evaluation of exercises on sensor-equipped weight training machines. Procedia Engineering 34, 562–567 (2012)

Author Index